# Global Issues and Innovative Solutions in Healthcare, Culture, and the Environment

Mika Merviö
*Kibi International University, Okayama, Japan*

A volume in the Advances in Human Services and Public Health (AHSPH) Book Series

Published in the United States of America by
IGI Global
Information Science Reference (an imprint of IGI Global)
701 E. Chocolate Avenue
Hershey PA, USA 17033
Tel: 717-533-8845
Fax: 717-533-8661
E-mail: cust@igi-global.com
Web site: http://www.igi-global.com

Copyright © 2020 by IGI Global. All rights reserved. No part of this publication may be reproduced, stored or distributed in any form or by any means, electronic or mechanical, including photocopying, without written permission from the publisher. Product or company names used in this set are for identification purposes only. Inclusion of the names of the products or companies does not indicate a claim of ownership by IGI Global of the trademark or registered trademark.

Library of Congress Cataloging-in-Publication Data

Names: Mervio, Mika Markus, 1961- editor.
Title: Global issues and innovative solutions in healthcare, culture, and
   the environment / Mika Merviö, editor.
Description: Hershey, PA : Information Science Reference, an imprint of IGI
   Global, [2020] | Includes bibliographical references and index. |
   Summary: "This book discusses worldwide conflicts within healthcare and
   environmental development as well as modern resolutions that are being
   implemented"-- Provided by publisher.
Identifiers: LCCN 2019055635 (print) | LCCN 2019055636 (ebook) | ISBN
   9781799835769 (hardcover) | ISBN 9781799835776 (paperback) | ISBN
   9781799835783 (ebook)
Subjects: LCSH: Environmental protection. | Medical technology. | Energy
   conservation.
Classification: LCC TD170.3 .G65 2020  (print) | LCC TD170.3  (ebook) | DDC
   362.1--dc23
LC record available at https://lccn.loc.gov/2019055635
LC ebook record available at https://lccn.loc.gov/2019055636

This book is published in the IGI Global book series Advances in Human Services and Public Health (AHSPH) (ISSN: 2475-6571; eISSN: 2475-658X)

British Cataloguing in Publication Data
A Cataloguing in Publication record for this book is available from the British Library.

The views expressed in this book are those of the authors, but not necessarily of the publisher.

For electronic access to this publication, please contact: eresources@igi-global.com.

# Advances in Human Services and Public Health (AHSPH) Book Series

Jennifer Martin
RMIT University, Australia

ISSN:2475-6571
EISSN:2475-658X

### Mission

The well-being of the general public should be a primary concern for any modern civilization. Ongoing research in the field of human services and public healthcare is necessary to evaluate, manage, and respond to the health and social needs of the global population.

The **Advances in Human Services and Public Health (AHSPH)** book series aims to publish high-quality reference publications focused on the latest methodologies, tools, issues, and strategies for managing the health and social welfare of the public. The AHSPH book series will be especially relevant for healthcare professionals, policy makers, government officials, and students seeking the latest research in this field.

### Coverage

- Access to Healthcare Services
- Social Welfare Policy
- Social Work
- Health Policy
- Public Funding
- Youth Development
- Assistance Programs
- Healthcare Reform
- Poverty
- Domestic Violence

IGI Global is currently accepting manuscripts for publication within this series. To submit a proposal for a volume in this series, please contact our Acquisition Editors at Acquisitions@igi-global.com or visit: http://www.igi-global.com/publish/.

The Advances in Human Services and Public Health (AHSPH) Book Series (ISSN 2475-6571) is published by IGI Global, 701 E. Chocolate Avenue, Hershey, PA 17033-1240, USA, www.igi-global.com. This series is composed of titles available for purchase individually; each title is edited to be contextually exclusive from any other title within the series. For pricing and ordering information please visit http://www.igi-global.com/book-series/advances-human-services-public-health/102256. Postmaster: Send all address changes to above address. Copyright © 2020 IGI Global. All rights, including translation in other languages reserved by the publisher. No part of this series may be reproduced or used in any form or by any means – graphics, electronic, or mechanical, including photocopying, recording, taping, or information and retrieval systems – without written permission from the publisher, except for non commercial, educational use, including classroom teaching purposes. The views expressed in this series are those of the authors, but not necessarily of IGI Global.

# Titles in this Series

*For a list of additional titles in this series, please visit:*
http://www.igi-global.com/book-series/advances-human-services-public-health/102256

*Handbook of Research on Leadership and Advocacy for Children and Families in Rural Poverty*
H. Carol Greene (East Carolina University, USA) Bryan S. Zugelder (James Madison University, USA) and Jane C. Manner (East Carolina Universty, USA)
Information Science Reference • © 2020 • 525pp • H/C (ISBN: 9781799827870) • US $265.00

*Emerging Trends in Indigenous Language Media, Communication, Gender, and Health*
Kehinde Opeyemi Oyesomi (Covenant University, Nigeria) and Abiodun Salawu (North-West University, Sout Africa)
Medical Information Science Reference • © 2020 • 320pp • H/C (ISBN: 9781799820918) • US $285.00

*Impact of Textile Dyes on Public Health and the Environment*
Khursheed Ahmad Wani (Government Degree College, Bijbehara, India) Nirmala Kumari Jangid (Banasthali Vidyapith, India) and Ajmal Rashid Bhat (Government Degree College, Bijbehara, India)
Information Science Reference • © 2020 • 335pp • H/C (ISBN: 9781799803119) • US $245.00

*Handbook of Research on Health Systems and Organizations for an Aging Society*
César Fonseca (Universidade de Évora, Portugal) Manuel José Lopes (Universidade de Évora, Portugal) David Mendes (Universidade de Évora, Portugal) Felismina Mendes (Universidade de Évora, Portugal) and José García-Alonso (University of Coimbra, Spain)
Medical Information Science Reference • © 2020 • 323pp • H/C (ISBN: 9781522598183) • US $365.00

*Cases on Cross-Cultural Counseling Strategies*
Bonnie C. King (Midwestern State University, USA) and Tiffany A. Stewart (Midwestern State University, USA)
Medical Information Science Reference • © 2020 • 380pp • H/C (ISBN: 9781799800224) • US $245.00

*Environmental Exposures and Human Health Challenges*
Paraskevi Papadopoulou (Deree – The American College of Greece, Greece) Christina Marouli (Deree – The American College of Greece, Greece) and Anastasia Misseyanni (Deree – The American College of Greece, Greece)
Medical Information Science Reference • © 2019 • 449pp • H/C (ISBN: 9781522576358) • US $255.00

*Forecasting and Managing Risk in the Health and Safety Sectors*
Luisa Dall'Acqua (Scientific Lyceum TCO, Italy & University of Bologna, Italy)
Medical Information Science Reference • © 2019 • 304pp • H/C (ISBN: 9781522579038) • US $245.00

701 East Chocolate Avenue, Hershey, PA 17033, USA
Tel: 717-533-8845 x100 • Fax: 717-533-8661
E-Mail: cust@igi-global.com • www.igi-global.com

# Table of Contents

**Preface** ................................................................................................................................................. xiv

**Chapter 1**
Toward Leadership Agility ....................................................................................................................... 1
    *Simon Cleveland, Georgetown University, USA*
    *Marisa Cleveland, Northeastern University, USA*

**Chapter 2**
Transition of Ecosystem Services Based on Urban Agro Ecology ........................................................ 14
    *José G. Vargas-Hernández, University Center for Economic and Managerial Sciences,*
        *University of Guadalajara, Mexico*

**Chapter 3**
Corralling Contract and the Land Use Relationship Between Pastoralists and Agriculturalists in
Nigeria.................................................................................................................................................... 34
    *Regina Hoi Yee Fu, Senshu University, Japan*

**Chapter 4**
Neuroeconomics: Connecting Health Science and Ecology.................................................................. 64
    *Torben Larsen, University of Southern Denmark, Denmark*

**Chapter 5**
Reading, Literature, and Literacy in the Mobile Digital Age ................................................................ 84
    *John Fawsitt, Kibi International University, Japan*

**Chapter 6**
Strategic Transformational Transition of Green Economy, Green Growth, and Sustainable
Development .......................................................................................................................................... 99
    *José G. Vargas-Hernández, University Center for Economic and Managerial Sciences,*
        *University of Guadalajara, Mexico*

**Chapter 7**
Cinematic Works-Based Affective Urban Design Atmospheres ................................................................ 120
    *Hisham Abusaada, Housing and Building National Research Center (HBRC), Egypt*

**Chapter 8**
Environmental Problems of Delhi and Governmental Concern ............................................................... 133
    *Mohd. Yousuf Bhat, Government Degree College, Pulwama, India*

**Chapter 9**
Hydrogen: An Environmental Remediation .............................................................................................. 168
    *Athule Ngqalakwezi, University of the Witwatersrand, South Africa*
    *Diakanua Bevon Nkazi, University of the Witwatersrand, South Africa*
    *Siwela Jeffrey Baloyi, Mintek, South Africa*
    *Thabang Abraham Ntho, Mintek, South Africa*

**Chapter 10**
Healthcare in India: Challenges and Innovative Solutions ....................................................................... 195
    *Reenu Kumari, MIET, India*

**Chapter 11**
Medical Tourism Patient Mortality: Considerations From a 10-Year Review of Global News
Media Representations ............................................................................................................................. 206
    *Alicia Mason, Pittsburg State University, USA*
    *Sakshi Bhati, Pittsburg State University, USA*
    *Ran Jiang, Soochow University, China*
    *Elizabeth A. Spencer, University of Kentucky, USA*

**Chapter 12**
Magnetocaloric as Solid-State Cooling Technique for Energy Saving ..................................................... 226
    *Ciro Aprea, University of Salerno, Italy*
    *Adriana Greco, University of Naples Federico II, Italy*
    *Angelo Maiorino, University of Salerno, Italy*
    *Claudia Masselli, University of Salerno, Italy*

**Chapter 13**
Global Racial, Ethnic, and Religious Subculture and Its Impact on Healthcare Organizations ............... 253
    *Rohit Singh Tomar, Amity University Madhya Pradesh, India*
    *Meenal Kulkarni, Symbiosis Institute of Health Science, India*

**Chapter 14**
Assessment of Healthcare Service Quality: Tertiary Care Hospitals of Dhaka City ................................ 271
    *Segufta Dilshad, North South University, Bangladesh*
    *Afsana Akhtar, BRAC University, Bangladesh*
    *S. S. M. Sadrul Huda, East West University, Bangladesh*
    *Nandeeta Samad, North South University, Bangladesh*

**Chapter 15**
Application of Nanotechnology in Global Issues ............................................................................... 292
    *Ranjit Barua, OmDayal Group of Institutions, India*
    *Sudipto Datta, Indian Institute of Engineering Science and Technology, Shibpur, India*
    *Jonali Das, Raja Peary Mohan College, Calcutta University, India*

**Chapter 16**
The Social Significance of Visual Arts: State, Nation, and Visual Art in Japan and Finland ............ 301
    *Mika Markus Merviö, Kibi International University, Japan*

**Compilation of References** ............................................................................................................ 317

**About the Contributors** ................................................................................................................. 361

**Index** ................................................................................................................................................ 365

# Detailed Table of Contents

**Preface** ............................................................................................................................................. xiv

**Chapter 1**
Toward Leadership Agility ................................................................................................................... 1
    *Simon Cleveland, Georgetown University, USA*
    *Marisa Cleveland, Northeastern University, USA*

Companies are often challenged by the cultural diversity of the growing workforce. As a result, organizational leaders should develop culturally agile competencies in order to engage and motivate their employees. Leadership development programs exist to contribute to an individual's and an organization's success; however, there is a lack of studies that examine how such programs contribute to the development of cultural agility in leaders. Culturally agile leaders are more inclusive in their hiring practices and more open to encouraging more diversity within their own leadership network. Such leaders value collaboration and understand how culturally-grounded traditions and preferences effect transactions. This chapter addresses the roles of positionality and cultural agility, leadership development programs, and capacity and responsibility in building culturally agile leaders. The chapter also proposes how leaders have the capacity and the responsibility to develop other leaders through a relational leadership approach to promote inclusion and diversity.

**Chapter 2**
Transition of Ecosystem Services Based on Urban Agro Ecology ...................................................... 14
    *José G. Vargas-Hernández, University Center for Economic and Managerial Sciences,*
        *University of Guadalajara, Mexico*

This chapter analyzes the implications of the transition of ecosystem services based on urban agro-ecology. It advances on the debate over the negative effects of the traditional and industrial-oriented agricultural production on the ecosystem services, food systems, climate change, etc. and analyses the principles, methods, and some practices that support the transition to urban agro-ecology. The method employed is the analytical of the theoretical and empirical literature review. It concludes that a transition from traditional and industrial-oriented agriculture towards more urban agro-ecology is inevitable to improve the ecological and environmental services, the economic efficiency, the social equity and justice, and the environmental sustainability of cities.

## Chapter 3
Corralling Contract and the Land Use Relationship Between Pastoralists and Agriculturalists in Nigeria............................................................................................................................................34
*Regina Hoi Yee Fu, Senshu University, Japan*

"Corralling contract" is the indigenous fertilization system commonly practiced in the African Sahel and its southern periphery. In this chapter, the practice of the corralling contract between Fulani pastoralists and Nupe agriculturalists in the Bida region of Niger State of central Nigeria is examined. The study attempts to find out how the farmers and herders arrange the corralling contract, how they utilize this instrument, and how it influences their social relationship. Findings suggest that pastoral Fulani groups have different strategies to maintain socioeconomic relations with specific villages through the adoption of corralling contract in order to ensure resources entitlement. While some groups can well manipulate the relationships with various villages through the adoption of the corralling contract, some groups prefer a more stable situation and just get the minimum advantages. Higher social status, larger herd size and longer history of interaction that allow trust to be built are the factors contributing to the popularity and bargaining power of a pastoral group.

## Chapter 4
Neuroeconomics: Connecting Health Science and Ecology....................................................................64
*Torben Larsen, University of Southern Denmark, Denmark*

A neuroeconomic psychology establishes behavioral economics as a positivist discipline connecting health science and ecology. This makes "economic ecology" and "neuroeconomics" important new sub-disciplines of economics. Pigovian carbon emission tax (CET) is the most effective intervention towards the "green-house effect." However, in the short-term, democratic center-coalitions are too weak to implement CET due asymmetric levels of economic knowledge between economists and the public. However, neuroeconomic research indicates that training can improve decision-making about complex issues as CET. A second-best strategy is presented as green hybrid multilevel transition (GHMT). Besides GHMT preparing doe CET, neuroeconomic psychology delivers a series of science-based advices on effective behaviors: a consumer pattern that integrates individual satisfaction with environmental conscience, efficacious business behavior like an entrepreneur, long-term political suspension of relative poverty by universal basic income, and cognitive training by meditative in-depth relaxation.

## Chapter 5
Reading, Literature, and Literacy in the Mobile Digital Age ...................................................................84
*John Fawsitt, Kibi International University, Japan*

Reading and literature are struggling for relevance an environment where attention and the data they provide are seen as key motivators for commercial actors, and there is great pressure for those actors to provide engaging media to secure a meaningful market share. Thus, this media has to attract and keep user attention as quickly and as continuously as possible. The only limiting factors being those of time and energy of the user. Leisure hours that allowed periods for unbroken concentration and perusal of written texts are now devoted to online activities. What is not debated is that the effort and focus required to engage with the writer of fiction or other longer texts cannot be as automatically assumed now as it was before the digital age. Therefore, how can or should reading and literature and our notion of them and their purposes change?

**Chapter 6**
Strategic Transformational Transition of Green Economy, Green Growth, and Sustainable
Development ............................................................................................................................... 99
 *José G. Vargas-Hernández, University Center for Economic and Managerial Sciences,*
  *University of Guadalajara, Mexico*

This chapter aims to analyze a strategic transformational transition of green economy, green growth, and sustainable development from the institutional perspective. The analysis begins questioning the implications of the concepts and principles of green economy, green growth, and sustainable development from different perspectives in the transformational transition considering the investment, trade, and capacity building though the design and implementation of strategies and policies as well as measures from an institutional analysis. The methodology followed was the analytical review of the literature to derive inferences, challenges, proposals, and conclusions. It is concluded that the green economy concept addresses current challenges delivering economic development opportunities and multiple benefits for the welfare of all human beings.

**Chapter 7**
Cinematic Works-Based Affective Urban Design Atmospheres ....................................................... 120
 *Hisham Abusaada, Housing and Building National Research Center (HBRC), Egypt*

This chapter examines the dilemma of using the term "atmospheres" related to architectural history. It theorises the nature of this relationship, developing an analytical framework creating the architecture of the city as similar to artwork. In this chapter, the authors investigated through the aspects of cinematic works—ideas, themes, and dramatic text—and overarching effects of the technical elements. The question is: How can the urban designer use the artworks in the field of urban design? This chapter discusses the atmospheres in many artworks of Western and Egyptian thought to explore the effect of the architecture of cities in creating the atmospheres of the cities.

**Chapter 8**
Environmental Problems of Delhi and Governmental Concern ....................................................... 133
 *Mohd. Yousuf Bhat, Government Degree College, Pulwama, India*

Delhi, the capital city of India, which is the concern of this chapter, has its own significance as it is the seat of governance, learning, and the healthcare service provider. Capital cities though inhabit people from every region and tend to be overcrowded, but in Delhi, the situation is not only the nature of a capital city, but also the industrial and commercial centre of high order in the northern zone of India, which is creating a number of environmental problems, such as air and water pollution, slum development, congested housing, etc. The chapter discusses all causes of Delhi's environmental problems like atrophy of political will, mismanaged urbanisation, court interventions, etc., and finally, the chapter tries to find out possible solutions in a detailed manner keeping in view the measures taken by other countries like China to deal with such problems.

**Chapter 9**
Hydrogen: An Environmental Remediation.................................................................................. 168
    *Athule Ngqalakwezi, University of the Witwatersrand, South Africa*
    *Diakanua Bevon Nkazi, University of the Witwatersrand, South Africa*
    *Siwela Jeffrey Baloyi, Mintek, South Africa*
    *Thabang Abraham Ntho, Mintek, South Africa*

Global warming is a pertinent issue and is quintessential of the environmental issues that the world is facing, and thereby, remedial actions and technologies that aim to alleviate this issue are of paramount importance. In this chapter, hydrogen has been discussed as an alternative energy that can potentially replace traditional fuels such as diesel and gasoline. The storage of hydrogen as a gas, liquid, and solid was discussed. The key issues in hydrogen storage were also highlighted. Furthermore, regulations and legislations concerning the emission of greenhouse gases from fossil fuels-based sources were discussed.

**Chapter 10**
Healthcare in India: Challenges and Innovative Solutions ............................................................ 195
    *Reenu Kumari, MIET, India*

India facing some complex challenges in healthcare sectors like adaptation of hi tech (digital health), financing, approachability, and adaptability. Therefore, this chapter explores an overview of health and healthcare, source of financing in Indian healthcare, challenges and innovative solutions for the problems. This chapter also discussed the healthcare industry (global and India perspective) and found that total healthcare industry size will be increased US$ 160 Billion (2017) and US$ 372 billion (2022). Health financing is by a number of sources: (1) the tax-based public sector that comprises local, state, and central governments, in addition to numerous autonomous public sector bodies and the private sector. There are major challenges in Indian healthcare: population, non-commutation disease, new technology, inaccessibility to primary healthcare, and medical education. This chapter concluded by providing appropriate knowledge to the multidisciplinary groups such as researchers, policymakers, administrators, and politician.

**Chapter 11**
Medical Tourism Patient Mortality: Considerations From a 10-Year Review of Global News
Media Representations.................................................................................................................. 206
    *Alicia Mason, Pittsburg State University, USA*
    *Sakshi Bhati, Pittsburg State University, USA*
    *Ran Jiang, Soochow University, China*
    *Elizabeth A. Spencer, University of Kentucky, USA*

Medical tourism is a process in which a consumer travels from one's place of residence and receives medical treatment, thus becoming a patient. Patients Beyond Borders (PBB) forecasts some 1.9 million Americans will travel outside the United States for medical care in 2019. This chapter explores media representations of patient mortality associated with medical tourism within the global news media occurring between 2009-2019. A qualitative content analysis of 50 patient mortality cases found that (1) a majority of media representations of medical tourism patient death are of middle-class, minority females between 25-55 years of age who seek cosmetic surgery internationally; (2) sudden death, grief, and bereavement counseling is noticeably absent from medical tourism providers (MTPs); and (3) risk information from authority figures within the media reports is often vague and abstract. A detailed list of health communication recommendations and considerations for future medical tourists and their social support systems are provided.

## Chapter 12
Magnetocaloric as Solid-State Cooling Technique for Energy Saving ............................................... 226
    *Ciro Aprea, University of Salerno, Italy*
    *Adriana Greco, University of Naples Federico II, Italy*
    *Angelo Maiorino, University of Salerno, Italy*
    *Claudia Masselli, University of Salerno, Italy*

Magnetocaloric is an emerging cooling technology arisen as alternative to vapor compression. The main novelty introduced is the employment of solid-state materials as refrigerants that experiment magnetocaloric effect, an intrinsic property of changing their temperature because of the application of an external magnetic field under adiabatic conditions. The reference thermodynamic cycle is called active magnetocaloric regenerative refrigeration cycle, and it is Brayton-based with active regeneration. In this chapter, this cooling technology is introduced from the fundamental principles up to a description of the state of the art and the goals achieved by researches and investigations.

## Chapter 13
Global Racial, Ethnic, and Religious Subculture and Its Impact on Healthcare Organizations ......... 253
    *Rohit Singh Tomar, Amity University Madhya Pradesh, India*
    *Meenal Kulkarni, Symbiosis Institute of Health Science, India*

This chapter deals with the racial, ethnic, and religious subculture and its impact on the global healthcare organizations and their practices. This study is exploratory in nature where secondary sources have been analyzed to find out the answers of the selected objectives. A discussion-based approach has been used to compare and contrast various information regarding healthcare practice and the role of ethnicity and religion in affecting it. Healthcare seems to be affected by race, ethnicity, and religion, but there is a huge scope of quantitative analysis to get a detailed and comprehensive result.

## Chapter 14
Assessment of Healthcare Service Quality: Tertiary Care Hospitals of Dhaka City ......................... 271
    *Segufta Dilshad, North South University, Bangladesh*
    *Afsana Akhtar, BRAC University, Bangladesh*
    *S. S. M. Sadrul Huda, East West University, Bangladesh*
    *Nandeeta Samad, North South University, Bangladesh*

The service quality measurement of healthcare services is always a big concern for the hospitals, patient rights activists, regulators, and general patients. This study deals with quality assessment of healthcare facilities concerning the private and public health facilities of Dhaka, Bangladesh. This study follows the survey research approach. Using the purposive sampling method, the individuals have been selected from households who have received healthcare services from public or private hospitals of Dhaka city in last year. The study collected data among 410 respondents. Standard statistical software (i.e., SPSS and STATA) have been used to analyze the data. This study confirms existing evidence that Bangladeshi patients have a growing concern with lower level of satisfaction in public healthcare services. The respondents faced multi-dimensional problems, characterized by a low level of overall service quality, interpersonal service quality, and technical or treatment-related quality at public hospitals. Further research is recommended to analyze the issues further.

**Chapter 15**
Application of Nanotechnology in Global Issues .............................................................................. 292
    *Ranjit Barua, OmDayal Group of Institutions, India*
    *Sudipto Datta, Indian Institute of Engineering Science and Technology, Shibpur, India*
    *Jonali Das, Raja Peary Mohan College, Calcutta University, India*

Nanotechnology basically means any kind of technology in a nanoscale, which can be applied in the existent world. It is a comparatively new research field, but it is not a completely new area and the research draws insights from many other research areas. It is generally considered that nanotechnology makes possible the coming of the new Industrial Renaissance since it has the potential for a reflective impact on modern society and economy in the early 21st century, similar to that of information technology (IT), electronics technology, especially in semiconductor technology or molecular and cellular biology. The purpose of this chapter is to look into the present aspects of nanotechnology. In this chapter, the authors discuss a variety of applications of nanotechnology in recent decades like modern engineering, robotics, food technology, medicine, etc., and also they indicate the current and potential uses of nanoscience and nanotechnologies. Social and ethical impacts as well as health and environmental impacts will be highlighted.

**Chapter 16**
The Social Significance of Visual Arts: State, Nation, and Visual Art in Japan and Finland ............ 301
    *Mika Markus Merviö, Kibi International University, Japan*

This chapter focuses on visual arts in Japan and Finland and the ways that social and political developments have been linked to reconstruction of cultural traditions. In particular, the most important turning point appears to be the establishment of modern state. This chapter follows the circumstances that prevented Japanese state for a long time during the era of feudal society from giving culture and visual arts more freedom and larger role in the process that leads to modernization and the nation state. After this, the author moves to discuss the parallel developments in Finland, where the process is very different. The idea of this chapter is to look these parallel developments both in terms of contrasts and similarities. In both societies, the visual arts have a long history, and often in national history writing, the cultural traditions are presented as a long unbroken narrative. However, the social and political developments have been greatly affected by cultural developments, but also culture has played a major role in social development.

**Compilation of References** ............................................................................................................ 317

**About the Contributors** ................................................................................................................. 361

**Index** ................................................................................................................................................ 365

# Preface

This book uses as its inspiration The *International Journal of Public and Private Perspectives on Healthcare, Culture, and the Environment (IJPPPHCE)*, one of IGI Global's Core Journals. It is a genuinely international journal with an innovative and interdisciplinary approach to issues of Healthcare, Culture, and the Environment. It is particularly these issues that have proved to be at the core of social transformation of modern societies and areas where there are very strong linkages between different issue areas. There is a particular need for both interdisciplinary and transdisciplinary approaches in finding solutions to issues that tend to defy disciplinary and other borders. The Journal also bears the dichotomy of 'Public and Private Perspectives' to indicate an openness to analyze social and cultural phenomena from the point of view of different actors and stakeholders. Public policy remains important, but it is important to be able to bring all the relevant perspectives to the analysis. The global approach in the journal means that we are serious about inviting scholars from all parts to the world to contribute fresh and original research that in many different ways may well turn out to challenge the borders of our disciplines and many other borders and limitations that still exist in the world. This book continues the best traditions of our journal and maintains its interdisciplinary and transdisciplinary approach as well as the global perspective.

This book consists of chapters that have been invited from selected authors among the scholars who have contributed their articles to the journal and asked them to enhance and update their articles to turn them into book chapters. Furthermore, this book also includes several articles that have not yet been published elsewhere and which are very fresh. What is common to all these works is that they fit into the general theme and provide innovative approaches that could well be termed as being transdisciplinary, moving beyond fixed disciplines and combining in an innovative ways tools to analyze topics that are central for contemporary societies.

Our experience among the authors has been that healthcare, welfare, culture and the environment are probably the most important and contested issue areas in international discourses concerning the contemporary societies. It is exactly these areas where the modern state as well as the civil society and all individuals are facing the biggest challenges to adjust to new realities, in terms of politics, culture, economy or ethics. These areas are often overlapping and in many countries there are already scholars working in some of these interdisciplinary fields and there surely is a need to encourage better communication across cultural and disciplinary borders as well as moving to all new directions.

To summarize about the issue areas of this book I will introduce briefly the main areas of study and how we have approached them.

*Preface*

1) **Healthcare:** According to Aristotle, the court physician and a son of a physician, proper balance or proportion makes for health, lack of it for disease. Too much food, or too much exercise, are bad for health, just as too little food or exercise are. The same holds, in particular, in ethical matters. The Doctrine of the Mean can't be used as the theoretical foundation of every scholarly thesis, but it certainly makes clear that 'health' as a concept has a long history of having been seen intricately entangled with a plethora of human activities. Healthcare as a system is directly connected with politics, administration, education, ethics, religion, culture and economy. Healthcare has significantly improved our quality of life and it is no wonder that its role in societies keeps growing. In short, health care is not only an industry directly covering more than ten percent of GDP in most industrialized countries, even without starting to discuss the close relationship between health and welfare. This book will give some of the answers what healthcare and welfare are all about and where they are moving within the societies that are rapidly changing.

2) **Culture:** While 'health' easily leads us to analyze from within the framework of social sciences and natural/ medical sciences 'culture' is more elusive and requires an understanding of philosophy and humanities. However, in our global society cultural differences and socially and culturally based forms of understanding/ construction are central to social analysis. Including "Culture" and encouraging research on the topic opens the discussion to cultural differences concerning the public and private as well as the "culture" and cultural understanding itself and their relevance to policy and society. In short, "Culture" is included to encourage new kinds of innovative pieces of research both in terms of developing the theory and paying attention to cultural diversity and the importance of culture in every society, as well as to the historical roots of social issues and forms of understanding in different societies. Culture is particularly suitable when it is combined with other issue areas that the journal covers. The result both in terms of issue areas as well as the theoretical approach often becomes transdisciplinary and helps to make new fresh contributions to international research and theory in various fields. Of course, the most important contribution will be the relevance of research for finding solutions to acute social problems.

3) **Environment** is the physical and cultural setting of human endeavors and the stage where everything else happens. In this era of our reflexive relationship with the threatened and beleaguered natural environment it is clear that the sheer survival of all living things is at stake and that their health, culture and everything they hold important is closely connected to the environment. Environmental issues are of primary importance for all the living things, including all the humans on this planet. Therefore, it is most opportune to discuss environmental issues in the context of social issues of our time. We hope that our efforts will help to find solutions to environmental and social problems and we do recognize that all these efforts are closely linked.

## SCOPE AND APPROACH OF THE BOOK

This book introduces discourses and analytical discussions of many different related topics, all adding to understanding on changing realities within the realms of healthcare, culture and the environment. In this book the idea has been to link discussions of in social, environmental and cultural sphreres in new ways to be able to understand the issues better both in the global and their own social and cultural context. This book aims to transcend all kind of boundaries, including the national boundaries and boundaries of academic disciplines to search explanations to developments in contemporary world. Area-specific or culture-specific discussions, of course, are closely linked to developments elsewhere, but there is also a need to analyze the impact of local cultures and traditions – without being blind to changes in these traditions.

Transdisciplinary approach is used here to promote depth of understanding as well as adaptability to skills needed to succeed in our changing world. In short, the problems that we are facing in contemporary world require both deeper understanding and a broader set of skills. The idea is to include multiple disciplines with a human centered goals. In the end the analysis not only integrates the perspectives of multiple disciplines but helps to find fresh and exciting perspectives.

## Introduction of Contents by Highlighting How They Relate the Current Situation and Future Trends in the Fields of Healthcare, Culture, and the Environment

In Chapter 1, "Toward Leadership Agility," Simon Cleveland and Marisa Cleveland introduce valuable ideas and insights about the importance of cultural agility in decision-making, collaboration and leadership. Flexibility and an ability of adapt have always been important for effective leadership, but with the digital landscape and globalization of the marketplace culturally agile leaders surely will recognize how cultural diversity makes a difference in how employees act and react. This chapter addresses three main areas of interest: positionality and cultural agility, leadership development programs, and capacity and responsibility. The positionality and level of cultural agility always inform and limit a leader's ability to make an impact on their organization. Leaders must reflect about themselves in relation to others and acknowledge the multiple roles, identities, and positions that each member of the organization contributes to the organization. The authors introduce a variety of leadership programs that give a good idea on how the leadership development can be promoted throughout the organization. Effective leaders are found throughout an organization, not just at the top, and organizations are dedicating significant amounts of time and money to develop and train emerging leaders in their organizations. As companies expand and compete in today's digital marketplace, leaders are focusing resources on training emerging leaders in a combination of technical, conceptual, and human skills.

On the other hand, the chapter analyses the failings or shadows of leadership that are all too common and make the leadership fail. These shadows include such issues as unfair increases in employee workload, humiliating and bullying employees, and claiming credit for the work of others. Faulty decision-making is often the work of psychopath leaders. Misuse of power, abuse of privilege, misinformation, inconsistency, misplaced loyalties and irresponsibility are all results of organizational leadership that fails. The authors next move on to analyze the types of shadow casters or leaders who turn to monsters. Behind these types are such issues as unhealthy motivation, narcissism, faulty decision-making, lack of moral imagination and moral disengagement.

After analyzing the failed leadership the authors turn to demonstrate how the relational leadership emphasizes the social process of connecting individuals, and enhancing social networks is an integral part of leadership development. Relationships matter, and engaged organizations are built following a relational leadership approach. One important relational leadership competency for any global leader is inspiration. When it comes to a vision that inspires the followers, global leaders should be cognizant of the set of cultural values in order to build high performing teams. To influence the behaviors of in-

*Preface*

dividuals with diverse backgrounds, these leaders can incentivize them in order to construct a unified mindset focused on the success of the organization. The inspiring leader is perceived by others to be someone who has knowledge and sensitivity to the problems that need to be addressed. Furthermore, a relational leader is not someone who micromanages and forces others to follow; a relational leader is the one who guides. Relational leaders are above all passionate about helping others in their organizations. In addition to inspiration, there is also the trait of passion to help others. This type of quality is associated more closely with relational servant leadership. Another relational leadership trait is competence. Competence is a kind of thirst for knowledge and self-improvement that demonstrates skills needed to overcome challenges.

In their conclusion the authors introduce the needs of modern leadership and how culturally agile leadership meets the demands of our time. Leaders who recognize and support inclusive discourse not only respect the rights of others but also reflect a concern for justice for all. Through continuing education, leaders are able to enhance their leadership abilities, share perspectives on relevant issues, and work toward solutions in the public interest. Corporations are no longer confined to one local area, and many businesses are unable to operate without the use of technology, which widens the digital landscape across cultures and time zones. By supporting culturally agile leaders through a relational leadership development approach, organizations will organically create holistic leaders beyond the traditional competencies to include networks and relationships. This chapter has its strengths both in analyzing in detail the failings of ethically and culturally inadept leadership and showing how culturally agile leadership can be achieved.

Chapter 2, "Transition of Ecosystem Services Based on Urban Agro Ecology," by José G Vargas-Hernández focuses on the darker sides of modern negative industrial model of agricultural production. There is clearly a need to seek transition to other alternative forms capable to provide ecosystem services and produce food. Vargas-Hernández introduces the concept of 'Agro ecology' as a response to the problems associated with the industrial model of agricultural production. The simplified form of agroecosystem is altered with the input substitution that occurs in industrial agro systems. The industrial agriculture framed in the free market economic model has a destructive impact on sustainable natural resources, biodiversity, food security, environmental services and climate change, leading to ecosystem disruption and human hunger. Conventional agriculture is not sustainable for the rarefaction of inputs, global competition, loss of biodiversity and benefits from ecosystem services, climate change, etc.

The current consumption model based on industrial agro culture is disconnected from food production leads towards poor consumption patterns and choices of diets beyond the requirements of urban land planning and development. Furthermore, there is a gap in the agro ecological transitions between the agro ecological practices and the ecosystem services delivery. The relationship between agro ecology and nature is relevant in the current crisis is framed by the land used for agriculture and the land dedicated to conservation of biodiversity and ecosystem services.

The agro ecological struggles leading to demands for right to food, land redistribution, concerns for pollution of soil, land and water, provision of ecosystem services, corporate dismantling and access to fair markets of organic products, etc., are some of the emerging issues in urban environments. To break this pervasive relationship, it is required the incorporation of an integral conservation programs framed by minimizing land use dedicated to agriculture and more to conservation of agroforest ecological biodiversity and other ecosystem services.

Agro ecology and food production systems face some economic, social and environmental challenges when economies are organized around a competitive export orientation of crops that have impacts on the ecosystem services integrity, food quality, public health, and disruption of traditional rural livelihoods. The autonomy and solidarity between peoples of rural and urban agro ecology displaces the control of global markets. This chapter is seeking solutions and alternatives to one of the most basic areas of human life and societies, environment and food production.

In Chapter 3, "Corralling Contract and the Land Use Relationship Between Pastoralists and Agriculturalists in Nigeria," Regina Hoi Yee Y. Fu explores the social and environmental dimensions of the 'corralling contract', the indigenous fertilization system commonly practiced in the African Sahel area. The chapter examines the practice of the corralling contract between Fulani pastoralists and Nupe agriculturalists in the Bida region of Niger State of central Nigeria. This study follows carefully how the farmers and herders arrange the corralling contract, how they utilize this instrument and how it influences their social relationship. Findings suggest that pastoral Fulani groups have different strategies to maintain socioeconomic relations with specific villages through the adoption of corralling contract in order to ensure resources entitlement. Meanwhile, land sharing and pastoral camp rotation enable multiple Nupe village members to have access to manured fields. Contrary to the conventional description, findings do not show a strong tendency of increasing payment to herders for the corralling contract.

This study clearly has broader significance since it demonstrates how two groups of people with very different lifestyles and traditions have formed contracts that have proved to be a very beneficial and flexible instrument that has helped also to maintain good relations between the two groups. The world has many other examples of less fortunate relationships between groups that have different lifestyles. Symbiosis and cooperation are the means to solve the problems of the world and the Fulani and Nupe have a lesson to teach the world. Furthermore, the corralling contract itself has still great potential especially in areas where social and environmental conditions require careful balancing and creative solutions that take into consideration different needs of people.

Chapter 4, "Neuroeconomics: Connecting Health Science and Ecology," by Torben Larsen is a truly transdisciplinary work combining in a very innovative ways economics, ecology, brain science and political science in order to produce a model that would better meet the demands of our time, especially the influences of greenhouse effect.

The chapter first illustrates the history of Invisible Market Hand in economic theory and moves on to discuss the basic ideas of market economy, economic growth and profits. After this discussion the author seeks to combine economic theory with ecological realities of our time as well as with ideas on neuroeconomic psychology. To counteract the Green-house effect scientific databases are searched for Bioecological economic model. A Bioecological Model of Bronfenbrenner and Ceci focuses on human development as the interaction of genes and environment which is not specific to economics. More models of economic ecology are available. The best fit is the model of "Dynamic Integration of Climate and Economy" as currently updated. DICE is an equation system centering a Cobb-Douglas function enriched with both variables on climate costs and differentiated expectations on development of consumption and production. The target function of DICE is a social welfare function based on historic data on consumer behavior, specified in additional equations. Despite the complexity of the target equations, they display an *ex ante* limitation of human adaptability that doesn`t meet good standards of cybernetic target functions. From brain science the author uses ideas to demonstrate how anxiety and rationality are very much present in our societies and politics. According to the author the economic solution to the Green-house effect is remarkable simple and known among economists for 100 years as Pigovian tax.

*Preface*

The author argues that Carbon Emission Tax (CET) should be collected at the source where carbon enters the economy, at the wellhead, mine shaft, or import terminal. A full CET would cover more than 80 percent of the economy's total emission of greenhouse gas. The importance of collection at the source is illustrated by the flight-sector. CET at the import terminal gives flight companies a strong incentive to search for aircrafts with a lower consumption of petrol or even to search for non-fossil alternatives such as biofuel while a passenger tariff is simply a public money machine. As charter tourism is only about 25% of total civil air transport, the social lopsidedness of CET on civil flights is limited. Protection of agriculture is typical a strong national political interest. A reason for excluding livestock from CET is the growing world population that requires more agricultural products including livestock to prevent serious hunger problems, especially in Africa South of Sahara. Recently, new technologies have emerged that enable agriculture to become $CO_2$-neutral, too.

The author argues that carbon emission tax (CET) is the most effective intervention towards the Green-house effect and continues to argue that the short-term, democratic center-coalitions are too weak to implement CET due asymmetric levels of economic knowledge between the profession of economists and the public. Neuroeconomic psychology explains the dichotomy in complex decision-making as that of the difference between rationalist and pragmatic patterns of decision-making. However, educational experiences with intertemporal choice show that simple courses significantly improve the over-all quality of decision-making. This indicates that an accelerated public dissemination of Bioecological economics can improve decision-making in the complex public economic issues.

It is clear that there is a need to seek solutions to climate change by combining new ideas and approaches. This chapter does exactly this and hopefully helps to persuade others to join the debate and develop innovative strategies to tackle the plethora of challenges we are facing.

In Chapter 5, "Reading, Literature, and Literacy in the Mobile Digital Age," John Fawsitt analyzes the basic issues of human intellectual pursuits and how the society and educational policies have responded to changes taking place in 'reading' and 'thinking'. Reading, accompanied by discussion and writing, has long been the staple of a humanist education. It is felt that exposure to and the process of trying to understand other thinkers and their ideas enriches and trains the human mind to comprehend and evaluate their existences, world, and values and to use this knowledge to live a richer and better life whatever they judge that to be. The changes brought about by the advance of science have altered our perceptions and needs so much that there has been a change in what people seek and expect to achieve when they attend university or when they receive of education of any kind. However it is not yet clear how much continuity will be exhibited and in which direction our relationship with technology will evolve. Due to technological and social changes together with digitization, reading is becoming more intermittent and fragmented, and long-form reading is in decline. Engagement with various types of literature and written text has been seen as providing both information and intellectual exercise for all of the faculties including those of reasoning, memory, analysis, and comparison. In contrast what comes to our attention online especially on mobile platforms is more likely to have been chosen for us by algorithms on the basis of our own previously expressed preferences. Whether that expression was given consciously or not. Overwhelmingly the material is of a current, personalized, immediate and usually directly informative or entertaining nature. The algorithm's processes are opaque and we can't be sure as to whether they align with our values rather than our tastes.

The changes in 'reading' have revolutionized our education. Universities among other institutions are affected by the conditions in their surrounding societies, and indeed they are not remaining untouched by the wave of technological developments in the digital area. The motivations for the incorporation of many of these inventions and innovations can be described as not purely pedagogic. Factors taken into consideration include those of efficiency, finance, administration, and appearance. E-learning is now well established and seen as integral to the educational experience. M-learning (mobile learning) is now in the course of becoming part of the curriculum. In M-learning undergraduates are enabled and encouraged to use tablets and smart phones to participate in their courses. Furthermore, literature faces even more fundamental changes than 'reading'. It is unrealistic and unwise to presume that no response is necessary to the digital challenge. Deep reading can be deployed towards any type of text that requires it to gain benefit, such as longer informative articles or treatises and academic texts of whatever nature, however most commonly it is deployed in the area of literature. It is this field which depends on it for its existence. For it has no further purpose than to be read. A literature without reading has lost its connection to the world. In terms of education, literature often requires interpretation, in fact often a different way to interpret that provides a different angle from or through which to understand. No one field of research or study can alone claim to be able to fully analyze or advise. Economics has always assumed people as rational actors who will make choices according to their incentives. People's actual psychology remains far more complicated and unknown than this. Sacrifice, heroism and cruelty remain features of human nature beyond the ken of most schools of economic and scientific thought. Understanding comes in many ways. One of the most important is the ability to identify with the other's anxieties, hopes, and fears. Most disciplines approach their subjects from above and merely seek to convey understanding, of various depths, on an intellectual level.

It is in this way literature will help students become people capable of understanding on multiple levels and facets. In an era when careers even the shape of society cannot be predicted, literature and reading can help build the literacy necessary to create a cohesive society that acknowledges difference even our own, and sees that within each difference is a fellow soul. It seems that we have to expand our vision of literature and literacy from the written to encompass anything that deepens the understanding of the world and opens our minds to different levels of meaning.

The author analyzes the consequences of the new technology that has already fundamentally changed the ways that younger generations 'read' and 'think' and how especially the emotional, attentional and cognitive demensions of 'reading' have changed. There seems to be no way back to the old world. In short, it is time to wake up and realize that there are huge consequences for educational policies and academic teaching – as well as for life in general.

Chapter 6, "Strategic Transformational Transition of Green Economy, Green Growth, and Sustainable Development," by José G Vargas-Hernández focuses on green area innovation. The idea is that public and private city spaces are developed towards social good and increasing their environmental value. This process is fostered by innovation and new technologies. In the last decades, several contradictions have been generated in terms of urban socio-spatial relations, and the authoritarian urban space and democratic urban space. The logic of the management of the dominant urban space is to create a space that, being public, is authoritatively constructed, that is, it is a restrictive and selective public. The urban space itself makes any articulation of the social movement difficult. The social movements in Latin America developed a very strong presence during the 1990s in such a way that they forced a reconfiguration of the political spectrum dominated by leftist governments. They arise in a context of a democratic opening that takes place after military dictatorships, within a tradition that is an oligarchic tradition and of much social inequality.

*Preface*

This social inequality meant that in the beginning any social struggle had to be very organized, it had to be very strong, because the social inequality was so great that the oligarchic classes were going to defend themselves by all means, they had defended themselves with the dictatorship and they were going to defend themselves with democracy. Authoritarian spatial conceptions develop in large part at a time when social polarization and social inequality began to endanger governance. Authoritarian urban spaces are the dominant urban spaces that try to defend themselves from a popular reaction, private urbanizations are exactly a good example, among others.

A sustainable city is supported by an urban planning and designing paradigm. Urban design and planning supported by local governments are urged to spur technical innovations, innovative practices and contribute to develop new business models and marketing strategies aimed to find green resource-efficient solutions for climate change adaptation and mitigation while restructuring and reinventing local economies.

Integrating environmental sustainability issues into green areas innovation strategy and greening the innovation process are becoming a strategic opportunity for organizations, communities, local governments, neighborhoods. Environmental sustainability is an integral part of green area innovation requiring greater demands of ecological performance out of eco-efficient green technologies with reduction and mitigation of green negative environmental impacts. An innovation environment also creates new business opportunities. The process of urban planning is aimed to create and develop an inclusive economy centered on innovative improvements at local level. The green urban planning has three priorities: the economic growth and development, social equity and the environmental sustainability. Local governments at the same time that other levels of federal and state authorities and agencies can take different approaches based on environmental economics and technologies innovations and political commitment to incentive and motivate involvement of actors and stakeholders to find specific innovative solutions to urban planning. Transforming green infrastructure models into action require new approaches of innovative planning and policymaking of programs.

Vargas-Hernández advocates that the urban public space must be reconstructed with a sense of collectivity, as an urban space of coexistence, of emotion, of trust, of looking, and it is the space of embracing. They are all urban spaces that must be built and, therefore, that urban space is a great achievement at this time. Furthermore, local communities and governments should initiate selectively some green area projects with different degrees of radicalness dimensions following a profile according to a strategic urban planning. Development of sustainable communities in smart cities can be supported by design, transfer and implementation through collaborative urban planning of innovative urban policies.

Future priorities for research in urban innovation should identify and develop principles of sustainable urban green planning and development to provide support to policy makers in local governments to design and implement mechanisms for more resilient communities and neighborhoods in cities. Green resilience innovation of urban life in contemporary cities can be supported in digital cities with the implementation of cyberspace to cities.

Chapter 7, "Cinematic Works-Based Affective Urban Design Atmospheres," by Hisham G. Abusaada uses the concept of 'Atmospheres' in very innovative and informative way to tackle cinematic works, literature, architecture and urban design and continues to show how the teaching in the realm of architecture and design could benefit from deeper understanding of 'Atmospheres'. While the concepts of atmospheric architecture, atmospheric perception, and aesthetics atmospheres are well-known, especially in regard to environmental design, the disregard of the concept of the atmospheres occurs mostly during the process of urban design – sometimes with stark consequences. In individual buildings typical examples of atmospheric elements include the use of climatic effects, colors, shade, light, and sound. The purpose of this chapter is to prove that the creation of the atmospheres of cities is as plausible as in artworks; both having similar aspects for making; which are ideas, themes, and dramatic text, as well as the technical elements. This work focuses on revealing the contribution of wide range of art (in particular, music, novels, and cinema) to deepen understanding on the impact of societal transformations on everyday events in different urban settings and places.

The chapter introduces the history of 'Atmospheres' as a concept as well as provides a wealth of information about the practices and prevailing forms of thought within epistemic communities related to architecture and urban design, especially in Egypt, but since the chapter follows the long journey of 'Atmospheres' in the history of thought and arts, the chapter really has a global significance. The novels of Naguib Mahfouz prove out to be particularly perceptive in terms of atmospheres and many of them have also been turned into cinematic works. After the stimulating journey among the past experiences of atmospheres the chapter turns to practical concerns of present professional practices—particularly in the field of urban design—which all too often fall short of paying full attention to the atmosphere. The goal, isn't to restore the atmospheres of the past but to find out how to create great atmospheres based on the design of the city. This goal starts with the process of improving teaching methods to achieve this in architectural education institutions and applying them in practice.

Chapter 8, "Environmental Problems of Delhi and Governmental Concern," by Mohammad Yousuf Bha focuses on the environmental issues and the issues of governance by analysing the case of Delhi, the capital city of India. Delhi is also a major industrial and commercial center with a number of environmental problems, such as air and water pollution, slum development and congested housing. The chapter follows the historical narrative of the city and focuses one by one on interesting related issues. Geography, economy and population growth are closely related. Delhi is not an ideal location for industrial development. Yet, for variety of socio-political reasons, it has become a key industrial center. This not only causes pollution, it also acts as a magnet for migration. Relocation of industries from Delhi should be one of the major objectives, as this research indicates.

As for the environment, pollution is one of the most critical problems facing the public and concerned authorities. According to the World Health Organization, Delhi is the most polluted city in the world in terms of suspended particulate matter. The growing pollution is responsible for increasing health problems. The deteriorating environment is the result of population pressure and haphazard growth as industrial development has largely been haphazard and unplanned. Though in 2002 metro rail service was initiated as safe and pollution free transport system the first phase got completed only in year 2006, but literally it has not sufficed the growing transportation needs of the huge population. Road transport has remained the sole mode of public transport over the years; there has been a phenomenal increase in vehicle population. Data accessed from the transport department of Delhi government puts the total number of registered vehicles at 1,05,67,712 till May 25, 2017 among them 31,72,842 are registered cars. However, motor cycles and scooters which make biggest chunk of the registered vehicles i.e. 66,48,730 are major air polluters due to poor emission standards.

As for the water issues, the city of Delhi has very little water sources of its own. The groundwater that is available is not adequate to serve the needs of the population of Delhi. Already 67% of the ground water resources has been utilized and the remaining water is of poor quality and further exploitation may result in serious degradation. Results of extensive chemical tests indicate that except for isolated wells,

*Preface*

the ground water of Delhi area is generally brackish and unfit for drinking or irrigation. Furthermore, domestic and industrial sewage generated within the Delhi capital area is the main source of pollution of the river Yamuna during its passage through the metropolitan area.

As for air pollution, it is estimated that about 3,000 metric tons of air pollutants are emitted every day in Delhi. The sources of air pollution in Delhi are: emissions from vehicles (67%), coal based thermal power plants (13%), industrial units (12%)) and domestic use (8%). All these environmental problems, such as air and water pollution, slum development and congested housing together and separately are creating gastrointestinal diseases, respiratory ailments, heart and various microbial infections abortions, miscarriage, stillborn, primary and secondary sterility, psychosomatic disorders, mental retardation, schizophrenia epilepsy and learning disabilities among school going children. In case of Delhi Apex Court seems to be playing predominant role to control pollution rather executive and legislature should have played proactive role in making Delhi a model city. This chapter strongly advocates an Action Plan for a particular phase of time to introduce tough anti-pollution measure. It is quite imperative that an effective policy decision is needed for appropriate allocation of land and control over unsustainable developmental process, that will result in meaningful impact on the existing environment and it is also vital for sustaining the desirable environmental quality in Delhi.

Chapter 9, "Hydrogen: An Environmental Remediation," by Athule Ngqalakwezi, Diakanua Bevon Nkazi, Siwela Jeffrey Baloyi, and Thabang Abraham Ntho focuses on the hydrogen from the point of view of environmental remediation. Global warming, in particular, requires quick remedial actions and therefore technologies that aim to alleviate this issue are of paramount importance. In this chapter, hydrogen is discussed as an alternative energy that can potentially replace traditional fuels such as diesel and gasoline. In short, this chapter explores the economic and technological viability of hydrogen for automotive applications but also shows that hydrogen economy potentially has also many other benefits for the economy, environment, and energy security. However, to make the use of hydrogen a true and popular alternative the storage of hydrogen as a gas, liquid and solid need to improved. The key issues in hydrogen storage are also highlighted to show how hydrogen actually has plenty of potential as a practical alternative. Furthermore, regulations and legislations concerning the emission of greenhouse gases from fossil fuels based sources have also been discussed as a background for finding alternatives. The environmental remediation standards themselves have been a subject to interest and research in different parts of the world, as the chapter explains. This chapter demonstrates that hydrogen, indeed, has plenty of potential for environmental remediation and that progress in this issue requires broad cooperation among people. The hydrogen economy and its storage for on-board applications still largely remain in the development phase. However, great progress has already been made in this field. The collective efforts by the governments, corporate businesses, and research and development spheres across the globe have already made a valuable impact in the fight on climate change and the magnitude of problem or risk for the whole planet surely requires that all available alternatives must be examined without delay.

Chapter 10, "Healthcare in India: Challenges and Innovative Solutions," by Reenu Kumari addresses the problems of future prospects of health care and health care systems in India. The health care system in India currently is multi-layered and complex, which makes it difficult to understand its true potential and appropriate quality of services. In India health care system is very approachable, but there remains many differences in quality of health care system between rural and urban areas. Most of the private hospitals are located in urban areas while public hospitals predominantly exist in rural areas. This is the main reason behind the unequal access to healthcare system in India. Private hospitals try to maintain their reputation in market through providing low cost medicine and high quality of services. On the other hand, the public hospitals provide low cost medicine, but are often short of basic hygiene and cleanness as well as such basic facilities and services as servant services, compounders, qualified doctors, appropriate infrastructure and equipment. Furthermore, a large number of private hospitals are not able to provide modern health care, because they are not fully aware and trained in modern medicine and, instead, rely heavily on traditional medicine and such practices as the *ayurveda, siddha, unani* and homœopathy.

Kumari points out that India needs to adopt new strategies that will ensure the protection of health of all individuals. Due to health financing problem in India there is marked health inequity, unequal accessibility to proper equipment and poor quality in terms of hospital services. In rural areas expenses of private hospitals are out of pocket and, on the other hand, in urban areas people can better afford such services. Moreover, rural citizens are not even aware about facilities and different types of medical insurances. So, every year government of India has tried to increase the amount of money allocated to health care (expenditure on health care) but these policies are not implemented in systematic and effective enough way. If the government increases public expenditure (including health) it should also be able to contribute to batter quality in the public and private health care sector. The health care also includes many other aspects such as the clinical trials, medical devices, telemedicine, health insurance, medical tourism and medical equipment. The Indian government faces a real challenge in bringing quality and good governance to all areas of public health.

Chapter 11, "Medical Tourism Patient Mortality: Considerations From a 10-Year Review of Global News Media Representations," by Alicia Mason, Sakshi Bhati, Ran Jiang, and Elizabeth A Spencer focuses on different forms of medical tourism and provides an analysis of patient mortality. As an industry medical tourism involves both the treatment of illness and the facilitation of wellness, with travel. Those who engage in the process of medical tourism are called 'medical tourists' and they do so for a variety of reasons. Many seek access to advanced medical technology, higher quality of care, and quicker access to novel or restricted medical treatments and procedures in their home countries. Medical tourists are motivated by several factors. Lower-cost procedures and discretionary cosmetic operations represent only a small segment of the global patient base. The chapter cites a report according to which 72% of respondents sought orthopedic/spine and oncology/cancer care, 60% are those seeking cosmetic/plastic surgery, 54% are seeking cardiovascular treatment, 52% neurology, while 40% are seeking IVF/fertility treatments. Critics maintain medical tourism is a poorly regulated healthcare industry, that increases patient risk due to weak pre-operative counseling and poorly coordinated post-operative treatment plans. Critics also argue that in some destinations the process creates preferential treatment centers, or dual delivery healthcare systems, that exacerbate current health disparities in local populations. Affordability is a driver. To create a sample for analysis, three researchers identified patient mortality cases using a variety of digital and database search strategies. The strategic focus was media representations of patient deaths and our research efforts concentrated on identifying cases presented in U.S., India, and Chinese media outlets. The final dataset included 50 cases, of these 80% are females ($n=40$) and 20% are males ($n=10$) who died as a result of medical tourism during the January 2009 to March 2019 time period. Twenty percent of the cases ($n=10$) resulted from domestic medical tourism, 80% ($n=40$) were international medical tourists. Findings show the most dominant global media representation of medical tourism patient mortality occurring between 2009 and 2019 is framed as middle class minority females between 25-55 years of age who seek cosmetic surgery internationally. Improved warning and advisory systems are needed to alert patients in high-volume, high-risk areas of cosmetic medical tourism destinations. Death is the ultimate price, and many seekers do not die, but are instead left with

*Preface*

scars, disfigurements, or remain in need of long-term care. Patient morbidity, in contrast to mortality, is a domain of exploration that may also yield novel considerations and recommendations for those considering international healthcare treatments.

Chapter 12, "Magnetocaloric as Solid-State Cooling Technique for Energy Saving," by Claudia Masselli, Ciro Aprea, Adriana Greco, and Angelo Maiorino focus on magnetocaloric technique as a potential energy saving and global warming remedy. The refrigeration is responsible of more than 20% of the overall energy consumption all over the world and most modern refrigeration units are based on Vapor Compression Plants (VCP). The traditional refrigerant fluids employing VCP, i.e. ChloroFluoro-Carbons (CFCs) and HydroChloroFluoroCarbons (HCFCs), have been banned by the Montreal Protocol in 1987, because of their contribution to the disruption of the stratospheric ozone layer. Currently, the most mature, eco-friendly, solid-state caloric technique is Magnetocaloric Refrigeration (MR), whose main innovation consists in employing solid materials with magnetic properties as refrigerants, able to increase or decrease their temperature when interacting with a magnetic field. Instead of the fluid refrigerants, proper of vapour compression, a magnetic refrigerant is a solid and therefore it has essentially zero vapour pressure, which means no direct Ozone-Depleting Potential and zero Global Warming Potential. Most of the criticism lays in the high costs in magnetic field generating and the expensiveness of the most promising magnetocaloric refrigerants. In this chapter magnetocaloric cooling is introduced from the fundamental principles to the present state of research. The authors believe that the magnetocaloric refrigeration and heat pumping technology have great potential and competitiveness for a large-scale implementation and for major positive environmental impact.

Chapter 13, "Global Racial, Ethnic, and Religious Subculture and Its Impact on Healthcare Organizations," by Rohit Singh Tomar raises with a transdisciplinary approach issues of ethnicity, culture and religion together with issues of health. Chapter provides examples and cases where innovative practices improved the behavior of staff towards patients. In many tribes and religion certain diseases are considered as taboo and treated with the help of traditional practices. Spiritual and religious practices have played an important role in the treatment of the psychological and mental illness. Much evidence is presented in research to support the role of spirituality and religion in the patient care and recovery from the diseases. However some studies do not support that religion and spirituality plays an important role in reducing human health related problems. Some studies don't even find any correlation in health care and religion. Health care seems to be affected by race, ethnicity and religion but there is a huge scope of quantitative analysis to get a detailed and comprehensive result. Study needs to be conducted across the world based upon primary data where it could compare various culture, sub-culture, ethnicity, races and religion and its impact on healthcare industry and patients.

Religious practice brings positivity in human life by promoting good health practices. This positive energy improves physical and mental health. Visiting religious place and praying brings in happiness by reducing depression and improving self-esteem. Religious practice improves physical strength and increases longevity; it improves an individual's chances of recovering from illness, and lessens the incidence of many diseases. Perhaps religion connects an individual with god, which is nothing but a manifestation of faith in life. Most religion condemns self-destruction of any kind. Beside that other followers serves as a support network, which unites an individual with family and friends. All religions suggests regimented pattern of living life where due importance is attached to cleanliness, eating habits, nutrition, exercise or any physical activity, way to live life and to be abstemious. This results in reduced mortality risk from infectious diseases and diabetes.

Chapter 14, "Assessment of Healthcare Service Quality: Tertiary Care Hospitals of Dhaka City," by Segufta Dilshad, Afsana Akhtar, Sadrul Huda, and Nandeeta Samad, provides a comprehensive analysis of healthcare services and tertiary care hospitals in Dhaka, Bangladesh. This study deals with quality assessment of health care facilities concerning the private and public health facilities of Dhaka, Bangladesh. An informative introduction of the health care situation in Bangladesh is followed by a survey research. Using the purposive sampling method, the individuals have been selected from households, who have received healthcare services from public or private hospitals of Dhaka city in last one year.

The results illustrate patient's level of satisfaction with the public and private health care services they received from different hospitals Dhaka city. For instance, among the male respondents 60.27% and among the female respondents 60.27% of them were satisfied with the services of healthcare services of Bangladesh and dissatisfaction was high among public hospitals comparing with private hospitals. The evidence further reveals that most of the participants had experienced facing deteriorated health condition at the public hospitals. Among the respondents, 76.2% of them were satisfied with the service quality of private hospitals and on the other hand, 76.4% of the respondents are not satisfied with the service quality of the public hospitals. Regarding treatment service satisfaction, the public hospitals services were less satisfactory comparing private hospitals. Despite some positive experiences, majority participants opine that they are less satisfied with public services specially the ways the services are operated and delivered to them. The situation of dissatisfaction arises due to many factors, which include administrative, interpersonal, environmental and technical service quality. The common expectations of patients are to get better quality of care, supportive staff members, and standard physical environment of hospitals.

Chapter 15, "Application of Nanotechnology in Global Issues," by Ranjit Barua, Sudipto Datta, and Jonali Das, raises the issue of present and potential significance of nanotechnology in different walks of life. Nanotechnology is already serving significant role in fueling progress and transformation in numerous fields of technology, natural sciences and engineering divisions such as energy, information and communication technology, medicine, environmental technology, home security, transportation, and food safety, with lots of others. Existing nanotechnology has greatly transformed physics, chemistry, biotechnology and material science and as a result we already know how to make innovative materials that have distinctive characteristics as their arrangements are resolute on the nanometer scale. Science and technology continue to progress forward in creating the fabrication of micro and nano-devices and systems that have major significance for a variety of consumer, biomedical and industrial applications.

It is widely felt that nanotechnology will be the next industrial revolution. Today, many of world's most creative scientists and engineers are finding new ways to use nanotechnology to improve the world in which we live. We see doctors detecting disease at its earliest stages and treating illness such as heart disease, cancer and diabetes with further effective and safer medicines. Although there are many research challenges ahead, nanotechnology already is producing a wide range of beneficial materials and pointing to breakthrough in many fields.

Chapter 16, "The Social Significance of Visual Arts: State, Nation, and Visual Art in Japan and Finland," by Mika Merviö, focuses on visual arts in Japan and Finland and the ways that social and political developments have been tied to reconstruction of cultural traditions. In particular, the most important turning point appears to be the establishment of modern state. However, the particular strains of nationalisms in Japan and Finland were very different reflecting different historical and social backgrounds. In both cases the state and political elite played an important role when they tried to influence the development of visual art traditions. In the Finnish case the art was widely understood to be a very important tool of

## Preface

social development. In Japan, the political elite for a long time was suspicious of freedom of expression, which certainly had a major impact on the development of Japanese visual arts.

The idea of this chapter was to look at parallel developments in these two countries both in terms of contrasts and similarities. In both societies the visual arts have a long history and often in national history writing the cultural traditions are presented as a long unbroken narrative. The reality is seldom so simple as the narratives of development of visual arts is these two societies clearly illustrate. However, the social and political developments have been greatly affected by cultural developments, but also culture has played a major role in social development. A more careful analysis or more nuanced reconstruction of past developments in these areas indicates that in both societies individual artists played a major role in changing the direction of art. In both societies art has survived very different challenges from strict censorship to ignorance of aesthetic values.

# Chapter 1
# Toward Leadership Agility

**Simon Cleveland**
https://orcid.org/0000-0001-9293-3905
*Georgetown University, USA*

**Marisa Cleveland**
*Northeastern University, USA*

## ABSTRACT

*Companies are often challenged by the cultural diversity of the growing workforce. As a result, organizational leaders should develop culturally agile competencies in order to engage and motivate their employees. Leadership development programs exist to contribute to an individual's and an organization's success; however, there is a lack of studies that examine how such programs contribute to the development of cultural agility in leaders. Culturally agile leaders are more inclusive in their hiring practices and more open to encouraging more diversity within their own leadership network. Such leaders value collaboration and understand how culturally-grounded traditions and preferences effect transactions. This chapter addresses the roles of positionality and cultural agility, leadership development programs, and capacity and responsibility in building culturally agile leaders. The chapter also proposes how leaders have the capacity and the responsibility to develop other leaders through a relational leadership approach to promote inclusion and diversity.*

## INTRODUCTION

Leaders in today's interconnected world must add cultural agility to their competencies. How leaders respond in unpredictable circumstances, whether they are flexible and able to adapt to unique situations, will contribute to their effectiveness as a leader. With the digital landscape and globalization of the marketplace, culturally agile leaders recognize how cultural diversity makes a difference in how employees act and react. Culturally agile leaders value collaboration and understand how culturally-grounded traditions and preferences effect transactions. Of significant relevance, culturally agile leaders will be more inclusive in their hiring practices and more open to encouraging more diversity within their own leadership network. As global corporations continue to hire employees from around the world,

DOI: 10.4018/978-1-7998-3576-9.ch001

individual cultures are merging with organizational ones. Adults already in the workforce are learning to adapt to the cross-cultural engagements, but without the proper education or training to promote the cultural adaptability, efforts to produce positive corporate results may fail.

This concept paper aims to address three main areas of interest: positionality and cultural agility, leadership development programs, and capacity and responsibility. The paper will begin with an examination of the ways a leader's positionality and level of cultural agility inform and limit their ability to make an impact on their organization. Next, contemporary issues in leadership development programs will be explored. The paper will conclude with a reflection on how leaders have the capacity and the responsibility to develop other leaders through a relational leadership approach to promote inclusion and diversity.

## CULTURAL AGILITY

Cultural agility inform and limit a leader's ability to make an impact on their organization. According to Milner (2007), "The nature of reality or of our truths shapes and guides our ways and systems of knowing and our epistemological systems of knowing" (p. 395). The default leadership studies and characterizations revolved around white males; however, within the last two decades, researchers have examined leadership from a cross-cultural perspective and included women and marginalized individuals (Kezar & Lester, 2010). Within the workplace, culturally diverse teams find it challenging to achieve open communication. This lack of communication can hinder a team's creativity and their ability to be productive.

Because who a person is shapes how that person perceives the world, leaders must recognize how their own subjectivity controls their perspectives and how, for the most part, their perspectives are limited to the constraints placed upon them by the perceived societal norms and the role models who raised them to understand these norms. Because today's societies are far more diverse than in the past, race plays a large factor in how leaders will need to approach actively pursuing a more inclusive workplace.

Kezar and Lester (2010) found studies examining leadership qualities focusing on gender, race, ethnicity, and the intersectionality of race and gender. In analyzing studies exclusively on women leaders, Kezar and Lester (2010) found women leadership is more participatory, relational, and interpersonal. Many "women leaders tend to conceptualize leadership as collective rather than individualistic, emphasize responsibility toward others, empower others to act within the organization, and deemphasize hierarchical relationships" (Kezar & Lester, 2010, p. 164).

Leaders must reflect about themselves in relation to others and acknowledge the multiple roles, identities, and positions that each member of the organization contributes to the organization. Positionality plays an important role in professional practice and research. A scholar-practitioner's positionality informs the problems identified and the questions asked about the respective contexts. A leader's identity informs the positionality, and there is an interconnectedness between positionality and power.

## LEADERSHIP DEVELOPMENT

Extant literature suggests that leadership is a process of influence and collaboration (Cleveland & Cleveland, 2018; Liang & Sandmann, 2015; VanVactor, 2012). Furthermore, collaboration is regarded as critical to success and "occurs when multiple individuals work together toward a mutual benefit"

(VanVacor, 2012, p. 560). Within an organization, leaders must encourage collaboration among their employees, through teams and shared company goals. With the rise of global corporations, and the increasing diversity within the workforce, leaders have a wider field of influence than previous leaders in previous centuries.

Leadership development is an important issue in corporate America (Galloway, 1997). Organizations are spending more time and money on leadership development to develop leaders (Cullen-Lester, Maupin, & Carter, 2017; Galloway, 1997; Know, 2014; Muyia & Kacirek, 2009). The W.K. Kellogg Foundation categorized 55 leadership development programs into outcomes: individual, organizational, community, field (communities of practice), and systemic effects (Nissen, Merrigan, & Kraft, 2005). Larsson, Sandahl, Söderhjelm, Sjövold, and Zander (2017) recognized leader development and leadership development differences and similarities and presented a theory-based, longitudinal study of leaders and their subordinates.

Eich (2008) used a grounded theory and studied attributes of leadership programs. The research identified three clusters of attributes for high-quality leadership programs: participants engaged in building and sustaining a learning community, student-centered experiential learning experiences, and research-grounded continuous program development. The study also showed that organizations of all sizes face challenges in regards to leadership training and implementing successful leadership development programs.

After analyzing a group of 30 CEOs and human resource executives, Groves (2007) introduced a best practices model for leadership development that is inclusive for succession planning in organizations. Groves (2007) notes evidence that supports the idea of using mentors in developing networks and exposing high potential employees to multiple stakeholders.

## Leadership Development

Tubbs and Schulz (2006) define leadership as "Influencing others to accomplish organizational goals," (p. 29). Leadership is essential for the success of any organization and without it, companies would not meet their objectives to deliver products and services to their customers. Most theorists recognize leadership as a process of influence (Cleveland & Cleveland, 2018; Sinclair, 2013). People become more culturally agile when they seek to acquire knowledge beyond their own assumptions. Caliguiri (2013) suggests several approaches to knowledge acquisition, experiential learning, and personal reflection: 1) question your own assumptions about culture, 2) learn how to learn about specific cultural differences, 3) build deep knowledge about one or more other cultures, 4) become comfortable being uncomfortable, 5) ignore your passport stamps and frequent flyer miles, 6) get some passport stamps and frequent flyer miles, and 7) be honest with yourself.

The rise of management education, business schools, and leadership development elicits the question of whether leadership can be taught (Daloz Parks, 2005; Day, 2000; Sinclair, 2009). Pigg (1999) acknowledges the importance of influencing behaviors in leadership development. The focus of leadership and leadership development is expanding beyond competencies to include networks and relationships. According to Yang, Yang, Zhang, and Spyrou (2010), "People's movement behaviors are strongly affected by their social interactions with each other" (p. 6). Cullen-Lester et al. (2017) present a conceptual model explaining how network-enhancing leadership development improves the leadership capacity of individuals and collectives.

Effective leaders are found throughout an organization, not just at the top, and organizations are dedicating significant amounts of time and money to develop and train emerging leaders in their organizations (Muyia & Kacirek, 2009). As companies expand and compete in today's digital marketplace, leaders are focusing resources on training emerging leaders in a combination of technical, conceptual, and human skills (Muyia & Kacirek, 2009). Terry and Liller (2014) argued that a leadership action plan is used to enhance and refine a leader's abilities. Such an action plan can serve as a roadmap to guide a leader toward a greater focus on the objectives while charting a course to navigate any challenges.

## Leadership Shadows

Organizational leadership is a challenging job. Leaders often have to overcome a number of shadows, which when ignored can often lead to organizational failures. Johnson (2017) proposes a number of leadership behaviors that cast a dark shadow in the ethical space. Some of these include increases in employee workload, while disregarding the interests of the key stakeholders, higher rates of absenteeism, humiliating and bullying employees, and claiming credit for the work of others. Johnson terms these the behaviors of psychopaths and attributes them to faulty decision making.

### Misuse of Power

The first shadow described by Johnson is the shadow of misuse of power. The author divides the power into two categories: soft and hard (Nye, 2008). The soft power is used by leaders to drive vision and to formulate relationships with employees. In contrast, the hard power is used to allocate tangible deliverables such as terminations or monetary rewards. Additionally, Johnson proposes five power biases: coercive, reward, legitimate, expect, and referent (French & Raven, 1959). Leaders in organizations exercise each of these biases to accomplish specific goals.

### Abuse of Privilege

The second shadow is the one of abuse of privilege. Johnson argues that the higher the power of the leader, the larger the leader's reward. Leaders are often rewarded hundreds of times greater than their employees simply because of their positionality in the organization (Sandler, 2015).

### Misinformation

The third shadow is the one of misinformation. Johnson argues that leaders have greater access to information which they can use for their personal gains (O'Connor, 2015). Information can be leveraged to subdue certain employees, or to reward others. For example, leaders may choose to hide the truth about the financial situation of the company in order to prevent certain employees from leaving prior to a company merger.

### Inconsistency

The fourth shadow is the one of inconsistency. Typically, inconsistencies most often arise when leaders depart in their decision making and actions from the status quo. For example, company policies may prohibit fraternizing with subordinates and nepotism; however, leaders are often found guilty of breaking such policies and engaging in behaviors that are in direct conflict with these policies (Frolich, Sauter & Stebbins, 2015).

## Misplaced loyalties

The fifth shadow is that of misplaced loyalties. Johnson argues that leaders frequently betray their employees and shareholders. One such example is the gross mismanagement of the pensions and investments of the Enron's employees and shareholders by the senior team who placed their own self-interests before those of others (Berman, 2008).

## Irresponsibility

The last shadow is that of irresponsibility. Leaders often engage in irresponsible behaviors because they rely on their power to shift their responsibilities and mistakes to others. There are countless examples of this shadow, such as the irrational behaviors of the leaders of WorldCom, Tyco, AIG and others during the Dot Com era.

# Shadow Casters

Johnson (2017) proposes five shadow casters that deserve recognition as should be studied by leaders in order to prevent them from occurring.

## Unhealthy Motivation

The first one is unhealthy motivation. This particular shadow caster is typically based on flawed personal characters. Johnson calls these monsters and subdivides them into six groups: 1) insecurity – associated with lack of personal belief in their own capabilities to lead; 2) battleground mentality – associated with the idea that life is series of battles that should be won, thus approaching each situation as a win or a loss; 3) functional atheism – the belief that the leader is the one carrying all responsibilities for failure of their organization; 4) personal fear; 5) denial of death – associated with fear to end failing projects and initiatives; 6) pure evil (consider the Holocaust).

## Personality Disorders

The second shadow is that of personality disorders. Johnson discusses narcissism, the idea of self-reliance and self-confidence, as one of the most resonating features of a leader's personality disorder (Higgs, 2009).

## Faulty Decision Making

The third shadow is that of faulty decision making. Some of the most frequent occurrences of this shadow includes planning fallacy and optimism bias (Flybjerg, Garbuio & Lovallo, 2009).

### Moral Imagination

The fourth shadow is the failure of moral imagination (Caldwell & Moberg, 2007). Johnson proposed three components in this shadow: 1) ethical sensitivity; 2) consideration for others; and 3) innovative solutions. Johnson argued that when leaders fail in each of these components, they open the doors for further organizational failures.

### Moral Disengagement

Finally, the fifth shadow is moral disengagement. Here, while leaders may have the knowledge about right and wrong, they may choose to act contrary to this knowledge, thus disengaging from the moral baseline (Bandura, 1999).

## Relational Leadership

Relational leadership emphasizes the social process of connecting individuals, and enhancing social networks is an integral part of leadership development (Cullen-Lester et al., 2017; Wasserman & Faust, 1994). The quality of the relationship between leaders and followers should be the focus with regards to leadership development education (Pigg, 1999). Maznevski and DiStefano (2000) proposed that global teams are more complex than domestic ones. This is because such teams consist of individuals with diverse cultural backgrounds. These cultural differences often lead to diverse relationship forming or decision making. As such, leaders of global teams should be cognizant of the cultural differences in order to build high-performing teams. These leaders need to be able to influence the teams in order to achieve the goals of the organization (Tubbs & Schulz, 2006).

Many contemporary approaches to leadership lack the embedded realm of everyday experience (Cunliffe & Eriksen, 2011). When investigating how to support culturally agile leaders, a relational leadership approach allows for conceptual and practical applications. Relational leadership is both a way of theorizing leadership and being a leader, noticing subtleties of relationships, engaging in dialogic conversations, respecting others, a way of viewing the world as intersubjective, emerging in our relationships with others, understanding the importance of conversations, (Cunliffe & Eriksen, 2011).

Cunliffe and Eriksen (2011) examined leadership as an embedded experience and in relationships, and a review of extant literature revealed three main themes: relationships between network elements, the social construct of leadership, and post-heroic leadership as the collective social practices of people. These three themes guide leaders toward a heightened level of awareness when recognizing relationships, using specific language, and thinking about collective activities (Cunliffe & Eriksen, 2011). If individuals are shaped by their social experience, then relational leading is focused on the kinds of relationships between individuals. Relational leaders use dialog to communicate with others in a meaningful and purposeful way (Cunliffe & Eriksen, 2011).

Relationships matter, and engaged organizations are built following a relational leadership approach. Carroll and Simpson (2012) examined leadership development as it relates to "the social capital required to build relationships that promote cooperative work" (p. 1284). As recognized by Carroll and Simpson (2012) and Day (2001), leadership development focuses on the relational dimensions of leadership practice. Williams and Wade (2002) recognize that leadership is a shared relationship.

"Influence occurs when individuals make behavior changes to be consistent with their peer network" (Valente & Pitts, 2017, p. 105). Relational leadership views leadership as a social construction (Uhl-Bien, 2006). Limited access to networks can impede a woman's career advancement (Athanasopoulou, Moss-Cowan, Smets, & Morris, 2018). Athanasopoulou et al. (2018) found women developed networks and viewed those networks as potential sources of information that would help them do their current job better, whereas men viewed their networks as ways to get a better job. Simosi and Xenikou (2010) obtained empirical evidence from 300 employees working in a large Greek service company to explore the nature of relationships between leaders and the commitment of employees. The findings concluded that the relationship of organizational leaders through social exchange with their followers have an effect on the followers' emotional attachment.

"Issues related to interpersonal development... concern the capacity for individuals to build relationships with others, such as enhancing a group's social capital and engaging in authentic leadership with followers" (Subramony, Segers, Chadwick, & Shayamsunder, 2018, p. 121). Individuals with diverse backgrounds form strategic networks where new opportunities become more visible (Upson, Damaraju, Anderson, & Barney, 2017). Groves' (2005) found a relationship between charismatic leaders, their followers, and their willingness to follow specific types of leaders.

## Relational Leadership Competencies

### Inspiration

One important relational leadership competency for any global leader is inspiration. When it comes to a vision that inspires the followers, global leaders should be cognizant of the set of cultural values in order to build high performing teams. To influence the behaviors of individuals with diverse backgrounds, these leaders can incentivize them in order to construct a unified mindset focused on the success of the organization. Bass (1988) argued that the inspiring leader is perceived by others to be someone who has knowledge and sensitivity to the problems that need to be addressed. Furthermore, a relational leader is not someone who micromanages and forces others to follow; a relational leader is the one who guides.

### Passion

Relational leaders are above all passionate about helping others in their organizations. In addition to inspiration, there is also the trait of passion to help others. This type of quality is associated more closely with relational servant leadership. According to Albright (2016), "to lead effectively, we must first serve, and that means the legitimate needs of others is the essence of what it means to serve" (Albright, 2016, p. 19). This quality focuses on the employees and their development. Leading is a form of guiding and supporting, understanding, being genuine, developing relationships, and building a community.

### Servitude

Albright (2016) identifies servitude as key trait in leaders who invest in their employees. The researcher breaks down servant leadership into four key values: inspiration, equality, community incorporation, and guidance. By investing in employees, servant leaders encourage and support their employees to hone their successes, instead of the leader's success. In organizations, servant leaders would be effective in helping their employees adapt to change by learning how the employees process things, and together, work on ways to integrate better within the organization. The practice of servant leadership is often characterized with positive results, such as decreased employee turnover (a major inhibitor for company success), improved job satisfaction, and employee loyalty. Inspirational leaders who practice such qualities will ensure that their organizations create a safe place where employees remain committed to the mission of the organization and motivated to follow its vision.

## Competence

A third relational leadership trait is competence. Competence is a kind of thirst for knowledge and self-improvement that demonstrates skills needed to overcome challenges. De Pree (2002) argued that competent leaders are transforming leaders. Transformational leaders stimulate their followers' efforts "to be innovative and creative by questioning assumptions, reframing problems, and approaching old situations in new ways," (Avolio & Bass, 2002, p. 2). Transformational leadership focuses on transforming employees since such "leaders who truly excel are those who transform results, performance, and culture" (Galloway, 2016).

## Transformation

Transformational leadership focuses on tasks that help understand what motivates and influences employees, how to coach employees to improve performance, how to help employees understand strategy, how to measure what employees want, monitors employee progress, and assesses value rather than numbers" (Galloway, 2016). This seems to be an approach of "transforming" the employee to "value" organizational goals. Transitional leaders "want to help others see value in the goals and leverage what others are interested in or motivated by," (Galloway, 2016).

## Mentorship

A fifth relational leadership trait is mentorship. Knapp (2015) argued that "Leaders are mentors who focus on developing strong relationships with organization members," (p. 856). This trait instills in a leader to guide, mentor, and provide feedback to the manager to help him/her see their actions from another perspective. Furthermore, this quality helps a leader to build trust with the employees through feedback and guidance.

## Adaptability

The final relational leadership trait is adaptability. Leaders who possess this trait can adapt to the situation at hand in order to mobilize employees to address difficult scenarios and succeed (Heifetz, Grashow, & Linsky, 2009). Such leaders are persistent, methodical, and able to accept disequilibrium and discomfort. Hult and Sivanesan (2013) also argued that such leaders are focused on the mission. They noted that the "mission when clearly articulated builds a shared sense of purpose throughout the cyber security function and extends beyond into the wider organization," (p. 116). Finally, Karaman, Çatalkaya, and Aybar (2016) argued that when it comes to crisis response, adaptable leaders strive to build a resilient command and control structure in order to minimize harm.

## CAPACITY AND RESPONSIBILITY

When leadership is viewed as a process of influence, it can be executed by individuals throughout an organization (Cleveland & Cleveland, 2018; Sinclair, 2013). Today's leaders face complex and daunting challenges. Leaders who empower their teams will enhance their organizations. However, barriers exist in building an inclusive organization. Culturally agile leaders are self-aware and use their own biases to develop their immunity to cultural stereotyping. By recognizing their own views, they are better able to promote a workplace that embraces inclusion and benefits from a diverse team.

When leaders recognize their responsibility to themselves and to their organizations to lead as authentically as possible while still maintaining the flexibility to deal with other cultures, whether in their own workplace or negotiating with outside vendors, companies, and countries, genuine relationships will develop. Culturally agile teams will respect cultural differences and capitalize on how best to use those differences to create the best possible solutions for their organizations.

## CONCLUSION

Based on the literature surveyed, it is clear that finding common ground is essential to organizational communication. Even with seemingly increased fragmentation, there needs to be productive discussions toward equitable solutions. Today's leaders must be culturally agile in order to engage in civil discourse because today's interconnectedness impact humanity globally and requires the ability to connect across cultures. Where once the default in corporate America was the white, middle-class man (Franklin, 2014), that normative standard is no longer a given.

In this digital age, people are able to connect with a wider range of cultures and influences throughout the world. A global economy means a more complex networking system. Advancements in technology mean more awareness about politics and foreign issues, and leaders must develop the competencies to handle the challenges facing cross-cultural teams. According to Savelsbergh, Poell, and van der Heijden (2015), experimental research proves the positive effects of maintaining a stable team. When team members form an understanding of each other's capabilities, it results in improved coordination of their actions.

Cross-cultural exercises prove successful in increasing students' cultural intelligence and will ease interactions among students from different cultures and countries (Duus & Cooray, 2014; Taras et al., 2013). Currently, colleges and universities both strive to foster competition through a meritocracy and create civil dialogue among the diverse population (Ewald, 2001). Leaders across college campuses understand the importance civility plays in effective leadership (Ewald, 2001).

According to Fairhurst (2016), the current political rhetoric is at opposite ends of the dichotomy; however, there is a movement toward improving governance through civil discourse. Blog posts and websites are dedicated to educating and promoting inter-group relations and building civility (Fairhurst, 2016), recognizing the need to promote cultural agility among each other.

Leaders who recognize and support inclusive discourse not only respect the rights of others but also reflect a concern for justice for all (Kesler, 2000). Through continuing education, leaders are able to enhance their leadership abilities, share perspectives on relevant issues, and work toward solutions in the public interest. With a decline in civic and social connectedness over the past forty years (Wituk, Ealey, Clark, Heiny, & Meissen, 2005), organizations face challenges where their leaders lack the capacities to address issues through civil discourse.

Gurin, Dey, Hurtado, and Gurin (2002) confirmed that racial and ethnic diversity promotes active thinking skills and intellectual engagement. Diversity among teams and organizations with culturally agile members and leaders are better able to meet the global challenges facing today's organizations (Caligiuri, 2013). The differences in cultural values affect how organizations communicate. Given the rise in global corporations, there is another aspect of cross-cultural awareness that finds its way into not only corporate America but also our classrooms and our political arena. Developing cultural agility is only one piece of the challenge when operating within today's digital landscape and cross-culturally populated organizations.

According to Chen and Hamilton, "It's important to note that, although a group may have high numeric representation of minorities, its policies or culture could lead racial minorities to feel alienated or undervalued," (p. 587). Within the realm of leadership development, leaders must learn to be culturally agile in order to exceed the demands of the growing global marketplace.

Corporations are no longer confined to one local area, and many businesses are unable to operate without the use of technology, which widens the digital landscape across cultures and time zones. By supporting culturally agile leaders through a relational leadership development approach, organizations will organically create holistic leaders beyond the traditional competencies to include networks and relationships.

Extant literature posits that organizational members input their skills and effort with the expectation of a return which will further their personal goals (Simosi & Xenikou, 2010). By understanding the importance of becoming culturally agile individuals, leaders are better able to understand how to communicate civilly with each other. Future research should focus on identifying most effective strategies to increase cultural agility among leaders, while building strong awareness about civil discourse.

## REFERENCES

Albright, M. (2016). Servant leadership: Not just buzzwords. *Strategic Finance, 98*(4), 19-20.

Athanasopoulou, A., Moss-Cowan, A., Smets, M., & Morris, T. (2018). Claiming the Corner Office: Female CEO Careers and Implications for Leadership Development. *Human Resource Management Journal, 57*(2), 617–639. doi:10.1002/hrm.21887

Avolio, B. J., & Bass, B. M. (Eds.). (2001). *Developing potential across a full range of Leadership Tm: Cases on transactional and transformational leadership.* Psychology Press. doi:10.4324/9781410603975

Bandura, A. (1999). Moral disengagement in the perpetration of inhumanities. *Personality and Social Psychology Review, 3*(3), 193–209. doi:10.120715327957pspr0303_3 PMID:15661671

Bass, B. (1988). The inspirational processes of leadership. *Journal of Management Development, 7*(5), 21–31. doi:10.1108/eb051688

Berman, D. K. (2008, October 28). The game: Post-Enron crackdown comes up woefully short. *The Wall Street Journal*, p. C2.

Caldwell, D. F., & Moberg, D. (2007). An exploratory investigation of the effect of ethical culture in activating moral imagination. *Journal of Business Ethics, 73*(2), 193–204. doi:10.100710551-006-9190-6

Caligiuri, P. (2013, March). Develop your cultural agility. *TD Magazine.* Retrieved from https://www.td.org/magazines/td-magazine/develop-your-cultural-agility

Chen, J. M., & Hamilton, D. L. (2015). Understanding diversity: The importance of social acceptance. *Personality and Social Psychology Bulletin, 41*(4), 586–598. doi:10.1177/0146167215573495 PMID:25713169

Cleveland, M., & And Cleveland, S. (2018). Building engaged communities—A collaborative leadership approach. *Smart Cities, 1*(1), 155–162. doi:10.3390martcities1010009

Cullen-Lester, K. L., Maupin, C. K., & Carter, D. R. (2017). Incorporating social networks into leadership development: A conceptual model and evaluation of research and practice. *The Leadership Quarterly, 28*(1), 130–152. doi:10.1016/j.leaqua.2016.10.005

Cunliffe, A. L., & Eriksen, M. (2011). Relational leadership. *Human Relations, 64*(11), 1425–1449. doi:10.1177/0018726711418388

Daloz Parks, S. (2005). *Leadership Can be Taught.* Boston, MA: Harvard Business School Press.

Day, D. (2000). Leadership development: A review in context. *The Leadership Quarterly, 11*(4), 581–611. doi:10.1016/S1048-9843(00)00061-8

De Pree, M. (2002). Servant-leadership: Three things necessary. *Focus on leadership: Servant leadership for the, 21*, 89-100.

Duus, R., & Cooray, M. (2014). Together we Innovate: Cross-cultural teamwork through virtual platforms. *Journal of Marketing Education, 26*(3), 244–257. doi:10.1177/0273475314535783

Eich, D. (2008). A Grounded Theory of High-Quality Leadership Program: Perspectives from Student Leadership Development Programs in Higher Education. *Journal of Leadership & Organizational Studies, 15*, 176–187. doi:10.1177/1548051808324099

Fairhurst, G. (2016). Reflections on Leadership and Ethics in Complex Times. *Atlantic Journal of Communication, 24*(1). . doi:10.1080/15456870.2016.1113612

Flyvbjerg, B., Garbuio, M., & Lovallo, D. (2009). Delusion and Deception in Large Infrastructure Projects: Two Models for Explaining and Preventing Executive Disaster (February 2009). *California Management Review, 51*(2), 170–193. doi:10.2307/41166485

French, R. P., & Raven, B. (1959). The bases of social power. In D. Cartwright (Ed.), *Studies in social power* (pp. 150–167). Ann Arbor: University of Michigan, Institute for Social Research.

Frolich, T. C., Sauter, M. B., & Stebbins, S. (2015, June 29). The worst companies to work for. *24/7 Wall St/Yahoo Finance.*

Galloway, R. F. (1997, Summer). Community Leadership Programs: New Implications for Local Leadership Enhancement, Economic Development, and Benefits for Regional Industries. *Economic Devekopment Review, 15*(2), 6–9.

Galloway, S.M. (2016). What makes a leader transformational? *Leadership Excellence, 33*(5), 20-21.

Groves, K. (2007). Integrating leadership development and succession planning best practices. *Journal of Management Development*, 26(3), 239–260. doi:10.1108/02621710710732146

Gurin, P., Dey, E., Hurtado, S., & Gurin, G. (2002). Diversity and higher education: Theory and impact on educational outcomes. *Harvard Educational Review*, 72(3), 330–367. doi:10.17763/haer.72.3.01151786u134n051

Heifetz, R. A., Grashow, A., & Linsky, M. (2009). *The practice of adaptive leadership: Tools and tactics for changing your organization and the world*. Harvard Business Press.

Higgs, M. (2009). The good, the bad and the ugly: Leadership and narcissism. *Journal of Change Management*, 9(2), 165–178. doi:10.1080/14697010902879111

Hult, F., & Sivanesan, G. (2014). What good cyber resilience looks like. *Journal of Business Continuity & Emergency Planning*, 7(2), 112–125. PMID:24457323

Johnson, C. E. (2017). *Meeting the ethical challenges of leadership: Casting light or shadow*. Sage Publications.

Karaman, M., Çatalkaya, H., & Aybar, C. (2016). Institutional Cybersecurity from Military Perspective. *International Journal of Information Security Science*, 5(1), 1–7.

Kegan, R., Lahey, L. L., Miller, M. L., & Fleming, A. (2016). *An everyone culture*. Becoming a Deliberately Developmental Organization.

Kezar, A., & Lester, J. (2010). Breaking the barriers of essentialism in leadership research: Positionality as a promising approach. *Feminist Formations*, 22(1), 163–185. doi:10.1353/nwsa.0.0121

Knapp, S. (2015). Lean Six Sigma implementation and organizational culture. *International Journal of Health Care Quality Assurance*, 28(8), 855–863. doi:10.1108/IJHCQA-06-2015-0079 PMID:26440487

Larsson, G., Sandahl, C., Söderhjelm, T., Sjövold, E., & Zander, A. (2017). Leadership behavior changes following a theory-based development intervention: A longitudinal study of subordinates' and leaders' evaluations. *Scandinavian Journal of Psychology*, 58(1), 62–68. doi:10.1111jop.12337 PMID:27859309

Liang, J. G., & Sandmann, L. R. (2015). Leadership for community engagement: A distributive leadership perspective. *Journal of Higher Education Outreach & Engagement*, 19(1), 35–64.

Maznevski, M. L., & DiStefano, J. J. (2000). Global Leaders are Team Players: Developing Global Leaders through Membership on Global Teams. *Human Resource Management*, 39(2 & 3), 195–208. doi:10.1002/1099-050X(200022/23)39:2/3<195::AID-HRM9>3.0.CO;2-I

Milner, H. R. IV. (2007). Race, culture, and researcher positionality: Working through dangers seen, unseen, and unforeseen. *Educational Researcher*, 36(7), 388–400. doi:10.3102/0013189X07309471

Muyia, H., & Kacirek, K. (2009, December). An Empirical Study of a Leadership Development Training Program and Its Impact on Emotional Intelligence Quotient (EQ) Scores. *Advances in Developing Human Resources*, 11(6), 703–718. doi:10.1177/1523422309360844

Nissen, L., Merrigan, D., & Kraft, M. K. (2005, March). Moving mountains together: Strategic community leadership and systems change. *Child Welfare*, 84(2), 123–140. PMID:15828404

Nye, J. S. (2008). *The powers to lead*. Oxford, UK: Oxford University Press.

O'Connor, A. (2015, August 9). Coca-Cola funds scientists who shift blame for obesity away from bad diets. *The New York Times*, p. A1.

Pigg, K. E. (1999). Community leadership and community theory: A practical synthesis. *Journal of the Community Development Society*, *30*(2), 196–212. doi:10.1080/15575339909489721

Sandler, M. (2015, August 8). CEO pay soars at top not-for-profits. *Modern Healthcare*. PMID:26642549

Savelsbergh, C., Poell, R. F., & van der Heijden, B. (2015). Does team stability mediate the relationship between leadership and team learning? An empirical study among Dutch project teams. *International Journal of Project Management*, *33*(2), 406–418. doi:10.1016/j.ijproman.2014.08.008

Simosi, M., & Xenikou, A. (2010). The role of organizational culture in the relationship between leadership and organizational commitment: An empirical study in a Greek organization. *International Journal of Human Resource Management*, *21*(10), 1598–1616. doi:10.1080/09585192.2010.500485

Sinclair, A. (2009). Seducing leadership: Stories of leadership development. *Gender, Work and Organization*, *16*(2), 266–284. doi:10.1111/j.1468-0432.2009.00441.x

Sinclair, A. (2013). Not just "adding women in": Women re-making leadership. *Seizing the Initiative: Australian Women Leaders in Politics, Workplaces and Communities*, 15-34.

Subramony, M., Segers, J., Chadwick, C., & Shayamsunder, A. (2018). Leadership development practice bundles and organizational performance: The mediating role of human capital and social capital. *Journal of Business Research*, *83*(C), 120–129. doi:10.1016/j.jbusres.2017.09.044

Taras, V., Caprar, D. V., Rottig, D., Sarala, R. M., Zakaria, N., Zhao, F., & Huang, V. Z. (2013). A global classroom? Evaluating the effectiveness of global virtual collaboration as a teaching tool in management education. *Academy of Management Learning & Education*, *12*(3), 414–435. doi:10.5465/amle.2012.0195

Terry, H., & Liller, K. D. (2014). The Doctoral student leadership institute: Learning to lead for the future. *Journal of Leadership Education*, *13*(1), 126–135. doi:10.12806/V13/I1/IB2

Tubbs, S. L., & Schulz, E. (2006). Exploring a taxonomy of global leadership competencies and meta competencies. *The Journal of American Academy of Business, Cambridge*, *8*(2), 29–34.

Uhl-Bien, M. (2006). Relational Leadership Theory: Exploring the social processes of leadership and organizing. *The Leadership Quarterly*, *17*(6), 654–676. doi:10.1016/j.leaqua.2006.10.007

Upson, J. W., Damaraju, N. L., Anderson, J. R., & Barney, J. B. (2017). Strategic networks of discovery and creation entrepreneurs. *European Management Journal*, *35*(2), 198–210. doi:10.1016/j.emj.2017.01.001

VanVactor, J. D. (2012). Collaborative leadership model in the management of health care. *Journal of Business Research*, *65*(4), 555–561. doi:10.1016/j.jbusres.2011.02.021

# Chapter 2
# Transition of Ecosystem Services Based on Urban Agro Ecology

**José G. Vargas-Hernández**
https://orcid.org/0000-0003-0938-4197
*University Center for Economic and Managerial Sciences, University of Guadalajara, Mexico*

## ABSTRACT

*This chapter analyzes the implications of the transition of ecosystem services based on urban agro-ecology. It advances on the debate over the negative effects of the traditional and industrial-oriented agricultural production on the ecosystem services, food systems, climate change, etc. and analyses the principles, methods, and some practices that support the transition to urban agro-ecology. The method employed is the analytical of the theoretical and empirical literature review. It concludes that a transition from traditional and industrial-oriented agriculture towards more urban agro-ecology is inevitable to improve the ecological and environmental services, the economic efficiency, the social equity and justice, and the environmental sustainability of cities.*

## INTRODUCTION

A debate over the negative impacts of industrial input-intensive agriculture is leading to consider a feasible transition to other alternative forms capable to provide ecosystem services and produce food. Agro ecology emerges in response to the problems associated with the industrial model of agricultural production. The simplified form of agroecosystem is altered with the input substitution that occurs in industrial agro systems. The industrial agriculture framed in the free market economic model has a destructive impact on sustainable natural resources, biodiversity, food security, environmental services and climate change, leading to ecosystem disruption and human hunger. Conventional agriculture is not sustainable for the rarefaction of inputs, global competition, loss of biodiversity and benefits from ecosystem services, climate change, etc.

DOI: 10.4018/978-1-7998-3576-9.ch002

The current consumption model based on industrial agro culture is disconnected from food production leads towards poor consumption patterns and choices of diets beyond the requirements of urban land planning and development.

There is a gap in the agro ecological transitions between the agro ecological practices and the ecosystem services delivery. The relationship between agro ecology and nature is relevant in the current crisis is framed by the land used for agriculture and the land dedicated to conservation of biodiversity and ecosystem services.

The agro ecological struggles leading to demands for right to food, land redistribution, concerns for pollution of soil, land and water, provision of ecosystem services, corporate dismantling and access to fair markets of organic products, etc., are some of the emerging issues in urban environments. To break this pervasive relationship, it is required the incorporation of an integral conservation programs framed by minimizing land use dedicated to agriculture and more to conservation of agroforest ecological biodiversity and other ecosystem services.

Agro ecology and food production systems face some economic, social and environmental challenges when economies are organized around a competitive export orientation of crops that have impacts on the ecosystem services integrity, food quality, public health, and disruption of traditional rural livelihoods. The autonomy and solidarity between peoples of rural and urban agro ecology displaces the control of global markets.

Research on agro ecology is based on the assumption that there is a positive relationship between biodiversity and the rate of ecosystem services. Agro ecological research tends to enhance agroecosystem functions of biodiversity systems based on ecological principles such as facilitation, niche partitioning, competition, etc. Sustainable urban agro ecology take into consideration the efficient competition of resources between the strategies developed by actors in the urban environment to enhance benefits and securing access to land use.

## ECOSYSTEM SERVICES

Ecosystem services are "the direct and indirect contributions of nature to human wellbeing" (The Economics of Ecosystems and Biodiversity (TEEB) 2010). Urban agro ecology has direct impacts on ecosystem services, environmental sustainability, food security and socio economic development in urban and peri-urban areas. The impact of urban agro ecology in the ecosystem services in urban environment, economic viability and productivity has been given little attention by organizations and institutions than the peri-urban commercial farms. Agro ecological techniques can improve yields and increase productivity while improving the supply of critical natural resources and environmental services (Pretty et al. 2006).

To enhance this situation, it is recommended to provide more extension, education, training and support for the agro economic services. The urban economies benefit from urban agro ecology if local governments develop institutions with a policy framework to provide agro ecological services, promote sustainable urban development and urban food security aimed to enable urban agro ecological participants to unlock supporting financial and technical services.

The analysis of agro ecology theory and practice may contribute to the discussion to pursue to the institutionalization of agro ecological development based on a farm system with investments and services supported by networking interactions between the state, society and inclusive participatory organizations.

A multilevel methodology to analyze the impact of agro ecological practices on environmental services flows and processes to assess the feasibility of the agro ecological transition.

The urban agro ecological systems have sustainable impacts to the beneficiaries on the delivery of environmental and ecosystem services. Agro ecology provides food, fuel, fiber and a wide range of ecosystem services. Ecosystem services is a natural process resulting of interdependencies and interactions between nature, economy and society that has an effect of human dependency. Urban agro ecology is spreading across urban marginal and vacant land in some cities around the world, embraced by local governments and civil society as source of heathy food, ecosystems services and job creation.

Urban and suburban agro ecological farms is an asset for the residents of the city by providing micronutrients and sustenance during lean seasons, supporting the health of ecosystems and habitat for wild plants, genetic diversity and pollination, supplying and regulating air and water. Urban agro ecological farming preserves the urban ecosystem biodiversity and the quality of local seeds. Urban agro ecology is an inextricable component of biodiversity conservation and environmental services in the urbanization processes evident in the industrialization, which drives complicated socio economic, political, environmental and ecological dynamics.

Small-scale urban agro ecological farming increases the incomes of households across the poor population segments and their demand for locally traded goods and ecosystem services beneficial to other economic sectors. Agro ecological practices of smallholder and cooperative farmers must develop the supportive infrastructure and ecosystem services to use their own resources and have access to local distribution channels in markets.

Sustainable agro ecological urban farming is a healthy food production to guarantee nutritious and accessible for everyone while natural resources are managed to maintain ecosystem functions and ecological services aimed to support current as well as future human needs. Usually, urban agro ecological fresh production is for self-consumption and where there are surpluses, these are traded in a market oriented economic agents or processed and packaged for later self-consumption or sale.

Sustainable agro ecology depends on ecological services such as topsoil and pollination of crops (Zhang et al. 2007). Plots managed agro ecologically have less erosion, retain the vegetation, have lower economic losses and enhance agro ecosystems resilience by improving the adaptive capacity and reducing vulnerability to climate changes, natural disasters and economic and environmental crisis and stresses.

The concept of integrated environmental and ecosystem services valuation can be an operational tool to steer the sustainable agroecosystems transition. (Duru and Thérond, 2015). An integrated ecosystem services valuation has to make explicit the feasibility of the expectations to be fulfilled (Cote and Nightingale 2012, Davies et al. 2015). The valuations of ecosystem services in the agro ecological context and agro environmentally are theoretically limited (Holt et al. 2016; Tscharntke et al. 2005, Prager et al. 2012, Merckx and Pereira 2015).

Ecosystem services assessments leads to increase the social learning and delivery but could not lead to more systemic changes in economic efficiency, social equity and inclusion and environmental sustainability in agro ecosystems.

Social-agro-ecological ecosystems management has some inherent trade-offs with ecosystem services delivery (Armitage et al. 2008, Galafassi et al. 2017) under distinct agro ecological scenarios although what is missing is the implementation of transition actions. Urban agro ecological systems requires more interdisciplinary and integrated analysis of sustainable urban planning, urban land rights, urban food planning, ecological systems and services, Etc. (Bohn and Viljoen 2005). Agroecosystem planning develops design approaches to take advantage of contributions from the incorporation of permaculture principles into farmer-oriented participatory research and development to support design solutions and site analysis by stimulate creative problem solving.

Regional spatial urban agro ecology follows a combination of economic scale and diversity patterns of urban developments with implications for urban spatial planning and governance. The spatial planning integration of peri-urban agro ecology can implement the use of participatory forms of land use and cultural heritage to improve ecosystem services and urban sustainable development (Simon Rojo et al., 2014; Swensen and Jerpåsen, 2008). Urban and peri-urban agro ecology becomes part of urban planning and management systems adopting production methods link up to ecosystem services sustainable natural resources, waste and eco-sanitation, recreational and leisure services, nature reserves, etc.

A common ground attained by consultation among the diverse viewpoints of the stakeholders becomes a basis for a normative vision of the expected agro ecosystem. The systemic vision of the agro ecological system facilitates the trajectories of change in the transitional space of ecosystem services. A model of agro ecosystems is a representation of a complex system of the stakeholders' perceptions integrating ecosystem services and their value biophysical, economic and sociocultural domains (Martín-López et al. 2014) to develop a systematic approach to agro ecological analysis framework.

Urban agro ecology includes production, inputs, processing, marketing and ecosystem services delivery activities closely related in space and time by the geographical proximity of resources flows. Agro ecology is supported by low cost and knowledge-intensive technology and innovations adapted by producers to advance food productivity, social equity and environmental sustainability and to encourage ecosystem maintenance and resilience. Urban agro ecology requires technology to perform under specific conditions of limited spaces and high values of land, use of natural and urban resources and wastes, direct contacts between producer and consumer.

Technology of urban agro ecology enterprises is still developing with the tendency towards more advanced and intensive. Available agro ecological technologies are adapted whilst new technologies are being locally developed undertaking participatory action-research to respond to the specific stakeholders involved in urban needs.

Urban agro ecological production requires the formulation and implementation of strategies aimed to minimize the risks and enhance biodiversity and ecosystem services opportunities to improve climate change, leisure, education, tourism, culture, etc. Agro ecological production and processing should be adjusted to mitigate climate change and other ecosystem stresses.

The agro ecological transition assessment has to measure its impacts and feasibility in the different elements of both agro ecological and conventional farming systems on the socio-ecosystem services (Mouchet et al. 2014; Crouzat et al. 2015). The agro environmental scheme is one example of payment for ecosystem services in environmental governance (Engel et al. 2008), although limited by the influence and polarization processes from powerful groups of pressure and pervasive strategic behaviors (Banerjee et al. 2013). Agro ecological systems requires to implement large scale institutional changes to integrate ecosystem services assessments to support local communities to steer transitions for the sustainability of agro ecological systems.

Food agro forestry has the potential to contribute to ecosystems services besides food provisioning which can be maximized by implementing sustainable urban planning and ecological designs using specific techniques in permaculture design and forest gardening to promote symbiotic polycultures planting that maximize yield (Rivera et al. 2004; Ellis 1998; Bostonian et al. 2004; Pretty 2008, Mollison 1988, 1979, Jacke and Toensmeier 2005, and Crawford 2010).

Low-resource agro ecology systems are based on the use of local resources and modest use of external inputs, which produces the majority of grains, roots, tuber, and crops. Organic and agro ecology farming systems that tend to rely more on local than external inputs and foreign certification seals, tend to be less dependent on foreign and volatile markets. However, keeping small organic and agro ecological farmers' dependent on input substitution does not motivate them to redesign agro ecosystems and tends to perpetuate dependence. In planning and maintaining complex agroecosystems, the permaculture offers risks and challenges.

## URBAN AGRO ECOLOGY SYSTEMS

Urban agro ecology as a theoretical and practical approach studies and applies agricultural systems from ecological concepts and principles in a socio-economic perspective to design, manage and implement sustainable agro-ecosystems to increase the sustainability of agro food systems. The concept of agro-ecosystem was introduced by the ecologist Odum in 1969 to represent the interaction of ecological, technological and socio-economic factors, expanding the scope of agro ecology (Wezel, et al., 2009; Gliessman, 2007).

Urban agro ecology is defined as growing of plants and raising of animals within and around cities. The cultural meaning of urban agro ecology experience in its spatial dimension values the place as a public right to have access to power dynamics. Urban and peri-urban agro ecology systems are the cultivation, grow, crop, process and distribution of agro ecological products for feeding local residents. Urban agro ecology often can be confused with urban agriculture, gardening, homesteading and subsistence farming. What distinguish urban agro ecology from urban agriculture is that the growing of the product is more dedicated for personal consumption or sharing and not always for commercial activities.

Urban agro ecology is more than technical and social in focus, it is explicitly rooted in radical political thought and action. Urban agro ecology has become part of the fabric of a sustainable urban society. Urban agro ecology hinges on the economic, social and environmental development of urban areas that determines their sustainability. Sustainable urban agro ecology contributes to the development of a sustainable productive and resilient, socially inclusive, food-secure, and environmentally-healthy city.

Any city around the world recognizes the relevance of urban agro ecology because its contribution to feed residents with more organic sustainable approaches. Urban agro ecology is located inside the inner cities or in the peri-urban spaces taking place on-plot or off-plot the private, public or semi-public residences and land. Urban and peri-urban vacant soils and lands located near industrial areas and roadways may be contaminated and the organic waste may contain substances environmentally risky (Nehls et al, 2015; Säumel et al., 2012; Schwarz et al., 2016).

Development of sustainable urban agro ecology can recycle nutrients and wastes in line with use of local natural resources and agro ecological principles for the production of compost and seeds and organic plant protection. Urban agro ecology is part of the urban environmental and ecological system solving problems such as re using urban wastes as productive resources and collecting fresh organic waste, which may cause health and environmental problems. Urban agro ecology can be genetic sanctuary islands to serve as the source of GMO-free seeds and avoid the advance of contaminated transgenic agriculture. Urban agro ecology is an urban source of organic food production, an issue of national security and environmental sustainability. Methods used in organic urban agro ecology are supported by the urban small-scale gardening movement.

An agro ecological model based on small-scale production employing ecological methods may not compete with and industrial agriculture model but provides improvements for the environments and the living conditions of those producers employed. Urban agro ecology tends to involve frequently urban residents who have access to urban land, natural resources and water and have time available combined with other activities and tasks related to the household and difficult to combine with full jobs that require traveling. Scaling-up agro ecology extends knowledge awareness of principles, methods and systems used by producers to design and implement policies and strategies aimed to transform the socio economic context and food system.

The emergence of urban agro ecology was burnout of urban gardening that was perceived as a sign of underdeveloped and poor people who cultivated at small-scale with little experience, and diverse. Urban agro ecology as a strategy reduces food insecurity and urban poverty while enhancing urban environmental sustainability management. Making urban agro ecology appealing and profitable may lift millions out of poverty offering a promising approach to feed the growing population. Urban agro ecology is a strategy to have social impacts in social inclusion, integration and poverty alleviation of disadvantaged populations.

Urban agro ecology surpasses the dimensions of what is vegetable gardening, becoming a significant source of fresh and healthy produce for the urban populations (Companioni et al., 1997). The agro ecology has multifunctional functions and dimensions to provide nutritious, healthy and fresh diet facilitate equitable and environmental sustainability, increasing ecological resilience. Agro ecology in urban areas improves nutrition, food security, sovereignty and offers economic, social, health and environmental benefits. Therapeutic activities related to urban agro ecology and gardening in communities accessible by urban green innovation areas encourage cooperative and inclusive human interactions, knowledge exchange, strengthens mental health, reduce poverty, etc.

Urban agro ecology can be used to restore degraded land and develop it into a productive green innovation area or public space to foster inclusive and sustainable socio-economic activities. Agro ecology is a green turn in local self-sufficient agriculture aimed to food security to defeat the agro food industry and fair use of the natural resources. Agro ecological crops and food production infrastructure in urban green areas and community spaces may serve to distinct and multiple purposes, needs and interests of all the stakeholders involved. Community-led urban agro ecology food systems play a relevant role to rebuild vacant spaces in cities and re-skill employment opportunities.

International organizations, local governments and sectoral organizations support intra and peri urban agro ecology is because it contributes to healthy nutrition and food security, improves the quality of the food intakes, increases the efficiency of the food system, the socio economic impacts on development, self-employment creation, household income and earnings, promotes production of agro ecological inputs, processing, packaging and marketing.

Urban agro ecology has concerns about sustaining food security and climate change in urban areas lowering temperatures and food-related greenhouse gas (GHG) emissions. The agro ecological approach is well suited to withstand economic and environmental stresses and pressures of climate change by avoiding or reducing the synthetic chemical inputs in cultivations such as fossil fuels, fertilizers and pesticides poisoning consumers, workers and communities. Consumers of agro ecological products gain access to healthy and nutritious local food at a fair cost, while supporting urban agro ecological farming. Urban agro ecological and workplace configuration affects the labor efficiency (Taghavi and Murat 2011; Venkatadri et al. 1997; Becker and Steele 1995; Burbidge 1971).

Urban agro ecological crops are a response on-farm carbon sequestration to climate change and contributes to anthropogenic greenhouse gas emission, biodiversity and habitat provisioning and conservation, biogeochemical maintenance, water cycling and complex trophic interactions, improvement of community livelihoods, food sovereignty and security. Urban agro ecology focusing on food sovereignty is a concept connected to political methodologies based on radical framework that can be used in any context between the practical and the political in urban agriculture.

Agro ecology as a way of life and the language of nature is based on ecological principles. Agro ecological principles are adopted by food social movements linked to struggles for food sovereignty, which is described as the human right to healthy, culturally appropriate food produced through ecologically sustainable methods, and to deðne the agro ecological systems. Agro ecology is a mix between principles of agricultural sciences and ecology focusing on the diversification of agricultural systems for the benefit of agroecosystem functions and services to enhance the environment and mitigate the negative social externalities of the industrial intensified agricultural model.

The principles of agroecosystem configuration implicit in land use functions appear reasonable supported, but some of them such as the analysis of spatial relationships are notoriously absent (Cavazza 1996; Veldkamp et al. 2001; Hatfield 2007; Osty 2008; cited in Benoit et al. 2012).

Urban agro ecology is devoted to food production and has economic, social and environmental multifunctional benefits and other non-economic and non-quantifiable benefits (Hampwaye, 2013). Urban agro ecological food production is becoming more familiar in its location and spatial organization in the urban spaces and distribution of activities, feasibility, and durability. The core principles of agro ecology as an alternative food system are sustained on the diversified and complementary use and recycle of nutrients and energy of the ecosystems with the aim to create a sustainable environment of resilient biodiversity. Use of agro ecological design principles can maximize food provisioning and ecosystem services.

Urban agro ecology meets only one part of the global food demand not enduring food self-sufficiency for inhabitants in cities. Urban agro ecology is part of a widely globalized food system experiencing a dramatic rise of trade in food and agricultural products (Busch and Bain, 2004; Wiskerke, 2016). Urban agro ecology requires one third of total urban area to meet global demand for food and only a small proportion is consumed by low-income food-insecure and most vulnerable people who have not access to land and lack skills to produce it.

Urban agro ecology includes food and non-food products from different types of crops, herbs, plants and animals or a combination of these, often high valued but perishable. Urban agro ecology revitalizes starchy roots traditional crops expanding production, playing a relevant role in the food system reducing dependencies on foreign food sources, making it widely available through distribution, logistics and consumer networks and improving the socio economic conditions of large urban areas. To reduce agro ecological food systems carbon footprint, the producer-consumer links of quality standards and integration of ecological issues, must be strengthening though direct sales, community supported agro ecology and value chain (Lev and Stevenson 2011; Stevenson et al. 2011).

The methods of agro ecology as an agro ecosystem are based on diversified and synergies of biological interactions to improve productivity of food, soil fertility and resilience through participatory frameworks while achieving high crops production, economic efficiency, sustainable environment and social inclusion. A food system model based on agroecosystems may address the vulnerability of national food systems

Agro ecology is conceived as the application of ecological science to the study, design and management of sustainable agroecosystems. Agro ecology as a science provides a framework to apply ecological principles to design and manage sustainable agricultural ecosystems. The science of agro ecology is based on several disciplines to evaluate the diversity of farm interacting with some agro ecological functions such as the impact on biodiversity, carbon sequestration, climate change, gas emissions, ecosystems resilience, food security, etc. Agro ecology is an aspirational and practical approach that encompasses agricultural systems thinking aimed to support local economies, strengthening ecological biodiversity, ecosystem resilience, social justice and food sovereignty.

Agro ecological biodiversity is covered by research on alternative agro ecology. Urban agro ecology research is at a crossroads. Advanced research has surfaced for urban agro ecology and highlighted spatial trends and organization of agro-food systems in urban and peri-urban areas developing countries. Urban agro-ecology research should involve democratic and participatory dialogues between scientists, urban planners, farmers, communities, agribusiness, and other relevant stakeholders. The research potential of urban agro ecology goes beyond its food production dimension and rising concerns for societal and environmental changes. Food sovereignty uses natural resources in diverse agro-ecological production and harvesting methods to maximize the contribution of ecosystems and improve their resilience and adaptation. Agro ecological approached can optimize sustainability of agroecosystems and become the basis of food sovereignty.

Urban agro ecology may increase and becomes part of the integral urban system when the city grows. The agro ecology brings significant economic, environmental, social and political benefits to farmers and urban populations. The economic scenario is associated to market oriented urban agro ecology with economic impact and high profitability. The social scenario of urban agro ecology is associated with the livelihood strategies of low income households who seek out sources of survival. The ecological scenario is related to urban agro ecology that is multifunctional in environmental management. Urban agro ecology is a viable the access to healthy food, environmental protection and enhanced human dignity.

The agroecosystem configuration drives multiple functions such as environmental, land and labor productivity of diversified farming systems. Agro ecosystems benefits from some soil rich in organic matter, which improves fertility and increases agro ecological production, providing food to society and income to producers (Adhikari and Hartemink 2016). Agro ecology provides nutrition and food security benefits for people and for maintaining the healthy environment and ecosystems. Urban agro ecology enhances food security taking into consideration the supplying and distributing food costs from rural production.

The advantages in urban agro ecology are related to produce that can be sold fresh due to the fact that cultivations are near consumers. Urban agro ecology changes how far healthy and nutritious food travels to a better informed consumer, increasing the access to local grown food and taking the right place for the food system and reintroducing the culture of how food grows seasonally in the region.

Traditional agroecosystems are temporal and spatial arrangements with several varieties of crops becoming less vulnerable to environmental catastrophes. Agro ecology can achieve the transition to low-carbon and the preservation of natural resources by improving the farming productivity and preserving the ecosystems. Agro ecological system transition implementation is supported by socio economic and ecological drivers based on biodiversity resources and ecosystem services to be delivered. Steering urban agro ecological transitions need to be supported by an ecosystem services framework to assess the implementation process depending of the urban context.

Agro ecology and permaculture are alternatives to the food systems that are changing the density of the social fabric in rural-inner cities-peri-urban areas to articulate diversity in food provision and sovereignty, food and environmental justice movements, etc. Around one billion people around the globe grow their own vegetables, fruits or raise animals, in urban inner and peri areas, representing one fifth of the agro ecological food production of the world and creating economic, social, environmental and health benefits. Households benefit by consuming accessible fresh and healthy food, promoting healthy lifestyles, saving money and controlling production and consumption.

Permaculture is being positive and nonspecific associated (Leakey 2012; Deb et al. 2008; Lovell et al. 2010) with agrobiodiversity, agroecosystem design, agroforestry and other polycultures (Francis and Porter 2011; Torre Ugarte and Hellwinckel 2010). Permaculture is defined as an integrated and evolving system of perennial and self-perpetuating plant and animal species useful to man, forming an agricultural ecosystem, (Mollison and Holmgren 1978, p. 1). Permaculture is identified more as an alternative agriculture more than explicitly questioned agro ecological framework (Gomiero et al. 2011; Pretty 2006; Bavec et al. 2009; Pretty 2005). The agro ecological content of the permaculture is the diverse and multifunctional site design at multiple scales to increase the productivity based on heritage agroecosystems, water management and global germplasm to achieve sustainability.

Permaculture is a pertinent approach to agro ecological transition and to agroecosystem design and configuration, diversity and perennially. The assessment of permaculture overlaps agro ecology across different sectors with an uneven relative prominence.

Urban agro forestry is an innovative tool bridging gaps between urban planning and agro ecology in the contexts of urban sustainability, increasing resilience, biodiversity and capacity of carbon sequestration. Agro ecological practices of agroforestry enhance ecosystem service provision. Agroforestry combines the crop agriculture and woody cultivation recognized as agro ecological practice for addressing sustainability and community resilience (Nair 1993; Kumar 2006; Nair 2007). Urban agro-forestry ecology is an urban resilience strategy to protect urban food and energy supplies (Altieri et al. 1999; Lovell 2010; Dubbeling et al. 2009; De Zeeuw et al. 2011) providing a set of economic, social and ecological services (Konijnendijk and Gauthier 2006).

Urban agro ecological food forestry as an interdisciplinary and multifunctional strategy to improve urban sustainability development, ecosystem services, ecological resilience, food security, climate change, poverty, public health, etc., by increasing the consumption of more nutrient-dense foods. Shade tree in agro ecological forestry protects crop plants against fluctuations in soil and temperature in a microclimate, which controls evaporation, exposure to sunlight, wind velocity and rain.

Social change and innovation in urban and peri-urban agro ecology do not come from government but from small urban farmer families willing to transform the living conditions towards a healthier life style. Urban agro ecology tends to expand involving innovation and adaptation as the capacities of the food systems expand creating more nutritional and healthy resources and more employment opportunities. Technological innovation of agro ecology requires the contribution of social and production techniques, organization and diffusion.

Coherent management urban agro ecological policies set the standards of a sustainable environment for cropping techniques and treatment of natural resources. Agro ecological management in urban and peri-urban areas must be supported by production, environmental protection, quality and marketing policies integrated to local governance and cultural heritage.

Policy approaches on urban agro ecology are related to economic, social and ecological scenarios. Agro ecological management of urban and peri-urban areas is a complex process of development and governance strategies and policies to support investments and safe guard the self-sufficiency of the ecosystem. Agro ecological funded projects tend to focus on linkages between agro ecological biodiversity, functioning, improvement of intercropping and agro ecological systems, agroecosystem management, agro ecological science and management.

The implementation of sectorial policies in urban agro ecology requires urban institutions to operate in the urban sphere such as an urban agricultural department or office and an interdepartmental committee to take care of urban food production, supply and consumption. Peri-urban agro-tourism is being developed and promoted by farmers offering tourist ecological services in the forms of larger agro-recreational parks and family-based agro-tourism including accommodation, food, fresh products, horse riding, etc.

The agro ecological system should include policies to combine knowledge, to secure access to land and to self–organize in solidarity towards food sovereignty and justice. Urban agro ecological farms rely more on grant funding, donations, social and volunteer services and other off-farm income to support their activities. The social and solidarity economy (SSE) refers to cooperatives, enterprises, organizations, mutual benefit societies, associations, foundations and social enterprises producing goods, environmental services and knowledge while pursuing economic, social and environmental aims and fostering solidarity.

Urban agro ecology properly managed may reduce the negative effects and externalities of industrial agriculture on public health and environment.

## Agro Ecological Transition

Agro ecological transition is a multi-sectorial project, complex of operating through the involvement of multiple temporal and spatial scales and diverse constituencies (Geels and Kemp 2007; Marques 2010; Piraux et al. 2010). Agro ecological literature addresses the transition to diversified and multifunctional production systems (De Schutter 2010, Kremen et al. 2012; Wilson 2008). The agro ecological transition is a complex process that requires the evaluation of resources and opportunities supported by disciplines and movements outside the boundaries.

The transition from industrial agriculture to agro ecological production requires experiential and scientific knowledge from the collaboration between research institutions, traditional producers, consumers, agro ecological movements, and other stakeholders. Stakeholders relate their socio economic values to an agro ecological transition by increasing their equity and reducing their power asymmetries in their assessment and decision processes (Felipe-Lucia et al. 2015). Among the various stakeholders of urban agro ecology should be stimulated the dialogue and cooperation relationships through the implementation of a multi-actor working team to organize joint analysis in urban development and to coordinate policymaking and sustainable urban planning.

The multidimensional nature of planning and policy development depends on urban agro ecology and nature management, among others. Urban agro ecology is a new method of urban planning based on a sustainable land exploitation and food production.

An agricultural research opens the potential for addressing agro ecological transition and diversified production systems of staple and complementary crops (Shepard 2013). The ongoing suburbanization, peri-urban transitional areas, and urban sprawl processes challenges the urban-rural development dichotomy and related phenomena of land uses of agro ecology mixed in a fragmented land mosaic.

The dominant model of agriculture is based on corporate control of chemical-intensive agro-industrial complexes of growing monocultures for export is decreasing healthy food quality, exacerbating global climate change, disrupting traditional rural livelihoods and eroding arable land. Transition to more sustainable agro ecological systems has the goal to achieve the resilience than the conventional agricultural systems has not achieved because the vertical interdependence links easy to break and fail (Servigne and Stevens 2015).

The trend toward agro ecological farming and the transition of agro ecological practices is global. A transition to more agro ecologically farming systems should be able to substitute agricultural chemicals by more ecological inputs. Gliessman and Rosemeyer (2009). describe the transition from conventional agriculture to agro ecological practices in a continuum by defining four steps. Agro ecology that is reliant on locally-produced inputs, increases the incomes for small-scale farmers. Agro ecological small farming systems are more food productive and more resilient to natural disasters. The self-sufficiency local agenda is the transition initiative that underestimates the benefits of international cooperation in the agro ecological food (Mason and Whitehead 2012).

Agro ecology is a participative action research for practicing science with stakeholders involved in agro ecological transitions (CuéllarPadilla and Calle-Collado 2011; Levidow et al. 2014, Méndez et al. 2017). Actors involved in urban agro ecology such as the urban planners, local governments, researchers, producers, etc. are facing some challenges in urban economic, socio-demographic and environmental changes resulting from the rural post-productivist transition to urban agro ecological food production (Wilson, 2009; Almstead et al., 2014; Roche and Argent, 2015). Food consumption patterns, linked to nutrition transition are decreasing the availability of water, which is utilized for irrigation in agriculture (Collette *et al.* 2011).

The multi functionality of earthworks for water control and harvesting are used and traditional agriculture systems and may be used in urban agro ecological systems to improve the productivity in urban land contexts (Evenari et al. 1982; Bruins et al. 1986; Boyd and Gross 2000; Mussery et al. 2013, Holt-Gimenez 2006, Prein 2002; Smukler et al. 2010). Traditional agriculture and agro ecological systems use earthworks for water control and harvesting despite the frequency (Lancaster and Marshall 2008; Frey 2011) and the risks posed by vulnerable and dispersive soils (Sherard et al. 1976).

Individual residents and land users can contribute to agro ecological transition by using simple techniques, practices and depoliticize the agro ecological transition (de Molina 2012; Lovell et al. 2010; Rosset and Martínez-Torres 2012), simplifying the sociological complexity of the challenges and processes.

Negotiation of the agro ecological transition between the different interests of the stakeholders is a normative political process involving emotional dimension (Wezel et al. 2009). To deliver the agro ecology principles in a transition process towards a new paradigm requires political actions.

The agro ecological transition is an analytical framework, which may be used to analyze discussion of permaculture from a holistic perspective of biophysical and social factors. Permaculture or permanent agriculture is an international alternative agro ecological movement and ecological design system potentially contributing to agro ecological transition. The agro ecological transition challenges the movement of permaculture driven by the transfer, dissemination, practices and application of knowledge and beliefs that acknowledges the sustainability (Jordan et al. 2008; Berkes et al. 2000). Permaculture is a design orientation used in agro ecological system research with a limited potential for transition. Agro ecological movements are critical agro ecological transition (Nelson et al. 2009; Altieri and Toledo 2011; Fernandez et al. 2012; Petersen et al. 2012).

Permaculture grows overlapping with agro ecology focusing on ecology and agriculture production in association with environmental movements toward agro ecological transition. Permaculture is an agro ecological movement with potential contribution to agro ecological transition by mobilizing social support for sustainability. The movement of permaculture has contributions constrained by cultural factors affecting the impact on the agro ecological transition despite that mobilizes social support in diverse geographical locations. Besides, the permaculture movements have been accused of inflating land and labor in perennial and polyculture systems.

In the forest transition model, increased agricultural intensification due to the current food crisis, is related to deforestation of land to be used for cultivation.

Popular agro ecological movements and networks advance traditional agriculture transition towards agro ecology through social and political support (Nelson et al. 2009; Ferguson and Morales 2010; Rosset et al. 2011; Altieri and Toledo 2011).

## CONCLUSION

There is an urgent need in favor of agro ecology for changes in global sustainable food systems. The new global sustainable food system aims to protect the agro ecosystems that provide the food and the ecosystem services based on equity, participation, democracy and justice. Urban farmers can actually take some steps for converting from industrial or conventional agroecosystems to have an agro ecological farm, such as reduce use and consumption of environmentally damaging inputs, substitute for alternative practices and redesign on the basis of new agro ecological processes.

Economic policy promoting urban agro ecological production should crate regulatory and financial mechanisms and provide incentives to urban farmers and communities to favor market conditions for food production and ecosystem services. Agro ecological production requires the action of local governments and supportive policies, communities, markets and citizens to spread the practices and benefits of crop and livestock production systems and managed as ecosystem services.

The cultivation of productive urban agro ecology in urban green spaces, open space locations and home production systems is an intervention strategy to be carried out to achieve economic, social and environmental benefits, to alleviate urban poverty and hunger, and enhance urban environmental services.

An ecosystem services valuation framework may help in steering agro ecological transitions. The evaluation of the impact of urban agro ecology activities in ecosystem services and food justice may vary depending of their own merits because the diversity in goals, scope, scale, type of access and participants.

## REFERENCES

Adhikari, K., & Hartemink, A. E. (2016). Linking soils to ecosystem services— A global review. *Geoderma*, *262*, 101–111. doi:10.1016/j.geoderma.2015.08.009

Almstead, A., Brouder, P., Karlsson, S., & Lundmark, L. (2014). Beyond post-productivism: From rural policy discourse to rural diversity. *European Countryside*, *4*(4), 297–306. doi:10.2478/euco-2014-0016

Altieri, M. A., Companioni, N., Cañizares, K., Murphy, C., Rosset, P., Bourque, M., & Nicholls, C. I. (1999). The greening of the "barrios": Urban agriculture for food security in Cuba. *Agriculture and Human Values*, *16*(2), 131–140. doi:10.1023/A:1007545304561

Altieri, M. A., & Toledo, V. M. (2011). The agroecological revolution in Latin America: Rescuing nature, ensuring food sovereignty and empowering peasants. *The Journal of Peasant Studies*, *38*(3), 587–612. doi:10.1080/03066150.2011.582947

Armitage, D., Marschke, M., & Plummer, R. (2008). Adaptive comanagement and the paradox of learning. *Global Environmental Change, 18*(1), 86–98. doi:10.1016/j.gloenvcha.2007.07.002

Banerjee, S., Secchi, S., Fargione, J., Polasky, S., & Kraft, S. (2013). How to sell ecosystem services: A guide for designing new markets. *Frontiers in Ecology and the Environment, 11*(6), 297–304. doi:10.1890/120044

Bavec, M., Mlakar, S. G., Rozman, C., Pazek, K., & Bavec, F. (2009). Sustainable agriculture based on integrated and organic guidelines: Understanding terms—the case of Slovenian development and strategy. *Outlook on Agriculture, 38*(1), 89–95. doi:10.5367/000000009787762824

Becker, F. D., & Steele, F. (1995). *Workplace by design: mapping the high-performance workscape*. New York: Jossey-Bass.

Benoit, M., Rizzo, D., Marraccini, E., Moonen, A. C., Galli, M., Lardon, S., ... Bonari, E. (2012). Landscape agronomy: A new field for addressing agricultural landscape dynamics. *Landscape Ecology, 27*(10), 1385–1394. doi:10.100710980-012-9802-8

Berkes, F., Colding, J., & Folke, C. (2000). Rediscovery of traditional ecological knowledge as adaptive management. *Ecological Applications, 10*(5), 1251–1262. doi:10.1890/1051-0761(2000)010[1251:ROTEKA]2.0.CO;2

Bohn, K., & Viljoen, A. (2005). More city with less space: Vision for lifestyle. In A. Viljoen (Ed.), *Continuous Productive Urban Landscapes: Designing Urban Agriculture for Sustainable Cities* (pp. 251–264). Burlington, MA: Architectural Press.

Bostonian, N. J., Goulet, H., O'Hara, J., Masner, L., & Racette, G. (2004). Towards insecticide free orchards: Flowering plants to attract beneficial arthropods. *Biocontrol Science and Technology, 14*(1), 25–37. doi:10.1080/09583150310001606570

Boyd, C. E., & Gross, A. (2000). Water use and conservation for inland aquaculture ponds. *Fisheries Management and Ecology, 7*(1-2), 55–63. doi:10.1046/j.1365-2400.2000.00181.x

Bruins, H. J., Evenari, M., & Nessler, U. (1986). Rainwater-harvesting agriculture for food production in arid zones: The challenge of the African famine. *Applied Geography (Sevenoaks, England), 6*(1), 13–32. doi:10.1016/0143-6228(86)90026-3

Burbidge, J. L. (1971). Production flow analysis. *Production Engineering, 50*(4-5), 139–152. doi:10.1049/tpe.1971.0022

Busch, L., & Bain, C. (2004). New! Improved? The Transformation of the Global Agrifood System. *Rural Sociology, 69*(3), 321–346. doi:10.1526/0036011041730527

Cavazza, L. (1996). Agronomia aziendale e agronomia del territorio. *Riv Agron, 30*, 310–319.

Collette. (2011). Thunnus obesus The IUCN Red List of Threatened Species 2011: e.T21859A9329255. doi:10.2305/IUCN.UK.2011-2

Companioni, N., Rodríguez Nodals, A., Carrión, M., Alonso, R. M., Ojeda, Y., & Peña, E. (1997). La Agricultura Urbana en Cuba. Su participación en la seguridad alimentaria. Conferencias. III Encuentro Nacional de Agricultura Orgánica. Villa Clara, Cuba: Central University of Las Villas.

Cote, M., & Nightingale, A. J. (2012). Resilience thinking meets social theory: Situating social change in socio-ecological systems (SES) research. *Progress in Human Geography*, *36*(4), 475–489. doi:10.1177/0309132511425708

Crawford, M. (2010). *Creating a forest garden: working with nature to grow edible crops*. Green Books.

Crouzat, E., Mouchet, M., Turkelboom, F., Byczek, C., Meersmans, J., Berger, F., & Lavorel, S. (2015). Assessing bundles of ecosystem services from regional to landscape scale: Insights from the French Alps. *Journal of Applied Ecology*, *52*(5), 1145–1155. doi:10.1111/1365-2664.12502

Cuéllar-Padilla, M., & Calle-Collado, Á. (2011). Can we find solutions with people? Participatory action research with small organic producers in Andalusia. *Journal of Rural Studies*, *27*(4), 372–383. doi:10.1016/j.jrurstud.2011.08.004

Davies, K. K., Fisher, K. T., Dickson, M. E., Thrush, S. F., & Le Heron, R. (2015). Improving ecosystem service frameworks to address wicked problems. *Ecology and Society*, *20*(2), art37. doi:10.5751/ES-07581-200237

De Molina, M. G. (2012). *Agro ecology and politics. How to get sustainability?* About the necessity for a political agroecologist. *Agroecology and Sustainable Food Systems*, *37*, 45–59. doi:10.1080/10440046.2012.705810

De Schutter, O. (2010). *Report submitted by the special rapporteur on the right to food*. United Nations Human Rights Council.

De Zeeuw, H., Van Veenhuizen, R., & Dubbeling, M. (2011). The role of urban agriculture in building resilient cities in developing countries. *J Agric Sci*, *149*(S1), 153–163. doi:10.1017/S0021859610001279

Deb, S., Barbhuiya, A. R., Arunachalam, A., & Arunachalam, K. (2008). Ecological analysis of traditional agroforest and tropical forest in the foothills of Indian eastern Himalaya: Vegetation, soil and microbial biomass. *Tropical Ecology*, *49*, 73–78.

Dubbeling, M., Campbell, M. C., Hoekstra, F., & van Veenhuizen, R. (2009). Building resilient cities. *Urban Agriculture Magazine*, *22*, 3–11.

Duru, M., Thérond, O., & Fares, M. (2015). Designing agroecological transitions: A review. *Agronomy for Sustainable Development*, *35*(4), 1237–1257. doi:10.100713593-015-0318-x

Ellis, M. A., Ferree, D. C., Funt, R. C., & Madden, L. V. (1998). Effects of an apple scab-resistant cultivar on use patterns of inorganic and organic fungicides and economics of disease control. *Plant Disease*, *82*(4), 428–433. doi:10.1094/PDIS.1998.82.4.428 PMID:30856893

Evenari, M., Shanan, L., & Tadmor, N. (1982). *The Negev: the challenge of a desert*. Cambridge: Harvard University Press. doi:10.4159/harvard.9780674419254

Felipe-Lucia, M. R., Martín-López, B., Lavorel, S., Berraquero-Díaz, L., Escalera-Reyes, J., & Comín, F. A. (2015). Ecosystem services flows: Why stakeholders' power relationships matter. *PLoS One*, *10*(7), e0132232. doi:10.1371/journal.pone.0132232 PMID:26201000

Ferguson, B. G., & Morales, H. (2010). Latin American agroecologists build a powerful scientific and social movement. *Journal of Sustainable Agriculture*, *34*(4), 339–341. doi:10.1080/10440041003680049

Fernandez, M., Goodall, K., Olson, M., & Mendez, E. (2012). Agroecology and alternative agrifood movements in the United States: Towards a sustainable agrifood system. *Agroecology and Sustainable Food Systems*, *37*, 115–126. doi:10.1080/10440046.2012.735633

Francis, C. A., & Porter, P. (2011). Ecology in sustainable agriculture practices and systems. *Critical Reviews in Plant Sciences*, *30*(1-2), 64–73. doi:10.1080/07352689.2011.554353

Frey, D. (2011). *Bioshelter market garden: a permaculture farm*. New Society.

Galafassi, D., Daw, T., Munyi, L., Brown, K., Barnaud, C., & Fazey, I. (2017). Learning about social-ecological trade-offs. *Ecology and Society*, *22*(1), 2. doi:10.5751/ES-08920-220102

Geels, F. W., & Kemp, R. (2007). Dynamics in socio-technical systems: Typology of change processes and contrasting case studies. *Technology in Society*, *29*(4), 441–455. doi:10.1016/j.techsoc.2007.08.009

Gliessman, S. R. (2007). *Agroecology: The ecology of sustainable food systems* (2nd ed.). New York: CRC Press.

Gliessman, S. R., & Rosemeyer, M. (Eds.). (2009). *The conversion to sustainable agriculture: principles, processes, and practices*. Boca Raton, FL: CRC Press; doi:10.1201/9781420003598

Gomiero, T., Pimentel, D., & Paoletti, M. G. (2011). Is there a need for a more sustainable agriculture? *Critical Reviews in Plant Sciences*, *30*(1-2), 6–23. doi:10.1080/07352689.2011.553515

Hampwaye, G. (2013). Benefits of urban agriculture: Reality of illusion? *Geoforum*, *49*, R7–R8. doi:10.1016/j.geoforum.2013.03.008

Hatfield, J. (2007) Beyond the edge of the field. *Arch. Soil Sci. Soc. Am.* https://www.soils.org/about-society/presidents-message/archive/13

Holt, A. R., Alix, A., Thompson, A., & Maltby, L. (2016). Food production, ecosystem services and biodiversity: We can't have it all everywhere. *The Science of the Total Environment*, *573*, 1422–1429. doi:10.1016/j.scitotenv.2016.07.139 PMID:27539820

Holt-Gimenez, E. (2006). *Campesino a campesino: voices from Latin America's farmer to farmer movement for sustainable agriculture*. Oakland: Food First.

Jacke, D., & Toensmeier, E. (2005). *Edible forest gardens: ecological design and practice for temperate climate permaculture* (Vol. 2). Chelsea Green Publishing Company.

Jacke, D., & Toensmeier, E. (2005). *Edible forest gardens: ecological design and practice for temperate-climate permaculture*. White River Junction: Chelsea Green.

Jordan, N. R., Bawden, R. J., & Bergmann, L. (2008). Pedagogy for addressing the worldview challenge in sustainable development of agriculture. *Journal of Natural Resources and Life Sciences Education*, *37*, 92–99.

Konijnendijk, C., & Gauthier, M. (2006) Urban forestry for multifunctional land use. In *Cities farming for the future: urban agriculture for green and productive cities*. International Development Research Centre, Ottawa. https://www.idrc.ca/en/ev-103884-201-1-DO_TOPIC.html.

Kremen, C., & Miles, A. (2012). Ecosystem services in biologically diversified versus conventional farming systems: Benefits, externalities, and trade-offs. *Ecology and Society*, *17*(4), 40. doi:10.5751/ES-05035-170440

Kumar, B. M. (2006). Agroforestry: The new old paradigm for Asian food security. *Journal of Tropical Agriculture*, *44*, 1–14.

Lancaster, B., & Marshall, J. (2008). *Water-harvesting earthworks*. Tucson: Rainsource.

Leakey, R. R. B. (2012). *Multifunctional agriculture and opportunities for agroforestry: implications of IAASTD. In Agroforestry: the future of global land use* (pp. 203–214). Dordrecht: Springer. doi:10.1007/978-94-007-4676-3_13

Lev, L., & Stevenson, G. W. (2011). Acting collectively to develop midscale food value chains. *Journal of Agriculture, Food Systems, and Community Development*, *1*(4), 119–128. doi:10.5304/jafscd.2011.014.014

Levidow, L., Pimbert, M., & Vanloqueren, G. (2014). Agroecological research: conforming—or transforming the dominant agro-food regime? *Agroecology and Sustainable Food Systems, 38*(10), 1127–1155. doi:10.1080/21683565.2014.951459

Lovell, S. (2010). Multifunctional urban agro ecology for sustainable land use planning in the United States. *Sustainability*, *2*(8), 2499–2522. doi:10.3390u2082499

Lovell, S. T., DeSantis, S., Nathan, C. A., Olson, M. B., Méndez, V. E., Hisashi, C., & ... . (2010). Integrating agroecology and landscape multi-functionality in Vermont: An evolving framework to evaluate the design of agroeco-systems. *Agricultural Systems*, *103*(5), 327–341. doi:10.1016/j.agsy.2010.03.003

Marques, F. (2010) Constructing sociotechnical transitions toward sustainable agriculture. In *Proc. Symp. Innov. Sustain. Dev. Agric*. Food ISDA.

Martín-López, B., Gómez-Baggethun, E., García-Llorente, M., & Montes, C. (2014). Trade-offs across value-domains in ecosystem services assessment. *Ecological Indicators*, *37*, 220–228. doi:10.1016/j.ecolind.2013.03.003

Méndez, V. E., Bacon, C. M., & Cohen, R. (2013). Agroecology as a transdisciplinary, participa-tory, and action-oriented approach. *Agroecology and Sustainable Food Systems*, *37*(1), 3–18.

Merckx, T., & Pereira, H. M. (2015). Reshaping agrienvironmental subsidies: From marginal farming to large-scale rewilding. *Basic and Applied Ecology*, *16*(2), 95–103. doi:10.1016/j.baae.2014.12.003

Mollison, B. (1979). *Permaculture: a designer's manual*. Tagari Publications.

Mollison, B. (1988). *Permaculture: a designer's manual*. Tagari.

Mollison, B., & Holmgren, D. (1978). *Permaculture one: A perennial agricultural system for human settlements*. Tagari.

Mouchet, M. A., Lamarque, P., Martín-López, B., Crouzat, E., Gos, P., Byczek, C., & Lavorel, S. (2014). An interdisciplinary methodological guide for quantifying associations between ecosystem services. *Global Environmental Change, 28*, 298–308. doi:10.1016/j.gloenvcha.2014.07.012

Mussery, A., Leu, S., Lensky, I., & Budovsky, A. (2013). The effect of planting techniques on arid ecosystems in the northern Negev. *Arid Land Research and Management, 27*(1), 90–100. doi:10.1080/15324982.2012.719574

Nair, P. K. R. (1993). *An introduction to agroforestry*. Kluwer Academic Publishers. doi:10.1007/978-94-011-1608-4

Nair, P. K. R. (1993). *An introduction to agroforestry*. Dordrecht: Kluwer. doi:10.1007/978-94-011-1608-4

Nair, P. K. R. (2007). The coming age of agroforestry. *Journal of the Science of Food and Agriculture, 87*(9), 1613–1619. doi:10.1002/jsfa.2897

Nehls, T., Jiang, Y., Dennehy, C., Zhan, X., & Beesley, L. (2015). From waste to value: urban agriculture enables cycling of resources in cities. In Urban Agriculture Europe (pp. 170–173). Berlin: Jovis.

Nelson, E., Scott, S., Cukier, J., & Galán, Á. L. (2009). Institutionalizing agroecology: Successes and challenges in Cuba. *Agriculture and Human Values, 26*(3), 233–243. doi:10.100710460-008-9156-7

Osty, P.-L., Le Ber, F., & Lieber, J. (2008). Raisonnement à partir de cas et agronomie des territoires. *Rev Anthr Connaissances, 2*(2), 169–193. doi:10.3917/rac.004.0169

Petersen, P., Mussoi, E. M., & Dalsoglio, F. (2012). Institutionalization of the agroecological approach in Brazil: Advances and challenges. *Journal of Sustainable Agriculture, 37*, 103–114. doi:10.1080/10440046.2012.735632

Piraux, M., Silveira, L., Diniz, P., Duque, G., Coudel, E., Devautour, H., ... Hubert, B. (2010) Agroecological transition as a socio-territorial innovation: the case of the territory of Borborema in Brazilian semi-arid. *Proc. Symp. Innov. Sustain. Dev. Agric.*

Prager, Reed, & Scott. (2012). Encouraging collaboration for the provision of ecosystem services at a landscape scale—rethinking agri-environmental payments. *Land Use Policy, 29*(1), 244–249. .landusepol.2011.06.012 doi:10.1016/j

Prein, P. (2002). Integration of aquaculture into crop–animal systems in Asia. *Agricultural Systems, 71*(1-2), 127–146. doi:10.1016/S0308-521X(01)00040-3

Pretty, J. (2005). Sustainability in agriculture: Recent progress and emergent challenges. Issues. *Environmental Science & Technology, 21*, 1–15.

Pretty, J. (2006). *Agroecological approaches to agricultural development*. Washington, DC: World Bank.

Pretty, J. (2008). Agricultural sustainability: Concepts, principles and evidence. *Philosoph Trans Royal Soc. Botanical Sciences, 363*(1491), 447–465. PMID:17652074

Rivera, M. A., Quigley, M. F., & Scheerens, J. C. (2004). Performance of component species in three apple-berry polyculture systems. *HortScience, 39*(7), 1601–1606. doi:10.21273/HORTSCI.39.7.1601

Roche, M., & Argent, N. (2012). The fall and rise of agricultural productivism? An Antipodean viewpoint. *Progress in Human Geography, 39*(5), 621–635. doi:10.1177/0309132515582058

Rosset, P. M., & Martínez-Torres, M. E. (2012). Rural social movements and agroecology: Context, theory, and process. *Ecology and Society, 17*(3), 17. doi:10.5751/ES-05000-170317

Rosset, P. M., Sosa, B. M., Jaime, A. M. R., & Lozano, D. R. Á. (2011). The campesino-to-campesino agroecology movement of ANAP in Cuba: Social process methodology in the construction of sustainable peasant agriculture and food sovereignty. *The Journal of Peasant Studies, 38*(1), 161–191. doi:10.1080/03066150.2010.538584 PMID:21284238

Säumel, I., Kotsyuk, I., Hölscher, M., Lenkereit, C., Weber, F., & Kowarik, I. (2012). How healthy is urban horticulture in high traffic areas? Trace metal concentrations in vegetable crops from plantings within inner city neighbourhoods in Berlin, Germany. *Environmental Pollution, 165*, 124–132. doi:10.1016/j.envpol.2012.02.019 PMID:22445920

Schwarz, K., Cutts, B. B., London, J. K., & Cadenasso, M. L. (2016). Growing gardens in shrinking cities: A Solution to the Soil Lead Problem? *Sustainability, 8*(2), 1–11. doi:10.3390u8020141

Servigne, P., & Stevens, R. (2015). *Comment tout peut s'effondrer. Petit manuel de collapsologie à l'usage des générations présentes*. Paris, France: Seuil.

Shepard, M. (2013). *Restoration agriculture: real world permaculture for farmers*. Austin: Acres U.S.A.

Sherard, J. L., Decker, R. S., & Dunnigan, L. P. (1976). Identification and nature of dispersive soils. *Journal of the Geotechnical Engineering Division, 102*, 287–301.

Simon Rojo, M., Moratalla, A. Z., Alonso, N. M., & Jimenez, V. H. (2014). Pathways towards the integration of peri-urban agrarian ecosystems into the spatial planning system. *Ecological Processes, 3*(13), 16.

Smukler, S., Sánchez-Moreno, S., Fonte, S., Ferris, H., Klonsky, K., O'Geen, A., ... Jackson, L. (2010). Biodiversity and multiple ecosystem functions in an organic farmscape. *Agriculture, Ecosystems & Environment, 139*(1-2), 80–97. doi:10.1016/j.agee.2010.07.004

Stevenson, G. W., Clancy, K., King, R., Lev, L., Ostrom, M., & Smith, S. (2011). Midscale food value chains: An introduction. *Journal of Agriculture, Food Systems, and Community Development, 1*(4), 1–8. doi:10.5304/jafscd.2011.014.007

Swensen, G., & Jerpåsen, G. P. (2008). Cultural heritage in suburban landscape planning. A case study in Southern Norway. *Landscape and Urban Planning, 87*(4), 289–300. doi:10.1016/j.landurbplan.2008.07.001

Taghavi, A., & Murat, A. (2011). A heuristic procedure for the integrated facility layout design and flow assignment problem. *Computers & Industrial Engineering, 61*(1), 55–63. doi:10.1016/j.cie.2011.02.011

The Economics of Ecosystems and Biodiversity (TEEB). (2010). *Mainstreaming the economics of nature: a synthesis of the approach, conclusions and recommendations of TEEB*. Geneva, Switzerland: TEEB.

Torre Ugarte, D. G., & Hellwinckel, C. C. (2010). The problem is the solution: the role of biofuels in the transition to a regenerative agriculture. In P. Mascia, J. Scheffran, & J. Widholm (Eds.), *Plant biotechnology for sustainable production of energy and co-products* (pp. 365–384). New York: Springer. doi:10.1007/978-3-642-13440-1_14

Tscharntke, T., Klein, A. M., Kruess, A., Steffan-Dewenter, I., & Thies, C. (2005). Landscape perspectives on agricultural intensification and biodiversity-ecosystem service management. *Ecology Letters*, *8*(8), 857–874. doi:10.1111/j.1461-0248.2005.00782.x

Veldkamp, A., Kok, K., De Koning, G. H. J., Schoorl, J. M., Sonneveld, M. P. W., & Verburg, P. H. (2001). Multi-scale system approaches in agronomic research at the landscape level. *Soil & Tillage Research*, *58*(3-4), 129–140. doi:10.1016/S0167-1987(00)00163-X

Venkatadri, U., Rardin, R. L., & Montreuil, B. (1997). A design methodology for fractal layout organization. *IIE Transactions*, *29*(10), 911–924. doi:10.1080/07408179708966411

Wezel, A., & Soldat, V. (2009). A quantitative and qualitative historical analysis of the scientific discipline agroecology. *International Journal of Agricultural Sustainability*, *7*(1), 3–18. doi:10.3763/ijas.2009.0400

Wilson, G. A. (2008). From "weak" to "strong" multifunctionality: Conceptualising farm-level multifunctional transitional pathways. *Journal of Rural Studies*, *24*(3), 367–383. doi:10.1016/j.jrurstud.2007.12.010

Wilson, G. A. (2009). Post-Productivist and multifunctional agriculture. International Encyclopedia of Human Geography, 379–386. doi:10.1016/B978-008044910-4.00895-6

Wiskerke, J. S. C. (2016). Urban food systems. In Cities and Agriculture. Developing resilient urban food systems (pp. 1–26). New York: Routledge.

Zhang, W., Ricketts, T. H., Kremen, C., Carney, K., & Swinton, S. M. (2007). Ecosystem services and dis-services to agriculture. *Ecological Economics*, *64*(2), 253–260. doi:10.1016/j.ecolecon.2007.02.024

## KEY TERMS AND DEFINITIONS

**Agro-Ecology:** The discipline that is responsible for administering the ecological principles of the production of food, fuels, fibers and pharmaceuticals. This encompasses a wide range of approaches and they consider it a science and a way of seeing life, whether organic, conventional, intensive or extensive.

**Ecosystem Services:** These are resources or processes of natural ecosystems (goods and services) that benefit human beings. It includes products such as clean drinking water and processes such as waste decomposition.

**Environmental Development:** An economic and social development that respects the environment. The objective of sustainable development is to define viable projects and reconcile the economic, social, and environmental aspects of human activities; It is about making progress in these areas without having to destroy the environment.

**Food System:** Refers to food produced, processed, distributed and consumed locally.

**Transition:** Step or change from one state, way of being, etc., to another. Intermediate state between an older one and another that is reached in a change.

**Urban Agro-Ecology:** Covers agroecological practices that are developed in or near cities. ... Urban agroecology usually develops on the roofs of buildings, on the walls of houses and on the balconies and terraces of buildings.

# Chapter 3
# Corralling Contract and the Land Use Relationship Between Pastoralists and Agriculturalists in Nigeria

**Regina Hoi Yee Fu**
*Senshu University, Japan*

## ABSTRACT

*"Corralling contract" is the indigenous fertilization system commonly practiced in the African Sahel and its southern periphery. In this chapter, the practice of the corralling contract between Fulani pastoralists and Nupe agriculturalists in the Bida region of Niger State of central Nigeria is examined. The study attempts to find out how the farmers and herders arrange the corralling contract, how they utilize this instrument, and how it influences their social relationship. Findings suggest that pastoral Fulani groups have different strategies to maintain socioeconomic relations with specific villages through the adoption of corralling contract in order to ensure resources entitlement. While some groups can well manipulate the relationships with various villages through the adoption of the corralling contract, some groups prefer a more stable situation and just get the minimum advantages. Higher social status, larger herd size and longer history of interaction that allow trust to be built are the factors contributing to the popularity and bargaining power of a pastoral group.*

## INTRODUCTION

Coexistence of farmers and herders in the semi-arid Africa has been described as symbiotic. Although confrontation occasionally occurred, in most cases they could be regulated in such a way that the peaceful cohabitation of the groups as a whole was not endangered. In West Africa however, conflicts over the use of scarce natural resources between farmers and herders are said to be on the increase in recent years. The occurrence of such conflicts is generally attributed to two factors: the changing patterns of resource use that lead to increasing competition for resources; and the breakdown of traditional mecha-

DOI: 10.4018/978-1-7998-3576-9.ch003

nisms governing resource management and conflict resolution. The generalization of increasing conflict gives an impression that the traditional mutual dependent and mutual beneficial forms of farmer-herder interaction that well-functioned in the past does not work anymore now. This perspective justifies direct interventions and implies new structures for new institutions for the co-operative management of natural resource use and conflict management.

Observations have been made in respect to the Nupe farmers and Fulani herdsmen in the Niger State of Nigeria. Case materials suggest, at least with the specific case of the Nupe farmers and the Fulani pastoralists in the field site, a perspective that contrary to the increasing-conflict view. Even though limited natural resources are shared and their production system is gradually converging, tension seems to be absent and their relationship shows no sign of deterioration. The farmer-herder interactions are frequent and mostly cordial. The traditional institutions governing natural resource use and conflict resolution are being preserved and are still functioning fine.

This paper focuses on the corralling contract which is one of the most important traditional institutions between farmers and herdsmen that have been practiced down through the ages. Corralling contract refers to the contractual agreement between farmers and herders to maintain livestock on croplands for a specified time period. Following the great reduction since mid-1980s and finally the withdrawal in 1997 of fertilizer subsidies by the Nigerian government, the corralling contract has become more important for resource-poor farmers who cannot afford fertilizer. Meanwhile, the decreasing availability of grazing resource due to the extension of cultivated area outpacing population growth also make herders rely more on the corralling contract as the tool to ensure access to resources. The corralling contact has gained more attention in recent years. Although it is not a new phenomenon, some scholars described it as a newly emerging traditional institution. Most researches focused on the ecological impacts of manure on soil fertility; very few examined the socio-economic implications of the corralling contract. While the contract is an institution that requires the agreements of both sides, most researches took only the farmers' perspective. Nevertheless, field observations revealed that farmers are rather passive in the adoption of corralling contract. The preconditions for them to adopt the practice are the presence of herders in nearby area and the location of village that is in an environment suitable for cattle stay when herders come for the season. Therefore, the perspectives of herders are indispensable for the thorough understanding of the corralling contract. However, there is no research the author could find so far that explains how the farmers and herders reach to the corralling contract and how the details are being arranged. This paper intends to provide a detail account of this important institution.

Research has been done on both sides to investigate the implementation of this traditional institution. The main questions here are: how do the two groups arrange the corralling contract; how do they utilize this instrument and how does it influence their socioeconomic relationship? Findings suggest that Fulani groups adopt different strategies to maintain social relations with specific villages in order to ensure resources entitlement. Their "popularity stakes" and the amounts of payment they can get through the contract vary greatly from each other. The competition for Nupe farmers to host a Fulani group is keen and costly therefore villagers have to combine collective efforts. Contrary to traditional depiction, richer and influential farmers do not necessarily benefit more from corralling contracts and there is no significant sign that Fulani herders claim more payment in cash or in kind than in former years.

## CORRALLING CONTRACT

### Ecological Benefits Brought by Corralling Contract

Corralling contract, or manure contract, is an indigenous fertilization system commonly practiced in the semi-arid area of West Africa (Asanuma, 2004; Neef, 2001). It is also known as "parcage system" in the French literatures and locally as "*hoggo* system". *Hoggo* in Fulani language means the cattle enclosure where cattle herds are kept overnight. When individual or group of farmer and herder enter into the contractual agreement, the herder has to corral his cattle overnight on the farmer's field for a specific period of time at the farmer's request. In return the farmer pays the herder in cash or in kind and allows livestock to graze on the crop residues on his fields.

Land scarcity and degradation from insufficient nutrient cycling increase the demand for manure in sub-Saharan Africa. In Nigeria, the lack of accessibility to good quality and affordable fertilizer and the unavailability of fertilizer in time of need make farmers rely on cattle manure. Demand for manure increased especially after the gradual reduction of the fertilizer subsidy since the mid 1980s and the liberalization of the fertilizer sector in 1997 (Nagy and Edun, 2002; Shimada, 1999). Fertilizer use declined sharply from a peak of 461,000 nutrient tones in 1994 to 173,000 nutrient tones in 2000. To ensure the availability of fertilizer for farmers, the federal and the state governments still procured and subsidized fertilizer in an ad hoc manner. However, the problem of lack of access to subsidized fertilizer for farmers still persisted. A substantial amount of subsidized fertilizers were sold on the black market (Nagy and Edun, 2002; USAID, 2007). Farmers have to rely on cattle manure to retrieve the productivity of their lands when fallow system for long-period is difficult. The benefits of the use of manure in crop production are the improvement in soil physical properties and the provision of N. P. K. and other mineral nutrients. The application of livestock manure increases soil organic matter content, which leads to improved water infiltration and water holding capacity as well as increased cation exchange capacity. Farmer access to manure requires either a decision to invest in animals or to enter into a corralling contract with someone who keeps livestock, usually the professional herders. The latter is more commonly in use in rural Nigeria not only because many farmers cannot afford to own livestock, but also because the corralling contract can bring a better efficiency in fertilizing a larger area of farmland with lower cost. Evidence suggested that in Niger, the fields manured through corralling contract received 5 to 13 times more manure than average land (Hiernaux et al., 1997).

Many researches have proven the effectiveness of corralling livestock on cropland for improving soil fertility (Schlecht et al, 2004; Sangarè et al, 2002; Achard & Banoin, 2003). It is more effective in maximizing nutrient cycling of soil comparing with merely applying manure transferred from other places. Based on the report of TropSoils (1991), the ecological benefits from manure applied by corralling animals can last for 10 years, which is much longer than that of transported manure which can last for only 3 years. The corresponding crop yields are also significantly higher. The difference is proven to be brought by cattle urine, which is difficult to be transported (Powell & Williams, 1993). Urine and manure together can effectively raise the PH level of soil and accelerate the decomposition of organic matter and termite mounds[1] (Brouwer and Powell, 1995; 1998). Many farmers regard the corralling contract with herders a better mean to fertilize their fields than the application of fertilizer by themselves.

## Social Impacts of Corralling Contracts

Corralling contract can be regarded as an exchange of services between herders and farmers: the service to fertilize croplands in exchange for the right to settle on fallow lands and to graze on crop residuals. Besides, exchange of farm products and milk products between farmers and herders is very common when herders are settling on the land of farmers (Grayzel, 1990; Wilson, 1984: Ogawa, 1998). Corralling contract is an important economic arrangement that facilitate the complimentary relationship of the two groups. Nevertheless, changes caused by economical, environmental and political factors are making the contract less accessible to some farmers. The corralling contract is no longer just a simple economical agreement. Some scholars point out that it has turned into tools and symbols in broader struggles among communities over access to land for field and pasture (Heasley and Delehanty, 1996).

The research of Neef (1997) in south-west Niger found out that richer and more influential farmers obtained greater access to manure through corralling contract than poor farmers. In many parts of sub-Saharan Africa, herders are claiming more and more payment in cash or in kind than in former years. Evidence in Southwest Niger suggested that tenants of short-term use rights used animal manure to a significantly lesser extent compared with landowners and tenants with medium-term use rights. Some of them feared that the landowner would reclaim the land back if he noticed that soil fertility was improved (Neef, 2001).

On the other hand, higher demand for manure enhances the bargaining power of herders and enables them to get a better position in the politics of manure. Many pastoralists use the contract as a trump in case of land conflicts (Loofboro, 1993) and as a strategy to obtain and secure permanent land use rights from private landowners or local leaders (Neef, 1997). Heasley and Delehanty (1996) illustrated the case study of four villages in southwestern Niger to demonstrate how the access to manure has become a signal point of entry into the political economy of agropastoral production emerging in the Sahel. In two of the villages studied, the pastoral Fulani could threaten to withhold or even boycott corralling contract to enforce claims to ownership of lands and secure free passage to grazing resources. However in another two villages studied, the Fulani herders were in weaker positions that manure could only ensure their temporary access to lands. Contrary to conventional depiction, access to manure was not guaranteed for wealthier farmers who have livestock ownership. The control of manure was rather likely to reside with a professional herder entrusted with the farmer's stock. These case studies showed that beneficiaries of manure contracts were not necessarily determined by wealth ranks, but increasingly by the vagaries of the shifting local politics of ecology control. Manure has become a potent political tool because the rules and procedures governing its accessibility are undergoing transition. Such transition is due to the changing production systems from strictly crop or livestock based into more Agropastoral based.

## THE FAILURE OF GRAZING RESERVE POLICY IN NIGERIA

The Fulani[2] are the most numerous and probably the most prominent of all the pastoral groups in West Africa. They expanded eastwards from the Gambia River over the last thousand years and stretched across the entire West Africa sub-region. Among the estimated 30 million of pastoralist in Africa, 10 million are found in Nigeria. The Fulani is the largest group of pastoralists in Nigeria which constitute about 95% of nomadic herders in the country. The presence of Fulani was recorded in the Hausaland of northern Nigeria as early as the thirteenth century (Awogbade, 1983:3). A number of classic monographs described the Nigerian Fulani, most notably St. Croix (1972), Hopen (1958), and Stenning (1959). They studied the pastoral clans in the semi-arid areas. More recent researches were from Awogbade (1983) who described the Fulani on the Jos Plateau and Gefu (1992) who studied the Fulani of Udubo Grazing Reserve. The study on pastoral Fulani in the humid and sub-humid regions of Nigeria is still limited. Some of the papers in Kaufmann, Chater & Blench (1986) studied the Fulani in southern Zaria. Omotayo (2002) and Fabusoro (2006) explored the land related issues of Fulani in Southwestern Nigeria.

The pastoral Fulani in Nigeria, same as other nomadic pastoralists in Africa in general, have for several centuries concentrated their activities in the dry savanna and arid regions where farming activities were limited and competition for resources with other forms of land use were practically non-existent (Tonah, 2002). During the twentieth century, Fulani herders in Nigeria began to migrate through and settle in whole zones that were previously inaccessible to pastoralists. Ecological change and population increase has reduced the tsetse challenge for the non-trypanotolerant cattle owned by Fulani. This has removed the major barrier that stopped their southern expansion in previous era (Blench, 2010). The conventional stereotypes of the Fulani as living in Northern Nigeria are becoming less and less true, year after year. Now it is not surprising to find Fulani pastoralists settling even in the costal states in southern Nigeria.

Since independence, the Nigerian government has placed emphasis on the sedentarization of nomadic pastoralists in its effort to develop the livestock sector. It was assumed initially that the intensive western ranching models could be introduced to replace the traditional Fulani systems of production. After several unsuccessful attempts at making the westernized ranges work locally, they were dropped in favor of improving the traditional livestock production systems. One of the suggestions was the need to protect and improve grazing areas and stock routes so as to stabilize the Fulani mode of production. In 1965 the federal government passed the "Grazing Reserve Law" which intended to provide grazing rights and all-year resources to the pastoralists (Powell, 1992). The idea of establishing grazing reserves was to provide grazing land on which nomadic pastoralists could settle permanently with the expectation that this would lead to empowerment and equitable property rights for the Fulani pastoralists, improved standards of living, improved cattle production and elimination of conflict between them and sedentary crop farmers (Omotayo, 2002; Fabusoro, 2006).

In the Third National Development Plan of 1975-80, the establishment of a total of 22 million hectares of grazing reserves for the exclusive use of nomadic pastoralists was proposed (Gefu, 1989:23-25). The long-term objective of the policy was to enable the herders to settle down and adopt modern technologies of livestock production. The provision of infrastructural facilities such as watering points, improved pastures, treatment centers and feed store was to be embarked upon by the government as part of the strategies to develop the grazing reserves (Olomola, 1998). Nevertheless, despite of what has been written on proposal, in reality very little has been accomplished beyond the demarcation of some identified lands. By 2003, only 2.8 million hectares, which were only 13% of what has been proposed, had been acquired by the government for the purpose of grazing reserves in the northern states. Out of these acquired lands only about 10% were legally gazetted (National Livestock Development Project, 2003:5; 2007:15). Due to the fact that these grazing reserves often located in very remote districts with bad access to transportation and market, the numbers of pastoralists voluntarily settling in has been limited. Basic infrastructures were often not available in these reserves. Even among the pastoralists who have settled, there were few signs of improved production and living standard. Serious problems of overstocking and range deterioration have been encountered (Suleiman, 1989: 42-43). Crop farmers and other users have encroached upon almost all the reserves (National Livestock Development Project, 2003:5). As a result, most of the grazing reserves have indeed been abandoned (Fabusoro, 2006:55). Regarding the Niger

State, attempt was made to establish grazing reserves at strategic locations to reduce transhumance by pastoralists. Eighteen grazing reserves in total of 104,309 hectares were designated, but only two reserves in total of 44,302 hectares have ever been gazetted (National Livestock Development Project, 2007:73; NYSC, 2010). Although governmental documents indicated that three grazing reserves were locating right in the Bida region, none of the interviewed pastoral Fulani, even those of the ruling Fulani council, acknowledged the existence of these grazing reserves. This poses a question of the actual status of grazing reserves listed on paper: whether they have ever actually existed, or have been abandoned and then encroached by farms since long time ago. Obviously, Nigerian pastoralists do not and cannot rely on the nation's grazing reserve policy for securing adequate land resources for their herds. The majority of them continue the nomadic production system, and to maintain a cordial cooperative relation with their hosting communities remains as the most important method to secure resources access.

## STUDY AREA AND METHODOLOGY

The research was conducted in the Niger State of central Nigeria where Nupe is the dominant ethnic group. Nupe agricultural villages and pastoral Fulani groups in the suburb of Bida were studied (figure 1). Bida is the second largest city in the Niger State. It was the capital town of the old Nupe kingdom in the early 19th century. At present it is still the political and cultural center of the current Bida Emirate where the highest level traditional chiefs, such as the Bida *Emir* and the Fulani *Dikko* are stationed. The study area locates about 150 km upstream of the confluence of the Niger and Benue rivers. It is surrounded by rivers the Kaduna to the west, the Gbako to the east and the Niger to the south.

The vegetation of the study area belongs to the Guinea savanna zone with yearly precipitation of about 1,100mm. There are two distinct seasons – the rainy season from April to October and the dry season from November to March. However, the pastoral Fulani divide the year in a different way. They divide a year into six seasons and their herding activities change accordingly. For simplicity the six seasons are grouped into two in this study: December to May is regarded as the dry season and June to November is regarded as the rainy season. The study area can be divided into uplands and lowlands roughly by the counter line of 250 feet, which is approximately 75-80 meters. This peculiar topography and the availability of water in surrounding river basins throughout the dry season allow the pastoral Fulani not to migrate in long distance between seasons. The activities of pastoral Fulani concentrate on the uplands during the wet season from June to November. In the dry season from December to May, the river valleys turn into important grazing resource for the pastoral Fulani and other nomadic Fulani which migrate through or settle in from the north.

The research is based on fieldworks carried out during September to October in 2005 and September in 2006. Similar to many researches on pastoralists, much time was needed to build up relationship with pastoral Fulani. The author first contacted the Fulani in September 2004 during another fieldwork. Some preliminary researches were done during December 2004 to January 2005 with two major pastoral Fulani groups. In order to investigate the migration pattern and the practice of corralling contact of more Fulani groups, permission was obtained from *Dikko Bida* before more extensive research could be carried out. Statistical procedure to select samples for interviews was impossible because the two groups were rarely studied and census data did not exist. The author was introduced to the Nupe and Fulani informants through the Fulani officer of *Dikko Bida* council and the extension staff of Bida Agriculture Development Project. With the basic information gathered during the preliminary research, a

*Figure 1. Bida and its environs (counter lines shown in feet)*

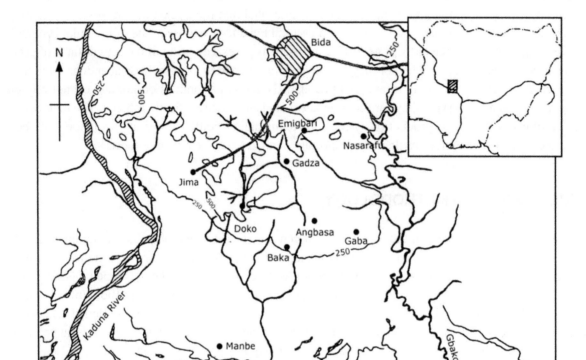

list of questions was designed for the semi-structured interview conducted during September to October 2005. Interviews were carried out mostly in the Fulani camp with the group heads. However as many of the heads gathered in market or Nupe village during day time, some of the interviews were carried out in market or village. Supplementary interviews were also conducted with other male household heads (*baade head*). Information of seventeen pastoral Fulani groups was gathered. For the farmers' perspectives, another list of questions was also designed for the semi-structured interviews with Nupe villagers. Farmers from sixteen Nupe villages who have hosted Fulani groups were questioned. For additional information about the relationship between pastoral Fulani and Nupe, dispute settlement, grazing reserve and traditional administration, interviews were conducted with the Nupe Village Area heads, the *Bida Dikko*, the assistant of *Bida Emir*, the officer of the Bida Agriculture Development Project and the officer of the Niger State Ministry of Agriculture. Fulani camp sites and farmers' manured fields were surveyed in details and maps were drawn out of the survey data. A digital area-line meter (PLANIX EX of Tayama Technics Inc.) was used to measure areas based on the survey maps.

## NUPE AND FULANI IN THE STUDY AREA

### Nupe Society

The Nupe people live in the heart of Nigeria in the low basin formed by the villages of the Niger and Kaduna Rivers. It is located between 9° 30´ and 8° 30´ north, in an area roughly 17,920 square kilometers in extent (Forde, 1970:17). Nupe is the dominant group in Niger and Kwara States. They were first described in detail by ethnographer Siegfried Nadel, whose book *Black Byzantium*, remains as anthropological classic. Accounts of the Nupe society can also be found in Forde (1970), Ibrahim (1992), Mason (1981) and Ismaila (2002). There are probably about a million Nupe, principally in Niger State[3]. They are primarily Muslims. Christianity was brought into the area since the mid-19th Century. Traditional religion still exists but is weak. The Nupe trace their origin to *Tsoede* who fled the court of Idah and established a loose confederation of towns along the River Niger in the 15th century. Nadel refers to *Tsoede* as the culture-hero and mythical founder of the Nupe kingdom. The Nupe were converted to Islam at the end of the eighteenth century by Mallam Dendo, a wandering Fulani preacher, and were incorporated into the Fulani Empire established by the *Jihad* led by Usman dan Fodio after 1804. Mallam Dendo's second son, Usman Zaki became *Etsu Nupe* (King of Nupe) in 1832 and the Fulani conquerors have been ruling the Nupe of the Bida Emirate since then. The city of Bida fell to the colonistist British forces in 1897 but the traditional administration of Bida Emirate has been preserved until now. The Fulani ruling class were assimilated with the Nupe by intermarriage and have lost their Fulani identity.

Extended family system of the Nupe is accompanied with patrilineal, patri-local and Islamic polygyny. Descent and succession to offices and the inheritance of rights to land and other property are patrilneal. The domestic unit is the elementary or polygynous family. The domestic family normally forms part of a larger social unit - the "*emi*" which means house. *Emi* is both a kinship and a territorial group and may also be referred to as *katamba*, a term for the common gateway or entrance to an enclosed compound. Nupe village normally composed of several *emi* which descended from one ancestor. Marriage of Nupe is largely endogamous, especially among the aristocracy. People normally marry within the same social class and the Nupe class endogamous marriage is characterized by a much lower marriage payment than exogamous marriage. Interethnic marriage is uncommon. The Nupe in Bida never inter-marry with the pastoral Fulani, but in 2005 there was the first case of a Nupe man marrying a Fulani woman.

### The Nupe Production System

The Nupe are subsistence farmers. Their means of livelihood slightly differ according to ecological division. In the uplands, they depend on shift cultivation during the rainy season, growing sorghum, millets, maize, melon, cassava, legumes and some vegetables. In the lowland watersheds that are floodplains of small rivers, they depend on rice farming in the rainy season and some of them grow off-season crops like sugar cane, cassava, sweet potato and vegetables after harvesting rice in the dry season. In the huge floodplains along the Niger, Kaduna and Benue Rivers, the Nupe rely on traditional rice farming in the rainy season and fishery (Hirose, 2002:187). Some farmers keep domestic livestock like chicken and goat, but only very few can afford investment in cattle fattening. Women do not own farmland and do not involve in farm work, but food processing is entirely done by them. Moreover, marketing of farm produce is normally in the hands of women. Because of increasing population density, land is becoming scarce. Fallow period is largely shortened in the uplands and lowland fields are cultivated annually. According to the AICAF (1994) report on a Nupe village called Gadza, the cultivated area of lowland rice is about 2 ha and upland fields are about 2.2 ha per family[4]. Few farmers use inputs such as chemical fertilizers, insecticides, pesticides, improved seeds or imported agricultural equipment. In 2006, a bag of 50kg fertilizer (NPK:15-15-15) was estimated at NGN 3, 000 (USAID, 2007), which was expensive relative to return and credit for purchase was unavailable. Like many parts of rural Nigeria, many young farmers have left their villages for education or better income in towns and cities. Their farms are usually managed by other household members and they just come back to work on their farms during long vacation. Since few years ago, motorbike-taxi driver in Bida town has become a very popular occupation for young Nupe farmers. Older Nupe farmers complained to the author that there is labor shortage especially for the community farms and younger farmers do not manage their farms as good as before. This situation also causes dissatisfaction of the pastoral Fulani as it is sometimes difficult to distinct farm from fallow land and farm encroachment may happen unintentionally.

## Land Tenure in Nupe Agricultural Communities

Nigeria is a former British colony and Africa's most populous country. It has about 120 million people, and 98 million hectares of land of which about three-quarters is arable. Prior to colonization, customary norms and laws governing the use of land evolved differently in various part of the country with a fundamental conception that land belongs to a vast family of which many are dead, few are living and countless yet unborn. In the customary land holding system, land is owned communally. It is the responsibility of the community heads to allocate the land to members of the community. Individuals within the community are entitled to portions of communal land for personal use and are expected to hold such land in trust for coming generations. Under the customary law, non-indigenes or strangers and migrants do not have guaranteed access to land.

Despite of the introduction of the Land Use Act in 1980 which took over the legal ownership of all land to the state, communal land right still prevail in most part of Nigeria including the study area. Due to the history of Fulani conquest in the 19th century, the customary land tenure system of the farming communities in Bida area is complicated. There is a three-layered structure of pattern of control over land and land related activities (Masuda, 2002). On the top level it is the traditional ruler, *Bida Emir*. In principle all the territories of the Bida Emirate are under the control of *Bida Emir*. In case of land dispute, the *Emir* is always in the supreme position of the traditional judicial system to arbitrage. Under the *Emir* there are primary landlords which were created by the feudalistic system of the kingdom. It consists of absentee or sometimes resident landlords. The absentee landlords in most cases are the privilege class of town Fulani living in Bida whose ancestors were important warriors of the *jihad* or dominant vassals of the *Emir*. On the other hand, resident landlords were mostly originated from the subordinates or slaves of noble Fulani in the distant past. Farmers regard the primary landlords as the "real landlords" who have the right to take over land from a community and to appoint secondary landlords. Primary landlords usually own large area of lands which cover the area of a few villages. Some of them hire Fulani herdsmen to manage their cattle herds and let them settle on their lands. At the bottle of the structure there are secondary landlords who are usually referred to as the "land managers" by farmers. The secondary landlords are usually the village chiefs. Their powers over village lands are restricted to allocation of farmland and management of vacant land, while each individual, group, or family can exercise exclusive right on the land on their own farming. Under the customary land system, pastoral Fulani do not have guaranteed access to land as they are still regarded as strangers even though some of them have been cohabitating with Nupe in the area for over a century. When pastoral Fulani decide to set up their camp on the land of a particular village, they must get the permission of the secondary landlord. Grazing lands in the study area are regarded as open access resources. The peaceful coexistence and mutual understanding with their Nupe hosts grant pastoral Fulani unrestricted access to any fallow land either during the rainy or dry season. Even for farmers who are hostile toward some Fulani, especially those migrate from far north during the dry season, they do not have the legitimate right to expel herdsmen from their fallow lands and harvested farmlands under the customary land system.

## Pastoral Fulani in Bida

Fulani are highly differentiated, not only according to clans but also by their economic pursuits and way of life. Fulani can be divided into two main types. The first type is called *"Fulbe siire"* in the Fulani language *Fulfulde,* meaning "town Fulani". They are Fulani urban dwellers who may or may not own cattle. They are mostly engaged in commerce, administration and education. The Fulani aristocrats who are the ruling class living in town belong to this type. Most of them have long abandoned the traditional lifestyle of Fulani, and they do not speak the Fulani language *Fulfulde*. The second type is the "pastoral Fulani", which in *Fulfulde* are called *"Fulbe na'i"* – "cow Fulani", or *"Fulbe ladde"* – "bush Fulani". The one important distinguishing feature differentiating the "pastoral Fulani" from the "town Fulani" is their close relationship with their cattle. This is the second group, the pastoral Fulani who reside in the bush and farmlands of the rural area that is the concern of this paper. Research on the pastoral Fulani in the Niger State is very limited. The ecological anthropological study conducted by Shikano (2002) in the mid-90s is the only account. There is no affinity between the pastoral Fulani and the town Fulani in the research area. Their clan organizations do not cross and they do not intermarriage. It has to be emphasized that pastoral Fulani is not a homogenous group. There are *"Fulbe wuro"*, the semi-settled or settled transhumant Fulani having permanent homestead; and *"Fulbe bororo*[5]*"*, the highly nomadic Fulani who still maintain a close system. *Fulbe bororo*, who are the true nomads, constitute only a small fraction of the total Fulani population in West Africa. The pastoral Fulani studied in this paper are *"Fulbe wuro"*.

The date when the pastoral Fulani first reached the land of the Nupe is unknown. It was estimated that semi-nomadic pastoralists made their appearance at a very early stage, but long-term group probably did not take place until much later and even after they settled it was only on a small scale (Johnston, 1967:135). Even by the time of the *jihad* conquest, Nadel (1942) estimated that the total number of Fulani, including the leading Fulani preachers and warriors, plus their cattle Fulani followers and Hausa mercenary soldiers, was not more than 1,000 or 1,500.

At about the seventeenth century, Fulani mallams and Fulani cattle owners began to settle in the land of the Nupe (Ismaila, 2002). The first group of pastoral Fulani that settled in the Bida region was the *Dindima'em, Juuliranko'em* group led by *Abdul-Maliki*. They migrated from an area named Machina which located somewhere north-east to the Sokoto country near Niger. During the colonial era, Fulani from the *Dindima'em, Juuliranko'em* group was selected by the *Emir* as *Dikko* - the chief of all pastoral Fulani in the emirate for the convenience of cattle tax collection. Apart from collecting cattle tax during

*Figure 2. View of hoggo of a Fulani camp in rainy season (DB group, October 2005)*

the rainy season, other major functions of *Dikko* are to settle disputes, to arbitrate divorce, to attend the transferal of cattle ownership and to represent the interests of his people in the national association of Fulani, Miyetti Allah Cattle Breeders Association (MACBAN). When the *Dindima'em, Juuliranko'em* group first settled in the Nupeland, they were just four in persons. After the migration of these four pioneers, pastoral Fulani from Machina and other regions in the north gradually infiltrated into the Bida Emirate. *Dikko Bida* estimated that by 2005 there were about 1,450 Fulani groups under his domain in the whole Bida Emirate, within which about 350 groups resided in the Bida region. The main pastoral Fulani lineage groups that are now settling in the region are the *Dindima'em*, the *Boodi* and the *Fittoji*. Pastoral Fulani in the region sustain their subsistence by raising cattle, sheep and chicken. Majority of them are pure pastoralist that they do not farm at all, but in recent years there is a trend for Fulani to borrow farm plots from the Nupe for very small-scale upland farming.

## Pastoral Fulani Camp

Pastoral Fulani form small group compose of several families and live in cooperation with one another. They call their camp as *wuro*. A pastoral Fulani camp in the Bida region normally composes of several *baade*, which refers to a household headed by a married man with an independent herd of cattle. The most senior male member of the whole group usually becomes the group head, *moudo wuro*. The seventeen Fulani groups studied vary greatly in size as indicated in table 1. The average number of people

*Figure 3. View of a Fulani camp in the dry season (AA group, January 2005)*

is 30 per group, but the smallest group just consists of 6 people while the largest one has 112 members. The average herd size is 252.5 per group, but the smallest group just owns 25 cattle while the biggest herd size of a group is 900 heads. The spatial structure of the homestead of a pastoral Fulani group in the region is shown in figure 4. In general the pastoral Fulani camp in the region is long and narrow rectangular in shape extending from south to north. This rectangular shape and orderly arrangement is related to the practice of corralling contract that Nupe farmers turn these camp sites into farms after the Fulani have moved.

The pastoral Fulani camp consists of the residential section for Fulani people and the enclosure for their cattle herd which is called *hoggo*. In the rainy season, *hoggo* is enclosed with logs, but in the dry season log is not necessary because there is no crops around the camp. After the Fulani move out, the manure and soil inside the *hoggo* are spread over the whole camp site, very often even beyond the camp site. The camp site of an average Fulani camp in the region is about 8,548m[6]. However as mentioned there is great variation among groups that the smallest camp site is just 2557m$^2$ while the largest one is almost 2 hectares in size.

*Table 1. Information of the pastoral Fulani groups studied*

| Group | Lineage | Year of residing in Bida Emirate | Place of origin | No. of household | No. of people | No. of cattle | No. of sheep |
|---|---|---|---|---|---|---|---|
| AA | D.J. | ~200 | Maasina | 6 | 51 | 353 | 162 |
| AJ | D.J. | ~200 | Maasina | 2 | 21 | 75 | 45 |
| AK | D.J. | ~200 | Maasina | 1 | 11 | 45 | 30 |
| DB | D.J. | ~200 | Maasina | 2 | 50 | 600 | 30 |
| GA | D.B. | 46 | Sokoto | 1 | 6 | 35 | 0 |
| KA | D.B. | 18 | Nararuka | 1 | 14 | 25 | 15 |
| MK | D.B. | 60 | Sokoto | 1 | 10 | 60 | 30 |
| AB | D.B. | 75 | Kano | 6 | 45 | 151 | 15 |
| DU | D.B. | 35 | Sokoto | 3 | 21 | 90 | 15 |
| AE | D.B. | 30 | Massina | 5 | 34 | 285 | - |
| SA | H.A. | 27 | Sokoto | 1 | 12 | 25 | - |
| WA | D.J. | 20 | Minna | 3 | 37 | 375 | 81 |
| AI | B.O. | 51 | Sokoto | 1 | 9 | 32 | 15 |
| DA | B.O. | 51 | Sokoto | 7 | 45 | 145 | 35 |
| MN | B.O. | 51 | Sokoto | 1 | 35 | 500 | 0 |
| RU | B.O. | 51 | Sokoto | 1 | 13 | 900 | 40 |
| IS | D.S. | 50 | Lapai | 9 | 112 | 427 | 87 |
| *Average* | | | | *3.0* | *30.9* | *242.5* | *40* |

Source: Fieldwork (September –October 2005).
Note for lineage: D.J.: Dindima'em, Juuliranko'em; D.B.: Dindima'em, Baasamanko'em;
H.A.: Hausaji; B.O.: Boodi; D.S.: Dindima'em, Sattanko'em.

*Figure 4. Spatial layout of a pastoral Fulani camp studied (The camp of AA group, September 2005)*

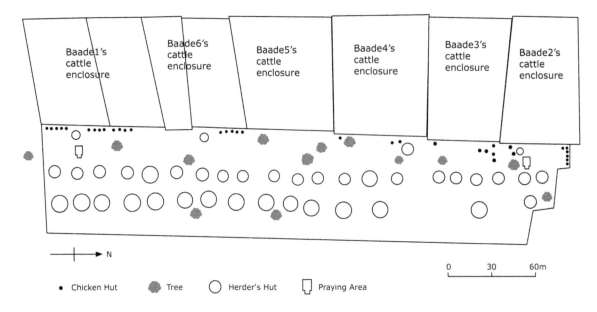

## CORRALLING CONTRACT BETWEEN NUPE AND PASTORAL FULANI

The brief record of the practice of corralling contract in the study area can be found in the ethnographies of Nadel (1942:206) and Shikano (2002:353). Nadel described it as *"an interesting cooperation"* between villagers or landlord and nomadic Fulani herdsmen. Presents of food, cash and assistance in the building of the camp were given to induce the pastoral group. Shikano even observed that invitation had to be done more than one year ahead. For the benefit of such cooperation, Nadel stated that *"I have myself seen the enormous difference in the growth of the crops between a plot on which the Fulani had made their camp and other, ordinary farm-plots"*. He also mentioned that it was an accepted arrangement among the Bida landlords to place one's fallow land at the disposal of the Fulani herdsmen previous to leasing it to a new tenant. The landlord could then obtain a much higher price for his land. The record of Nadel and Shikano were very brief and no further information was provided, but they proved that the corralling contract has been an arrangement being practiced at least for half a century.

### Invitation to Pastoral Fulani

Pastoral Fulani in the research site migrate two times in a year; in June they migrate to the drier uplands to avoid tsetse fly and to get closer to the markets in Bida town; in late October they move to the *fadama* lowlands to secure water and pasture for their cattle in the dry season. As mentioned before, the topography of the region benefits the Fulani that seasonal migration distance is relatively short as compared with pastoralists in other regions. Interviewed pastoral Fulani groups settled on uplands around the Bida town in the rainy season; and in the dry season they migrated about 10-20km west to the river basin of the Gbako River, or about 20-30km southwest to the large floodplain of the Niger River. It was similar to the case of the Fulani in Jos Plateau whose seasonal migration distance was 20km (Awogbade, 1983). On the contrary, Hopen (1958) estimated that the average one way distance for transhumance was about 100km (ranging from 10km to 303km) in Sokoto. In southwestern Nigeria the dry season grazing radii ranged from 32km to 125km (Fabusoro, 2006). Comparing to these figures, pastoral Fulani in the research site are carrying out their pastoral activities in a rather limited district. It allows Nupe farmers to easily stay in contact with the pastoral Fulani even after they migrate away. They can observe the behavior of the Fulani, find those they trusted, and to frequently visit the groups that they are targeting to host.

The battle for inviting popular Fulani groups begins few months to a year before the seasonal migration. Some groups receive invitation from several villages in every season. The number of invitation a group receive depends on its popularity stakes and strategy to be linked with various villages or to stay free. Likewise, Nupe farmers also have their options to invite a few groups simultaneously or to tightly target a particular group. Among the groups studied, five of them got invitations from three or more villages for each season, but more of them, that were nine, got invitation from just one village for each season. Nupe farmers need to formally declare the wish to host the group at least few months before the season changes by visiting the group with kola nut and gifts. When the village is already hosting a Fulani group, farmers need to express welcome for the group to come again next year before they move away. Fulani group head is not the only person who can accept invitation; other male household heads also can be the contact point of villagers. There is an unspoken rule among the pastoral Fulani in the region that before the decision is made; no gift other than kola nut should be taken from any village to avoid conflict. Village representatives normally visit the group at least three to four times before the decision is made. They gather information about their competitors and the amount of gifts they propose to offer. Some villages offer more and more gifts every time they visit the Fulani camp in order to out beat other villages.

## Cost to Host a Group of Pastoral Fulani

There is great variation in the cost to host a pastoral Fulani group. The amount of gifts farmers need to offer mainly depends on the expectation and the size of the group. Generally speaking, farmers need to pay cash and kola nut once their invitation is honored. After the Fulani has settled in the village, farmers then need to offer sorghum, rice, salt and so on. Moreover, in recent years popular groups additionally request for truck money to move their belongings. Farmers need to pay 50% or even all of the transportation cost which can sometimes reach a few thousands Nigerian Naira. Nevertheless, not all the pastoral Fulani use the corralling contract to take financial benefit from farmers. It depends on their bargaining power and the strategy adopted. As shown in table 2, nine out of the studied groups received nothing from farmers in the dry season of 2005. Most of them settled on uncultivated area of floodplains for the dry season. Based on the information obtained from the Fulani group studied, the estimated average amount of gifts that a Fulani group received for the corralling contract for the dry season of 2005 was NGN 2,803 (about US$21), and for the rainy season of 2005 was NGN 4,295 (about US$33)[7]. This amount probably underestimated the actual amount because Fulani tended to tell a lower amount dur-

*Table 2. Estimated amount of gifts received by the Fulani groups studied   (n=17)*

| Amount in Nigerian Naira | Number of groups (Percentage shown in parenthesis) | |
|---|---|---|
| | 2005 Dry season | 2005 Rainy season |
| 0 | 9 (53%) | 3 (18%) |
| 1-1,000 | 1 (6%) | 4 (24%) |
| 1,001-5,000 | 4 (24%) | 5 (29%) |
| 5,001-10,000 | 1 (6%) | 3 (18%) |
| >10,000 | 2 (12%) | 2 (12%) |

Source: Fieldwork.

ing interview. While it can cost some villages almost nothing to host a smaller group, it can cost some villages over NGN 10,000 to host a big or popular group. In the rainy season of 2005, the biggest group in the area received gifts equivalent to NGN   23,156 (about US$177) for the corralling contract.

Apart from gifts, there are many other things that the hosts need to provide to their Fulani guests. They need to supply labor. Younger farmers normally need to clear the land, cut tree, set up hoggo, and to assist in building shelters after the Fulani has selected their camp site. Female villagers need to provide labor and assist in cooking when their Fulani guest held ceremony. The hosts have to allow the guests to access to resources like water, firewood, building materials for shelter and most importantly, the grazing resources like crop residuals on harvested fields and pasture grasses on pre-cultivated fields and fallow lands. Thatching grass as the main building material for shelter is specially requested in the dry season. According to informants, it is a tedious job for farmers to gather all the materials necessary for the whole Fulani camp. Not only labor is costly but also the materials themselves are expensive if these are purchased in market. Pastoral Fulani therefore expects a lower cash gift and no salt is requested for dry season.

## Village Selection by Pastoral Fulani

Decision for selecting the village to set up camp site is made in a meeting of all *baade* heads. Pastoral Fulani well aware the value of their cattle manure and think to set up camp site in a particular village is a kind of service to "sit for" the village. The meeting for village selection is "*kauten hore bolwen hala hodde*", means gathering together to discuss about sitting. In the meeting, all *baade* heads can express their opinions and preferences. Decision is made with consensuses among all *baade* heads and final approval of the group head. When deciding where to sit for, the record that cattle reproduction was good in the village concerned is the most important consideration. Pastoral Fulani is deeply superstitious in this regard. Villagers can be quite sure that the Fulani will come again when they think their village is a place that brings luck. On the contrary, when cattle reproduction was bad, no matter how much the villagers are willing to pay the Fulani are not willing to go there again within few years. The second most important criterion is the availability of water and pasture especially during the dry season. The location of village is the third important consideration. Pastoral Fulani avoid villages close to rivers, streams and forests in the rainy season to prevent cattle from sickness. In the dry season they prefer villages closer to grazing resources on floodplains. Given similar conditions among villages, a good relationship between herdsmen and villagers is an important criterion. Pastoral Fulani avoid villages that always trouble them with farm encroachment and showed hatred when they settled before. Besides, they appreciate the good personal quality of farmers. For instance, some of them are more willing to sit for farmers who are hardworking and well utilize the sites they settled before.

No Fulani group admitted that the amount of payment affected their consideration. However, some of the informants had ever sat for certain villages when the payments offered were higher. Besides, there was a case that herdsmen did not sit for a village again because the farmers did not offer any gift other than kola nut. For the popular groups, to maintain fairness among villages and different households within a big village is important. They rotate among villages as well as different households of a village to avoid conflict among Nupe. All the groups studied had the experience that they could not sit for particular villages even when they wanted to because almost all the lands were under cultivation. Pastoral Fulani avoid settling in such kind of villages because it can easily result in unintended farm encroachment which harms the relationship with villagers. In case the group really wants to sit for a particular village for some reasons while the location of the village is unfavorable or there is lack of enough vacant land within the village, the Fulani may request farmers to lease a piece of land in another village to accommodate the Fulani camp. There were few cases like this recorded but in 2006 a village which used to lease land to other villages to host Fulani began to refuse the request of farmers. Therefore the group concerned sat for another village which was not their first choice. Female members of the group have no influence in village selection although the camp location greatly affects their well-being. They cannot complain even if the village is far from market or they need to trek longer for water and firewood. After the group has decided which village to sit for, a day is selected for the village representatives to present kola nut and cash to the Fulani group head. Once this ritual is done, the corralling contract is formally set up. A few *baade* heads of the group will visit the village and choose the camp site from a few pieces of land selected by villagers before the seasonal migration.

## CASE STUDIES OF DIVERSE STRATEGIES OF PASTORAL FULANI

All the seventeen pastoral Fulani groups studied have different migration patterns and different degree of closeness with Nupe villages. They also arranged the corralling contract in slightly different ways in response to their different conditions and needs. The adoption of the corralling contract for pastoral Fulani is not just a simple economic arrangement, but the most powerful tool for them to ensure access to resources and to maintain a harmonious relationship with the Nupe. They have different strategies with respect to the use of the corralling contract in accordance to their peculiar circumstances. Although generalization is difficult, their strategies could be roughly grouped into four different types. The four strategies are namely; the utilization of the corralling contract as a political tool, the utilization of the corralling contract as an economic tool, the passive adoption of corralling contract and the adoption of corralling contract with an exclusive village. Each of the strategies is illustrated below with a representative case study.

### Strategy One: Corralling Contact as a Political Tool

As the chief of all pastoral Fulani in Bida Emirate, the group of *Dikko Bida* needs to act as a role model regarding the practice to sit for Nupe villages. *Dikko Bida*, *Alhaji Adamu Dikko*, owns one of the largest herds in the area. He is a highly respected elder who persists in maintaining the traditional nomadic lifestyle of Pastoral Fulani. Although he earns a monthly salary and owns a house in Bida town as office, this ninety years old chief still lives in his simple shelter with his family members in the group. The group of *Dikko Bida* began to sit for villages north to Bida town during the rainy season about sixteen years ago. He moved northward as the *Dindima'em* group expanded so he moved slightly northward to explore new grazing resources. As the chief of Pastoral Fulani, his group was soon welcomed by villages. For rainy season, the group rotates among four Nupe villages, namely Kologa, Bube, Akote and Emigbari. Meanwhile for the dry season, the group has been sitting for just one village – the Eyagi village, for all the last 80 years. For *Dikko Bida*, corralling contact is not a tool to get economic benefit, but a political tool to symbolize the harmonious social relationship between Pastoral Fulani herdsmen and Nupe farmers, as well as to maintain the linkage with the Emir. Eyagi village was the birthplace of the mother of the late Emir. In addition, the village head of Eyagi has always been the ward head, *Etsu Yenkpa*, who is responsible for dispute settlement for villages under his management. Sitting for Eyagi can be regarded as an annual virtual to acknowledge the allegiance of Pastoral Fulani to the Bida Emirate. Regarding the four villages for rainy season stay, unlike other popular groups, *Dikko Bida*'s group does not take any cash gift from villagers; rather cash is always given to village heads whenever they come to greet Dikko. The group only receives kola nut as the ceremonial gift. However, in recent years villagers volunteer to offer money for the group to hire truck to move their belongings. Exchange of gift between herders and villagers is more often comparing with other groups studied. Produce like yam, sweet potato, maize, rice and sorghum are always given to the group. In return, village heads receive cheese, milk, chicken and money. The exchange is unbalanced; Dikko always offers more gifts to show generosity and to gain prestige. Corralling contract with villages does not bring economic benefits to *Dikko Bida*'s group, but it has an important political meaning for the maintenance of the cordial social relationship between Nupe and pastoral Fulani.

## Strategy 2: Corralling Contract as an Economic Tool

Some informants from more popular groups which always receive many invitations described the corralling contract as a kind of "exchange". They regarded it as a kind of service provided for farmers to achieve better yield, and in return they could take some advantage from it. As mentioned before, these popular groups receive higher payments and more gifts from farmers for the contract. They sit for different villages in each season. However, they are not absolutely utilitarian in their consideration for village selection. The monetary benefit they receive is just one of the conditions that they expect farmers to fulfill. The long term harmonious relationship with various villages is a more important consideration for them. The groups with higher popularity usually have higher social status, such as belonging to the ruling house or led by respectful Islamic mallam. Besides, their groups are usually bigger in size, and it is highly welcomed by many villages because they can get more benefit of manure at once.

The group of *Aliyu Abdullahi* is one of the most popular groups in the area. The group belongs to the same lineage group of *Dikko Bida*. The group has 51 people and 353 heads of cattle. It is the second largest group in the area in term of population. The group sits for various villages south to Bida town in the rainy season. In the dry season, the group migrates to the basin of River Gbako and rotates among a few villages. Table 3 lists the locations of the camp site and the amount of gift the group received from 1996 to 2006.

*Table 3. Camp sites and amount of gift received of Aliyu Abdullahi group[8]*

| Year | Dry season (Dec – May) Village | Amt. of gift (NGN) | Rainy season (Jun – Nov) Village | Amt. of gift (NGN) |
|---|---|---|---|---|
| 2006 | Nasarafu | 3, 200 | Alukusu Tako | 7, 350 |
| 2005 | Tswatagi | 5, 267 | Emigbari | 7, 650 |
| 2004 | Nasarafu | 5, 367 | Emigbari | 8, 528 |
| 2003 | Shabamaliki & Nasarafu | 5, 600 | Lemuta | 7, 578 |
| 2002 | Nasarafu | 4, 267 | Patishin | 16, 628 |
| 2001 | Nasarafu | 4, 667 | Ekota | 10,122 |
| 2000 | Shabamaliki | 7, 480 | Emigbari | 7, 428 |
| 1999 | Nasarafu | 4, 500 | Lemuta | 6, 778 |
| 1998 | Tswatagi | 2, 867 | Lemuta | 6, 378 |
| 1997 | Nasarafu | 3, 867 | Emigbari | 6, 578 |
| 1996 | Tswatagitako | 3, 933 | Emigbari | 6, 378 |

Source: Fieldwork.

The amount of gift received in certain years deserves some elucidations. In the dry season of 2006, only 150 pieces of kola nut, 40 kg of sorghum and NGN1,000 were requested by the group. It was because of the lack of rainfall in 2005 that herders were willing to accept a lower payment from farmers. On the other hand, in the rainy season of 2002, a record-breaking high amount of gift and payment were received from a farmer of Patishin. Moreover, to avoid the cattle disease that prevailed in the area during that time, the farmer accepted the request to rent a piece of land in Ndaceko village in order to accommodate the Fulani camp. Although the group did benefit a lot financially that year, they did not continue the contract with that farmer because they did not want to provoke other villages. As listed in table 3, the group did not sit for a village continuously for over two years. The informant of the group pointed out that, *"It is good to maintain relationship with various villages because it gives you more freedom and bargaining power."* When deciding which village to select, informant said, *"You cannot follow money, you need to follow cattle."* Financial benefit is just one of their considerations, what really matter to them are the welfare of their cattle and the good relationship with villages which makes grazing on farmers' land an easier task. To avoid conflict among villages, the group needs to rotate. In the dry season of 2003, there was a special case that farmers of two villages needed to share the site. The group originally promised to sit for a Nasarafu villager but they wrongly chose the land belonged to a Shabamaliki farmer. None of them were willing to give up so finally herders requested them to share the site equally. For large village like Nasarafu and Shabamaliki, the group even needs to rotate among different households in order to avoid conflict within village. To run the corralling contract is similar to running a "business" to those popular groups. They care about financial benefit, but they also need to maintain "customer relations" with various "clients" and to prevent them from "fighting among themselves".

## Strategy 3: Passive Corralling Contract

Some pastoral Fulani groups are not so eager to engage in corralling contract with villages. They value freedom of mobility higher than the close relationship with certain Nupe communities. *Adamu Iya* belonged to the *Boodie* lineage group which began to settle in the Bida area in the 1930s from the Sokoto region. His group is small; it has just 9 people and 32 heads of cattle. Unlike the *Dindima'em* group, most of the *Boodie* groups studied do not formally engage in corralling contract with Nupe villages. Even though they do not get so many benefits from the corralling contract comparing with the two types of group mentioned above, they are less nomadic in the sense that they do not need to move to different villages every year. They usually settle on a particular village continuously for several years during the same season. For example, until 2005, *Adamu Iya* has been setting up his camp in Fakunba village during the rainy season for 5 consecutive years, and in Gaba village for the dry season for 10 consecutive years. However, he never sets up his camp on the same spot for two consecutive years. When he sits for a small village, he can choose a plot of fallow land as he wishes and let villagers later share the land among themselves. When the village he settles is large, he needs to follow the advice of the village head and rotates among the lands of different households. *Adamu Iya* does not actively engage in corralling contract with villages although a few villages always show welcome to host his group. He does not receive payment from villagers, but village head of each village usually give him 100 pieces of kola nut and some grains for gratitude and to express welcome for his coming back. Nevertheless, *Adamu Iya* usually does not give promise that he must come again. He prefers to keep his flexibility. If he wants to come again in the following year, he just walk-in and gets permission to settle from village head. Villagers still need to provide some basic services, such as clearing the land for his camp and assisting in building camp. There is usually no specific reason when *Adamu Iya* decides not to return to a village. He prefers to be flexible so that he can explore new environment for his cattle any time he wants. The precondition for him to stay in a village is that the villagers have shown welcome and have ever forgiven him for minor destruction caused by farm encroachment.

## Strategy 4: Fixed Corralling Contract

There were a few groups studied that do not carry out seasonal migration. They settle both in the rainy and dry seasons in a village for years. Their life-style can be regarded as semi-settled, but they do not own permanent shelters and need to move their cattle enclosure frequently within the village following the request of their hosts. These groups are usually smaller and own a smaller herd. The villages they stay are usually larger in scale, with large area of vacant land or fallow land. Besides, there must be water resource available even during the dry season. Groups prefer not to migrate but just sit for a particular village because it is "too much suffering" to move around villages. They do not get any payment for their cattle manure, but they do receive kola nut and grain sometimes from villagers for courtesy. The group of *Aliyu* moved into the Bida area from the Sokoto region about 45 years ago. They have 21 people and 90 heads of cattle in the group. *Aliyu* has never moved out of Gbanchitako village for over a decade. The stable relationship with the village enables him to get a relatively large plot to do his own farming. He also gets a plot in a nearby village. Although he gets no payment from farmers, he gets land to farm and the right to use the cattle manure exclusively for his own farms during the dry season. *Aliyu* move his cattle enclosure following the wishes of farmers in the rainy season. In the dry season, he can let his cattle to stay on his two farms for two months respectively. *Aliyu* is not interested in getting financial benefits by corralling contract. Stability is more valued and by sitting for a village all year round for long term, he is able to sustain a semi-settled life.

## UTILIZATION OF FULANI CAMP SITE

Cattle manures are accumulated inside the *hoggo* during the season. Calves, sheep and other animals are not corralled inside the *hoggo* and are left free in the camp at night. Every morning the Fulani women and children sweep the residential section and drop the animal faeces in the *hoggo*. Pastoral Fulani do not use cattle manure as fuel or construction material. All the animal faeces are concentrated inside the *hoggo*. In the next rainy season after the group have moved away, Nupe farmers will spread the faeces all over the previous camp site and the area is usually extended. They often transfer part of the manure to their other farms as well. The size of such manured field depends on the size of the camp. Based on the 24 surveyed manured fields (table 4), the average size of such manured field is 14,016m$^2$. These fields are extended on average by 199% beyond the original size of the Fulani camp site.

*Table 4. Summary of the manured fields surveyed*

| No. of field surveyed | Smallest field size | Largest field size | Average field size | Lowest portion to size of camp | Highest portion to size of camp | Average portion to size of camp |
|---|---|---|---|---|---|---|
| 24 | 2,542m$^2$ | 38,312m$^2$ | 14,016m$^2$ | 81% | 605% | 199% |

Source: Fieldwork.

The arrangement for the corralling contract of Nupe farmers can be categorized into two: hosting by collective effort of the whole village and hosting by single household of the village. The former is far more common in the study area, especially for smaller scale villages that host Fulani group during the wet season. The latter is only practiced in larger village or during the dry season when small group of pastoral Fulani, often just one *baade*, migrate into the region for relatively short period of stay. For the region is close to the great floodplain of the Niger River, many pastoral Fulani basing in the north pass through the research area during the dry season for reaching the floodplain for water and pasture. They move in small groups and stay on the lands of Nupe villages for short period like a few weeks.

For villages that host pastoral Fulani during the wet season, not only the very field that is utilized as camp site, but also the fields surrounding the camp site, that the farming activities for the year have to be suspended because of inevitable cattle destruction. If the benefits brought by the corralling contract are not shared among all the villagers whose farming activities have been affected, tension can easily be generated. For villages that host Fulani group during the dry season, there is no such trouble because no farming activity will be affected.

When arrangement is done by collective effort of the whole village, the land for past Fulani camp site will be divided into many small plots and distributed to village members according to the norms and rules of the village. For a group of about 1 hectare, it is normally divided into eight to nine plots. There was a case that a Fulani camp site of 1.5 hectare was turned into a field of about 2.6 hectares and was then divided into twenty-three small plots. Fourteen manured fields achieved through collective effort were surveyed and the average area of such divided small plot is just about 1,447m$^2$. Dividing the field into so many long and narrow small plots may not comply with the principle of economics of scale, but for many Nupe farmers the notion of fairness in the community is highly important. That is the reason why larger group is more desirable by Nupe farmers. By hosting a bigger Fulani camp, they can ensure members of every household of the village can get a share of the manured land.

Village head often has the first priority to choose the plots he prefers and the remaining are then distributed to other members usually according to seniority. According to farmers, the plots in the middle and of the two ends are most wanted, because farmers think the plots in the middle have more faeces accumulated and the plots at the two ended can easily be extended into larger size. As the most senior member of the village and usually the secondary landlords, the village heads have the privilege to get a bigger portion. In the measurement it is found that the plots of village heads are 146% bigger than plots of ordinary villagers.

The cost incurred for the corralling contract is divided among members of the village. If the cost is not shared, it would be a very heavy burden for farmers as it can easily cost a few thousands Nigeria Naira to host a popular group. In most cases the cost is divided unequally. Senior and wealthier members, especially the village head, of the village usually contribute more money and grains. This justifies their bigger shares of the manured land. Younger farmers can contribute less money, but they need to provide physical labor in assisting the clearing of land and building of camp in order to entitle to the manured land.

Figure 5 shows a typical example of a measured manured field which the unit of host was the whole village. It used to be the camp site of the group of Aliyu Abdullahi in Emigbari village for the rainy season of 2005. The Emigbari village has a population of fifty-six. Villagers are earnest about the corralling contract with pastoral Fulani for maintaining the fertility of their land. In 2006, there were totally nine fields made out of previous pastoral Fulani camp site in the whole village. Twenty-seven percent of their land was manured through the corralling contract with pastoral Fulani groups. The field in figure 5 was divided into 13 long and narrow plots. All the six household heads of the village obtained their shares. The distribution was uneven: 25% and 29% of the field were taken by the village head and the deputy village head respectively while the remaining 47% were shared by the other four household heads.

It is noteworthy that six plots out of the thirteen plots, about one-fifth of the whole field, were occupied by non-villagers. In the research area it is common for the owner of such manured fields to sell the use right of part of their plots for good income. Ten plots with the use right sold to non-villagers were identified. The lump-sum price for the use right of 5-7 years for a plot of 1130m$^2$ was NGN 4,100 on average. Every year the tenants also need to submit 10% of the harvest to their landlords as rent. In this example, three farmers from other villages have bought the use right for four plots. One of the household heads sold the use right of 70% of his plot to his father-in-law of another village. In addition, two non-villagers obtained the plots for no lump-sum cost through affinity relationship and friendship.

When the unit of host is a household, the distribution of land is much simpler. Figure 6 shows the layout of a manured field which the unit of host was a household of a larger village. It used to be the camp site of the group of *Aliyu Abdullahi* in Nasarafu village for the dry season of 2003. The previous group camp site was largely extended and then distributed among the three brothers of the household. Plot size of the manured fields achieved through individual household effort is often much larger. For the ten surveyed fields of such, the average plot size was 7,645m$^2$ in area. It is probably because as the household do not need to share the land with other villagers, they can arrange the group to set up camp site in the midst of their family land and then easily spread the manure all over on the land when the Fulani has gone. Instead of dividing the plot into smaller size and selling the use right, farmers of such fields seem to prefer more extensive farming. In the study area, only the large and populous villages on the floodplains of river basins allow their villagers to invite a Fulani group on their own. To avoid competition among villagers, coordination of village head is necessary but villagers still compete on time and gift to invite the same group sometimes. Pastoral Fulani usually intentionally rotate among different households when sitting for such large villages in order to avoid conflict among villagers.

## FARMERS' ACCESS TO MANURED LAND

In the conventional depiction, richer farmers and cattle herders are often described as the major beneficiaries of the corralling contract. Resources poor farmers with less access to manure, suffer more on land degradation as they have to provide pasture for cattle grazing but get nothing in return to retrieve the nutrient cycling of their lands. In this research, evidences from case studies suggested that it is not necessarily the case at least with respect to the Nupe farmers and pastoral Fulani herders. Under the same local settings, different pastoral Fulani groups have different strategies regarding the adoption of corralling contract with Nupe villages. The classical description that herders are the dominative beneficiaries of the politics of manure was only true to some herdsmen. The corralling contract was the most powerful instrument that assisted in the access to resources for all the groups studied, but what they could get from the arrangement varied greatly from each other. While some groups could well manipulate the relationships with various villages through the adoption of the corralling contract to their advantages, some groups preferred a more stable situation and just got the minimum advantages out of the contract. Findings revealed that higher social status, larger herd size and long history of interaction which allowed trust to be built were the factors contributing to the popularity of a group. Each of the strategies illustrated before has its merits and demerits if comparisons are to be made. But the important message is that different strategies have been evolved and adopted by the pastoral Fulani in accordance to their particular circumstances and needs. Under the customary land system of the study area, the corralling

*Figure 5. Layout of a manured field in Emigbari village (measured in September 2006)*

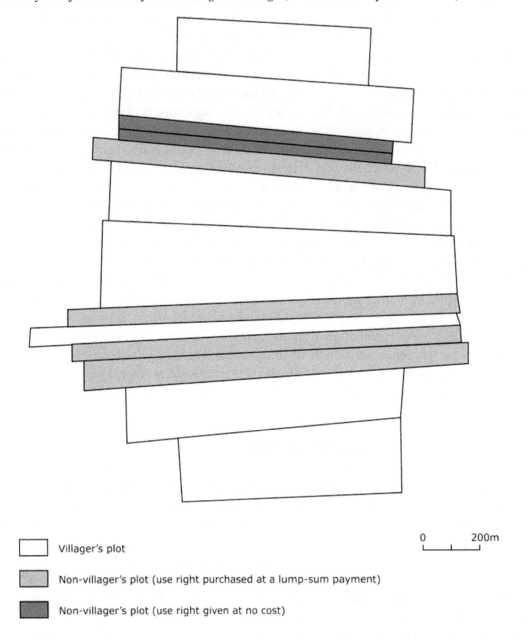

contract plays a pivotal role that facilitates the interdependence of the two groups. This well-functioning traditional institution allows limited resources to be shared and balance to be maintained.

Concerning the farmers, although village chiefs usually benefit more, the sharing and rotation practices allows members of the whole village to get their shares of manured fields. Village heads and village elderly usually contribute the major part of the payment to invite herders. They are also the ones responsible for the gifts to herders in ordinary time and during their stays in order to maintain the relationship. Therefore, younger farmers justify their bigger share of the manured land. All the interviewed farmers confirmed the higher yield brought by the manure. They pointed out that the benefits of the previous cattle corral

*Figure 6. Layout of a manured field in Nasarafu village (measured in September 2005)*

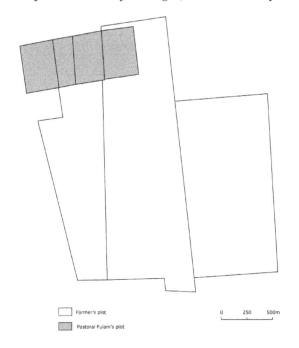

can last for six to ten years and the performance is at least three times better than the chemical fertilizer that they can get in the market. The manured plots can be a source of cash income when the farmers sell the usufructuary rights of their plots. When a village fails to invite any Fulani group, farmers can still access to such manured fields through affinity relationship, friendship or by purchasing the right of usufructuary. There was no evidence that access to manure was concentrated only to wealthier farmers. Although the competition for corralling contract sometimes created tensions between villages or among villagers, such tensions were never serious as farmers knew the norm that pastoral Fulani would rotate among them and they expected that they would get their chance sooner or later. Villages do cooperate occasionally by jointing efforts to host a group together.

Contrary to the conventional description, findings did not show a strong tendency of increasing payment to herders for the manuring service. Table 5 illustrates the change in the value of gift received for the pastoral Fulani studied groups during 2001 to 2005. The value increased for NGN 69.8 for the dry season and just NGN 7.83 for the rainy season during those four years. On the other hand, some groups even received fewer amount of gifts than previous years.

As illustrated in the examples, some pastoral Fulani groups preferred not to get formal invitation and not to get payment in order to remain flexible. It was only the popular groups whose payments received had shown a slightly increase over the last ten years. The major increase of financial burden to farmers was the truck hiring cost to move the belongings of herders. This burden had partly or totally shifted from popular Fulani groups to farmers. However, as mentioned before, herders did consider the affordability of farmers and the amount of payment was never their only consideration in village selection. With respect to the Nupe farmers and the pastoral Fulani herders of the study areas, the notion that herders are asking for more payment and making manure only accessible to wealthier farmers is not applicable. Nevertheless, informants reviewed that prior to the mid-1990s, pastoral Fulani seldom received any payment for

*Table 5. Change in value of gift received for the pastoral Fulani groups during 2001 to 2005*

| Value of gift received | Dry Season | | Rainy Season | |
|---|---|---|---|---|
| | No. of group | Amt. of change | No. of group | Amt. of change |
| Increased | 4 | NGN 341.45 | 6 | NGN 531.44 |
| Unchanged | 10 | | 6 | |
| Decreased | 3 | - NGN 59.72 | 5 | - NGN 611.12 |
| Net change | 17 | NGN 69.80 | 17 | NGN 7.83 |

Source: Fieldwork.

the manure service. Farmers might just take care of lodging and food when the herders stayed on their fields. Herders even needed to move their camp sites two times in each season to sit for four villages in a year. The great improvement of the pastoral Fulani's term of benefits from the corralling contract should be related to the changes of fertilizer policy of the Nigeria government as well as the degradation of land due to population increase and excessive farm expansion.

## CONCLUSION

The corralling contract has enhanced the mutual dependence of the Nupe farmers and the pastoral Fulani herders. It is especially essential to pastoral Fulani who, under the customary land system, has no guaranteed access to land. With the failure of the government in providing grazing reserve, the corralling contract has remained as their most important asset that assists them to access to resources. It is also the most important antifriction for the social relations between the two groups. Most of the interviewed farmers answered that they would forgive pastoral Fulani for minor crop encroachment for the sake of the cattle manure. Village heads were expected to assist their Fulani guests when they have disputes with other villages. In the study area, most of the disputes caused by cattle encroachment into farms could be settled by village heads, just few cases needed to be settled by *Dikko* or the *Emir*.

Despite all the merits mentioned, the corralling contract is not without constraint. First of all, farmers have limited power to manipulate the adoption and the arrangement of the corralling contract. When a village fails to establish relationship with the Fulani groups in the surrounding area, it is difficult for villagers to invite Fulani groups from other areas. There are implicit rules governing the territory of a Fulani group, new comers normally do not just infiltrate into the area without the consensus of the existing groups. It is to avoid competition on resources and unnecessary social conflict. A village has less access to corralling contract when it is located in an unfavorable environment for cattle. Because of these limitations, farmers can hardly plan for the fertilization proactively. Secondly, excessive farm expansion due to population growth and decreasing productivity of land, have created great limitation not only for herders but also for farmers. Many herders revealed that they could not sit for certain villages even if they wanted to because almost all the lands were under cultivation. When a village hosts a Fulani group during the rainy season, not only the spot where the camp is set up, but also the surrounding farmlands are expected to be sacrificed due to inevitable farm encroachment by cattle and sheep. It is a serious constraint that stops pastoral Fulani from sitting for some Nupe villages and it may lead to a vicious circle for the poor farmers. Thirdly, the corralling contract prevents the pastoral Fulani from settling down permanently on a place and limits their progress in development. All interviewed herders pointed out that if they would stop providing manure for Nupe farmers someday, there would be war between them. The corralling contract helps them to access to resources, but on the other side of the coin, they do not have the chance to develop a more stable life because they cannot break the expectation of farmers that they would move and rotate. Although some groups could begin to get lands for farming from their hosts, they were not supposed to farm on the same lands continuously after they have moved out. Pastoral Fulani well acknowledged the responsibility that they should help poor farmers to fertilize their lands, but they also expressed their wishes for a more stable and secure life.

Despite of the history of the Fulani conquer in the early nineteenth century, Nupe farmers generally are not antagonistic toward the pastoral Fulani. Although they do not form martial relations, the Nupe and the pastoral Fulani have a wide range of social interaction. The camp sites of pastoral Fulani are generally close to the homestead of villagers. Fulani men often gather in village and pray in the mosque with their Nupe fellows. The corralling contract has a very positive impact on the social relation of the two groups. They see each other as partner: the Nupe need the Fulani for manure while the Fulani need the Nupe for land and fodder. The corralling contract is not just a casual arrangement, but a dynamic and well-functioning traditional institution that facilitate the collaboration of the two groups. It is an important example of local adaptation and innovation that allow balance to be maintained when limited resources are being shared. By contrast, statutory efforts to draw territorial distinctions between agriculture and livestock production have created social rifts in many regions. Technological solutions, such as chemical fertilizers, have not halted the decline in agricultural productivity. The corralling contract should be advanced as part of the complex set of social and biophysical conditions in agropastoral regions. In spite of working in vain to simplify the system with statutory and technological solutions which separate agriculture from livestock production, efforts should be focused on removing the constraints of the corralling contract on both side and facilitating it in order to enhance the association between agricultural and livestock production.

## ACKNOWLEDGMENT

The main fieldwork for this paper was funded by the Sasakawa Scientific Research Grants of the Japan Science Society (Project number: 17-055). It was also a part of the "Local economics subsistence as the substantial base for human security" research project (JSPS KAKENHI Grant Number 23653063). The author is sincerely grateful to Emeritus Professor Toshiyuki Wakatsuki at Shimane University who gave the author the first opportunity to begin research in Nigeria under the "Watershed Ecological Engineering for Sustainable Increase of Food Production and Restoration of Degraded Environment in West Africa" research project (JSPS KAKENHI Grant Number 15101002). Special thanks are due to Mr. Mohammad Abdullahi, Mr. Solomon, Mr. Joshua Aliyu and Dr. Rotimi Fashola for their earnest assistances for the fieldworks. The follow up research of this paper was funded by JSPS KAKENHI Grant Number 16K20982 under the project "Vulnerability of pastoralists' livelihoods to climate change".

This paper is also funded by a Senshu University research grant in 2020 and is based on part of the results of a project entitled "Local innovation for agricultural development of Africa".

## REFERENCES

Achard, F., & Banoin, M. (2003). Fallows, forage production and nutrient transfers by livestock in Niger. *Nutrient Cycling in Agroecosystems, 65*(2), 183–189. doi:10.1023/A:1022111117516

AICAF. (1994). *Report of the study for formulation of sustainable agriculture development plans for Africa – Valley bottom rice farming in West Africa (second phrase)*. Association for International Co-operation of Agriculture and Forestry.

Asanuma. S. (2004). Desertification in Burkina Faso in West Africa and Farmer's Coping Strategy – Life-scale Analysis. *Expert bulletin for International Cooperation of Agriculture and Forestry, 25*(2), 1-18. (in Japanese)

Awogbade, M. O. (1983). *Fulani Pastoralism – Jos Case Study*. Ahmade Bello University Press Limited.

Blench, R. (2010). *The transformation of conflict between pastoralists and cultivators in Nigeria, Review Paper prepared for DFID, Nigeria*. Review Paper prepared for DFID, Nigeria. Unpublished paper. Online. http://www.rogerblench.info/Development/Nigeria/Pastoralism/Fadama%20II%20paper.pdf

Brourwer, J., & Powell, J. M. (1995). Soil aspects of nutrient cycling in a manure experiment in Niger. In *Livestock and sustainable nutrient cycling in mixed farming systems of sub-Saharan Africa. Vol. 11: Technical papers. Proceedings of an International Conference held in Addis Ababa, Ethiopia, 22-26 November 1993*. ILCA.

Brourwer, J., & Powell, J. M. (1998). Micro-topography, water balance, millet yield and nutrient leaching in a manuring experiment on sandy soil in south-west Niger. In *Soil fertility management in West African land use systems. Proceedings of a workshop held in Niamey, 4-8 March 1997*. Margarf.

Croix, St. (1972). *The Fulani of Northern Nigeria: Some general notes*. Gregg International Publishers Limited.

Fabusoro, E. (2006). *Property rights, access to natural resources and livelihood security among settled Fulani agro-pastoralists in Southwestern Nigeria* (Unpublished research report). University of Agriculture, Abeokuta, Nigeria.

Forde, D. (1970). The Nupe. In D. Forde (Ed.), *Peoples of the Niger-Benue Confluence* (pp. 15–52). London: International African Institute.

Gefu, J. O. (1989). The dynamics of Nigerian pastoralism: An overview. In *Pastoralism in Nigeria: Past, Present & Future. Proceedings of the national conference on pastoralism in Nigeria*. Ahmadu Bello University.

Gefu, J. O. (1992). *Pastoralist Perspectives in Nigeria: The Fulbe of Udubo Grazing Reserve*. Uppsala: Scandinavin Institute of African Studies.

Grayzel, J. (1990). Markets and migration: A Fulbè pastoral system in Mali. In J. Galaty & D. Johnson (Eds.), *The world of pastoralism* (pp. 35–67). New York: Guilford Press.

Heasley, L., & Delehanty, J. (1996). The Politics of Manure: Resource Tenure and the Agropastoral Economy in Southwestern Niger. *Society & Natural Resources, 9*(1), 31–46. doi:10.1080/08941929609380950

Hiernaux, P., Fernàndez-Rivera, S., Schlecht, E., Turner, M. D., & Williams, T. O. (1997). Les transferts de fertilité per la betail dans les agro-écosystèmes du Sahel (Nutrient transfers of animals in Sahelian agroecosystems). In *Proceedings of Regional Workshop on Soil fertility management in West Africa land use systems*. Margraf Verlag.

Hirose, S. (2002). Rice and upland farming in the Nupe community. In S. Hirose & T. Wakatsuki (Eds.), *Restoration of Inland Valley Ecosystems in West Africa* (pp. 183–212). Association of Agriculture and Forest Statistics.

Hopen, C. E. (1958). *The pastoral Fulbe Family in Gwandu*. London: OUP for IAI.

Ibrahim, S. (1992). *The Nupe and their neighbours from the 14th century*. Ibadan, Nigeria: Heinemann Educational Books.

Ismaila, D. (2002). *Nupe in history (1300 to date)*. Jos, Nigeria: Olawale Publishing Company Ltd.

Johnston, H. A. S. (1967). *The Fulani Empire of Sokoto*. London: Oxford University Press.

Loofboro, L. (1993). *Tenure Relations in Three Agropastoral Villages: A Framework for Analyzing Natural Resource Use and Environmental Change in the Arrondissement of Boboye, Niger*. Discussion Paper No. 4. Land Tenure Center.

Mason, M. (1981). *Foundations of the Bida Kingdom*. Zaria, Nigeria: Ahmadu Bello University Press.

Masuda, M. (2002). People and Forests in Guinea Savanna. In S. Hirose & T. Wakatsuki (Eds.), *Restoration of Inland Valley Ecosystems in West Africa* (pp. 233–302). Association of Agriculture and Forest Statistics.

Nadel, S. F. (1942). *A black Byzantium: The kingdom of Nupe in Nigeria*. London: Oxford University Press.

Nagy, J. G., & Edun, O. (2002). *Assessment of Nigerian government fertilizer policy and suggested alternative market-friendly policies*. Unpublished IFDC consultancy paper.

National Livestock Development Project. (2003). Accelerated development of grazing reserves and stock routes for settlement of pastoralists. Report submitted by the Sub-committee on grazing reserves and stock routes to the National Livestock Development Project, Kaduna, Nigeria.

National Livestock Development Project. (2007). *Report of the pastoral resolve (pare) committee on pastoral development and empowerment*. Kaduna, Nigeria: National Livestock Development Project.

Neef, A. (1997). Le contrat de parcage, fumure pour les riches? Une étude de cas au sud-ouest du Niger. In *Soil fertility management in West African land use systems. Proceedings of a workshop held in Niamey, 4-8 March 1997*. Margarf.

Neef, A. (2001). Land Tenure and Soil Conservation Practices – Evidence From West Africa and Southeast Asia. In *Sustaining the Global Farm. Selected papers from the 10th International Soil Conservation Organization Meeting held on May 24-29, 1999*. Purdue University and the USDA-ARS National Soil Erosion Research Laboratory.

NYSC. (2010). *Nigeria - Bida Agricultural Development in Niger State Project*. http://documents.worldbank.org/curated/en/309101468099257815/pdf/multi-page.pdf

Ogawa, R. (1998). Agriculture in Pastoral Fulbe Society – Re-thinking of the "traditional" production system. In Y. Takamura & Y. Shigeta (Eds.), *The problems of Agriculture in Africa* (pp. 88–113). Kyoto University Press. (in Japanese)

Olomola, A. S. (1998). *Pastoral Development and Grazing Resource Management in Nigerian Savannah Area*. Paper presented at "Crossing Boundaries, the 7th annual conference of the International Association for the Study of Common Property, Vancouver, Canada.

Omotayo, A. M. (2002). A land-use system and the challenge of sustainable agro-pastoral production in southwestern Nigeria. *International Journal of Sustainable Development and World Ecology*, 9(3), 369–382. doi:10.1080/13504500209470131

Powell, J. M., & Williams, T. O. (1993). Livestock, nutrient cycling and sustainable agriculture in the West African Sahel. IIED Gatekeeper Series No. 37. International Institute for Environment and Development.

Sangarè, M., Fernández-Rivera, S., Hiernaux, P., Bationo, A., & Pandey, V. (2002). Influence of dry season supplementation for cattle on soil fertility and millet (Pennisetum glaucum L.) yield in a mixed crop/livestock production system of the Sahel. *Nutrient Cycling in Agroecosystems*, 62(3), 209–217. doi:10.1023/A:1021237626450

Schlecht, E., Hiernaux, P., Achard, F., & Turner, D. (2004). Livestock related nutrient budgets within village territories in western Niger. *Nutrient Cycling in Agroecosystems*, 68(3), 199–211. doi:10.1023/B:FRES.0000019453.19364.70

Shikano, K. (2002). Ecological anthropological study on daily herding activities of pastoral Fulani in central Nigeria. In S. Hirose & T. Wakatsuki (Eds.), *Restoration of Inland Valley Ecosystems in West Africa* (pp. 303–369). Association of Agriculture and Forest Statistics.

Shimada, S. (1999). A study of increased food production in Nigeria: The effect of the Structural Adjustment Program of the local level. *African Study Monographs*, 20(4), 175–227.

Stenning, D. (1959). *Savannah nomads: A Study of the Woodabe Pastoral Fulani of Western Bornu Province Northern Region*. International African Institute.

Suleiman, H. (1989). Policy issue in pastoral development in Nigeria. In *Pastoralism in Nigeria: Past, Present & Future. Proceedings of the national conference on pastoralism in Nigeria*. Ahmadu Bello University.

TropSoils. (1991). *Integrated management of an agricultural watershed: Characterization of a research site near Hamdallaye, Niger*. TropSoils Bulletin No. 91-03.

USAID. (2007). *Nigeria food security update*. Famine early warning systems network paper May 2007. www.fews.net/south

von Kaufmann, R., Chater, S., & Blench, R. (1986). Livestock systems research in Nigeria's subhumid zone. In *Proceedings of the second ILCA/NAPRI symposium*. Addis Ababba: International Livestock Centre for Africa.

Wilson, W. (1984). *Resource management in a stratified Fulani community* (PhD dissertation). Howard University.

## ENDNOTES

[1] Termites are recognized as "ecosystem engineers" by scholars because they promote soil transformations by disturbance processes. They collect particles from different soil depths and deposit them in the mounds which can be regarded as soil nutrient reservoirs. The decomposition of termite mounds releases the nutrient back to soil.

[2] The Fulani are referred to variously as Fulbe, Peul, Fellah or Fula in literatures. They call themselves Fulbe. In this paper the Hause term Fulani is used as it is a more widely used term in Nigeria.

[3] World Christian Database estimated in 2008 that there were 1,197,139 Nupe, out of which 92% were Muslims, 5.2% Animists and 2.8% Christians. Online. https://worldchristiandatabase.org/wcd/. (Accessed on 1 April 2008). Nupe continue to account for about 1% of the total population of Nigeria.

[4] In Nadel's finding in 1942, the average farm size for a man is 2-2.5 acres.

[5] The term *Bororo* is Hausa word derived from *Bororo'em*, a Fulani name for a "special" class of *Fulbe ladde*. Despite its widespread use, it has slightly pejorative overtones and is not used by the people themselves.

[6] The camp sites of ten Fulani groups were surveyed in 2005 and the camp site of two major groups were surveyed again in 2006. From the twelve surveyed camp site records and the cattle data of the corresponding groups, the average camp site area per cattle which is 35.25m$^2$ is obtained. As the average head of cattle for the 17 groups is 242.5 heads, the average size of camp site is calculated as 35.25m$^2$ x 242.5, which is 8,548m$^2$.

[7] The values of rice and sorghum were calculated based on the official record of the average retail market price of the Bida market of the Niger State Agricultural Development Project. The values of kola nut and salt were given by Fulani informants while the values of yam and fresh maize were given by farmer informants. Regarding the average exchange rate of US dollar to Nigerian Naira, during 1 December 2004 to 31 May 2005 it was 1:132.35 and during 1 June 2005 to 30 November 2005 it was 1:130.58.

[8] For a more accurate estimation, track hiring cost is not included in the calculation of the gift value. However, it is one of the biggest burdens for farmers when hosting a popular group. The values of agricultural products calculated based on the retail price of the Bida Market in 2005 rainy and dry season. The official record was obtained from the Niger State Agricultural Development Project. For time-series comparison 2005 was used as the base year for value calculation.

# Chapter 4
# Neuroeconomics:
## Connecting Health Science and Ecology

**Torben Larsen**
https://orcid.org/0000-0002-5704-7753
*University of Southern Denmark, Denmark*

## ABSTRACT

*A neuroeconomic psychology establishes behavioral economics as a positivist discipline connecting health science and ecology. This makes "economic ecology" and "neuroeconomics" important new sub-disciplines of economics. Pigovian carbon emission tax (CET) is the most effective intervention towards the "green-house effect." However, in the short-term, democratic center-coalitions are too weak to implement CET due asymmetric levels of economic knowledge between economists and the public. However, neuroeconomic research indicates that training can improve decision-making about complex issues as CET. A second-best strategy is presented as green hybrid multilevel transition (GHMT). Besides GHMT preparing doe CET, neuroeconomic psychology delivers a series of science-based advices on effective behaviors: a consumer pattern that integrates individual satisfaction with environmental conscience, efficacious business behavior like an entrepreneur, long-term political suspension of relative poverty by universal basic income, and cognitive training by meditative in-depth relaxation.*

## INTRODUCTION

The vision of the "Invisible Market Hand" (IMH) as organizer of economic development is an unprecedented success in human development history (Smith, 1776). Lately, the unequivocal success story of IMH is turned upside-down by global threats e.g. the "Green-house effect". This study explores options to correct documented market failure departing from a SWOT-analysis.

DOI: 10.4018/978-1-7998-3576-9.ch004

## SWOT-ANALYSIS OF MARKET ECONOMy (IMH)

The SWOT-analysis is based on the economic tools of demand and supply as applied to the industrialization of G7 in 2.1. G7 includes USA, Canada, Germany, France, UK, Italy and Japan
with 10% of world population accounting for nearly half of the global gross domestic production (GDP). Industrialization originated in UK about 1750 and spread to Continental Europe and North America and after WWII to Japan.

### The Basic Function of Market Economy (IMH)

### What Is Economic Growth?

According to classical economics are human needs infinite, however with a decreasing marginal utility of consumption. This means that the higher income the higher total satisfaction of consumers. The price of goods and services limits the level of satisfaction that given the income and the relative prices determine which of alternative buying options is chosen. In-summary, income and relative prices are the most important determinants of consumption as confirmed by empirical investigation. This means that in general, consumer options rise whenever the prices decline, or the income raises. For a certain period in a specific region e.g. a year in a specific country, the aggregate value of foods and services is accounted as the gross domestic product (GDP) representing the consumer option of that specific country in that period. Economic growth is the case of relative growth in GDP compared to prices, the growth in consumer options. This growth thesis is in the simple case - without either foreign trade, financial transactions or changes in stocks - operated at the level of nations as changes in GDP corrected for eventual changes in the level of baseline prices.

GDP started to grow in UK about 1800. The quantitative comparison of the Pre-industrial, Feudal Village – economy with the Post-industrial, digitalized economy is not straight-out due substantial changes in the composition of consumption. A simple indicator of over-all economic growth in G7 is employment in agriculture. In the Pre-industrial economy, about 80% of the working population was poor copyholders in agriculture. In Post-industrial digitalized economies with advanced agriculture, <5% of the working force is required directly in agriculture and indirectly in the agricultural refinery industries to feed the population. This indicates a gross 15-fold increase in labor productivity which corresponds to an average economic growth during ups and downs as well as peace and war of about 1.2% pro anno since 1800.

### Economic Growth by Market Economy is Profit-Driven

At the social level, profit serves as a risk-premium for an extra effort to overcome initial obstacles and insecurity which is termed "Return on Investment". The thriven for profit, affects that production factors are applied to an extent, where the proportion between all production factors and the corresponding price is the same and minimal (Pareto, 1906). The trick of Pareto-optimal allocation is that simultaneously with optimal allocation of resources for society is production costs for enterprises minimized, wherefore collective and private interests progresses hand by hand in this line of development.

The Pareto-optimal combination of production factors represents the *technical aspect of economic effectiveness*. So, economics is basically a multidisciplinary process, presupposing free and informed communication between economists and a diversity of production specialists.

*Box 1.*

---
**Optimization of Technology to minimize unit production costs**
The optimal combination of multiple factors of production e.g. land, labor, capital and enterprise, is that of equal marginal cost per utility unit:

$$\text{Pareto Optimal Allocation: } \frac{dX1}{P1} = \frac{dX2}{P2} \ldots \frac{dXi}{Pi} \ldots \frac{dXn}{Pn} \quad X_i = \text{Production Factor}_i \quad P_i = \text{Price of i}$$

Minimal Variable Production Costs: $Vc = \Sigma X_i P_i$

---

Economic effectiveness requires more than technical efficiency, the final product must satisfy the needs of the consumer as good as possible, too. The marketing aspect of economics requires sensitivity to consumer preferences including assessment of an appropriate contribution margin (**M**). The consumer price must cover both variable ($V_c$) and fixed costs ($F_c$). Given production level (**Q**), the return of investment (**ROI**) is given in box 2:

*Box 2.*

---
(1) **ROI = Q*P*(M-1)*V_c - F_c**

($V_c = \Sigma X_i P_i$); $F_c$ = (Interest+depreciation of stationary facilities) + (fixed salaries to management)

---

## Automatic, Non-Bureaucratic Regulation of Markets by an Invisible Hand

Adam Smith (1776) launched a vision of the "Invisible Market hand" (IMH). Later, economists have explained, how IMH automatic balances demand and supply as illustrated in Figure 1. Demand increases as the price declines. In contrast, Supply declines when the price declines as the less effective producers no longer get a profit as explained by the ROI Formula. Empirical confirmation of declining marginal productivity was observed by economists in the 18th Century British farming where expansion beyond a certain level required cultivation of less fertile land.

In a period with high prices, supply increases while low prices increase demand until some equilibrium (E) is established. In the short-term, this adaptation process is implemented by fluctuations in the stock. In the long-term, price and stocks become stabilized by a simple, non-bureaucratic self-stabilization mechanism, IMH, appeared to work for local agricultural products in the early days of industrialism in the UK and explained the social benefit from new technologies as the "Spinning Jenny" and the "Steam Engine". This applies to ROI, too as markets with high ROI attract new investors until ROI is normalized..

The private accumulation of ROI (Profit) was not a political issue at the onset of industrialization where the dominant attitude among the leading economists of the Manchester School of Economics was admiration of entrepreneurs that were able to form large mechanized factories (Levy, 2014). The accounting of ROI as economic driver is a sophisticated operation that presupposes development of double-entry bookkeeping which appeared in the early 19th Century.

*Figure 1. Competitive market*

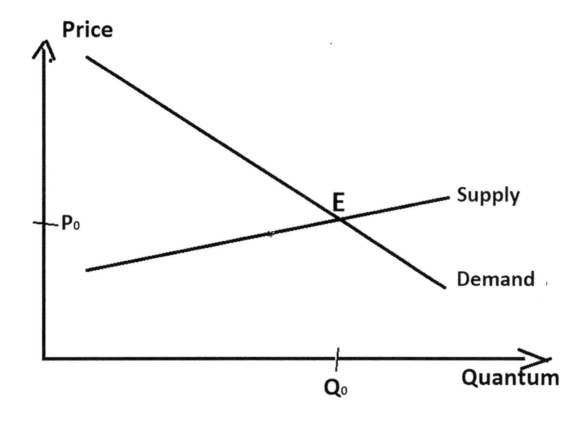

## SWOT-Analysis of IMH

Strength: A 15-Fold Productivity Gain in National GDP
Redoubling Life-Expectancy to 80 Years

How sure is it, that the 15-fold growth in GDP per capita in G7 is due to IMH? The opposite to IMH – central managed economy – has been tried out in Eastern Europe in the period from the Russian Revolution 1917 to the "Fall of the Iron Wall 1989". This large-scale social experiment with hundreds of millions of people during >70 years provides a unique option for comparison with IMH. Key figures on relative economic growth are summarized in Table 1. It shows that the central planned economy in Russia had a good initial growth in the Stalinist period up to World War II. This path of growth continued until mid 70es, where it decayed in the period where liberalist globalization by IMH gave pace to market economies. The history of Eastern Europe – dominated by USSR 1950-89 – is very parallel to that of USSR with the amendment that after the decline of USSR, Eastern Europe benefits from inclusion in the EU where it is the fastest growing region.- an extra argument of special growth capacity of IMH.

Further, Table 1 shows the surprising finding, that while private capitalism (IMH) is superior to socialism (USSR), the growth capacity of the Chinese state capitalism seems to be superior to that of private capitalism. In 50 years from the mid 70es Chinese economy has accomplished the same 15-fold growth that took 200 years in the private capitalist world.

*Table 1. GDP per capita growth rates across economic systems (%)*

| Country | 1928-40 | 1950-59 | 1959-76 | 1976-91 | 1991-98 | 2000-2017 |
|---|---|---|---|---|---|---|
| USSR/Russia | 3.8 | 2.9 | 3.2 | 0 | -7.5 | 1.7[1] |
| Eastern Europe | N/D | 3.2 | 3.5 | -0.8 | 1.8 | 4[2] |
| China | N/D | 5.2 | 1.2 | 5.6 | 7.0 | 9.4 |
| G7 | | | | 3.0 | | 1.7 |

Source: Maddison A (2003): The World Economy: Historical Statistics and OECD and World Bank.

Notes: [1] 2000-08 Russian economy grow by 8% p.a. which recovered the loss of productivity in the Decennium after the collapse of USSR Since the financial crisis Russian growth is at the G7 level.
[2] Countries in Eastern Europe joined the EU per 1.5.2004.

## Weakness: Reinforcing Concentration of Income/Wealth by Monopolist Capitalism

Until 1970, economic growth was distributed in a relative stable way between ROI and wages due the formation of social democratic parties rooted in strong labor unions. Since 1970, the rise of multinational companies (MNC) due to economies of scale by globalization far beyond simple size of production, has changed this trend in favor of ROI (Petty, 2014), see Table 2. Firstly, the demand of e.g. electronic products correlates positively with price e.g. due branding-effects (Erk et al., 2002), wherefore the inclination of the Demand curve typical is slightly positive. Secondly, the marginal production costs – determining Supply - decline due economies of scale which represents the most radical break with classical economics. The intersection point between supply and demand is not a self-sustaining equilibrium, but the point of reinforcing monopolistic gains (M).

A study by the International Monetary Fund (IMF) shows that the key driver in this new development trend is not MNC directly, but an associated formation of a new class of international super-managers with extreme high salaries/benefits. This class is known as the "Top 1%" in the income hierarchy possessing about 50% of global wealth (Francese and Mulas-Granados, 2015). To join the Top 1% in a worldwide perspective, personal wealth must approach one million USD. This basic market failure needs to be corrected by democratic measures.

## Opportunity: Suspension of Relative Poverty by Social Welfare Policy

Orbitofrontal Cortex centers *expected utility* (Levy and Glimcher, 2012). Orbitofrontal Cortex connects our Pro-mammal (Emotion) and Neocortical (Thinking) brains wherefore other primates as monkeys do represent a simplified model of Man`s needs/emotions. The needs of Monkeys can be classified as material, social and sexual:

- <u>Material needs</u>:
    - A diversity of food
    - Fresh water
    - A domicile for sleeping
    - Clothes as our skin is degenerated compared to monkeys
- <u>Special communication and organization for security and development</u>
- <u>Sex partners for reproduction</u>

*Table 2. Key figures on the hegemony of multinational corporations*

---
- MNC parents and their foreign affiliates accounted for 48% of all US Foreign Trade 2010
- MNC Parents accounted for 40% of all capital expenditures, but only 20% of the work force 2010
- MNC Parents accounted for 73% of all Research and Development expenses 2010
- MNC was slow some years after the 2008 financial crisis. From the beginning of the 1990´es through 2008 the average rate of their growth in Direct Foreign Investments (FDI) was 8% p.a. (more than the double of the rest of the economy)
- A most descriptive figure of the global dominance of MNCs, is that the Top 100 MNCs since 2009 increased their net capital with 12% p.a. while the growth of global GDP has been 2%

<u>This Hegemony relates to the following privileges</u>:

(1) *Ownership* advantages encompass the development and ownership of proprietary technology or widely recognized brands that competitors cannot use. Empirical analysis shows MNC are often technological leaders that invest heavily in developing new products, processes and brands, which are then kept confidential and are protected by intellectual property rights.

(2) *Localization* advantages refer to the benefits that come from locating near the final buyers or closer to more abundant and cheaper production factors, such as expert engineering or raw materials or tax shelter. The localization can consider tax levels, too.

(3) Finally, multinationals *internalize* benefits from owning a technology, brand, expertise or patent that is risky or unprofitable to rent or license to other corporation due difficulties enforcing international contracts.

Data Source: Barefoot 2012 / Bloomberg/PWC, 2017.

---

The basal costs of living are estimated to 5 USD/day by the World Bank (WB, 2015) which represent an absolute level of poverty to cover the material needs of survival. According to WB, 10% of the world population lives in absolute poverty. This represents a dramatic progress compared to 1980 where 50% lived in absolute poverty. In the industrialized world, the discussion focuses relative poverty, which beyond basal physical survival enables an acceptable social living. Relative poverty is often defined by economists as 50% of the median income. 2017, the median personal income in USA was about 32.000 USD indicating 16.000 USD as relative poverty threshold. Relative poverty is about 8 times more income than absolute poverty (16.000/(5*365)) USD. The concept of relative poverty is illustrated in Figure 3.

It is a classical economics that man has great or even vital utility of small quantities. As the quantity of consumption increases of specific goods, the marginal utility gradually declines which is expressed in the common saying "A little of everything is good while too much spoils everything". This presupposes a utility function which approaches an asymptotic ceiling in accordance with modern physiology of the digestion system. The marginal utility function (Y-axis in Figure 3) – as the first derivative of the utility function – is continuously declining from a high initial level as income grows rising the quantity consumed. Relative poverty (50% of median income) means that basal social needs are just covered wherefore further consumption beyond that level has a marginal utility <1 as marked in Figure 3. Thus, the bio-humanity-option of economic growth appears to be elimination of relative poverty.

*Figure 2. Market with increasing economies of scale*

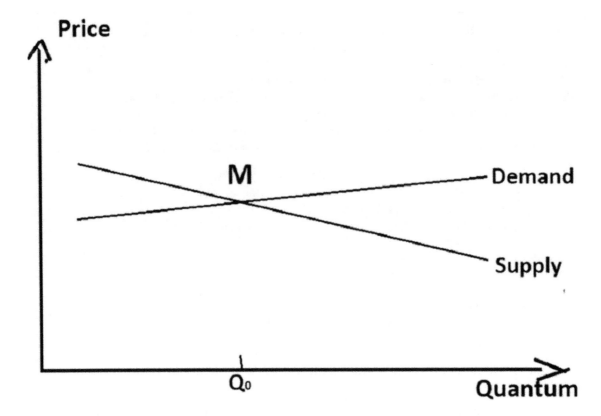

## Threats

Unfortunately, IMH produces global negative third-party effects to a third-party who did not choose to incur that cost or benefit. Education is an example of positive external benefits as individual acquisition of Know-How has positive spillovers on other persons.

### 1. Climatic Catastrophe With Enormous Rise in Sea Levels, Drought Periods, Acid Rain, Other Extreme Weather Phenomena, and Declining Biodiversity

The top-priority is counteraction of the "Green-house effect" as reported by Intergovernmental Panel on Climate Change (IPCC). Facing the "Green-house-effect" as an economic phenomenon requires guidance (rather than replacement) of IMH towards *technical efficiency*, see ROI Formula on Pareto-optimization. A Pigovian Tax (1920) is long known as relevant for such goal.

### 2. A Global Population Explosion

Economic growth is associated with reduced infant mortality due improved sanitation in households which leads to population growth. However, the reduction infant mortality is associated with family planning reducing the fertility rate, too. Table 3 shows data for evaluation of the global over-population problem. The global rate of fertility was as high as 4.7 in 1960. Since 1960 the global rate of fertility has

*Figure 3. Relative poverty as MI$_{50\%}$*

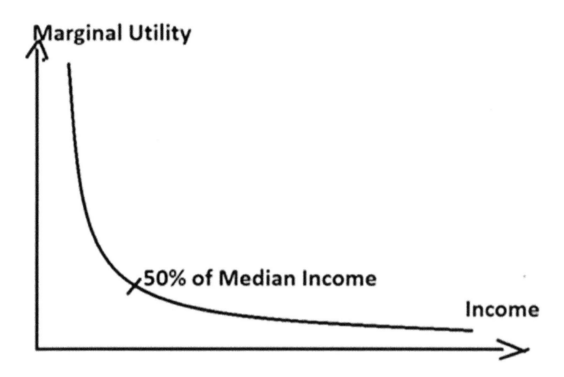

been more than halved to an actual level of 2.3. In a global view there is strong long-term correlation between the fertility rate and GDP per capita. A fertility rate of 2.15 is required to maintain a population. In many advanced industrialized economies, the fertility rate has declined far below 2. This applies to G7 as well as emerging economies (China, Indian, Brazil and South Africa) whereof China is state capitalist.

In summary, the global population actual grows by 1% pro anno. In 2030, the global fertility rate is expected to be 2.1 with a world population equilibrium of 8.6 billion. Some years after 2030 the global population declines below 8.6 billion. In relation to climate change, the growth of the world population is rather an intermediate than a long-term challenge.

*Table 3. Key figures on global population growth*

| Item | 1960 | 1980 | 2000 | 2020 | 2030-Prognosis |
|---|---|---|---|---|---|
| Global Fertility Rate | 4.7 | 3.6 | 2.7 | 2.3 | 2.1 |
| Global Population in Billions | 3 | 4.5 | 6.1 | 7.8 | 8.6 |

Source: World Bank Database.

*3. An Epidemic Rise in Job-Related Stress*

WHO warns that the world faces a mental health crisis by an epidemic raise in diagnosed depressions (Markus et al., 2012). In 2020, depression will be the second leading cause of world disability and 2030, it is expected to be the largest contributor to the burden of disease. More studies link epidemic depression to stresses related to economic crisis. On this background must economics – parallel to the "Green-house effect" – search cost-effective solutions to stress.

## ECONOMICS FOR THE 21st CENTURY

### A Framework of Embedded Economics

A modern approach to economics must recognize documented threats to continued liberalist market growth as job-related stress and global warming, illustrated like a living cell in Figure 4.

The Bioecological model of "embedded" economics in Figure 4 implicates important overlaps to related sub-disciplines as neuroeconomics and economic ecology.

*Figure 4. Bioecological economics*
Note: *The cell plasm is the Eco-system where the macroeconomic dynamics is constituted by Households as served by Corporations and coordinated by the Public Sector. The macroeconomic system is managed by man's ability to adapt as determined by the nervous system.*

### Neuroeconomic Psychology

Integration of Neuroeconomics and Psychology

A neuroeconomic model of complex choices identifies decision-making as integration of Reward-seeking - intrinsic controlled by the fear network - *and* Neocortical cognition. These three forces determine individual *risk-preference* (Larsen, 2017). In the continuum of risk-preference, rationality is the center balancing the peripherals, extraversion and risk-aversion.

Goldberg (1993) identifies by empirical study three degrees of center-oriented flexibility:

- Most flexible are Open-minded rationalists. They are inventive/curious vs. consistent/cautious, appreciates art, emotion, adventure, unusual ideas, curiosity, and variety of experience.
- Moderate flexible are Agreeable rationalists. They are friendly/compassionate vs. challenging/detached. High agreeableness is often seen as naive or submissive.
- Conscientious rationalists are inflexible. They are efficient/organized vs. easy-going/careless.

Figure 5 shows a neuroeconomic psychology integrating neuroeconomics and personality psychology by degree of risk-willingness. Figure 5 resembles the prospect theory in the way that center-oriented deci-

*Figure 5. Neuroeconomic psychology*
Note: *The Figure summarizes the interaction of the 3 basic neural forces: Reward system (RSS), Fear network and Cognition. The cognitive peak with maximal flexibility (variability) Is the center of the distribution of risk willingness. The poles represent emotional biases, either fear or passion. The critical break-even between fear and passion is between Open-minded and Agreeable according to a correlation study (Becker et al., 2012).*

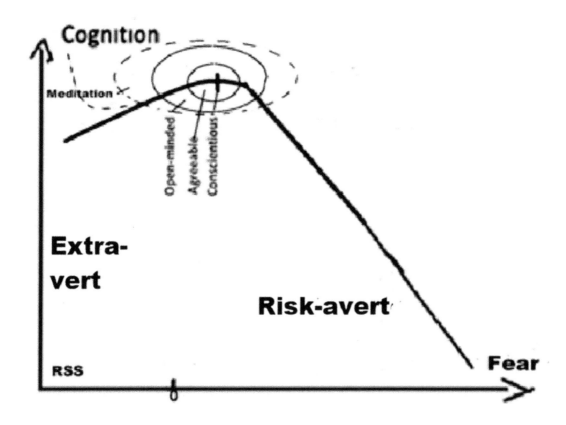

sions are the most rational while peripheral decisions have large positive respective negative emotional biases (Kahneman, 2012). However, neuroeconomic psychology relates these behavioral patterns to an individual personality trait that of risk willingness.

Risk-willingness is operated on a scale from 0 through 10 (Dohmen et al., 2012). Table 4 shows the population distribution by risk-willingness about 1980. A theory, described as "The Rise of the Creative Class" is based on statistics of the distribution of occupations (Florida, 2012). This theory is supported by the demographic distribution 2015 where open-minded rationalists redoubles in post-industrial countries due to liberal patterns of parenting and drastic rising levels of education as demonstrated by European follow-up research (Andersen and Lorentzen, 2005).

Academic Majors divide historically in the natural sciences e.g. physics and chemistry and the Humanities e.g. theology, language and psychology. Social science has developed in between the classical Majors from the beginning of the 19th Century. A qualitative study of psychological differences across Majors shows often moderate differences (about 0.5 standard deviation), but especially open-mindedness differentiates strongly among Majors (Vedel, 2016). Further data analysis indicates that differences

*Table 4. Demographic Redistribution of Economic Agents (%)*

| Group (Score) | Pragmatics (0-1) | Rationalists Conscientious (2-3) | Rationalists Agreeable (4-6) | Rationalists Open-minded (7-8) | Explorers (9-10) |
|---|---|---|---|---|---|
| Males | 3 | 7 | 22 | 15 | 3 |
| Females | 10 | 15 | 21 | 4 | 0 |
| All 1980 | 13 | 22 | 43 | 19 | 3 |
| All 2015 | | | 20 ? | 40 | |

Source: Recalculation of data from Dohmen et al. (2012): The changed orientation of "Agreeable" to "Open-minded" over time is due more liberal breeding and better education of the population.

primarily are due individual temper and only secondarily affected by socialization. The most relevant findings are (Vedel, 2016) are:

- Students of political science score high on open-minded rationality
- Students of economics, law, medicine, and political science score high on extraversion (Explorers)
- Students of science (including engineering) score high on agreeable rationality
- Students of psychology and arts/humanities students score high on both openness and Pragmatism

The most brilliant modern economist Keynes (1937) stated, that a good economist like Marshall should be "*as aloof and incorruptible as an artist, yet sometimes as near the earth as a politician*" (Economic Journal, 1924). The Keynesian appraisal of open-minded rationality is not representative to modern economists who are profiled as Extraverts. Students of psychology are significant more Open-minded than those of business/economics. An associated aspect of Extraversion is demonstrated in demonstrated in a follow-up study that finds that students of business/economics, too, have a significant higher frequency of the "Dark Triads" (Machiavellianism, Narcissism and Psychopathy) (Vedel and Thomsen, 2017). In-all, this indicates that economists could gain in Open-mindedness learning more psychology.

## IMPLICATIONS OF BIOECOLOGICAL ECONOMICS

### Neuroeconomic Psychology and Healthcare

More studies link epidemic depression to stress related to economic crisis (Markus et al., 2012). One of the simplest means to improved stress-management recommended by Markus et al. (2012) is meditation. Already meditation marks the creative class (Hurk et al., 2014). Meditation is a simple cognitive technique developed in the Indo-European tradition (Veda about 1500 B.C.). Repeating silently a sound-word (mantra) in a relaxed and quiet sitting position elicits an in-depth relaxation (Benson and Klipper, 1975). Regular repetition of such exercise gradually releases deeper and deeper stress and strain towards a state of reinforcing open-mindedness (Holen, 1976), see outer circle in Figure 6. A complement to increased cognitive endurance by physical fitness (Oaten et al., 2006). The long-term prospect of meditation is to turn-around the burden of stress into reinforcing mental development. A health technology assessment indicates meditation as a cost-effective health training (Larsen, 2007).

## Neuroeconomic Psychology and Behavioral Economics

### Consumer Behavior

Evidence from marketing of goods and services indicate that Open-minded rationalists have a key position as "First Movers" to have innovations disseminated to the broader society (Gountas and Corciari, 2010). Figure 6 shows open-minded rationalists as a core group of decision-makers in the dissemination of economic progress.

*Figure 6. Market penetration curve*

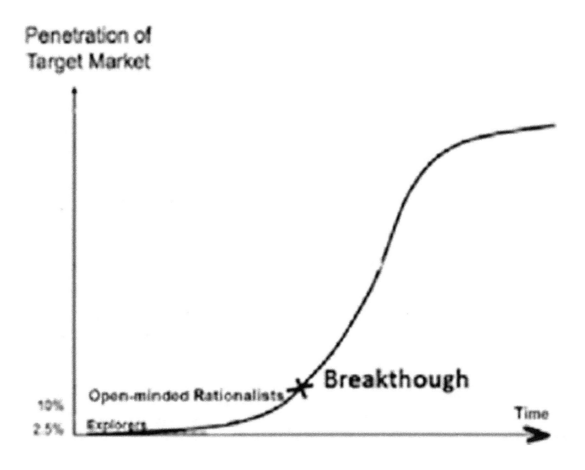

### Business Behavior

Entrepreneurs are more effective managers than other Academic educated business mangers[11]. On this background departs the description of production economy from a study of effective business management (Saraswathy, 2001). The study describes an approach to making decisions and performing actions in entrepreneurship processes, where you identify the next, best step by assessing the resources available in order to achieve your goals, while continuously balancing these goals with your resources and actions.

This novel view of effectuation is called *Pilot-in-the-plane*. It describes a future you can influence by your actions, i.e. you can create your own opportunities Effectuation differs from the causal logic, where there is a predetermined goal and the process to achieve it is carefully planned in accordance to a set of given resources. Causal logic is not suited for entrepreneurship processes that are inherently characterized by uncertainties and risks. Four principles of effectuation characterize *Pilot-in-a-plane*:

- **Bird-in-Hand**: You have to create solutions with the resources available here and now.
- **Lemonade Principle**: Mistakes and surprises are inevitable and can be used to look for new opportunities.
- **Crazy Quilt**: Entering into new partnerships can bring the project new funds and new directions.
  **Affordable Loss**: You should only invest as much as you are willing to lose.

The novel view and the four principles are used in entrepreneurship processes to plan and execute the next best step and to adjust the project's direction according to the outcome of actions. This characterizes an Open-minded with a good balance between rationality and innovation. Formula 1 above is a simple tool to supervise that the bird learns to fly.

## Neuroeconomics and Economic-Political Targets

The introductory phase of the social Welfare State has focused free and equal access to 1) Education, 2) Healthcare and 3) Social security. New Public Management (NPM) as advanced by Hoods (1991) serves to assure a market-like effectiveness of public welfare services:

- Replacing detailed classical budget control by strategic planning based on well-defined outcomes (utility units) that enables cost-effective analysis
- Advancing 1) Privatization and 2) Outsourcing.

Follow-up on NPM shows that it has failed in UK (Hoods and Dixon, 2015) but delivered significant productivity gains in Denmark (Petersen et al., 2015) where the limits is reached with negative side-effect on employee-stress (Andersen et al., 2016).

A series of distinguished economists recommend Universal Basic Income (UBI) at the level of relative poverty (50% of median income) for long-term equalization of income e.g. the Nobel Prize Laureats: Milton, Meade, Samuelson, Simon, Solow, Tinbergen and Tobin. Supposing UBI is restricted to nationalized citizens differentiated in accordance with: Youngsters 18-26 years: 50%), Pensioners >65:70% and Household size: Only 50% to the second or more member of the household. Within these conditions *UBI of 1000 USD for adult singles has a gross budget of 13% of GDP* in advanced industrialized economies. UBI can be financed by 1) Set-off in existing public cash-payments, 2) Abolition of the bottom-deduction in income tax and its proliferations and 3) Max. 3% rise in the bottom tax rate. Personal incomes <100,000 USD gains from UBI. The Neoliberalist argument against UBI is reduced willingness to work. However, recent trials in Finland and Canada indicates in accordance with Figure 5 that UBI helps Pragmatics as their improved social security increases motivation to seek normal jobs. Other benefits from UBI:

- The growing hegemony of multinational companies calls for new interventions to keep up with the status quo.
- The technological unemployment is expected to follow from intense digitalization e.g. robotics where UBI will enable a large low-income group to complement UBI with new types of low-skill jobs.

Common welfare by UBI must be complemented by stronger national account with multinational corporations (MNC) too, preventing further skewedness in the distribution of income and wealth.

## Economic Ecology

The general state of the Eco-system is indicated by in- and outgoing arrows of different thickness indicating imbalance between the amount of energy received on earth from the sun and the amount of energy reflected to the universe. This depends largely on three factors:

- The amount of energy absorbed by green plants and especially forests
- The amount of energy reflected by surface ice e.g. at the poles
- The atmospheric concentration of $CO_2$ determines the degree of reflection of sun energy back to space

Searching scientific databases for *Bioecological economic model* a Bioecological Model of Bronfenbrenner and Ceci (1994) focuses human development as the interaction of genes and environment which is not specific to economics. More models of economic ecology are available. The best fit is the model of "Dynamic Integration of Climate and Economy" (DICE, Nordhaus 2007) as currently updated. DICE is an equation system centering a Cobb-Douglas function enriched with both variables on climate costs and differentiated expectations on development of consumption and production. DICE has a social welfare function based on additional regression equations of consumer behavior. However, such ex ante limitation of human adaptability doesn`t meet good standards of cybernetic for complex adaptive systems.

### The Principle of Pigovian Tax

The economic solution to the "Green-house effect" (CET) is remarkable simple and known among economists for 100 years as Pigovian tax (Pigou, 1920). CET eliminates $CO_2$-emission equivalents reconstructing a new decentral Pareto-optimal allocation of production factors by IMH based on a new price structure including CET. The Nobel Prize in economics 2018 was awarded to an economist that has developed a calculation model for CET that of Dynmic Integration of Climate and Economy (Nordhaus, 2007). Further, massive international student demonstrations in 2019 indicate a generational value-change like that of 1968. As the environmental issue gains democratic top-priority in several EU-countries as demonstrated by the European Parliamentary Election 2019, the research interest in CET within the economic profession is elevated from moderate to top-priority. The action-mechanism of CET is illustrated in Figure 7. Due to the relative flatness (large price elasticity) of both supply and demand, due to the existence of non-fossil energy alternatives, is the substitution effect relative strong.

A review of 27 different implemented CET-projects located in both Europe and North America whereof the most important is from the Canadian Province "British Columbia" (Metcalf, 2019). A carbon tax is

*Figure 7. Pigovian $CO_2$-tax on electricity plants*

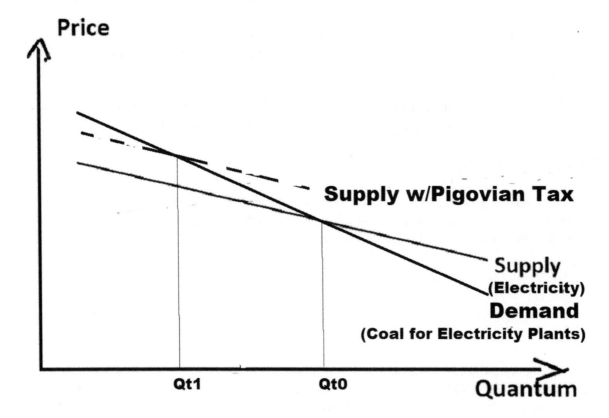

a cost-effective policy tool towards greenhouse gas emissions. Further, it is easy to implement, easy to administer, and straight forward for firms to comply to. It entails costs to society, but not large ones. We cannot clean up the environment for free. On the implementation:

## 1. Revenue Neutrality With Focus on Fairness

Often the motive for tariffs is the need for public revenue to finance public expenses. Regarding CET, it is going to affect goods and services for the broad population and therefore "Turns the heavy side downwards" like a general added-value tariff. This means that simple fairness implicates that a major part of the revenue from CET should be reserved for compensation of clients and businesses. The residual revenue should be used to public investments in related infrastructure to get the full transitional benefit of CET e.g. charging stations for electric cars.

## 2. Collection at the Source of Emission

CET should be *collected at the source* where carbon enters the economy, at the wellhead, mine shaft, or import terminal. A full CET would cover more than 80 percent of the economy's total emission of greenhouse gas. The importance of collection at the source is illustrated by the flight-sector. CET at the import terminal gives flight companies a strong incentive to search for aircrafts with a lower consumption of petrol or even to search non-fossil alternatives as biofuel while a passenger tariff is simply a public money machine. As charter tourism is only about 25% of total civil air transport, the social lopsidedness of CET on civil flights is limited.

Protection of agriculture is typical a strong national political interest. A reason for excluding livestock from CET is the growing world population that requires more agricultural products including livestock to prevent serious hunger problems, especially in Africa South of Sahara. Recently, new technologies have emerged that enable agriculture to become $CO_2$-neutral, too. The Danish Association of Agriculture has 2019 committed itself to $CO_2$-neutrality 2050 which very much weakens the argument for exclusion of agriculture to CET.

### 3 Stepwise Implementation

CET is not a stand-alone-project, it must include follow-up investments in new infrastructure. CET must be introduced as soon as possible and stepwise learn from benchmarks due the different state of available alternatives to fossil energy in different sectors:

- Transport alternatives e.g. electric cars are in an early state of technological development wherefore a very accelerated transition can be very costly compared to the situation in 5 years where the long-term technology for electrification of transport may be far better. Especially, fuel alternatives to long-distance flights are lacking, although biofuel is an option.
- The technology of a $CO_2$-neutral agriculture (farming and livestock) is even less developed and more complicated. In combination with continued expected growth of the world population and growing demand for agricultural products, the transition towards a $CO_2$-neutral agriculture must be somewhat slower than other sectors. Modern farming has specific ecological problems:
  ○ *Fertilizers* for accelerated growth has negative side-effects as pollution of the subsoil water, over-growth of seaweed with deoxidation and eventual health risks (nitrate)
  ○ *Pesticides* serving to fight noxious animals for crops have in some cases negative health effects as cancer

Ecological agriculture controls pollution and health risks, but the price is a lower level of crops. Reduction of $CO_2$-equivalents in livestock e.g. meat production represents a special challenge where further research is of crucial importance.

## Discussion of Implementation of CET

Implementation of CET is at least as complex as intertemporal choice investigated by neuroeconomics. The dichotomy between "Rationalist" and "Pragmatic" thinking regarding CET is expected to be even larger than in "laboratory" designs. On this background presupposes the implementation of CET a long-term build-up of enough public enlightenment to bridge the gap between rationalist and pragmatic thinking. At the level of individuals, a study shows that short courses in economics significantly improve the ability of college students to solve complex economic tasks as intertemporal choice (Paxton, 2019). This is a broad learning effect that is not specific correlated with hyperbolic discounters. Analogue, systematic public enlightenment on CET could create the necessary political base, too. Figure 5 presents a strategy for Green Hybrid Multilevel transition (GHMT) as a second-best solution based on actual frontrunner policies.

*Table 5. National checklist for a green hybrid multilevel transition (GHMT)*

> 1) The most effective means to a green economy is a Pigovian $CO_2$ Tax (**CET**)
> 2) Following the best pragmatic examples of both China and G7, national top-down ***Infant-energy-incentives*** are a second-best but far weaker alternative
> 3) A complementary top-down initiative is ***Governmental-private collaboration on research and development of especially carbon neutral farming and non-fossil airplane fuels***
> 4) Strong and close ***Meso-level collaboration between large-city mayors and business associations*** e.g. regarding infrastructure for electrified transports is more effective than expected due to much unknown internal preparation for carbon neutrality in large corporations
> 5) Consolidation of the political interventions in complementary relationship with a frontline of specialists in economic ecology respectively ecology. Such novel collaborative body between politicians and scientists is recently implemented in Denmark and British Columbia as an **independent climate council**
> 6) ***Facilitate bottom-up contributions*** as voting by the consumer basket e.g. in the format of:
>   - **Simple Living** as important element in the full green transition strategy
>   - Entrepreneurship in **ecological design for the 3R`s** (recirculation, reuse and reduction)
> 7) National protection plans for expected heating-effects in all levels of society e.g. rising sea levels are necessary. It seems already too late to moderate consequences of global warming. Even moderate global heating (1.5-2.0°C) have serious consequences in specific regions with poor countries that are unable to protect themselves e.g. Bangladesh. The seventh and final strategy element is **international economic solidarity** as previewed in the Paris Agreement.

## CONCLUSION

The development of a neuroeconomic psychology represents the establishment of behavioral economics as a positivist discipline between health science and ecology that replaces classical Homo Economicus. This makes "Economic Ecology" and "Neuroeconomics" important new sub-disciplines of economics. Regarding economic ecology, a Pigovian carbon emission tax (CET) is the most effective intervention towards the "Green-house effect". However, in the short-term, democratic center-coalitions are too weak to implement CET due asymmetric levels of economic knowledge between the profession of economists and the public. However, neuroeconomic research indicates that goal-directed training can improve decision-making about complex issues as CET. In order to raise the level of public understanding to implement CET a second-best strategy is Green Hybrid Multilevel Transition (GHMT).

Besides a strategy towards global heating, neuroeconomic psychology delivers a series of advices on effective behaviors:

- A consumer pattern that integrates individual satisfaction with environmental conscience
- Efficacious business behavior like an entrepreneur
- Long-term suspension of relative poverty by Universal Basic Income
- Cognitive training by positivist dialectics and meditative in-depth relaxation

## ACKNOWLEDGMENT

This research is that of a long-term journey with interchanging collaborative partners in more different projects. Especially, I am thankful for guidance over more decades from a series of long-term collaborators in the International Institute for Advanced Studies (IIAS). President, Professor George Lasker; Dr. Bernard Teiling, Director (retired) of Nestlé Headquarters; Professor Emeritus Kensei Hiwaki, Tokyo International University; Specialist in neurology, Dr. med. Helmut Sauer and Hugh Gash, Emeritus Associate Professor, Institute of Education, Dublin City University. Further, I am thankful to professors in behavioral economics Alexander Koch and Julia Nafziger, Aarhus BSS; To Professor Emeritus Terkel Christiansen, windmill-engineer Henrik Stiesdal and Professor of Economics, PhD, Kjeld Moeller Petersen, SDU. I am thankful to private guidance from PhD Erik Frandsen, long-term meditator Leif Justesen and MSc in Comparative Literature Anne-Stine Larsen.

## REFERENCES

Andersen, K. V., & Lorentzen, M. (2005). *The Geography of the Danish Creative Class – A Mapping and Analysis. Danish part of "Technology, Talent and Tolerance in European Cities: A comparative analysis"*. Cph Business Sch.

Andersen, L. B., Bjoernholt, B., Bro, L. B., & Holm-Petersen, C. (2016). Leadership and motivation: A qualitative study of transformational leadership and public service motivation. *Review of Adm. The Sciences*.

Barefoot, J. K. (2012). *US Multinational Companies – Operations of US Parents and their Affiliates in 2010*. Bur. US Survey of Current Business.

Benson, H., & Klipper, M. (1975). *The Relaxation Response*. W Morrow & Co.

Dohmen, T., & Falk, A. (2012). *Individual Risk Attitudes: Measurement, Determinants and Behavior. Consequences*. JEEA.

Erk, S., Spitzer, M., Wunderlich, A. P., Galley, L., & Walter, H. (2002). Cultural Objects modulate reward circuitry. *Neureport*, *13*(18), 2499–2503. doi:10.1097/00001756-200212200-00024 PMID:12499856

Florida, R. (2012). *The Rise of the Creative Class-Revisited:10th Anniversary Edition*. Basic Books.

Francese, M., & Mulas-Granados, C. (2015). *Functional Income Distribution and Its Role in Explaining Inequality*. IMF. doi:10.5089/9781513549828.001

Global Top 100. (2017). *Companies by market capitalisation. 31 March 2017 update*. PWC: www.pwc.com/top100

Goldberg, L. R. (1993). The structure of phenotypic personality traits. *The American Psychologist*, *48*(1), 26–34. doi:10.1037/0003-066X.48.1.26 PMID:8427480

Gountas. J, Ciorciari J (2010). Inside the Minds of Trendsetters. *Aus Magazine Sci/Tech*, 14-17.

Holen, A. (1976). *The Psychology of Silence*. Oslo: Dyade Publishing.

Hood, C. (1991). Public Management for All Seasons. *Public Administration*, 6(3).

Hood, C., & Dixon, R. (2015). What We Have to Show for 30 Years of New Public Management: Higher Costs, More Complaints. *Governance: An International Journal of Policy, Administration and Institutions*, 28(3), 265–267. doi:10.1111/gove.12150

Hurk, P.A.M., Wingens, T., & Giomni, F. (2011). On the Relationship Between the Practice of Meditation and Personality – an Exploratory Analysis of the Mediating Role of Mindfulness Skills. *MF*, 2, 194-200.

James, W. (1907). *Pragmatism's Conception of Truth. Lecture 6 in Pragmatism: A New Name for Some Old Ways of Thinking*. New York: Longman Green and Co. doi:10.1037/10851-000

Kahneman, D. (2003). Maps of bounded rationality: Psychology for behavioral economics. *The American Economic Review*, 93(5), 1449–1475. doi:10.1257/000282803322655392

Kahneman, D., Tversky, A. (1979). Prospect Theory: an analysis of decision under risk. *Econom*, 263-91.

Keynes, J. M. (1936). *The General Theory of Employment, Interest and Money*. Palgrave MacMillan.

Larsen, T. (2006). *Neuropsychology for Economists*. Lulu.com.

Larsen, T. (2017). Homo Neuroeconomicus – a Neuroeconomic Review of Functional Magnetic Resonance Imaging of Economic Choice. *IJUHD*, 7(1), 44–57.

Levy, D. J., & Glimcher, P. W. (2012). The root of all value: A neural common currency. *Current Opinion in Neurology*, 22(6), 1027–1038. doi:10.1016/j.conb.2012.06.001 PMID:22766486

Levy, J. (2014). *Accounting for ROI and the History of Capital*. U. of Chicago Press.

Maddison, A. (2003). *The World Economy: Historical Statistics*. OECD Data 1928-98.

Markus, M., Yasami, M. T., & Ommeren, M. (2012). *DEPRESSION – A Global Public Health Concern*. WHO. doi:10.1037/e517532013-004

Meadows, D.H., Meadows, D.L., Randers, J., Behrens, W.W. (1972). *The Limits to Growth*. Univ. Books.

Metcalf, G. E. (2019). On the Economics of a Carbon Tax for the United States. *BPEA Conference Draft*.

Nordhaus, W. (2007). *The Challenge of Global Warming: Economic Models and Environmental Policy*. Yale Univ.

Oaten, M., & Cheng, K. (2006). Longitudinal gains in self-regulation from regular physical exercise. *British Journal of Health Psychology*, 11(4), 717–733. doi:10.1348/135910706X96481 PMID:17032494

Paxton, J. (2019). Economics training and hyperbolic discounting: Training versus selection eff. *Applied Economics*, 51(55), 5891–5899. doi:10.1080/00036846.2019.1631439

Peck, J. (2005). Struggling with the creative class. *International Journal of Urban and Regional Research*, 29(4), 740–770. doi:10.1111/j.1468-2427.2005.00620.x

Petersen, O. H., Hjelmar, U., & Vrangbæk, K. (2015). Is contracting out of public services still the great panacea? A systematic review of international studies from 2000-2014. *Proceedings NormaCare Meeting*.

Petty, T. (2014). *Capital in the Twenty-First Century*. Belknap/Harvard University Press.

Pigou, A. C. (1920). *The Economics of Welfare*. Cambridge.

Sarasvathy, S., & Read, S. (2001). *Effectual Entrepreneurship*. Routledge.

Schutt, F. (2003). *The Importance of Human Capital for Economic Growth. IWIM*. University of Bremen.

Smith, A. (1776). *Inquiry into the Nature and Causes of the Wealth of Nations*. Oxford. doi:10.1093/oseo/instance.00043218

Summary for Policymakers. (2018). *Special Report on Global Warming of 1.5°C approved by governments*. IPCC.

Vedel, A., & Thomsen, D. K. (2017). The Personality of Academic Majors. *Personality and Individual Differences*, *116*, 86–91. doi:10.1016/j.paid.2017.04.030

# Chapter 5
# Reading, Literature, and Literacy in the Mobile Digital Age

**John Fawsitt**
*Kibi International University, Japan*

## ABSTRACT

*Reading and literature are struggling for relevance an environment where attention and the data they provide are seen as key motivators for commercial actors, and there is great pressure for those actors to provide engaging media to secure a meaningful market share. Thus, this media has to attract and keep user attention as quickly and as continuously as possible. The only limiting factors being those of time and energy of the user. Leisure hours that allowed periods for unbroken concentration and perusal of written texts are now devoted to online activities. What is not debated is that the effort and focus required to engage with the writer of fiction or other longer texts cannot be as automatically assumed now as it was before the digital age. Therefore, how can or should reading and literature and our notion of them and their purposes change?*

## INTRODUCTION

The ultimate focus of this article is on the place of reading and literature in the academic lives of university students in the face of widespread adoption of digital and M-learning. Reading, accompanied by discussion and writing, has long been the staple of a humanist education. It is felt that exposure to and the process of trying to understand other thinkers and their ideas enriches and trains the human mind to comprehend and evaluate their existences, world, and values and to use this knowledge to live a richer and better life whatever they judge that to be. It could be said to be the core of a classical humanist education. The word classical implying that it is no longer the only form of humanist education available. The changes brought about by the advance of science have altered our perceptions and needs so much that there has been a change in what people seek and expect to achieve when they attend university.

DOI: 10.4018/978-1-7998-3576-9.ch005

However, it is not yet clear how much continuity will be exhibited and in which direction our relationship with technology will evolve.

## What Is "Reading"

"Over time a crucial part of basic reading and literacy skills has come to be so-called "deep reading", that is, reading of long and potentially complex, linear texts requiring sustained mental focus over an extended period of time (e.g., essays; novels and short stories; articles;expository texts). Due to digitization, reading is becoming more intermittent and fragmented, and long-form reading is in decline" (Mangen 2014).

Reading in this context is taken to mean extended engagement with a longer text containing information that is offered as material for further thought or to be in some way related to other concepts or data. Examples would be the reading of literary works such as a novels, biographies, histories, journals, philosophical works, treatises etc. These works need not be exclusively of an artistic nature. Engagement with these types of work has been seen as providing both information and intellectual exercise for all of the faculties including those of reasoning, memory, analysis, and comparison. This provides material for expression and discussion based on experiences and opinions other than our own which we can evaluate in the light of what we know about ourselves and our world. The reverse is also true in that we can perceive our own world in the light of what we have read. While much literary work is of a subjective nature and sometimes can appear limited in scope; juxtaposed with other and our own interpretations it can provide a more objectively informed position to evaluate the choices we and others make or have made in our lives. Central to this activity is sustained engagement.

In contrast what comes to our attention online especially on mobile platforms is more likely to have been chosen for us by algorithms on the basis of our own previously expressed preferences. Whether that expression was give consciously or not. Overwhelmingly the material is of a current, personalised, immediate and usually directly informative or entertaining nature. The algorithm's processes are opaque and we can not be sure as to whether they align with our values rather than our tastes. If a person wants to read about great people. Then perhaps the algorithm can deduce that the person is a pale pink or orange male of a certain age and so provides material that deals with such people to the exclusion of other genders and colours. The purpose of the algorithm not being to educate and empower but to return a profit to its shareholders. This is not to belittle the sensitivity of algorithms which can be adjusted to suit the most exquisite of ethical outlooks. At present however it remains that the material read online is that which will ensure our continued use of the site, app, or platform.

## The University and Its Members

What then is the place of reading in this new age? While it will be a while before we know exactly. It can be said to have already suffered a drastic decline in the university sphere. Ken Robinson in his noted TED talk decrying the over cerebralisation of Education, said the world's education systems are geared towards the production of University Professors. People who live mostly inside one side of their heads and that in the future such focus will be less important and we should focus more on creative rather than intellectual ability. While this has great truth, he admits we can not be sure what qualities will actually be needed.

If we look to the past, it was in the 13th century that the universities supplanted the monasteries as the centres of culture, guardians of knowledge and drivers of change (Johnson 1990). They themselves were then transformed by the increasing availability of printed works from the 15th century onward. and the medium of instruction depended less absolutely on oral aural interaction and to a greater extent on the printed word and from a wider variety of sources and views; increasing the possibilities of specialisation in scholarship. In the 19th century science became a major preoccupation and in the 20th the humanities came under pressure to maintain their position. Thus it is unreasonable to expect that the style, content, and goals of a university education will remain unchanged.

It is important not to focus too exclusively on the impact of technology on university teaching. It of course acts alongside other developments which it has effected and been effected by, for example the increasing requirements for universities to be financially self-supporting and to meet the needs of the economy. There is another change and that has been the need to cater to the very desirable development of a far greater number of young people attending college. In the past the students who make up the bulk of this far greater number would due to social structures and constructions, and sometimes lack of aptitude, not been given the opportunity to further their education. Some observers might feel that this has led to an inevitable decline in standards, but for much of history enrollment in colleges was decided simply by the random circumstances of rank or wealth. There is also a tendency to idealise the universities of the past. Not every undergraduate was a Newton or a Marx. The spectrum of brilliance probably remains unchanged while only absolute numbers have increased. Now however, the less brilliant are not protected by rank or wealth or even the confidence of marketability when they graduate. Generally, once they graduate they will have to compete in their own separate marketplace. The marketplace reserved for the less brilliant and less fortunate financially. Their degrees are a ticket of entry to the race not a cheque to be redeemed at the cashiers window. Consequently, a premium has been placed on the marketability of skills. I do not want to say that acquiring practical skills is beneath the function of universities and while some may feel that 3rd level education is increasingly market driven it must be remembered that its original purpose was to supply lawyers.

It is for these less brilliant that the humanist education is most important. They are the most likely to be left in positions of disadvantage compared to the more academically successful. It is they who need most to be able to examine their lives and their worlds lest they fall into despondency and become the victims of demagogues. They are the demos in our democracies. The mesh of society is only as strong as its weakest strand. This is why their intellectual enrichment is so important. Unfortunately, they are also the most vulnerable to the enticements of digital media. They are also less likely to engage with reading for other than immediate purposes.

Nor have humans themselves remain unchanged. A preliterate human can be expected to have different and greater powers from the modern, perhaps their powers of memory and narration would have been greater. Our own literacy-based powers might have been useless in the preprinting ages. We can say that with the arrival of the digital age we may have arrived at just such another juncture. The skills and abilities that have served us over the past centuries will either become disused or adapted to the new age.

## Smartphones and Their Users

Perhaps the greatest change in Western or even global society over the last decade has been the advent of mobile communications devices, in particular the smart phone. " The whole world is engulfed in a revolution" (Robinson 2006). This is readily evident wherever we go their presence has made itself felt. There have even been health warnings about its use and overuse; from traffic safety to its effect on our spinal health through the type of habitual posture its use encourages. In such a short time Its effects on every part of society have been revolutionary. Even though it has just begun to be felt. That universities should expect to be an exception to this change is unrealistic. Already the tools of e and M learning have been taken up enthusiastically by college administrators and educational entrepreneurs keen to cut costs and ensure outcomes. As humanities teachers we should be proudly and irrevocably against measures which place education on the same level as efficiency and finance. A university as a physical place is now open to question.

However, it is not the question about what mobile technology may or may not do to or in Universities but what it has already done to the students who enrol. Our learners have been exposed to smartphone content constantly for at least 6 years prior to becoming undergraduates. And this exposure has not been random. Vast resources have been brought to bear to claim and reclaim the attention of young minds with the objective of understanding everything about them. So that they may be even more precisely targeted and controlled. Disturbing as this may be for the political and economic fields it is even more disturbing for the field of personal human development and autonomy.

Nicholas Carr in an Article in the Wall Street Journal sets out the scientific basis for concern over smart phone use by citing the findings of several experiments that show that " Not only do our phones shape our thoughts in deep and complicated ways, but the effects persist even when we aren't using the devices. As the brain grows dependent on the technology, the research suggests, the intellect weakens" (Carr 2017). I would go farther than this and say that not only the level of our intellect but also its stamina for concentration, and our ability to refrain from using the device decline.

Many of the tests Carr cites focus on performance and the perceived changes in the abilities of the students. Others take account of physiological symptoms displayed such as spikes in blood pressure when unable to check the phone. One other focused on the social effects whereby people without phones were able to achieve deeper personal connections with random partners over a period of ten minutes than people who had their phones with them.

*The irony of the smartphone is that the qualities we find most appealing—its constant connection to the net, its multiplicity of apps, its responsiveness, its portability—are the very ones that give it such sway over our minds. Phone makers like Apple and Samsung and app writers like Facebook and Google design their products to consume as much of our attention as possible during every one of our waking hours, and we thank them by buying millions of the gadgets and downloading billions of the apps every year. (Carr 2017).*

While most of the tests conducted have been related to effects on cognitive skills Carr does state that "The way a media device is designed and used exerts at least as much influence over our minds as does the information that the device unlocks." It must be remembered also that these tests were performed in the earlier years of the widespread adoption of smartphones upon people who at least at one stage in their lives did not possess one. If we think of the potential effect on the minds of people who were raised when the devices have been ubiquitous and where they are objects of attention, then we can only hope that the limits of their influence had been reached circa 2014-15 rather than continue their encroachment so as to approach the state of becoming almost hard-wired into our brains.

These concerns may be unfounded with respect to the level of possibility of our students' practical participation in the workforce if the contention or prediction made by Greg Satell (2017) that our increasingly automated economy will place greater emphasis on social rather than cognitive capability. In that our ability to relate to our fellow humans may become our most sought after skill and not the ability to solve problems or find order in events. However, this ability may also be damaged according to research by Przybylski and Weinstein's 2012 study, whose "results demonstrate that the presence of mobile phones can interfere with human relationships, an effect that is most clear when individuals are discussing personally meaningful topics."

## What Is the Relevance to Reading?

As previously stated reading has been seen as a core component of college education. It was here that students met the most influential and perhaps difficult texts that would serve challenge them and perhaps change their view of the world whether socially or morally. In the present day the smartphone is where students do most of their reading. The content is of course adapted for that medium, but it is also adjusted to build up a cycle of intermittent engagement, gratification, repletion or satiation, surcease, withdrawal and re-engagement; all in the shortest time-span possible. This is a contrast to the continuous engagement that any text of length requires.

Smartphones do not just act upon our brains while we are using them. The cycle of engagement remains while we are otherwise occupied. Our brains have been trained over years of use to expect the release of dopamine at regular intervals. Something an academic or challenging text will not supply in the quantity or periodicity established. If the brain is in a permanent state of expectancy than it is very difficult to build up a proficiency door extended application to another text.

Here is one Australian English Literature teacher, Tegan Bennett Daylight's assessment of his teacher training class's literary experience after dividing them into three groups according to their experience of reading. The reading back ground of one of the groups who have read even less than the groups who have read Harry Potter and the Hunger Game she sums up as having:

*read their and their friends' Facebook pages, their own news feed, and the occasional copy of a women's or a men's magazine. None, unless they have been made to by their high school English teacher, has read anything by an Australian author.*

Daylight then goes on to relate what the outcome of their experience was:
"This is what my students have learned: how to read more than 200 words of a text at a time.....To feel those ideas about literature, so angrily learned, change the way they see." (Daylight 2017). One of Daylight's themes is that difficulty is fundamentally part of the reading experience. This is echoed by Brown, Roediger, and McDaniel in "make it stick" (2014) where they argue that learning most effectively takes place under conditions that pull the learner towards their limits.

The fact then is not that a significant number of students do not read but that they can't read. The implanted cycle prevents them from engaging with the text. Physically the posture is very similar. The activity is superficially at least similar, but the form of attention and stimulation is extremely different. Smartphone usage is characterised by intermittency. it collects and presents information and data in readily digestible and bite-sized amounts

In my own experience after being primarily dependent on a smartphone for reading matter for several years. I found that I was unable to read as continuously as before. As soon as I came to a stimulating section of text I felt a sudden urge to change the activity by picking up my phone and playing a game or reading an online article. The same urge takes hold of me as I try to write this article.

Asking a student to read is asking them to stop doing something they enjoy, to act against their compulsions, to break a habit that has been carefully instilled over a long period of time during their formative years.

## THE FUTURE CURRICULUM

As stated earlier universities are affected by the conditions in their surrounding societies, and indeed they are not remaining untouched by the wave of technological developments in the digital area. The motivations for the incorporation of many of these inventions and innovations can be described as not purely pedagogic. Factors taken into consideration include those of efficiency, finance, administration, and appearance. In the last instance, the desire to appear to be a cutting edge institution to prospective students is extremely important. It is one area of competition between schools, along with campuses, sports teams, and leisure facilities.

E-learning is now well established and seen as integral to the educational experience. M-learning (mobile learning) is now in the course of becoming part of the curriculum. In M-learning undergraduates are enabled and encouraged to use tablets and smart phones to participate in their courses. An interesting aspect of the advent of M-learning is the way in which it is mostly contrasted with E-learning. Yesterday's slick and elegant PCs and laptops are now referred to as 'bulky' and 'inflexible'. It seems that once we make the decision to introduce technology into education the only sure outcome is its obsolescence and replacement. M-learning is going to go the same way. On a first glance the benefits of M-learning are easy to see. Students can access and submit materials wherever they are and whenever they want. Teachers can distribute and collect confident that the student will receive and be able to submit. Any difficulties that have been seen in the literature are mainly on the practical and technical side and deal with the reliability of connections, cross platform compatibility and security. Concerns have been shown about maintaining any distinction between and disruption of students' and teachers' personal and professional spaces. Educational issues are mentioned as existing in the areas of assessment and supervision.

Whether this adoption is more beneficial for the students or for the institution is not clear. Also unclear is whether account has been taken of the previously discussed issues regarding student cognition, intellect, and even mental and physical wellbeing, which should be at the forefront of discussion of any innovations that impact on the student body. It has been largely taken that this adoption is "a good thing" and the permeation of society by these devices should be reflected on campus. Whether they are fit for our purposes is left unanswered and perhaps unasked.

Nonetheless this is what is occurring and teachers of the humanities have to decide how they are going to respond to this process.

Ought we design our courses so as to accommodate students who are conditioned not to read anything longer than a social media post and to switch their attention frequently so as to find fresh stimulation, provide fresh data, and be exposed to fresh and more targeted advertising? While we decry this, nothing is being done and nothing will be done as this is the current commercial model. We as educators have to take the decision; are we going to aid and thereby reinforce this process or stand in its way? M-learning presents us with this choice.

It is perhaps inevitable that smartphones should make their way into education. This is for several reasons. Firstly, they are what the student is most accustomed to using for recording and executing their social obligations. Secondly, they present a quick and convenient method of distributing materials and ensuring the student has absolute access to them. It is assumed that students will work with the material in their free time. However, teaching staff seem not to have recognised the ways in which the really serious providers have trained their audience to behave. Moreover, we are entering the same arena as these professionals an arena that they dominate completely. We haven't got a chance.

There is a further issue. Most educational technology is designed to make "learning" as in memorisation and composition easier. However just as it has already been stated that real learning takes place when difficulties have to be overcome to master the material as posited in the book 'make it stick' (Brown, Roediger and McDaniel 2014).

One course for humanities teachers is to accept that we are now entering, or are already in, a new world that will be or is inhabited by a new type of human who does not feel the need to act without reference to the internet or to put it another way, is able to live and think in harmony with it. Who find more focus and sustenance in their relationships than in developing their own abilities. A useful simile might be that of a netbook computer that can function with lower specification hardware and resources because it stores most of its software and data online and only accesses it when it is needed. This person is more able to focus on their immediate surroundings and hopefully live a mindful existence. However, these new people will have no objective parameters by which to judge their own behaviour and attitudes beyond the reaction of others who share their platforms.

In this new world people may be more aware of things rather than knowledgeable about them. It may be that people become un-moored from their cultural heritage. Not that that is always detrimental. Many of the world's conflicts and tragedies are exacerbated by an over dependent and exclusionary attitude towards a specific culture to create a personal identity. However, will technology supply another culture? Companies have only one objective. Not to ensure and increase profits as quickly as possible is a breach of trust towards their shareholders/employers. In fact, the past decade has shown that the internet facilitates the separation of identities and the intensification of belief systems. It also provides an amplification and a prioritisation of the more and more confrontational and rigid. "social media is the most wonderful of communication, at its worst it has a multiplier effect of the nastiest and most negative comments possible" (Jeremy Corbyn in the Guardian 28 March 2018).

Reading provides for arriving at an individual conclusions or at least a differentiation of opinion between reader and author. The proactive nature of reading places the reader in control. However, they cannot control the course of the text. (They can however control the pace and intensity.) They can decide in their own minds the worth, practical or imaginative of what they see. A device makes it its business to detect and follow its user's inclinations. Personal growth is not seen as part of the business model.

As a non-digital medium paper requires some commitment to obtain. Consequently, the reader is likely to feel more invested in its utilisation, more likely to persevere. It is also far easier to abandon an online article than it is to ignore a book we have paid for are carrying in our bag. We must at least learn to ignore its existence on our desk and bookshelf.

Anne Mangen (2016) lists several paths of research to cover the various aspects or dimensions that reading appears to possess:

- **Ergonomic Dimension**: Reading is a physical, multisensory engagement with a device.
- **Attentional/Perceptual Dimension**: Reading is allocation of attentional resources; perceptual processing.
- **Cognitive Dimension**: Reading is cognitive, linguistic processing.
- **Emotional Dimension**: Reading is, potentially, an emotionally impactful experience.

*Reading, Literature, and Literacy in the Mobile Digital Age*

- **Phenomenological Dimension**: Reading is a personally meaningful activity.
- **Sociocultural Dimension**: Reading is a socio-culturally (and ideologically) appraised and historically contingent activity with sociocultural implications.
- **Cultural–Evolutionary Dimension**: Reading is an exocerebral extension of the brain (Bartra, 2014) developed under pressure of the increasing informational demands of an ever more sophisticated cultural habitat.

Without dismissing the importance of the other items, in the context of this paper, the attentional and cognitive dimensions appear to be the most relevant concerns. The cultural-evolutionary dimension also appears relevant in that the advent of the digital age signals a great transformation in the exocerebral extension. In that it has become much more user friendly in terms of speed and process of access. The journey has been greatly shortened. No longer do we have to find the way ourselves. We simply give the taxi the address we want to go to. We no longer need to know where the area is in relation to other areas. Where in the area our destination is or what type of neighbourhood it is.

Another option open to teachers is to swim against the tide and offer students the experience of 'raw' literature and hope that the 'angrily learned' experience Daylight (2017) mentions, empowers the student. Should we persist in the providing the same opportunities for engagement that were provided in the pre-digital pre-mobile age?

Reading on devices seems to be an obvious compromise The adoption of e-readers also seems like an option that provides space for focus. Devotees of the codex may take comfort in There seems to be a middle ground that provides the university teacher with both the benefits of the utility of digital media and the focus of conventional print.

However, to regain focus, reading generally is in decline. It is in the university context where reading is most extensively engaged in that this decline is most keenly felt. The increasing digitisation and incorporation of BYOD policies and M-learning are merely making it more difficult for students to engage in it. These students are also actively being pursued by commercial interests. As has been stated change is a constant, yet we do not know what we should discard. Certainly, the domination of the web by large businesses does not bode well for human autonomy. It is wise to maintain a level of critical faculty before we decide that the world of cyberspace will provide us with adequate and equivalent scope for our development. While we are still laying the ground rules for our digital future we should retain and develop our own faculties. It is possible that control will be wrested back for the individual but that is a far from certain outcome.

## Is There a Place for Reading?

As previously stated, the issue is not as much with the mobile technology itself as with the patterns of attention it has imprinted on people's minds. Patterns that are deep enough to cause people to neglect their responsibilities when driving heavy goods vehicles at high speeds. Office workers have been found to spend up to a third of their time either web surfing or on social media sites. Simply removing phones from physical sight has some effect. Changing the devices to more specialised ones without a multiplicity of features, such as E-readers, reverting to bound books etc. are all ameliorative measures that can be taken and are probably worthwhile. No victory is absolute, education is generally a game of inches and we can only do our best.

In a very concrete and simplistic way it would be possible to utilise the idea of incremental reading whereby students are presented with increments of a text at intervals judged to match the rthym of engagement engendered by their devices, to be read in steadily lengthening segments. Unfortunately, simply withdrawing the device for specific periods or attempting to wean students off them is unlikely to break the cycle of attention. Feelings of anxiety and unease will still penetrate the mind of the undergraduate reader. The will be as smokers longing to light up as they ascend from the subway. Some of them will do it on the stairs.

## A Fallback Position

It is perhaps best that we recast reading as an activity with an effect that we are consciously aiming for. In the past books were read for their content, for 'wisdom', and for practical benefit. The effects were assessed as being 'good'. To be well read was seen as an accomplishment. Now many of these benefits are no longer needed or are more instantaneously attainable. Universities humanities teachers may be well advised to open classes specifically designated as teaching a practice. One can read as one does yoga. Courses can be taken for therapeutic reasons as much as pedagogical reasons. The object of the teacher should be to release the mind from the habits and anxieties engendered through over-dependence on smartphones by helping students focus on longer texts for longer periods. The object of the student being to, by regular practice, to overcome any feelings of withdrawal, anxiety or depression, to achieve a greater balance in their daily lives. Reading as an activity can help younger people regain control of their lives.

Even Steven Jobs forbade his children from using the iPad within two years of introducing it to the world as a revolutionary and liberating device; recognising how damaging it could be. (Alter 2017) However, no matter how we feel about the mobile digital device revolution, until it is overtaken we have to accept that, "It's part of the future. We have to wake up to how potent an influence it really is. The answer is to learn to self-regulate around it. People don't yet know how to manage this substance in their hands" (Brown 2018). As teachers we should aid in this process of self regulation by doing what we have always done with the consciousness that we have acquired a further mission. That is to protect the integrity of young minds until they can do so themselves.

## LITERATURE

If reading is under such threat or undergoing such changes then we have to look at the body of work that requires such reading and assess whether it can or should continue in its present form. It is unrealistic and unwise to presume that no response is necessary to the digital challenge. Deep reading can be deployed towards any type of text that requires it to gain benefit, such as longer informative articles or treatises and academic texts of whatever nature, however most commonly it is deployed in the area of literature. It is this field which depends on it for its existence. For it has no further purpose than to be read. A literature without reading has lost its connection to the world.

Of course we need to define what we mean by "literature" just as much as we did with "reading". It turns out to have been assigned numerous and at times, incongruent definitions. For example the definitions given by the Collins Dictionaries of English and American English. C14: from Latin *litterātūra* writing; see letter.[1] 1. written material such as poetry, novels, essays, etc, esp works of imagination characterized by excellence of style and expression and by themes of general or enduring interest, 2. the body of written work of a particular culture or people Scandinavian literature, 3. written or printed matter of a particular type or on a particular subject, scientific literature, the literature of the violin, 4. printed material giving a particular type of information, sales literature, 5. the art or profession of a writer, 6. obsolete learning And in the American English dictionary: 1. the profession of an author; production of writings, esp. of imaginative prose, verse, etc. 2.a. all writings in prose or verse, esp. those of an imaginative or critical character, without regard to their excellence: often distinguished from scientific writing, news reporting, etc. b. all of such writings considered as having permanent value, excellence of form, great emotional effect, etc. c. all the writings of a particular time, country, region, etc., specif. those regarded as having lasting value because of their beauty, imagination, etc. American literature d. all the writings dealing with a particular subject the medical literature 3. all the compositions for a specific musical instrument, voice, or ensemble 4. printed matter of any kind, as advertising, campaign leaflets, etc. 5. Archaic acquaintance with books; literary knowledge. Copyright © HarperCollins Publishers

*Thus definitions of literacy vary from "all writings... without regard to their excellence", to "all of such writings considered of permanent value...". At the end of both sets of definitions, the ominous words "archaic" and "obsolete" appear. For this paper the tendency towards "value" and "excellence" is adopted. Having decided on this definition we are then forced to admit that "value" and "excellence" are subjective concepts which no one can claim to have decisive authority over. When we consider the forms people have found "value" and "excellence" in, we are indeed almost returned to the full breadth of definitions given above. For example no one can deny haiku as literature despite their brevity. However in the case of reading we must mean, that which takes more than a glance or more than a moment. We must mean the story as we see it whether it be short or long, works that are both high and low brow.*

The answer as to whether something is literature or not can not be definite and in the end it is up to what the user feels "literature" should provide. "Chewing gum for the mind" was a phrase used to disparage some of the more popular and accessible novel-form texts of the 20th century, such as those often referred to as "airport books". Large, sensational and titillating in nature, produced to a formula, and purchased to while away the time spent in uncongenial surroundings. Banal formulaic dialogue amid predictable situations caused by longings for sex, wealth and demonstrations of macho toughness. Would not have been seen as "serious" literature by those seriously engaged in it. "Literature" was meant to engage on a deeper level and leave the reader enriched and with an experience of something new. Still Steven King demonstrates the wide arc of literature and the impossibility of cordoning off areas as worthy or unworthy via the variety of his works from the "X men" to novels that explore forgotten incidents of his childhood. At first seen as merely material for fanzines they have become widely respected as works of literature. There are examples of "bad" literature this is literature whose values, (not style or topic) we disagree with. Their qualification to be considered as literature rests on their ability to influence. George Orwell's opinion that it should be possible, (to paraphrase him) "hold that this is a good book,(as in well written), and that it should be burnt by the public hangman". In her preface to Graham Greene's "The Quiet American" Zadie Smith holds that "English writers these days work in spasms, both in quantity and quality, and so keen are they to separate 'entertainments' from 'literature' that they end up writing neither." (Zadie Smith 2004 p xiii). She also commends the fertility and cross-disciplinary nature of Greene's works, "Reportage turned to novel turned to film". The perceived incongruity of being both marketable and good can be traced to a discomfort with sharing the common taste, with sharing space with more immediately commercial actors.

## Literature in Media

The following exchange from the script of the Novel "Being There": STIEGLER Mr. Gardiner, I'm Ronald Stiegler, of Harvard Books.

CHANCE *(a two-handed handshake) Hello, Ronald.*

STIEGLER *Mr. Gardiner, my editors and I have been wondering if you'd consider writing a book for us? Something on your political philosophy. What do you say?*

CHANCE *I can't write.*

STIEGLER *(smiles) Of course, who can nowadays? I have trouble writing a post card to my children! Look, we could give you a six figure advance, provide you with the very best ghostwriters, research assistants, proof readers...*

CHANCE *I can't read.*

STIEGLER *Of course not! No one has the time to read! One glances at things, watches television...*

CHANCE *Yes. I like to watch.*

STIEGLER *Sure you do! No one reads! ...Listen, book publishing isn't exactly a bed of roses these days...*

CHANCE *(mild interest) What sort of bed is it?*

The above is an example of literature supplied from a novel by Jerzy Kosiński. When it is changed to fit a film script is it still literature? When the words are spoken by the actors are they still literature? When they are recorded and played back are they still literature? When I copy and paste them into my text, are they still literature? Even if they are from a story not Kosiński's own are they still literature? Why are the words there? Because I remembered them and they became somewhat apposite to a task I have been given. I felt they had something to say to the topic I am treating. I am using them for support. To show not only I have felt this. They are indeed words from the mouth of a fictional character. But I know they have been written by someone who at least acknowledges the possibility that they may be expressed. It is as if the author has said, "No, you are not mad. I have seen it too." This is one of the powers of reading literature. We can find confirmation of our inner selves in text. It does not have to be in a work of art but feelings, instincts, and inclinations on certain subjects are not always reducible to a factual or demonstrable level. As humans we inhabit a world slightly different from that of the concrete and provable. Some works put our consciousness directly in touch with another consciousness. We are hailed across a void caused by the real, the outside, and the expressed.

## Literature From Media

Would these words be valid as literature if they had been written by an artificial intelligence? Now we have to say that we are in the midst of developing a simulated literature. An increasing amount of what we read is; anything that can be reduced to a formula probably can be. A great many of our most successful authors write to a formula, often against their own wishes but in response to their readers' publishers' and the rest of the market's demands. However texts will increasingly become machine written from templates, especially those that we consume as news and reportage. Whether from the sports stadium or the stock market. These cannot be called literature unless we come to see them as works of skill or beauty without personal meaning, perhaps as natural forms.

## Literature in Society

The role of literature vis a vis society in contrast to the internal is well expressed in an interview with the author Martin Cruz Smith commenting on his own very popular and profitable works.

*By looking at the underworld you see how mainstream society works. You can travel through a social fracture and, for a limited amount of time, you can behave differently and ask whatever embarrassing questions you like." During his time studying sociology, he says he was "interested in people who are marginalised and what they say about the wider society. I had no talent for academic sociology but I do have a talent to write a certain type of story that has something to do with the world and things that have happened to the world: the collapse of the Soviet state, Los Alamos, Pearl Harbor, Cuba and now Chernobyl. The ability to describe these things and what they might mean is a talent I enjoy. I don't want to preach, the story always has to come first. But if you think you've changed someone's outlook on the world for the better, that is obviously very satisfying as well. And to do it through crime fiction is just fine.*

(2005 Interview with Guardian Books)
https://www.theguardian.com/books/2005/mar/26/featuresreviews.guardianreview15

This quality of asking questions of society itself is not confined to newer works. It is interesting that some works and authors well-established as literature have already become "problematic". It highlights the chasm that has opened up between the generations of readers. Questions that are raised about these older works and authors, their attitudes and depictions of othered members of society, that are central to how our societies move forward are being hashed out in the lecture halls and literary columns rather than on the streets. Literature provides a sounding board for our ideas of how we want to live.

## LITERATURE AND LITERACY

Previously the English form of the word "literate" indicated an ability to decode the written word. However nowadays it is also used in phrases such as "computer literate" "science literate" these indicate an understanding of the subject's workings and an awareness of its ends (consequences). However it is questionable whether the average "user" of media today is really aware of the ends of the use. They are more likely to be focussed on the goals of their use. The ends of the use are those of the supplier of the platform or technology which is to keep the user' attention and harvest their data. There is a constant subtext of engagement which has nothing to do with the user's preferences. Use, even skilful use, therefore cannot be connoted with real literacy. Whatever the merits of the supplied material, it is important that we should have alternatives or at least knowledge of them. We cannot make decisions unless we have a choice. How can this type of literacy be taught? One way is through narratives which posit various, situations, outcomes, and emotions and how humans are affected by, and deal with them.

How do the effects on literacy differ for reading and for literature? In reading we would stress understanding whereas in literature we would prioritise feeling and impression. Reading provides information for us to work with, and the space for us to work with it. It has a relationship with literature and just as the story is the easiest place for children to become both interested in, and accustomed to literature. Narrative is also perhaps useful as a mode of learning perhaps the most effective mode. Popular knowledge of many fields whether those in space, time, or endeavour rests mostly on what has been related via means of a story. Generally only specialists venture into these areas further. For the ordinary person the pathway into these fields is one of narrative. Most people know stories that take place in those worlds whether in history or even science. We need personalities, emotions, successes, and tragedies to engage us. Many online game makers utilise narrative to engage their users' imaginations. Things are framed in terms of "Quest" and "Last Hope", "Armageddon" etc. in short, a narrative.

Among others, politicians have realised this and are becoming increasingly well versed in the provision of narratives to further their goals. A literacy or literature increase our abilities in analysing them. Political actors who cannot construct a narrative are generally ineffectual. The importance of narrative has been increased with its disappearance from modern daily life as lived in the cities. In this expanding world, older myths that sustained identities have been discarded to make way for further economic growth and progress. At least in theory the playing fields have been levelled and merged, and older distinctions and values that provided a framework of pride and meaning are now seen as barriers to development. This leaves many people thirsty for a narrative with which they can identify. Perhaps one of our failings has been allowing conventional literature to become a highbrow or artistic pastime. It has become the preserve of an alienated comfortable elite that struggles to understand how the common person sees the world.

In terms of education, literature often requires interpretation, in fact often a different way to interpret that provides a different angle from or through which to understand. No one field of research or study can alone claim to be able to fully analyse or advise. Economics has always assumed people as rational actors who will make choices according to their incentives. People's actual psychology remains far more complicated and unknown than this. Sacrifice, heroism and cruelty remain features of human nature beyond the ken of most schools of economic and scientific thought. Understanding comes in many ways. One of the most important is the ability to identify with the other's anxieties, hopes, and fears. Most disciplines approach their subjects from above and merely seek to convey understanding, of various depths, on an intellectual level.

It is in this way literature will help students become people capable of understanding on multiple levels and facets. In an era when careers even the shape of society cannot be predicted, literature and reading can help build the literacy necessary to create a cohesive society that acknowledges difference even our own, and sees that within each difference is a fellow soul. It seems that we have to expand our vision of literature and literacy from the written to encompass anything that deepens the understanding of the world and opens our minds to different levels of meaning.

By engaging our students with literature we can give them opportunity to build this wider literacy. We should actively seek out narrative works that deal with a wide variety of aspects of life, for example, the economic, professional and medical fields where participants need to not just understand but also empathise with people's situations. What we do know is that narratives support attention and that is what reading, literature, and thought are competing for.

## REFERENCES

Alter, A. (2017). *Irresistible: The Rise of Addictive Technology and the Business of Keeping Us Hooked*. Harmondsworth: Penguin.

Brown, J. (2018) Hypernatural monitoring: The real reason we're so addicted to our smartphone. *Independent*. https://www.independent.co.uk/news/long_reads/smartphone-addiction-hypernatural-monitoring-technology-nomophobia-social-media-a8253191.html

Brown, P.C., Roediger, H. L., & McDaniel M. A. (2014). *Make it stick*. Harvard University Press 2014.

Carr, N. (2017, Oct. 6). How Smartphones Hijack Our Minds. *Wall Street Journal*.

Daylight, B.T. (2017, Dec. 23). 'The difficulty is the point': Teaching spoon-fed students how to really read. *Reading Australia Guardian*.

Johnson, P. (1990). *A History of Christianity*. Harmondsworth: Penguin Books.

Kosiński, J. (1970). *Being There*. Harcourt Brace.

Kosiński, J., & Jones, R. C. (1979). *Being There* [Screenplay]. United Artists.

Mangen, A. (2014). *Evolution of Reading in the Age of Digitisation: European Cooperation in the field of Scientific and Technical Research*. COST IS1404 E-READ.

Mangen, A., & van der Weel, A. (2016). The evolution of reading in the age of digitization: An integrative framework for reading research. *Literacy*.

Nagy, G. (2020). A Minoan-Mycenaean scribal legacy for converting rough copies into fair copies. *3$^{rd}$ January 2020; Classical Inquiries*. Harvard's Center for Hellenic Studies. Harvard University. https://classical-inquiries.chs.harvard.edu/a-minoan-mycenaean-scribal-legacy-for-converting-rough-copies-into-fair-copies/

Perkins, A. (2018, Mar. 28). Jeremy Corbyn decries abuse of antisemitism protest MPs. *Guardian*.

Przybylski, A. K., & Weinstein, N. (2012). Can you connect with me now? How the presence of mobile communication technology influences face-to-face conversation quality. *Journal of Social and Personal Relationships*, *30*(3), 237–246. doi:10.1177/0265407512453827

Satell, G. (2017, Mar. 30). How to Win with Automation (Hint: It's Not Chasing Efficiency). *Harvard Business Review*.

Smith, Z. (2004). Preface. In *The Quiet American*. Penguin.

Wroe, N. (2005). Crime Pays. Cruz-Smith M. Interview (Wroe N.). *The Guardian*. https://www.theguardian.com/books/2005/mar/26/featuresreviews.guardianreview15

## ENDNOTE

[1] A deeper analysis of etymology reveals that *litterātūra* was adopted to Latin through Etruscan from Greek διφθέρα/ *diphthérā,* which in its early uses in Cyprus refers directly to leather and parchment. Word *diphtherāloiphós* means, etymologically, 'parchment-painter', derived from a combination of the noun *diphthérā* 'leather, parchment' with the verb *aleíphein* 'dab' in the sense of 'paint with a brush-pen'. Herodotus (5.58) also tells that the word διφθέρα/*diphthérā* was used by the Ionian Greeks of Asia Minor with reference not only to 'parchment' but also to 'papyrus'. In short, the term *diphthérā* in its early days was used to refer to different types of prestigeous scribal writing methods within the Greek cultural sphere and it can be traced back to the Mycenaean era (when writing was limited to a very small elite). Of course, the meaning of 'writing' or 'painting with a brush-pen' changed rapidly in Greek language, when the skill spread among the people and the Greeks developed a rich literary tradition. The material element was lost especially in translation quite early (Cf. Nagy 2020).

# Chapter 6
# Strategic Transformational Transition of Green Economy, Green Growth, and Sustainable Development

**José G. Vargas-Hernández**
https://orcid.org/0000-0003-0938-4197
*University Center for Economic and Managerial Sciences, University of Guadalajara, Mexico*

## ABSTRACT

*This chapter aims to analyze a strategic transformational transition of green economy, green growth, and sustainable development from the institutional perspective. The analysis begins questioning the implications of the concepts and principles of green economy, green growth, and sustainable development from different perspectives in the transformational transition considering the investment, trade, and capacity building though the design and implementation of strategies and policies as well as measures from an institutional analysis. The methodology followed was the analytical review of the literature to derive inferences, challenges, proposals, and conclusions. It is concluded that the green economy concept addresses current challenges delivering economic development opportunities and multiple benefits for the welfare of all human beings.*

## INTRODUCTION

Cities and urban ecosystems are facing important global challenges at local scale for a sustainable development. Academic research on green area innovation is a new shift in the environmental debate in public and private city spaces towards social good and environmental value fostered by innovation and new technologies (Berger et al., 2007).

In the last decades, several contradictions have been generated in terms of urban socio-spatial relations, and the authoritarian urban space and democratic urban space. The logic of the management of the dominant urban space is to create a space that, being public, is authoritatively constructed, that is, it is a restrictive and selective public. The urban space itself makes any articulation of the social movement difficult. The social movements in Latin America developed a very strong presence during the 1990s in such a way that they forced a reconfiguration of the political spectrum dominated by leftist governments. They arise in a context of a democratic opening that takes place after military dictatorships, within a tradition that is an oligarchic tradition and of much social inequality.

This social inequality meant that in the beginning any social struggle had to be very organized, it had to be very strong, because the social inequality was so great that the oligarchic classes were going to defend themselves by all means, they had defended themselves with the dictatorship and they were going to defend themselves with democracy. Authoritarian spatial conceptions develop in large part at a time when social polarization and social inequality began to endanger governance. Authoritarian urban spaces are the dominant urban spaces that try to defend themselves from a popular reaction, private urbanizations are exactly a good example, among others.

## ENVIRONMENTAL SUSTAINABILITY

The urban economic growth and development perspective considers the city as the location where production, distribution, consumption contributing to the process of urban innovation taking place. Sustainable urban planning policies must adopt innovative approaches to consider the anticipation of risks and rationalization of investments. The process of urban innovation is inherent to the practice of sustainable development. Environmental sustainable urban planning and development as a theoretical framework supported by transitional theory can be adapted to specific issues of urban land use and infrastructure. Environmental sustainability transitions may contribute to improve economic growth and development.

A sustainable city is supported by an urban planning and designing paradigm. Urban design and planning supported by local governments are urged to spur technical innovations, innovative practices and contribute to develop new business models and marketing strategies aimed to find green resource-efficient solutions for climate change adaptation and mitigation while restructuring and reinventing local economies.

Integrating environmental sustainability issues into green areas innovation strategy and greening the innovation process are becoming a strategic opportunity for organizations, communities, local governments, neighborhoods (Porter and Reinhardt, 2007). Research and green innovation resources are available and accessible in a network, which can be used adaptable to some specific demands, such as technologies, tools, data, experimentation and testing facilities, and user groups.

Environmental sustainability is an integral part of green area innovation requiring greater demands of ecological performance out of eco-efficient green technologies with reduction and mitigation of green negative environmental impacts. An innovation environment creates new business opportunities. Leon (2006) identifies the need to transform efficiency and quality of public infrastructures and services and the creation of a business environment as the necessary factors for a city to become an innovation hub (Komninos & Sefertzi, 2009).

The process of urban planning is aimed to create and develop an inclusive economy centered on innovative improvements at local level (Smart city Edinburgh, 2011). The green urban planning has three priorities: the economic growth and development, social equity and the environmental sustainability. Local governments at the same time that other levels of federal and state authorities and agencies can take different approaches based on environmental economics and technologies innovations and political commitment to incentive and motivate involvement of actors and stakeholders to find specific innovative solutions to urban planning. Transforming green infrastructure models into action require new approaches of innovative planning and policymaking of programs.

The question of urban space is measured exactly with the question of time. Urban public space is the space of long times, of coexistence, of trust, which is not created from one day to the next. It is created in several years, because urban spaces are often created and then it is said that people do not use it. It has to take some time for people to get used to and enjoy other conceptions of urban space. It's a long time, and if politicians ruled for a few years what is a short time that totally plays against any idea of urban public space.

The democratization of public spaces for citizens and for citizens includes domestic urban spaces, production and consumption spaces, the community, and the world space. Public property is state property, especially in terms of urban built spaces, there is no urban public non-state built space. Private property is the anchor of all modern law and all bourgeois democracy. In the movements of occupation of urban spaces, of reuse for community purposes, and of democratization of urban space as a vehicle, people speak of a "doing" of the community and a renewed sense of the collective. The democratization of urban space tries to reinvindicate privatization through various forms, such as real estate projects and as repressive responses to communities.

In the Western model only the urban space of the citizenship has been relatively democratized. Democracy works only at the level of the urban public space of citizenship, it is not in the family, it is not in the factory, it is not in consumption, it is not in the community, nor in world relations. Representative democracy that what you have. In the end, it is an island of democracy today very fragile, in an archipelago of despotisms, in the family, in the factory, in the street, in the community and in consumption. Therefore, to democratize democracy is to democratize those urban spaces and all of them have an urban public space dimension. That is, the family today cannot be understood as a private space, because it is publicly regulated.

The democracy that emerges in Latin America in the last thirty years arises together with neoliberalism, does not have a democratic social content. It is democracy plus the opening to the markets. Unlike European democracy, Latin American democracy is a democracy that does not have a concept of social democracy of inclusion nor the rights achieved for a long time. Democracy in Latin American countries was almost instantaneous and it was not necessary to organize movements in these areas because the State was somehow responding. In Europe, after the 70s, the great movements have been feminist movements and environmental movements. Obviously, after the worker's movements where that growth was organic, not a revolution, organized since the beginning of the 20th century and between two wars and after the student movements.

The democratic model of urban spaces has collapsed, not by decision of the countries, but by external decision. Organizational energies, collective consciousness, the culture of contestation and mobilization are not built from one day to the next. The reinvindicative tradition resides in the different cultures or democratic models that have been created, considering that there is a crisis of social movements that comes from the social-democratic policies and political cultures that were created over time.

All social relationships are urban social-spatial, but they are in different ways. Santos (2007) distinguishes six modes of production of power, knowledge and law: they are the domestic space, the space of production, the space of citizenship, the space of the community, the space of consumption and the world space. It is all these geo-spaces that must be democratized. A city where public space is very difficult because the physical spaces between the built-up parts are so vast, that far from favoring the creation of public spaces, create deserts of citizenship, socially and politically neutralizing zones that hinder organized social protesting mobilization.

The neo-liberal model went through the private space and left the public space. However, when the public space is abandoned, financial crises and ecological crises come into the home. That is to say, one does not earn much by taking refuge in the private space, because now there are many people without employment and then they eat poisoned products. The idea that private property is not touched has been created, which also forces us to rethink all other concepts of real estate property and even ownership of the land.

There is a great transposition of the conceptions of democracy. The solutions that they recommended as just for the people to develop, have made the advanced countries fall in their own trap, locked in their own ideology and has no solution to the problem. It is therefore a failed model, at the level of civilization, from the point of view of social inequality and from the point of view of social discrimination. Eurocentric critical theory, social and philosophical theory, has trained very well for critical denunciation, but not so well to formulate alternative proposals.

To address urban development challenges, innovative urban planning and design must have a green innovation orientation, interconnected and innovative development stages and processes (Adams 2006; Blewitt, 2008). The conceptual reference model of innovative green areas creates city identity of the urban planning and design by describing the interrelationships between smart urban environmental sustainability with innovative structural components and processes.

Innovation and design of urban green spaces must be linked to the city`s ecological, environmental and sustainable development transformed in high quality of life, safety and sharing of the space of local residents, economic growth and social development, fluidity of interactions between networks. The ecology of productivity implies having another concept of land productivity that is not merely based on the production cycle that promotes the negligent use of pesticides, herbicides and fungicides. It is a great transformation where the conquest of diversity and biodiversity are in the focus of a neoliberal development. New urban farming spaces used for agricultural production and recreational purposes may be based on new business models implementing shared knowledge on site-specific planning and organizing processes, new financial and marketing strategies supported by interdisciplinary networks.

Innovative data collection initiatives and management programs based on vacant properties can be used for purposes of innovative urban greening and vacant property. In the city there are many empty spaces, built buildings that have never been occupied. The urban occupation movements are not movements of the same dimension as the others, they are smaller movements, they are smaller organizations, they are sometimes what people today call spontaneous movements, where there has to be an aggregation, or that is to through social networks, but it is clear that there is another type of mobilization here whose political connotation is very difficult to identify, and which are even totally hostile to politics.

The urban greening as an established innovative programs based on vacant properties can lead to develop stable neighborhood environments by providing fresh food, training and jobs for residents. Democratic urban planning participation and managing green infrastructure can result in environmental changes benefiting all the social groups involved, including unrepresented and disadvantaged populations sectors.

Urban social innovation identifies urban resources for sustainable development, economic growth and social development. This created identity as an ability of the city blends social innovation in urban planning change and regeneration projects to develop a city socially inclusive among citizens and communities (Cozens, 2008; Belisent, 2010). A transition from niche innovations toward a more environmental sustainable development should be directed to change the composition of multi-segmented land

and infrastructure uses and urban mobility and transportation modes. Niche innovations redefine the limits to environmental and ecological development goals in terms that decoupling economic growth from environmental degradation can be achieved (Smith et al., 2010).

## URBAN ECOSYSTEMS

The city as an ecosystem is a concept that considers the political and cultural role of a living city which includes plants, animals, food production, etc. The cities have a potential role to become urban innovation environments. Innovative practices are prevalent in the environmental improvement in urban areas to create ecosystems services and ecological functions of urban land, which are in high demand. Land-use planning and urban development challenged the technological innovation, research, assessment, policy and strategy development and adaptation to the urban ecosystem.

The urban innovation ecosystems aim towards design and implementation of innovative environmental and sustainable development process in cities. Green urban innovation ecosystems management has the aim and the task to manage the resources, facilities and innovation assets portfolio embedded in regional and even national research and innovation projects and fostering fruitful interlinkages from the perspective of smart cities.

Urban innovation ecosystems are characterized according their social orientation towards environmental and sustainable development. This innovative green urban planning and design policies is the result of an interrelated concepts and issues of urban growth based on innovation ecosystems and environmental social sustainability built on green city.

The proxies related to innovation and economic wealth creation scale with urban size indicating that urban growth is driven by innovation and scale economies of energy consumption. This contributes to the attractiveness of the city in terms of location for investments and collaboration driving the emergence and development of social networks. Social networks and other forms of participation that facilitate the electronic means available, allow forms of electronic democracy, the result of a new reality that is there, of a virtual public space, which is a space with enormous potential.

The urban natural and environmental green resources and infrastructures involved in the innovation ecosystems may be capitalized and developed into new business models. Business models turning technological capabilities into innovations to offer useful green products and services require continuous flows of data and information. The outcome of this innovative urban planning is the generation of a business model based on public-private cooperation in urban innovation for smart growth opportunities (Barcelona 2011; Downs & Mohr, 1976).

Urban site design needs innovative planning approaches to integrate all the urban green areas in an urban green ecosystem overcoming several restrictions and constraints (Kuhns et al., 1985; Evans et al., 1990). Urban sites could be developed when available aimed to give green products and services to the population by creating new urban woodlands with innovative afforestation of native species (Harmer, 1999; Baines & Smart, 1991). A dense innovation ecosystem with knowledge interactions creates economic value and more rapid growth through the acquisition, processing, and use of information (Komninos 2008, a, b; Shapiro, 2003).

The fundamental transformations of these ecosystems build their own resilience based on diversity of urban systems driven by green innovation. Scale-crossing practices create unique pathways for a diversity of experiences breaking up closed group thinking (Oh et al. 2004, Scheffer & Westley 2007), fostering collective actions while nurturing innovation capabilities. To achieve this end, it is necessary to develop a comprehensive structural framework of urban innovation ecosystem as an approach to resolving ecological, environmental and sustainable issues.

Urban innovation ecosystem management aims to manage the portfolio of innovation assets, resources, facilities, information flows, knowledge, etc., through mechanisms such as the partnerships among economic agents and actors governing the access to users and developers. Urban innovation ecosystems should promote participation of citizens and organizations, firms, local governments in the planning and development of urban economic activities and utilities. Urban innovation can be harnessed in a highly politicized urban environment making urban governance more sensitive to urban green ecosystem dynamics facing social, ecological, environmental and sustainable uncertainties.

Social innovation of urban green areas improves the quality of life and attractiveness of the city and develops the ability to sustain human-dominated local ecosystem services. Community greening innovation is a community-based tool to promote social learning, adaptive management and resilience of the green ecosystems (Tidball & Krasny, 2009). Theoretical analytical transitions are the result of interactions and learning processes between the levels of niche-innovations, sociotechnical regimes and sociotechnical landscapes aimed to improve performance and support from groups.

Niche-innovation challenges develop processes to provoke changes at the sociotechnical landscape level creating pressure on the regime. Niche innovations in growth agenda may shift to current growth dynamics operating at the sociotechnical landscape level considered as merely exogenous to improve the mobility, infrastructure and building stock. The sociotechnical regimes and landscapes can become destabilized due to inner pressures and tensions originated from outside that may result in niche-innovations.

The ecosystem services approach is defined as the benefits that humans derive directly or indirectly, from the functioning of healthy ecosystems (Costanza et al., 1998). This concept is acknowledged in urban green space research 2 (Bolund & Hunhammar 1999; Elmqvist et al., 2004; Ernstson et al., 2008; Niemelä et al., 3 2010). The assumption behind this concept is that physical and mental health and wellbeing are reliant upon functional natural environments (Millenium Ecosystem Assessment, 2005). Wellbeing and health services must be inter-connected.

An adaptive method of urban planning and design using the concept of ecosystem services is a framework for promoting and supporting innovation in urban green areas. Adaptive systems are suited to "safe-to fail"- innovative design to be implemented and monitored, despite that the adaptive urban design framework "safe to fail" inhibits innovation decision making by tending to favor existing knowledge and performance of the specific intended ecosystems services (Holling, 1978; Kato & Ahern, 2008). Adaptive urban planning is open to innovative design and creativity, learning and gaining knowledge to be applied in future projects (Ahern, 2012; Lister, 2007; Rottle & Yocom, 2010).

Urban social innovation drives urban growth and structures local ecosystems services among the stakeholders through a process of recursive communication. The media play a relevant role in communicating the green innovations advancements although new methods have to be explored (Otto, 1998). Stakeholder's engagement in urban planning processes and encouragement of communities to initiate green infrastructure projects ensure more democratic outcomes. Wealthier and more democratic cities invest more on innovative capabilities to foster development than less democratic and poor cities in a context of the path dependency effect.

Emphasis on the assessment of urban ecosystem services does not motivate and support the innovations of specific ecosystem services in urban and infrastructure development activities (Aherna, Cilliers,

Niemeläc, 2014). In this urban ecosystem of innovation, innovative companies coexist with research institutes, training and tech transfer centers, urban green areas, infrastructure and facilities, etc.

Relationships of cooperation and partnership strategies are required among the stakeholders and user communities to share research and innovation resources and capabilities, technological and administrative know-how, methodologies, etc. Strategic cooperation and sustainable partnerships among the main stakeholders are required to share common resources and to establish urban and regional innovation ecosystems.

Urban areas are considered the object of innovation ecosystems able to empower citizens with capabilities to design, co-create and develop best urban working and living spaces. Innovation in urban green areas can provide solutions by defining quality of service and operational standards and results. An innovation approach based on environmental integration of urban areas combining housing, urban green areas and business activities in a kind of eco-neighborhoods could be supported by a public-private partnership. Urban innovation ecosystems may constitute Public-Private-People-Partnership ecosystems aimed to co-create, experiment and validate scenarios supported technology platforms user-driven involving and providing opportunities citizens, small, medium and large businesses, corporations, local governments, and any other stakeholder.

The smart city innovation ecosystem successfully implemented fosters innovation, cooperation and development into green innovative urban planning and design. An innovative urban policy interrelates issues of green ecosystems innovation, intelligent communities and environmental and sustainable development with urban growth. Waste management and sustainable public transportation using environmental-friendly fuels are only two areas where can be applied smart city innovations supported by proactive citizenship behavior and advanced technologies.

ICT-based applications in urban innovation deploying broadband infrastructure plays an important role to enhancing citizens' quality of life. Technology applications in urban innovation ecosystems in federated platforms but dependent on contexts and user locations can provide support for new urban e-services. ICT has a limited role at processing and integrating real-time information where processes are not based to any great extent on handling transactions in such areas as local public e- government communication between authorities and citizens, innovation, entrepreneurship and social inclusion, education, etc. (Gonzáles & Rossi, 2011).

E-democracy, E-government and digitizing the public administration are using ICT systems and tools aimed to improve the quality of services to citizens. These functions of the model Peripheries are identified in five archetypal smart urban settings: neighborhood, street, square, museum and park and City Hall (Schaffers, Garcia Guzmán, Navarro y Merz (eds.), 2010). Website created to support public awareness campaigns, local and regional government councils, city green planning and environmental organizations, business managers and leaders applying innovative and cost-effective solutions to harnessing the benefits of the green economy, can be conceptualized in general terms as green regions.

Technological innovations can create an efficient urban layout as a human dominate ecosystems in terms of consumption of natural resources, energy, land, water, etc., with the ability to sustain ecosystems services and the structure of social dynamics. There is a race for natural resources that press the earth and create a new conflict between those who want the land to extract natural resources, such as wood, and those who want to conserve them for the sustainability of the environment.

Innovation ecosystems can be fostered and developed through collaboration frameworks by future Internet testbeds integrated with elements of Living Lab environments. Cooperation frameworks include some elements of sharing access to experimentation facilities and technological knowledge resources.

The method of Living Lab concept is used as a model for organizing and conducting innovation projects, programs and experiments. Living labs are organized around experimental research projects driven by internet integrated in smart city programs with common resources for research in green innovation ecosystems and environments. Living labs provide product and services of real social innovation and improvements and lower the risks for future use based on public data of the city.

The social dynamics in the context of an open Living Lab ecosystem should have the purpose of involving citizens and stakeholders to ensure innovative solutions. Living Lab is an ecosystem supported by Public, Private and People Partnership (4 Ps) aiming to co-create innovative scenarios and provide opportunities to users/citizens based on technology platforms. Local urban growth and development partnerships and coalitions have the critical role of building bridges in planning structures of public-public adjacent areas for concerted urbanism, framing renewal and urban green innovation activities.

Development of green area innovations face several challenges although the empirical studies on this subject are very scant (Berchicci & Bodewes, 2005; Hall and Vredenburg, 2003; Ottman et al., 2006). Green innovation is one of the key factors of community development, government management and city governance to achieve economic growth, social development, environmental sustainability, and a better quality of life. Community initiatives are independent of government but can influence government actions to benefits for low-income groups under a framework of governance. Local governments supporting community and non-governmental organizations in a wide range of urban community programs contribute to improve basic services, housing, environmental sustainability management, infrastructure, micro-finance for enterprise development (Boonyabancha 2003, Stein 2001).

## GREEN INNOVATION

The unit of analysis in this study is green areas innovation projects. Theory and practice of green innovation has become a strategic priority for urban settlements and local communities. Multidisciplinary theoretical approaches combined may be used to analyze niche innovations, regimes and landscapes (Geels & Schot 2007). Measuring green areas innovation environmental impact and performance is a complex process. To identify key elements and dimensions such as the design strategy, process and performance integrated to green areas innovation involving micro and macro environmental issues, provides the multi-faceted nature of the phenomenon.

Innovation is regarded as more relevant in urban settings than in rural areas due to the concentration of population, availability of cognitive resources, greater accumulation of knowledge and experiences, etc. Innovation knowledge and practices is concentrated in urban undertakings have the property of transformability of uncertainties into opportunities by realignment of resources and organizational structures. Innovation knowledge can contribute to the resilience of a system by opportunities to realign organizational structures and resources at local urban undertakings.

Some of the issues relating to innovativeness of green areas related to cities developed are the motivations to engage in urban green areas innovation, community-level environmental projects, tools that local governments and communities have used to address sustainability issues, measurement of environmental performances, challenges and risks they face. Some available innovative tools for the analysis of spatiotemporal urban planning and growth patterns are the spatial metrics, remote sensing, urban growth models, etc.

Sources and spatial data of urban geography and urban modeling research and innovative technologies improve the analysis, representation and modeling of urban dynamics (Batty, 1989; Knox, 1994). Green area innovation is related to its physical life cycle (Gauthier, 2005) from the environmental impact perspective it has to population of the community and neighborhoods it is serving. Innovation and wealth creation tends to scale with city population size.

Other examples of green innovation programs are the participatory urban planning projects integrating technology applications for the design and development of an environmental sustainability, bio economy growth and social development by implementing viable programs to reduce urban carbon footprint. Large-scale participatory urban innovation processes are required to create and implement applications aimed to improve urban infrastructure linking cluster of business and organizations with their own activities.

Innovation among local NGOs are working on programs with other community organizations with and in absence of government and small private business entrepreneurs to benefit the urban poor (Hasan 1997, UN-Habitat 2003; Burra & Patel 2002). Combining the approaches for urban innovation not necessarily has altered the state's urban sprawl but may achieve substantive political progress to overcome the environmental-economic crisis.

Dynamic management systems can support the creation and development of natural green innovation for urban green spaces and the most efficient use of natural resources to provide the best practices to the urban green project. The empirical literature reports that the challenges of innovative urban green spaces are very scant (Berchicci & Bodewes, 2005; Hall & Vredenburg, 2003; Ottman et al., 2006). Any green project seeking and encouraging innovative solutions by landowners and governments must be coupled with high mitigation and remediation costs of most dysfunctionalities.

Innovation strategies of cooperation supported by a network and strategic alliances ensure long-term viability of urban projects (Belisent 2010). Farming as a green innovation requires interdisciplinary cooperative exchanges among networks of actors. Green areas innovation offers other relevant opportunities for environmental sustainability, improvement of living conditions, waste reduction, new business creation, etc., and to achieve this goals, it is required to set strategies, policies and targets to move forward and implement the projects. Green areas innovation should be supported by tools to measure environmental impacts such as life cycle analysis at each stage of development.

The conceptual framework for green urban area innovation is based on a multi-faceted process wherein key types of environmental development are in interaction, natural resources, materials, energy, and pollution, etc. which have an impact on the environment at different stages: Analysis, formulation, implementation, disposal and evaluation. Green urban areas innovation is becoming mainstream among the local governments and communities used to protect, strive and enhance the natural environment by conserving natural resources and energy and by reducing or eliminating pollution, waste and toxic agents (Roy et al., 1996).

Innovative forms of funding urban green areas innovation are the financial sources for support coming from the business sector, private donors, agencies promoting development, social organizations, non-governmental organizations, etc. To implement these innovative urban green space strategies, it is necessary to develop management strategies for financing maintenance and developing multifunctional urban green spaces, including technological and cultural innovations.

Green areas innovation is relevant to generate knowledge in transdisciplinary and adaptive innovation in urban planning and design processes such as the framework for "safe-to-fail" adaptive urban planning and design using innovative knowledge (Ahern, 2011; Lister, 2007). A relevant variable to assess the smart cities trends is the demographic density to facilitate contacts and social interactions through the innovative knowledge and ideas flows (Glaeser & Gottlieb, 2006).

A recent approach to this characterization considers the multiple dimensions of innovation as a continuous variable. Incremental green area innovations are more related to the increasing use of existing key green dimensions such as eco-efficiency, the use of materials and processes with a lower environmental impact (Hellström, 2007). Incremental green area innovations are characterized by small and incremental improvements of previous green area versions and their reliance on small changes on existing technologies, process, etc. Green areas research projects originating from business sources expected to respond to the market forces maybe more oriented by incremental innovation.

Radical green areas innovation contributes to achieve environmental sustainability objectives. Introducing urban radical green innovation in any stage of development to addressing urban environmental sustainability challenges across different dimensions such as natural and environmental resources, material selection, energy use, waste usage and pollution prevention can bring substantial benefits to the urban communities and neighborhoods. Green process innovation correlates positively to competitive advantage (Chen et al. 2006).

Open innovation can be user-driven to serve as a mediating, exploratory and democractic participative ITC platform to support urban innovation policies in demand-driven cycles of experimentation. An open innovation in smart cities should be governed by cooperation frameworks within an environment of diverse resources and assets accessible for users. Current projects on smart cities, future internet research and Living Labs have common technological challenges on the use of resources, methods, technologies, facilities and user communities for research and innovation. For example: open innovation models in procurement policies can be developed in order to create sustainable cooperation frameworks aligning technology to societal challenges.

Green innovations in urban farming development exhibit the characteristics of social innovations contributing to a sustainable urban environment and urban food movements. Urban farming is using diverse technologies such as hydroponics or aquaponics to promote green and environmental innovation programs. Urban farming production serves to offer processing, cooking and selling fresh food produce to the local market and enhancing the food business with high quality, sustainable and innovative approaches. Urban farming as an innovation is an incubator using special forms for promoting new concepts of organizing urban life and consumption around sustainable food production. These special forms can be diverse integrated designs to reproduce built environments such as modular containers and components, greenhouses, etc.

Urban farming can bring some ecological and environmental benefits besides the revenues. The innovative farming start-ups are supported by interdisciplinary academic knowledge, business experiences, urban developers, city and local government agencies and financial investors. Large scale high-tech commercial farming initiated by start-ups is facing specific urban land-use challenges regarding urban permitting, zoning, designing and constructing. Tacit knowledge is necessary for projects and contributes to the innovation (Nonaka & Takeuchi, 1995) and creativity to negotiate transition towards integrated risk management where the participants are less likely to negotiate from entrenched positions (Pahl-Wostl et al., 2007).

Integration of community gardens through an innovative approach facilitates opportunities for residents to participate more actively in urban green space planning processes to provide ecosystems services, environmental and sustainable education, alternative and accessible forms for physical activities, bridges interactions between different social groups, enhance local ecological and environmental outcomes.

The concept of smart cities is related to environments of open and user driven innovation respond to challenges. Some European Commission programs are aimed to mobilize urban areas as agents of change to experiment in the smart city concept in open and democratic innovation environments. The smart city concept involves an environment of open and user driven innovation enabled by ICT infrastructure supported by future internet enabled services for experimenting and validating research and innovation.

Future Internet technologies integrate augmented reality services in cultural heritage, safety and security with networks of video-cameras to monitor urban spaces. Future Internet technologies can give support to a platform for monitoring and governance processes of social interactions, development of mobility behaviors, participatory civic decisions, learning natural and cultural heritage and delivery of e-government services.

Urban planning and designing can use the Smart City reference model as an analytical tool to identify innovativeness of policies and processes not only in urban green areas innovation but also in tourism culture, finance, etc. The innovation urban plan must operate under and innovative green plan to include alternative energies, for $CO_2$ reduction, green building policies, green logistics and urban transport. Logistic services are connected with product development facilities with professional users.

Urban innovation ICT programs can stimulate citizens, communities, social organizations, business and other societal applications, scaling-up real-life deployment projects to large-scale levels. Innovative ICT-based services with user-driven innovation link smart cities with experimental infrastructure and facilities to design new applications and green services. The co-creation of green services in different areas requires the employment of innovative devices and customized sensors used by citizens.

Smart city solutions are currently more citizen oriented than local city government based solutions, but the best are the combination of both orientations. The real impact of smart city solutions, the funding mechanisms and business models already developed and implemented for their sustainability has not been demonstrated.

The smart city reference model is used by urban planners to describe the smart innovation elements and characteristics, leading by the conception of greening as urban infrastructure used for environment protection (Atkinson & Castro 2008; Belisent 2010) "interconnected" among the different economic agents (Bell, Jung and Zacharilla, 2009; Bizer, Heath, and Berners-Lee, 2009; Gillett, Lehr, & Osorio, 2004; Ergen, 2009) and "intelligent" as the capacity to produce added value based on the creative human capital (Chee-Yee, and Kumar 2003; Leon, 2006; Florida, 2003). Urban planners engage stakeholders through the formulation and implementation of innovation strategies ensuring democratic participation and outcomes (Oshun et Al 2011).

The trend based on the emergence of civic Internet information systems is an innovative tool socially constructed which can provide benefits for the design and implementation of green urban innovation projects towards facilitating the regeneration and economic development. On the one hand, it is beneficial to facilitating the structural environmental growth by contributing to have up-dated workforce and on the other, proving support to speed the communication marketing process.

The Future Internet Technology is an overlapping implementation of a mixture of technologies, paradigms and time-frames. It is important to consider the assumption that projects and tasks are supported by technical innovation with the latest release of short-lived software. Participation is a bottom-up approach to Future Internet technology integration. Future Internet technology is a complex technological and societal domain using driven processes of innovation, shaping and application for achieving socioeconomic and business benefits.

Radical green innovations include the use of new technologies, components, processes, etc., aimed to reduce and mitigate the overall environmental impact. Green areas innovations are radical if they are new and offer new features based on radical new technologies to the population of communities and neighborhoods. Local communities and governments can find and deal with local demands of citizens by responding with innovative projects in green areas in some specific local urban spaces, managing risks attached and providing greater benefits. More innovative projects are found to be associated with longer waiting times before having some results (Roberts & Hauptman 1987). In this sense, the radicalness of innovation in urban green areas needs to be conceptualized and operationalized in the varying degrees of dimensional urban innovation spaces (Roberts & Berry 1985).

Increasing the environmentally shares has an impact on housing and transportation green innovations in terms of green technological solutions aimed to improve land and energy efficiency, climate change adaption, etc., which contributes to the development of niches and replacing sociotechnical regimes. A radical innovative undertaking relies on technology innovation where the organization, community or local government has technical experience. Using technology in innovation process is related to develop or bring new technological knowledge available in the scientific community. Creation of technological knowledge is more than a technology problem and engaging in new practices to capitalize on the new technology's advantages (Tyre & Hauptman, 1992).

Innovative undertakings require working where the organization, community or local government has more experience to be regarded as radical innovation. Green area projects are more or less radical innovative undertakings on some differentiated dimensions such as the development degree of technology employed, technology costs, technological uncertainty, technical inexperience, business inexperience. Radical innovations in green areas taking less time to be terminated and less tied to market meds represent risks and therefore are less supported by communities and local governments (Souder, 1987).

Consistent theory on radical green areas innovation with empirical research experienced by practitioners is relevant for urban societies, economies and firms. Research on green innovation projects originated exclusively within R&D is more likely to be radical innovations. Green area projects originating from R & D in areas where there is less experience and practice greater uncertainty is considered more radical innovation. Radical innovation has implications for community, organizations, local governments and business development strategies pursued by firms opening new markets (Knight, 1967; Roberts and Berry, 1985).

There is not commonly accepted definition of radical innovation and also its validity and reliability of operationalization and measure is difficult to test by a manager, single-item judgment or a panel of experts. Radical innovation is defined in terms of the degree of innovations and newness (Abernathy, 1978; Hage, 1980). Radical innovation has multiple dimensions and factors as the technological uncertainty, technical experience, business knowledge and experience and technology costs. Radical innovation technologies in innovative undertakings of urban green areas and incur in high technology costs.

The situation that there is not commonly accepted definition of radical innovation makes difficult to compare results of innovation and accumulate knowledge (Downs and Mohr, 1976). According to the degree of change, Damanpour (1991) classifies innovations in different dimensions of radicalness, which in fact are difficult to operationalize. Radical innovation has greater amount of change and impact on the organizations and community (Daft & Becker 1978). The different dimensions of green radical innovation make contributions to explaining the effects upon local community and government.

Radical innovation of green areas through multinational representation of the construct influences other aspects such as innovation management. However, measurement of innovation radicalness is complex and differentiated considering the multiple dimensions, which may be more radical some of them and other less radical.

Radical innovation is best represented as a continuous variable. Description of innovation as radical or incremental is ambiguous and depends of the judgment. The continuo of radical-incremental innovation can be measured as a continuous variable more than restrict the range of the construct to dichotomize innovations only as radical or incremental. Technological innovation has implications on R&D in green areas taking into account that the reliable measures of different dimensions' radicalness provide diagnostic value to formulate and implement planning strategies of projects with the consequent benefits for the local communities and governments. Users of green areas are able to appreciate the benefits, evaluate the risks and face the challenges derived of engaging on innovation process.

Research on radical innovation has focused on technological innovation in contexts where radical and incremental innovation is a challenge (Ettlie, Bridges and O'Keefe, 1984). Technological innovation may be disruptive change because it is experiencing rapid development and have a greater impact on existing practices and experiences on organizational, community, local governments and societal systems (Tyre and Hauptman 1992). Technological innovation as part of the innovation process required to create new knowledge represents a challenge (Schoonhaven, Eisenhardt & Lyman, 1990). This innovation program creates long-term economic growth centered on entrepreneurship, high-tech networked collaboration (Barcelona, 2011, Downs & Mohr, 1976).

Organizations promoting radical innovation require new strategies and structures (Ettlie, Bridges & O'Keefe, 1984). Radical innovation demands knowledge and promotes greater changes upon the organization operation (Dewar & Dutton, 1986). A radical innovation green area project is more likely to fail than an incremental project (Souder, 1987). An innovative radical project's construct in technology management is a fundamental to understand innovation management; however, radical innovation efforts to increase product development are more likely to fail (Schoonhaven, Eisenhardt, & Lyman, 1990).

Radical technological innovation creates new knowledge and practices similar to administrative innovations. Radical innovations are usually more expensive and require more resources to successfully commercialize (Ettlie & Rubenstein, 1987). Innovation in green areas has more impact when the technology is more expensive and requires more technical experience. Radical innovation has a greater demand for resources and knowledge generation. However, radical innovations well financed are more likely to survive (Roberts & Hauptman, 1987).

Communities pursuing incremental and radical innovation have effects on performance (Damanpour, 1991; Meyer & Roberts, 1986) and for innovation management. Radical innovation relies in rapid developing technology knowledge making the innovative undertaking riskier if the agent and actor lack experience. The more they have inexperience, the more they are regarded as a radical innovation (McGeough and Newman 2004).

Radical innovation in green areas creates more change and has more impact on the community and neighborhoods. The amount of change attached to the green innovation projects where there is not extensive technological knowledge and experience make them very costly. Radical innovation in green areas brings significant changes such as making old such as emerging, transforming and disappearing the old and conventional forms and processes (Kaplan, 1999; Van de Ven et al., 1999).

Radical innovations create greater knowledge demands for the organizations, communities, local governments and the society in general. Radical innovations on green areas demand greater change in the existing practices are a source of change and have bigger impact on society (Damanpour, 1991).

Radical innovation incorporates technology that is a risky departure from existing practice of green areas where the magnitude or cost of change required may be sufficient (Ettlie, Bridges & O'Keefe, 1984).

Technology costs of radical innovative undertakings could be so high. The degree of radical innovation is related to the embryonic technology that is new and rapidly developing that represents a departure from existing practices with the corresponding financial risk. In radical innovation the green areas undertaking is costly and venturing into uncertainty because the lack or underdevelopment of knowledge. Radical innovation is riskier that incremental innovation.

Innovation environments should be open to develop collaboration models for sharing actual common resources used in research and innovation processes of living lab facilities and experimenting methods on future internet technologies for implementing urban innovation policies and ecosystems. Collaboration framework structures are developed for sharing research and innovation resources considering ownership and governance, access, transferability and interoperability of urban innovation projects.

Living Lab integrates services in urban contexts where users and citizens define and prioritize elements of urban cultural heritage and explore security and private issues for the safety of urban environments. The Living Lab-convergent service platforms are developing in a discovery-driven arena settings. Projects requiring radical technological innovation are associated with longer waiting times to complete and before there is a desired product (Roberts & Hauptman, 1987; Schoonhaven *et al.* 1990).

Living Labs Interfaces projects engage users in co-creation innovation processes whereas Future Internet Experimentation (FIRE) projects involve end users and communities in assessing the socio-economic impacts of technological changes and controlled innovation technologies. FIRE and Living Lab projects use some models for resources sharing in experimentation and innovation opportunities in user's communities.

Living Labs innovation projects share the use of some methodologies with the interface with FIRE experimentation approaches. The application of some technologies in the context of a Living Lab is leading to the adaptation of innovative solutions enabled by the emergence of breakthrough concepts, ideas, and scenarios. Living Labs Innovation and Future Internet Experiments requires infrastructure to create opportunities for urban green innovative services, wellbeing and mobility services in the context of urban space based on real-time digital data.

Collaboration between researchers, practitioners and policymakers to design and implement innovative urban planning can be sustained by a policy network to share information, solve problems and find policies. Adaptive design explores creative and innovative practices through design experiments. Adaptive design can be used as a framework for process to selected urban projects and open to design innovations and creativity, to gain innovative ecological knowledge, research design, practices and methods. An adaptive model in innovative urban green spaces strategies follows a systemic approach with inputs, processing and outputs to the local urban context. From the perspective of innovation, Bachelor and Butterworth (2008) the term learning alliance is used to define a group of individuals or organizations with a shared interest in innovation and the scaling-up of innovation.

The national systems of innovation approach sustain that the urban value creation system is shaped by physical and immaterial infrastructure, networks and collaboration, entrepreneurial climate and business networks, demand for services and availability of advanced end-users. Networks can influence policy processes and change the regime (Van Herk et al., 2011). The value creation system is affected by policy implemented to stimulate the networks, public-private partnerships, and the enhancement of innovative conditions (Porter, 1990).

Innovation is fostered through the collaborative innovation processes between the interaction of clusters and networks. It is important to develop collaboration in green areas innovation with the actors and agents of local social and ecological systems. Collaboration within the innovation is an ongoing interaction process between technology, research and applications development, validation and practical utilization. Cooperation relationships frameworks and synergy linkages should be developed between urban innovation policies, local government ICT and future internet research and open users. Future Internet research is a competitive offer proving its added value to citizens/users. Future Internet enables co-creation of innovative scenarios by users and citizens who may contribute a build new applications and public data to the open city.

## CONCLUSION

The analysis sheds light on environmental sustainability and urban ecosystems as dimensions of green innovation, which includes the management and use of natural resources, energy minimization, materials reduction, and pollution prevention. The interest of this analysis lies in providing support to urban settlements in managing the risks inherent in green area innovation, incremental or radical as a community's management would experience. Citizens and public managers must understand the risks inherent in pursuing either radical or incremental innovations.

Competitiveness based on research and innovation skills to promote knowledge bio economy, is one of the urban development challenges. Transition toward sustainable urban development innovations should be aimed to changing the composition of land use, infrastructure and transportation regimes. Cooperation strategic models aimed to provide access to shared research and innovation resources in urban areas ecosystems and environments are needed to create and develop more suitable working and living urban spaces.

Returning to the urban public space that needs to be reconquered. The urban public space must be reconstructed with a sense of collectivity, as an urban space of coexistence, of emotion, of trust, of looking, and it is the space of embracing. They are all urban spaces that must be built and, therefore, that urban space is a great achievement at this time.

Local communities and governments should initiate selectively some green area projects with different degrees of radicalness dimensions following a profile according to a strategic urban planning. Development of sustainable communities in smart cities can be supported by design, transfer and implementation through collaborative urban planning of innovative urban policies.

Future priorities for research in urban innovation should identify and develop principles of sustainable urban green planning and development to provide support to policy makers in local governments to design and implement mechanisms for more resilient communities and neighborhoods in cities. Green resilience innovation of urban life in contemporary cities can be supported in digital cities with the implementation of cyberspace to cities.

# REFERENCES

Abernathy, W. J. (1978). *The Productivity Dilemma*. Baltimore, MD: Johns Hopkins Univ. Press.

Adams, W. (2006). *The future of sustainability, re-thinking environment and development in the twentyfirst century*. The World Conservation Union.

Ahern, J. (2011). From fail-safe to safe-to-fail: Sustainability and resilience in the new urban world. *Landscape and Urban Planning*, *100*(4), 341–343. doi:10.1016/j.landurbplan.2011.02.021

Ahern, J. (2012). Urban landscape sustainability and resilience: The promise and challenges of integrating ecology with urban planning and design. *Landscape Ecology*, *28*(6), 1203-1212. doi:10.100710980-012-9799-z

Aherna, J., Cilliers, S., & Niemeläc, J. (2014). The concept of ecosystem services in adaptive urban planning and design: A framework for supporting innovation. *Landscape and Urban Planning*, *125*, 254–259. doi:10.1016/j.landurbplan.2014.01.020

Atkinson, R., & Castro, D. (2008). *Digital quality of life*. The Information Technology and Innovation Foundation.

Baines, C., & Smart, J. (1991). *A Guide to Habitat Creation*. London Ecology Unit.

Barcelona. (2011). *Smart city 22*. http://www.22barcelona.com

Batchelor, C., & Butterworth, J. (2008). *Learning Alliance Briefing Note 9: Visioning* (draft). http://www.switchurbanwater.eu/la_guidance.php

Batty, M. (2008). The size, scale, and shape of cities. *Science*, *319*(5864), 769–771. doi:10.1126cience.1151419 PMID:18258906

Belisent, J. (2010). *Getting clever about smart cities: new opportunities require new business models*. Forester for Ventor Strategy Professionals.

Bell, R., Jung, J., & Zacharilla, L. (2009). *Broadband economies: creating the community of the 21st century*. Intelligent Community Forum.

Berchicci, L., & Bodewes, W. (2005). Bridging Environmental Issues with New Product Development. *Business Strategy and the Environment*, *14*(5), 272–285. doi:10.1002/bse.488

Berger, I. E., Cunningham, P. H., & Drumwright, M. E. (2007). Mainstreaming Corporate Social Responsibility: Developing Markets for Virtue. *California Management Review*, *49*(4), 132–157. doi:10.2307/41166409

Bizer, C., Heath, T., & Berners-Lee, T. (2009). Linked data – the story so far. *International Journal on Semantic Web and Information Systems*. http://linkeddata.org/docs/ijswis-special-issue

Blewitt, J. (2008). *Understanding sustainable development*. London: Earthscan.

Bolund, P., & Hunhammar, S. (1999). Ecosystem services in urban areas. *Ecological Economics*, *29*(2), 293–301. doi:10.1016/S0921-8009(99)00013-0

Boonyabancha, S. (2003). A decade of change: From the Urban Community Development Office (UCDO) to the Community Organizations Development Institute (CODI) in Thailand. Poverty Reduction in Urban Areas working paper 12. International Institute for Environment and Development.

Burra, S., & Patel, S. (2002). Community toilets in Pune and other Indian Cities. *Participatory Learning and Action*, *44*, 43–45.

Chee-Yee, C., & Kumar, S. (2003). Sensor networks: Evolution, opportunities, and challenges. *Proceedings of the IEEE*, *91*(8), 1247–1256. doi:10.1109/JPROC.2003.814918

Chen, Y. S., Lai, S. B., & Wen, C. T. (2006). The influence of green innovation performance on corporate advantage in Taiwan. *Journal of Business Ethics*, *67*(4), 331–339. doi:10.100710551-006-9025-5

Costanza, R., d'Arge, R., De Groot, R., Farber, S., Grasso, M., Hannon, B., ... Paruelo, J. (1998). The value of the world's ecosystem 2 services and natural capital. *Ecological Economics*, *25*(1), 3-15.

Cozens, M. (2008). New urbanism, crime and the suburbs: A review of the evidence. *Urban Policy and Research*, *26*(4), 429–444. doi:10.1080/08111140802084759

Daft, R. L., & Becker, S. W. (1978). *The Innovative Organization*. New York: Elsevier.

Damanpour, F. (1991, September). Organizational innovation: A meta-analysis of effects of determinants and moderators. *Academy of Management Journal*, *34*, 555–590.

Dewar, R. D., & Dutton, J. E. (1986). The adoption of radical and incremental innovations: An empirical analysis. *Management Science*, *32*(11), 1422–1433. doi:10.1287/mnsc.32.11.1422

Downs, G. W., & Mohr, L. B. (1976). Conceptual issues in the study of innovation. Admin. Sci. Quart., 21.

Elmqvist, T., Colding, J., Barthel, S., Borgström, S., Duit, A., Lundberg, J., ... Bengtsson, J. (2004). The dynamics of social–ecological systems in urban landscapes: Stockholm and the National Urban Park, Sweden. *Annals of the New York Academy of Sciences*, *1023*(1), 308–322. doi:10.1196/annals.1319.017 PMID:15253913

Ergen, M. (2009). *Mobile broadband—including WiMAX and LTE*. Springer.

Ernstson, H., Sörlin, S., & Elmqvist, T. (2008). Social movements and ecosystem services–the role of social network structure in protecting and managing urban green areas in Stockholm. *Ecology and Society, 13*(2), 39. https://www.ecologyandsociety.org/vol13/iss2/art39/

Ettlie, J. E., Bridges, W. P., & O'Keefe, E. D. (1984, June). Organization strategy and structural differences for radical versus incremental innovation. *Management Science*, *30*(6), 682495. doi:10.1287/mnsc.30.6.682

Ettlie, J. E., & Rubenstein, A. H. (1987, June). Firm size and product innovation. *Journal of Product Innovation Management*, *4*(2), 89–108. doi:10.1111/1540-5885.420089

Evans, R. G., & Stoddart, G. L. (1990). Producing health, consuming health care. *Social Science & Medicine*, *31*(12), 1347–1363. doi:10.1016/0277-9536(90)90074-3 PMID:2126895

Florida, R. (2003). *The rise of the creative class and how it's transforming work, leisure, community and everyday life*. Perseus Books Group.

Gauthier, C. (2005). Measuring Corporate Social and Environmental Performance: The Extended Life-Cycle Assessment. *Journal of Business Ethics, 59*(1/2), 199–206. doi:10.100710551-005-3416-x

Geels, F. W., & Schot, J. (2007). Typology of sociotechnical transition pathways. *Research Policy, 36*(3), 399–417. doi:10.1016/j.respol.2007.01.003

Gillett, E., Lehr, H., & Osorio, C. (2004). Local government broadband initiatives. *Telecommunications Policy, 28*(7), 537–558. doi:10.1016/j.telpol.2004.05.001

Glaeser, E., & Gottlieb, J. D. (2006). Urban Resurgence and the Consumer City. *Urban Studies (Edinburgh, Scotland), 43*(8), 1275–1299. doi:10.1080/00420980600775683

González, J. A., & Rossi, A. (2011). *New trends for smart cities, open innovation Mechanism in Smart Cities. European commission within the ICT policy support programme.* http://opencities.net/sites/opencities.net/ðles/content-ðles/repository/D2.2.21%20New%20trends%20for%20Smart%20Cities.pdf

Hage, J. (1980). *Theories of Organization: Form, Process and Transformation*. New York: Wiley.

Hall, J., & Vredenburg, H. (2003, Fall). The Challenges of Innovating for Sustainable Development. *MIT Sloan Management Review*, 61–68.

Harmer, R. (1999). Creating New Native Woodlands: Turning Ideas into Reality. Forestry Commission Information Note 15.

Hasan, A. (1997). *Working with government: The story of OPP's collaboration with state agencies for replicating its Low Cost Sanitation Programme*. Karachi, Pakistan: City Press.

Hellström, T. (2007). Dimensions of Environmentally Sustainable Innovation: The Structure of Eco-Innovation Concepts. *Sustainable Development, 15*(3), 148–159. doi:10.1002d.309

Holling, C. S. (1978). *Adaptive environmental assessment and management*. New York: John Wiley.

Kaplan, S. (1999). Discontinuous Innovation and the Growth Paradox. *Strategy and Leadership, 27*(2), 16–21. doi:10.1108/eb054631

Kato, S., & Ahern, J. (2008). Learning by doing: Adaptive planning as a strategy to address uncertainty in planning. *Environment & Planning, 51*(4), 543–559. doi:10.1080/09640560802117028

Knight, K. E. (1967, October). A descnptive model of the intra-firm innovation process. *The Journal of Business, 40*(4), 478496. doi:10.1086/295013

Knox, P. L. (1994). *Urbanization: introduction to urban geography*. New Jersey: Prentice Hall.

Komninos, N. (2008a). *Intelligent Cities and Globalisation of Innovation Networks*. London: Routledge. doi:10.4324/9780203894491

Komninos, N. (2008b). *Intelligent cities and global innovation networks*. London: Routledge. doi:10.4324/9780203894491

Komninos, N., & Sefertzi, E. (2009). *Intelligent cities: R&D offshoring, web 2.0 product development and globalization of innovation systems*. Second Knowledge Cities.

Kuhn, E. P., Colberg, P. J., Schnoor, J. L., Wanner, O., Zehnder, A. J. B., & Schwarzenbach, R. P. (1985). Microbial transformation of substitute benzenes during infiltration of river water to groundwater: Laboratory column studies. *Environmental Science & Technology, 19*(10), 961–967. doi:10.1021/es00140a013

Leon, N. (2006). *The well connected city A report on municipal networks Supported by The Cloud*. Imperial College London.

Lister, N. M. (2007). Sustainable large parks: Ecological design or designer ecology? In G. Hargreaves & J. Czerniak (Eds.), *Large parks* (pp. 35–54). New York: Architectural Press.

McGeough, U., & Newman, D. (2004). *Model for sustainable urban design with expanded sections on distributed energy resources*. Sustainable Energy Planning Office Gas Technology Institute and Oak Ridge National Laboratory GTI Project # 30803-23/88018/65952.

Meyer, M. H., & Roberts, E. B. (1986). New product strategy in small cal issues and applications of stabstical techniques, technology-based firms A pilot study. *Management Science, 32*. doi:10.1287/mnsc.32.7.806

Millenium Ecosystem Assessment. (2005). Ecosystems and human well-being: synthesis; Mukul A. Lack of funds hampers social science research. *Times of India*. Retrieved from https://timesofindia.indiatimes.com/india/Lack-of-funds-hamperssocial-science-research/articleshow/10237494.cms

Niemelä, J., Saarela, S. -R., Söderman, T., Kopperoinen, L., Yli-Pelkonen, V., Väre, S., & Kotze, D. (2010). Using the ecosystem services approach for better planning and 13 conservation of urban green spaces: A Finland case study. *Biodiversity and Conservation, 19*(11), 3225-3243.

Nonaka, I., & Takeuchi, H. (1995). *The Knowledge-Creating Company*. Oxford, UK: Oxford University Press.

Oh, H., Chung, M. H., & Labianca, G. (2004). Group social capital and group effectiveness: The role of informal socializing ties. *Academy of Management Journal, 47*(6), 860–875. doi:10.5465/20159627

Oshun, M., Ardoin, N., & Ryan, S. (2011). *Use of the planning outreach liaison model in the neighborhood planning process: a case study in Seattle's Rainier Valley neighborhood*. Urban Stud Res.

Ottman, J. A., Stafford, E. R., & Hartman, C. L. (2006). Green Marketing Myopia. *Environment, 48*(5), 22–36. doi:10.3200/ENVT.48.5.22-36

Otto, U. (1998). Innovative Qualität statt neues Etikett: Erste Erfahrungen mit einer konsequenten Programmatik Bürgerschaftlichen Engagements. In J. Braun & O. Klemmert (Eds.), *ISAB-Schriftenreihe: Vol. 54. Selbsthilfeförderung und bürgerschaftliches Engagement in Städten und Kreisen. Fachtagung des Bundesministeriums für Familie, Senioren, Frauen und Jugend am 16./17. Februar 1998 in Bonn* (pp. 215–235). Köln: ISAB.

Pahl-Wostl, C., Craps, M., Dewulf, A., Mostert, E., Tabara, D., & Taillieu, T. (2007). Social learning and water resources management. *Ecology and Society, 12*(2), 5. https://www.ecologyandsociety.org/vol12/iss2/art5/

Porter, M. (1990). *The Competitive Advantage of Nations*. New York: Free Press. doi:10.1007/978-1-349-11336-1

Porter, M., & Reinhardt, F. L. (2007). A Strategic Approach to Climate. *Harvard Business Review*, *85*(10), 22–26.

Roberts, E. B., & Berry, C. A. (1985). *Entering new businesses: Selecting strategies for success. Sloan Manage Rev*, 3–17.

Roberts, E. B., & Hauptman, O. (1987, March). The financing threshold effect on success and failure of biomedical and pharmaceutical star-ups. *Management Science*, *33*(3), 381–394. doi:10.1287/mnsc.33.3.381

Rottle, N., & Yocom, K. (2010). *Ecological design*. Lausanne: AVA Publishing.

Roy, R., Wield, D., Gardiner, J. P., & Potter, S. (1996). *Innovative Product Development*. Milton Keynes: The Open University.

Santos, B. (2007). *Crítica da razão indolente: contra o desperdício da experiência*. São Paulo: Cortez.

Schaffers, H., Garcia Guzmán, J., Navarro, M., & Merz, C. (Eds.). (2010). *Living Labs for Rural Development*. Madrid: TRAGSA. http://www.c-rural.eu

Scheffer, M., & Westley, F. (2007). The evolutionary basis of rigidity: locks in cells, minds, and society. *Ecology and Society*, *12*(2), 36. https://www.ecologyandsociety.org/vol12/iss2/art36/

Schoonhaven, C. B., Eisenhardt, K. M., & Lyman, K. (1990, March). Speeding degrees in organizational studies from the Univerproducts to market Waiting time to first product introductions in new sity of San Francisco firms. *Administrative Science Quarterly*, *35*, 177–207. doi:10.2307/2393555

Shapiro, J. (2003). *Smart cities: explaining the relationship between city growth and human capital*. Harvard University.

Smart city Edinburgh. (2011). http://www.edinburgh.gov.uk

Smith, A., Voss, J. P., & Grin, J. (2010). Innovation studies and sustainability transitions: The allure of the multi-level perspective and its challenges. *Research Policy*, *39*(4), 435–448. doi:10.1016/j.respol.2010.01.023

Souder, W. (1987). *Managing New Product Innovation*. Lexington: MA Lex.

Stein, A. (2001). *Participation and sustainability in social projects: The experience of the Local Development Programme (PRODEL) in Nicaragua*. IIED working paper 3 on poverty reduction in urban areas. International Institute for Environment and Development, London.

Tidball, K. G., & Krasny, M. E. (2009). *From risk to resilience: what role for community greening and civic ecology in cities?* Wageningen: Wageningen Academic Publishers.

Tyre, M. J., & Hauptman, O. (1992). Effectiveness of organizational responses to technological change in the production process. *Organization Science*, *3*(3), 301–320. doi:10.1287/orsc.3.3.301

UN-Habitat. (2003). *Water and sanitation in the world's cities: Local action for global goals*. London: Earthscan.

Van de Ven, A. H., Polley, D. E., Garud, R., & Venkataraman, S. (1999). *The Innovation Journey*. New York: Oxford University Press.

Van Herk, S., Zevenbergen, C., Rijke, J., Ashley, R., (2011). Collaborative research to support transition towards integrating flood risk management in urban development. *Journal of Flood Risk Management*, *4*(4), 306 - 317.

## KEY TERMS AND DEFINITIONS

**Ecosystem:** Biological system constituted by a community of living beings and the natural environment in which they live.

**Environmental:** Of the environment or related to it.

**Green Innovation:** The creation and diffusion of technological means to remedy climate change.

**Sustainability:** It refers to biological systems that can preserve diversity and productivity over time.

**Sustainable City:** That city where there is adequate mobility, saving energy and water resources, reducing auditory pollution and creating pleasant public spaces where there are green areas with great functionality.

**Urban:** Of the city or related to it.

**Urban Space:** It is the proper space of a city, that is, of a population grouping of high density. It is characterized by having an infrastructure so that this large number of people can cope harmoniously in their daily lives.

# Chapter 7
# Cinematic Works–Based Affective Urban Design Atmospheres

**Hisham Abusaada**
https://orcid.org/0000-0001-6530-7714
*Housing and Building National Research Center (HBRC), Egypt*

## ABSTRACT

*This chapter examines the dilemma of using the term "atmospheres" related to architectural history. It theorises the nature of this relationship, developing an analytical framework creating the architecture of the city as similar to artwork. In this chapter, the authors investigated through the aspects of cinematic works—ideas, themes, and dramatic text—and overarching effects of the technical elements. The question is: How can the urban designer use the artworks in the field of urban design? This chapter discusses the atmospheres in many artworks of Western and Egyptian thought to explore the effect of the architecture of cities in creating the atmospheres of the cities.*

## INTRODUCTION

Many scholars in the fields of the architecture of the city create places based on the "time and not space" (Till, 2009, pp. 95-96), where some see time as "crucial" (Gehl, 1987; Gehl & Svarre, 2013, p. 101). In this vein, many scholars innovated the ideas such as serial vision (Cullen, 1961), situate essential events in the form of a narrative event (Tschumi, 1981; Tschumi, 1994), reconsidering an urban experience as a series of events, and collection of experiences (Norberg-Schulz, 1988). Further, based on respecting the characteristics of the rhythms; time emerges through "rhythmic repetition", "not repetition but change" (Lynch, 1972, p. 65), as well as "events recur without strict measure, repetition is not necessarily" (Lefebvre, 2004, p. 6). Any aesthetics experience has three characteristics completeness, uniqueness, and providing quality (Dewey, 2005, p. 44). These characteristics achieve based on "rhythm" that identifies with the regularity "of recurrence amid changing elements" (Dewey, 2005, p. 168). Also, Dewey sees "rhythm is rationality among qualities" (Dewey, 2005, p. 174).

DOI: 10.4018/978-1-7998-3576-9.ch007

*Cinematic Works-Based Affective Urban Design Atmospheres*

The progress of modern technology and the developments in human needs represents two matters to shape architecture (Böhme, 1993). Böhme is considered the daily experience a critical figure in the realms of 'atmospheric architecture' and 'urban ambiences'. The drama of life and everyday experience was creating a different atmosphere that reflects on people's mood. This way of thinking is emerging some of the similarity between architecture and artworks.

Urban design is inspired from a variety of experiences, consequently, stirring up the experiences is an essential part of education in this field. Therefore, the screening of some selected film scenes located the students in a different experience that he/she may never feel during his or her life. Through the multitude films scenes, students can carry inspiration from distinct experiences that are compatible with the nature of their designs, and more practical solutions can be achieved.

The manuscript objective is to confirm that the creation of the atmospheres of cities as plausible as it occurs in artworks; both having the same aspects for producing; which are ideas, themes, and dramatic text, as well as the technical elements. The next focuses on revealing the contribution of the artworks, mainly, in the cinematic works to understand the impact of societal transformations on everyday events according to different urban places. This understanding considers the base to teach students how to design urban spaces according to the scenario of events like what the directors make in the films. The impact of the atmospheres in literature and artworks is the core of this work, which discusses the validity of the hypothesis that an architecture of the city can create a differentiated or possibly immortal atmospheres. The appearance of the indicators that link between the atmospheres in four Egyptian cinematic works related to the change of architectural history is the central theme for the theoretical analysis. The analytical reading focuses on the narrative of the Egyptian novelist Naguib Mahfouz. The discussion will focus on the written novel and the film that appeared based on the stories.

## THE ATMOSPHERES IN THREE TRACKS

The following defines the concept of the atmospheres in Encyclopedias, dictionaries, and Western civilizations, such as Greek, Roman Empire, and at the beginning of modernity. It also shows the atmospheres in narrative artworks, as well as related to the urban design dimensions.

### In Encyclopedias, Dictionaries, and the Western Civilizations

Ambience is an expression in American English meaning 'atmosphere'. It appears in the metaphorical sense of mood, associated with air and the quality or condition of the surrounding space, internally or externally. It can be accepted as synonymous with 'la ambience' in French and 'ambience' in British English to show an excellent moral and a fantastic, impressive, or exciting atmosphere. Also, it uses to express on the affective persons, things, objects, spaces, places, or events. Some online English dictionaries show the development of the meaning of 'atmosphere' over time (Business Dictionary, 2018; Dictionary.com, 2018; Cayne, 1998; Dictionary.com, 2018; Merriam Webster, 2018; Stevenson, 2010; Simpson & Proffitt, 1997). The word ambient emerged "in Late 16th century: from French ambient or Latin ambient- 'going round'"; it relates to "the immediate surroundings of something" (Stevenson, 2010, p. 59). Also, it includes ambient greenhouse environment, surrounding environment, water environment, the temperature of surrounding sand, climatic conditions, ambient background noise, ambient temperature and surrounding music.

In the late 19th century, the word 'atmospheres' was derived from "French ambiance"; which appeared for the first time as an expression that means "surrounding", as well as used in art writing to express "arrangement". The atmosphere "look and feel of a place (mall, restaurant, tourist destination, etc.) that evokes a unique atmosphere or mood and makes it 'sticky' (compels people to come and stay around)". Ambient identified in the sense of a dominant tone or tone of expression to express a specific position or atmosphere surrounding a place, space, or object (Cayne, 1998).

The synonyms of the word atmosphere as an exceptional quality or impression according to *Merriam-Webster Dictionary* on the internet are air, aroma, aura, climate, flavor, halo, karma, mood, nimbus, note, odor, patina, smell, temper, vibration. Also, related words are aureole (or aureole), mystique, romance; genius loci; feel, feeling, sensation, sense, spirit; attribute, character, characteristic, image, mark, notion, peculiarity, picture, property, trait; color, illusion, overtone, semblance, suggestion, tone. Synonyms are circumstances, situations and things are the environment and the center and the context of speech and surrounding areas. The relevant words, in this vein, are: location, place, space, background, element, state, geography, habitat and small environment.

According to the Greek, the Roman Empire, and at the beginning of the modernity, space linked with the concept of the atmospheres in our contemporary life, such as the natural and built environment, the social world, and the aesthetic worlds that include the artefacts. The French use ambiances, meaning atmospheres to explain how the creation of specific situations is. A word ambience began in the time of the Latin Empire to describe the meaning of "to surround or go around" (Thibaud, 2002, pp. 1, 8). In the 17th-century, the word atmosphere is derived from Latin *Atmosphaera* that combines the Greek ἀτμός (atmos) meaning 'vapor' and σφαῖρα (sphaira) meaning 'sphere' (Gandy, 2017, p. 354). Steven Connor "traces the Greek atmos to the original Sanskrit atman", which explain the meaning of the atmospheres is "'breath', 'life', or 'soul'. John Wilkins is considered the first one who used the word 'atmosphere' in the year 1638 to be "the earliest recorded English usage is in a scientific tract" (p. 354). The atmosphere used as a term in physics; it appeared in the article that entitled "The Ambience of Ambience". In addition, Isaac Newton (1643-1727) called the great void of the universe that comprises the planets, stars, and galaxies through which orbits pass as "the atmosphere" (Jaaniste, 2010).

The phrase atmosphere used between 1860-1863, in Gottfried Semper' book that entitled *Style in the Technical and Tectonic Arts, Or, Practical Aesthetics,* translated by Harry Francis Mallgrave and Michael Robinson, with translation from Latin and Greek by Amir Baghdadchi. In addition, Simper, used the atmospheres to describe two entirely different things. The first, when he pointed out that it is a word derived from the meaning of the real atmospheres, which accompanied "the development of high Greek drama" at the time of "the birth of classical Greek temple" (Semper, 2004, p. 50). Semper described this case as "the true atmosphere of art" (Semper, 2004.). The second deals with the image of the surrounding spatial environment, conceived that it is acceptable in general to say, "a comet's tail is an atmosphere" created either "the partial evaporation or burning of the heavenly body when it is close to the sun" (p. 99). At the end footnote of Semper book, he wrote that every artistic creation and pleasure

*presumes a certain carnival spirit, or to express it in a modern way, the haze of carnival candles is the true atmosphere of art. (Semper, 2004, p. 438)*

Moreover, he believes that the atmosphere as an independent human creation needs "the destruction of reality, of the material [which occurs] ...if the form is to emerge as a meaningful symbol" (p. 349).

*Cinematic Works-Based Affective Urban Design Atmospheres*

The Modern movement focuses on the progress, development, and creativity, which is aimed at changing the patterns of thinking in the Enlightenment era. It is intended to create more dynamic and rational different atmospheres rather the depressed traditional image in the middle age, which was adopted a higher reproduction situation more than the past. At the mid of the nineteenth century, positively in 1851, as Jenner quoting, when Mary Merrifield used the atmosphere to describe the grand exhibition of the Crystal Palace in London (Jenner, 2013, p. 14). The atmosphere was palpable, and it was referring to harmony in colors. She describes the atmosphere as saying that the parts of the building are covered with blue-colored fog as if open to the air. It was contrasting with the warm color of the cloth and the ceiling with light blue beams, and the blue sky above the church wing produces a very satisfactory and at the same time very natural appearance. It is difficult to distinguish where art begins, and nature ends. In this context, Joseph Paxton (1869) designed 'the Crystal Palace' as an "icon of Empire exhibits par excellence" (Sturm & Turner, 2012, p. 26). Besides, the architect of the Frankfurt Opera House described the Crystal Palace as a "piece of the sculptural atmosphere" located in London. Followed by comparing between the floating elements of the Crystal Palace with the Turner's painting that "dematerialized the landscape and dissolved it into infinity" (Jenner, 2013, p. 14), as Sigfried Giedion compared the airy, and "through a humid atmosphere" (Jenner, 2013).

## In the Narrative Artworks Thought

In a dictionary of the field of the narrative that entitled *A Glossary of Literary Terms* review set of examples about the changes of the atmospheres. In the literature manuscripts it depends on two variables are the first deals with the changes of the scene, while the second focuses on the developing of the events and the characters (Abrams & Harpham, 2011, pp. 14-16). Based on this dictionary, the atmospheres are dealing with "the emotional tone pervading a section or the whole of a literary work". These tones "fosters in the reader expectations as to the course of events, whether happy or (more commonly) terrifying or disastrous" (p. 19). This dictionary includes many examples of how to create the atmospheres in the West relate to people and place, as follows: In William Shakespeare's tragic novel Hamlet, which was published in 1609, he sets up the spooky and foreboding atmosphere. This work tells the story of the prince of Denmark, who shows him the spectre of his father, the king, asking for revenge of his death; Shakespeare prepares a terrifying atmosphere through a nervous, convulsive, and delusional dialogue started by the guard as he expects the ghost to appear. In the short story *The Pit and the Pendulum* for Edgar Allan Poe (1842), a man is sentenced to death during the Spanish Inquisition. The first paragraph of the story sets up an atmosphere of horror, especially in the final line where the black-robed judges appear whiter than a sheet and "thin even to grotesqueness" (Poe, 2017, p. 114). The reader can only expect a tale of horror to ensure "for a while" (p. 114), a terrible atmosphere reflects the case of the absence of justice. Allan Poe wrote:

*They appeared to me white — whiter than the sheet upon which I trace these words — and thin even to grotesqueness; thin with the intensity of their expression of firmness— of immovable resolution— of stern contempt of human torture. (Poe, 2017, p. 114)*

This story focuses on the theme around arguments about the public role (and publicity value) of the penitentiary. The spectacle of punishment remains a central cultural goal and tool of the prison even as the physical punishment itself is hidden behind the prison walls (Haslam, 2008, p. 270). The atmospheres relate to place appears in two works: a) in what Coleridge induces a combination of religious and superstitious terror by his description of the first scene in the narrative poem *Christabel* (1816). b) in Hardy's novel *The Return of the Native* (1878) makes Egdon Heath an immense and brooding presence that reduces to pettiness and futility the human struggle for happiness for which is the setting. The atmospheres in the novel that turned to cinematic works create based on the visions of the director, and afterwards, mainly, based on the viewpoint of the production designer.

For instance, in a novel of "Great Gatsby" that was published for the first time in 1925, Kelsey Egan wrote in her article entitled 'Film Production Design: Case Study of The Great Gatsby', according to what the American writer Francs Fitzgerald wrote that this era is known as 'the Jazz era,' the production designer guides the work to create the atmospheres in the film (Egan, 2014, p. 7).

In 2013, the new film was appeared (Fisher, Knapman, Luhrmann, Martin, & Wick, 2013). In this version, Catherine Martin was the production designer; she tried to create the 1920s atmospheres based on developing the spatial spaces to change the context of the events carefully, mainly, by focusing on the realm of architecture styles. Catherine determined to revive the style of visual arts, architecture, and design called Art Deco, which first appeared in France before the First World War. Three essential elements characterize this style, and these are bold geometric shapes, luxurious decorations, and industrial aspects. In addition, Catherine's visual influences are those that dominated the machine age, and this is demonstrated in the Gatsby Palace. The theme of the novel revolves around the idea of the falsity of the upper class, and the neo-rich are portrayed as vulgar and disgraceful. To create the right atmospheres in the film, the set designers have the upper-class characters living in immense palaces with high ceilings and cold marble floors, while the lower-class characters live in lower-quality housing and apartments. In the film, Nick Karrawee's residence looks welcoming with the beautiful white-flowered corridor, the symmetry between the round tree benches, and the well-preserved garden. The living room is surrounded by low ceilings and beams of sawn oak, with stucco-eco furniture and terracotta tiles spread throughout the room surrounded by a hearth with oak furniture that gives a sense of nature and purity. Katharine Martin wrote everything is about how to find what we thought were the core Long Island jewels. The quality of the small house with the intimate atmospheres of private rooms creates a space of content that reflects the character of Nick Carraway.

There are a few previous types of research using the word "atmospheres" in urban design (Stefansdottir, 2017). These fields are addressing the relationships between people and place based on the visual, aesthetics, and perceptual dimensions It focuses on the non-physical aspects, such as sensual and spiritual quality. The fifth-dimension serves the aspect related to the phrase "the spirit of architecture"; it implies the sense of the term "atmospheres" (Abusada & Elshater, 2018). Although they are non-physical and invisible dimensions, both play an influential role in the design of the city. The atmosphere is the essence and stands side-by-side with the tenth dimension of urban design (the teleological dimension) and corresponds to the fifth dimension, achieving the aura of the object, the living spirit, the genius loci, and the sense of the ego. The principles of the atmospheres as the fifth dimension in urban design practice should be considered through the two determinants of Places and People, each of which should reinforce the other. I. Places: (a) planning; uses, functions, activities, and densities; (b) design; urban configuration, form, and formation (blocks, spaces, and circulation patterns). II. People: impression (image and tones; light, sound, texture, smell, and taste), sensation, and emotion.

## In the Urban Design Dimensions

John Dewey sees an experience in the artwork should have a satisfying emotional quality has internal integration and fulfilment reached through ordered and organized movement. This artistic structure may be immediately felt. In so far, it is aesthetic. Dewey thinks that there is "no experience of whatever sort is a unity unless it has [an] aesthetic quality" (Dewey, 2008, p. 40). Architecture is classified among artworks, both of which depend on the viewer to determine their effects on people's mood. These effects follow on three dimensions of the urban design (Carmona, Heath, Oc, & Tiesdell, 2003; Elshater & Abusaada, 2017): the aesthetic, perceptual, and sensual. Regarding the concept of atmospheres, Mikel Dufrenne (1953), Gernot Böhme (1993-2010), Juhani Pallasmaa (2014), and Tonino Griffero (2014-2016) focus on these three dimensions are as follows.

Regarding the aesthetic dimension, the atmospheres focus on the concept of "what the spectator perceives" (Dufrenne, 1973, p. 168), it is something aesthetic, and that every beauty is inherent in what the spectator sees. It means that the spectator is not the one who makes the beauty as much as a receiver for this work (p. 76). If the work is beautiful, the spectator senses it immediately. In this vein, the atmospheres are the "matter of a certain quality of objects or of beings" (p. 168). It is the aesthetic object "elicits a feeling or emotion in a spectator, viewer or listener which 'completes' the aesthetic object and 'surpasses' it (p. 521). Böhme (1993) sees the atmospheres emerge as an "aesthetics of reception" (p. 120). In this respect, he sees that the atmospheres depend on the "ecstasies" [an overwhelming feeling of great happiness or joyful excitement], as well as "the reality of the perceived as the sphere of its presence and the reality of the perceiver, insofar as in sensing the atmosphere" (p. 122). Böhme (1998) also believes that he atmospheres are ranging "between the subject and the object"; it "mediate between the aesthetics of reception and the aesthetics of the product or of production" (Böhme, 1998, p. 112). In this, Böhme supposed if the productive effort to create beauty is to have any meaning at all, then it must be supposed that our experiences of beauty are, at least to a certain extent, shared. Further, if somebody needs "to perceive something, that something must be there, it must be present; the subject, too, must be present, physically extant" (Ibid.). He also believes that the "analysis of the atmospheres or atmospherics produced by spatial professionals will thus yield understanding of what is required for the unfolding of atmospheres of a particular type –both through the producers' own actions and their acceptance by others" (Böhme, 2005, p. 139). Consequently, the aesthetics of atmospheres achieve "the beauty: the artist, the designer, the architect will want to know what he or she has to do to ensure that a public will experience his or her objects or arrangements as beautiful. And to say what the artist has to do would be the task of aesthetics" (Böhme, 2010, p. 23).

Regarding the perceptual dimension, the aesthetics of atmospheres "shifts attention away from the 'what' something represents, to the 'how' something is present" (Böhme, 1998, p. 114). Accordingly, "sensory perception as opposed to judgement is rehabilitated in aesthetics and the term 'aesthetic' is restored to its original meaning, namely the theory of perception"(ibid.). The quality of a space focuses on the daily experience and mindful physical presence. It is a holistic interplay between a range of physical and social design aspects (Böhme, 2013, p. 21). And, "the object pole of mindful physical presence in space" (p. 27). The atmospheres emerge as "something quasi-objective"; it is often those who define the meaning of the atmosphere who are emotionally affected by it (Böhme, 2014, p. 43). The atmospheres have strong impacts in "emotions and moods" (Pallasmaa, 2014, p. 20). In addition, atmospheric perception is "a holistic and emotional being-in-the-world" (Griffero, 2014, p. 15). Moreover, "an atmosphere finds in the perceiver a more or less clearly defined felt-bodily sounding-board" (Griffero, 2016, p. 27).

Arguably, this dimension follows the visual and behavioral dimensions, which depends "The habitual image of one's own body gained through experience "in particular vision and touch)" (Schmitz, 2017, p. 9). Notably, the effects of emotional atmospheres is based on the feeling of the body, and not only due to "the prior experiential knowledge" (p. 14).

Regarding the sensual dimension, the atmospheres constitute 'in- between' environmental qualities and human sensibilities (Böhme, 2000, p. 14). It is defined as a "multi-sensory in its very essence", "[t]he judgement of environmental character is a complex multisensory fusion of countless factors which are immediately and synthetically grasped as an overall atmosphere, ambience, feeling or mood" (Pallasmaa, 2014, p. 230). The atmospheres also defined as "a social situations" (p. 231), whereas, every cultural, regional, or national entity has a different atmosphere. Genius loci, and the spirit of place, which gives the place its unique perceptual character and identity is closely related with the atmospheres; which consider "a specific quality" (Pallasmaa, 2014, p. 231). Consequently, the atmospheres represent as "a qualitative sentimental prius, spatially proud out, of our sensible encounter with the world", "it means grasping a feeling in the surrounding space" (Griffero, 2014, p. 5).

## CINEMATIC ATMOSPHERES RELATED ARCHITECTURAL HISTORY

This section is answering the question of how one can use the artworks in urban design teaching related to the change of atmospheres. In this, this work is based on the reading of four novels for Nagibe Mahfouz turned to cinematic works. The Mahfouz novels in this article are Zuqaq al-Midaq [Midaq Alley] (1947), New Cairo (Known as Cairo 30) (1947), Qasr El-Shouq [Palace of Longing] (1957), and Miramar (1957). In these novels, the atmospheres range from the residential buildings (the palaces), the traditional 'Al-Hara' [the alley], the public realm (the squares), and the building facades. The next deals with the reading of the structural context of some known sites in Egypt during the fifty years between the 20s and the 70s. These reading are writing down a change in the atmospheres of the places related the ideas, themes, and dramatic text.

### In the 1920s: An Atmosphere of the Residential Building

The second book of the *Al-Thulathia* (Cairo Trilogy), entitled [*Qasr El-Shouq*] (Palace of Desire) (1947), presents the life of the family of Ahmed Abdel-Gawad (the father). The main events of the movie occurred after the death of his son, Fahmy, during the events of the 1919 revolution, and this part of the novel was ended by the death of the leader, Saad Zaghoul. The facts take place in the Al-Hussein area, especially in the area of Al-Gamaliya known as the Palace of Desire (Beattie, 2005, p. 126). This palace was set up in the era of Shajr al-Durr "tree of pearls" (1236-1240); it occupied the entire region as a palace, and its exclusive possession was for MS Shouq.

Afterwards, this palace turns into a residential district characterized by the buildings and the small narrow streets for pedestrians, both formal and informal. The existence of a patriarchy is one of the notions of this novel; this term appeared in the interior design of the house of the Abdel-Gawad family, where he was mediating between his male sons while eating and the girls waited for them until they were finished with their food. Also, the design of the individual buildings used internal courtyards and Putin Al-Mashrabiya to cover the windows to avoid pedestrians on the street from being able to see the

*Cinematic Works-Based Affective Urban Design Atmospheres*

female that sits behind it. In the novel (1957), and afterwards in the movie (1966), some changes in the atmospheres can be tracked by the daily life of Kamal, Al-Sayed's youngest son.

These changes appeared as he moved between the narrow streets of Al-Gamaliya, the traditional and old area in Cairo, to the vast roads in Al-Abbassia (the new Cairo suburb) to meet his friends. Once Kamal sees Aida, the sister of his aristocratic friend Husain Shaddad, he falls in love. And, because of the vast differences between their social classes, it seems that their love is without hope. This love reaches the point of worship, he tells Aida at one of their meetings. In this scene, the variations of the atmosphere emerged between the two places in Al-Gamaliya, and Al-Abbassia. When Kamal goes to visit his friend Shaddad, Aida decides to meet him on the stairs that lead to the garden. She seems to enjoy the love of youth; you can see in her eyes. In this scene, the production designer used the English-style garden kiosk and the beautiful landscape to show the style of the aristocratic era. Moreover, the architecture of the palace sits between the period of romanticism and the differentiation between classes in the 1920s.

## In the 1930s: An Atmosphere of the Public Realm

The novel of "New Cairo (Known as Cairo 30)" adopts the terms of social classes that were prevalent in the 1930s as a basis for the dramatic construction of the events; the nepotism is the basis for this artwork. The idea is based on the fact that Mahjoub Abdel-Daim, a young university graduate, gives up his dignity to get a job. His supervisor gives him advice, which is that he must accept marriage to a pretty girl who belongs to one of the Egyptian Pasha of the aristocratic class. Surprisingly, she was the girl that he had hoped to marry, and the biggest shock was that she would marry him now to make it easier for her to be the mistress of the Pasha in secret. This novel focuses on the offensive (sarcastic) style of dialogue in the indoor places. The film also showed ordinary external scenes, most of which explained the meaning of the public realm, which was crowded with a different group of people. The first public realm appeared in the opening scene; it was the city road located in front of the entrance of Cairo University.

By referring to the novel of Cairo 30, the heroine of the tale was living in the Hara but became an adulteress to sell herself. After this transformation, the girl moved to live in a residential building in one of the prestigious neighborhoods and became the mistress of the pasha, the aristocrat. But of course, this does not mean that all who live in the Hara are honorable and those who live in high-class neighborhoods are prostitutes. But it is purely a reading of the events of novels with their dramatic transformations. The first scene, between the main three actors after they graduated from the college, occurred on the sidewalk of the public street. This scene shows the majestic atmosphere of the students graduating. It is followed by footage showing Abdel Dayem riding a carriage dragged by a mule in the countryside; the atmosphere arises from the natural landscape in the poor rural area. In this time, the Pasha, the general boss of Abdel Daim, flirts (or tries to attract) the girl Ihsan Shehata, the heroine of the film, in the streets of the capital Cairo, while he is riding in a luxury automobile in this period. In the real public realm, in the big-hearted squares of the old city, the final footage shows Aly Taha, the political activist, the real lover of Ihsan, distributing the political publications.

## In the 1940s: An Atmosphere of the Traditional Hara [the Alley]

Arguably, the alley is the smallest planning unit in the structural hierarchy of the traffic routes in the ancient Egyptian city. The alley is similar to the closed-end roads in Western thought, which branch out from "Al-Hara [the alley]". This is slightly larger in width and allows not only pedestrian movement but also includes a small shop for commercial activities.

Like many Egyptian cafés in the 1940s, Kirsha's café in Midaq Alley denotes a center of social life. Those cafes were where the men spent most of the night drinking tea and coffee, smoking shisha, playing card games and dice, and most of all, they chatted about everything. A prominent theme in this novel is this unchanging quality of Egyptian society. Perhaps the characters of the film are one of the most influential influences on the atmosphere, each of them in his place, from Hamida the main character in the alley to the Kirsha the owner of the coffee shop, and even to Radwan al-Husseini, Al-Sheikh Darwish, and Zita, the beggar and pimp.

The neighborhood of Al-sanadiqyah [the boxes] is one of the oldest markets in the Al-Azhar area. It branches out of Zuqaq al-Midaq [Midaq Alley], the hero of Mahfouz's novel Zuqaq Al-Madak. Moreover, it was the centre of the process of packing and manufacturing boxes for the bride, along with incense, spices, and perfumes, in addition to natural fat (not artificial margarine). For instance, Mahfouz showed a vast space for the Egyptian 'Hara [the alley]' in his book _ *Zuqaq al-Midaq* [Midaq Alley] _ (1947). Its events take place in one of the most famous neighborhoods of Al-Azhar Al-Sharif in the historic capital of Fatimid. This site symbolized, at that time, the alley as the country of Egypt, and the people represented the middle class at the time.

## In the 1960's: An Atmosphere of the Building Facades

In one of the fabulous motels (pansions), through the life of a seven of pensioners (retired), living in Alexandria, the novel, "*Miramar*," tells about the Egyptian political situation in the 1960's. Marianna is the owner of the Miramar pansion, and for Tolba Marzuk, one of those who was affected by the revolution of the year 1952, the government reserved its property. Amer Wagdi, a retired journalist, Hosni Allam is an uneducated man rich in heredity (from the dignitaries), and Mansour Bahy feels guilty about his friends being arrested. Sarhan Al Beheiry, working in the government, thinks of himself as one of those who rule. Al Beheiry presented for Zahra (the protagonist) a promise by marriage; the girl that was drawn from the countryside, came Alexandria to flee from her hometown to refuse to marry a rich old man. In the vein, Mahfouz focuses on the atmosphere of violence against women through narrative, and that there are no differences between urban and rural violence. Zahra chose to express this violence when she pointed out that all her grandfather wanted was "to exploit me," and "to sell me." When she came to Alexandria, Sarhan Al Beheiry tried it with her.

In this novel, Mahfouz writes about the effects of the 1952 Revolution, such as the elimination of class discrimination among the Egyptian people, and of capital control in the areas of agricultural and industrial production. Mahfouz chose the Miramar Building, which was built by the Italian architect, Giacomo Alessandro Luria in 1926, and which brought together the Islamic and Italian styles. In this novel, Zahra enabled the popular struggle, and Mahfouz built an atmosphere of protection, using the idea of the female fight. The inside and outside the building are dominated by a feminized atmosphere. In the1969 film, the production designer chose the Miramar Building for its aesthetic qualities. He also showed this building and its architectural splendor in many scenes, and I think that linking the nature of the story, as well as the atmosphere achieved by the structure of Miramar, reflects how we too can build an atmosphere.

## CONCLUSION

The atmosphere/ambience (in noun) and atmospheric/ambient (in adjective) are classified as a human emotional condition and environmental entity as follows: a) the case that creates a proper environment for pleasure and conveys better impressions for people. It refers to an environment, milieu, or boundary [the environment surrounds it from its vast borders on all sides], surrounding influence, encircling, encompassing (surround and have or hold within/enclose), and environing (to encircle). b) These are entities that affect the environmental parameters, such as light and sound foremost; they stimulate feelings of inspiration. Arguably, simply, it is a status that occurs not only during three prime times of the day: morning, midday, and evening, but exists throughout the day, at any time, and in all places. c) It is the combination of the overall feeling created by the tone (the attitude that is expressed) and mood (a sense that the one gets from). The spirit cares about internal feelings and contributes to building up the entire atmosphere of a narrative that exists at a spot.

In this vein, the word atmospheres means a) the "surrounding" context in reality; b) arrangement (order or pattern) in the art writing; c) places that evoke a unique mood (look and feel); and d) aspects that compel people to come and stay around. Moreover, this is understood e) in the sense of a dominant tone or a tone of expression to express a specific position or atmosphere surrounding a place, space, or object; f) the sense of the surrounding spatial environment; and g) the environmental context at large and the place surrounding us in all the stages of our life. The atmospheres are emotional tone that fosters the course of events. There is a set of examples of the atmosphere in the field of the narrative, while the atmosphere changes in literature manuscripts depending on the changes of the scene and developing of the events and the characters. I hope this matter is possible in the field of architecture, so all of these feelings can be transferred to the user in the situation in which he lives. Hence, it can create a distinct atmosphere with the same methods and techniques that exist in the literature and films. Consequently, the question is: can the urban designer create urban atmospheres in the same way?

The issues that deal with the impact of the urban atmospheres on the criteria and principles of city design were not found. Accordingly, the urban designer should respect the design principles in any case, but many of them see the perceptual dimension has these principles, that depend on the five senses. Afterwards, the designer should use these principles to achieve the effect on feelings, not only the perceptual dimension but also the functional, social, visual, morphological, temporal, ecological and environmental dimensions.

The current work concluded with some lessons learned to determine how different atmospheres can create in ways close to those used by filmmakers. Moreover, these lessons provide suggestions focuses on how to take advantages from the history of architectural thought to generate well-defined atmospheres. These analytical studies should be included as an educational goal in the departments of urban, taking into consideration that the manifestations of these impacts occur through four stages:

1. Monitor the impacts of architecture on the atmospheres,
2. Ask students about the causes of their impacts of architecture,
3. Compare their opinions with some previous experiences, and
4. Record the impressions that match the experiences.

The elements that the urban designer can manage to influence the atmospheres are unity, mood, process, the visual elements, identity, and familiarity. It includes compatibility with the surrounding space, similar building materials, and the tension between interior and exterior. It also includes the temperature, lighting, shapes, levels of intimacy, the sound of space, surrounding objects, physical and psychological paths of light, and shadows.

In this respect, there is a set of theoretical, scientific rules that should be studied by architects to improve the atmosphere of spaces. However, the atmospheres that are explained in this paper go beyond these dimensions; it seeks to try to understand the characteristics of the urban environmental context for places based on their architectural styles, the scenario of events, social life, types of activities, and the nature of people.

## ACKNOWLEDGMENT

This chapter is an updated and enhanced version of the article "Revisiting the Word "Atmospheres" in the Urban Design Academic Field Based on the Artworks, In 'International Journal of Sustainable Entrepreneurship and Corporate Social Responsibility' (IJSECSR), 3(2), 22-43. doi:10.4018/IJPPPHCE.2019070102).

## REFERENCES

Abrams, M., & Harpham, G. (2011). *A Glossary of Literary Terms*. Boston: Cengage Learning.

Abusada, H., & Elshater, A. (2018). *The Fifth Dimension: Urban design- The atmospheres of the City*. Cairo, Egypt: Academic Bookshop. (in Arabic)

Beattie, A. (2005). *Cairo: A Cultural and Literary History*. Oxford: Signal Books.

Böhme, G. (1993). Atmosphere as the Fundamental Concept of a New Aesthetics. *Thesis Eleven, 36*, 113-126.

Böhme, G. (1998). Atmosphere as an aesthetic concept. In G. Confurius (Ed.), *Constructing Atmospheres* (Vol. 68, pp. 112–115). Daidalos.

Böhme, G. (2000). Acoustic Atmospheres: A Contribution to the Study of Ecological Aesthetics. *The Journal of Acoustic Ecology, 1*(1), 14–18.

Böhme, G. (2005). Atmosphere as the subject matter of architecture. In H. Meuron & P. Ursprung (Eds.), Natural History (pp. 398–407). London: Springer Science & Business Media.

Böhme, G. (2010). On beauty. *The Nordic Journal of Aesthetics, 39*, 22–33.

Böhme, G. (2013). Atmosphere as Mindful Physical Presence in Space. *OASE Journal for Architecture*, 21-32.

Böhme, G. (2014). Urban Atmospheres: Charting New Directions for Architecture and Urban Planning. In C. Borch (Ed.), *Architectur, Architectural Atmospheres – On the Experience and Politics of Architecture* (pp. 42–59). Basel: Birkhäuser. doi:10.1515/9783038211785.42

Business Dictionary. (2018). *Ambiance*. Retrieved from www.businessdictionary.com: http://www.businessdictionary.com/definition/ambiance.html

Carmona, M., Heath, T., Oc, T., & Tiesdell, S. (2003). *Public Places, Urban Spaces: The Dimensions of Urban Design*. Oxford: Architectural Press.

Cayne, B. S. (1998). *The New Lexicon Webster's Dictionary Of The English Language*. New York: Lexicon Publications Inc.

Cullen, G. (1961). *The Concise Townscape*. Architectural Press.

Dewey, J. (2005). *Art as experience* (1934 ed.). Penguin.

Dewey, J. (2008). *The Later Works of John Dewey, 1925-1953: 1934, Art as Experience* (A. Boydston, Ed.; Vol. 10). Carbondale, IL: Southern Illinois University Press.

Dictionary.com. (2018). *Ambience*. Retrieved from www.dictionary.com: https://www.dictionary.com/browse/ambience

Dufrenne, M. (1973). *The Phenomenology of Aesthetic Experience*. Evanston, IL: Northwestern University Press.

Egan, K. (2014). Film Production Design: Case Study of The Great Gatsby. *Elon Journal of Undergraduated Research in Communications*, *5*(1), 5–14.

Elshater, A., & Abusaada, H. (2017). *What Is Urban Design? Learning and Thought*. Partridge Publishing.

Fisher, L., Knapman, C., Luhrmann, B., Martin, C., Wick, D. (Producers), & Luhrmann, B. (Director). (2013). *The Great Gatsby* [Motion Picture]. Retrieved 10 30, 2018, from https://www.youtube.com/watch?v=PNLoFqVXoqg

Gandy, M. (2017). Urban atmospheres. *Cultural Geographies*, *24*(3), 353–374. doi:10.1177/1474474017712995 PMID:29278257

Gehl, J. (1987). *Life Between Buildings*. New York: Van Nostrand Reinhold.

Gehl, J., & Svarre, B. (2013). *How to Study Public Life*. Island Press. doi:10.5822/978-1-61091-525-0

Griffero, T. (2014). *Atmospheres: Aesthetics of Emotional Spaces* (S. D. Sanctis, Trans.). Ashgate Publishing Limited.

Griffero, T. (2016). Atmospheres and Felt-bodily Resonances. *Studi di estetica*, *49*(4), 1-49.

Haslam, J. (2008). Pits, Pendulums, and Penitentiaries: Reframing the Detained Subject. *Texas Studies in Literature and Language*, *50*(3), 268–284. doi:10.1353/tsl.0.0004

Jaaniste, L. O. (2010). The Ambience of Ambience. *A Journal of Media and Culture*, *13*(2). Retrieved from http://www.journal.media-culture.org.au/index.php/mcjournal/article/view/238

Jenner, R. (2013). Ambient atmospheres: Exhibiting the immaterial in works by Italian Rationalists Edoardo Persico and Franco Albini. *Interstices: Journal of Architecture and Related Arts*, 13-24.

Lefebvre, H. (2004). *Rhythmanalysis, space, time and everyday life*. London: Continuum.

Lynch, K. (1972). *What Time is this Place?* MIT Press.

Merriam Webster. (2018). *Ambience*. Retrieved from www.merriam-webster.com: https://www.merriam-webster.com/dictionary/ambience

Norberg-Schulz, C. (1988). *Architecture: Meaning and Place, Selected Essays (Architectural Documents)*. New York: Rizzoli International Publications.

Pallasmaa, J. (2014). Space, Place, and Atmosphere: Peripheral Perception in Existential Experience. In C. Borch (Ed.), *Architectural Atmospheres: On the Experience and Politics of Architecture* (pp. 18–41). Basel: Birkhauser. doi:10.1515/9783038211785.18

Poe, E. A. (2017). *Great Horror Stories*. Meniola, New York: Dover Publication.

Schmitz, H. (2017). The Felt Body and Embodied Communication. *Yearbook for Eastern and Western Philosophy*, 9-19. doi:10.1515/yewph-2017-0004

Semper, G. (2004). *Style in the Technical and Tectonic Arts, Or, Practical Aesthetics*. Los Angeles: Getty Research Institute.

Simpson, J., & Proffitt, M. (1997). *Oxford English Dictionary Additions Series* (Vol. 3). Oxford: Oxford University Press.

Stefansdottir, H. (2017). The role of urban atmosphere for non-work activity locations. *Journal of Urban Design*, 319–335. doi:10.1080/13574809.2017.1383150

Stevenson, A. (2010). *Oxford Dictionary of English*. London: OUP Oxford.

Sturm, S., & Turner, S. (2012). "Built Pedagogy": The University of Auckland Business School as Crystal Palace. *Interstices, 12*, 23–34.

Thibaud, J.-P. (2002). From Situated Perception to Urban Ambiences. In *First international Workshop on Architectural and Urban Ambient Environment* (pp. 1-11). Nantes, France: Centre de recherche m'ethodologique d'architecture; Ecole d'architecture de Nantes.

Till, J. (2009). *Architecture Depends*. MIT Press.

Tschumi, B. (1981). *The Manhattan Transcripts*. New York: St. Martin's Press.

Tschumi, B. (1994). *Bernard Tschumi, architecture and event: April 21-July 5, 1994*. The Museum of Modern Art.

Zumthor, P. (2006). *Atmospheres: Architectural Environments- Surrounding Objects*. Basel: Birkhauser.

# Chapter 8
# Environmental Problems of Delhi and Governmental Concern

**Mohd. Yousuf Bhat**
*Government Degree College, Pulwama, India*

## ABSTRACT

*Delhi, the capital city of India, which is the concern of this chapter, has its own significance as it is the seat of governance, learning, and the healthcare service provider. Capital cities though inhabit people from every region and tend to be overcrowded, but in Delhi, the situation is not only the nature of a capital city, but also the industrial and commercial centre of high order in the northern zone of India, which is creating a number of environmental problems, such as air and water pollution, slum development, congested housing, etc. The chapter discusses all causes of Delhi's environmental problems like atrophy of political will, mismanaged urbanisation, court interventions, etc., and finally, the chapter tries to find out possible solutions in a detailed manner keeping in view the measures taken by other countries like China to deal with such problems.*

## INTRODUCTION

The Constitution of India guarantees fundamental rights of its Citizens including Protection of Life and Liberty. Preservation of Environment is implicit in Article 21 of its constitution which guarantees right to life, the most fundamental of all rights. The environmental dimension of human rights is recognised by the ever-increasing support for right to a healthy environment particularly by the Landmark Stockholm Declaration (1972), which stated that "both aspects of man's environment, the natural and the man-made, are essential to the well-being and to the enjoyment of basic rights even the right to life itself", and that "man has the fundamental right to freedom, equality and adequate conditions of life, in an environment of quality that permits a life of dignity and well-being". The cause of preservation of environment is furthered by number of environmental laws and to deal with series of environmental problems particularly associated urban life style namely air and water pollution.

DOI: 10.4018/978-1-7998-3576-9.ch008

Delhi the capital city of India, which is the concern of this chapter, has its own significance as it is the seat of governance, learning, health care service provider. Capital cities though inhabit people from every region and tend to be overcrowded but in Delhi the situation is not only the nature of capital city, but also industrial and commercial centre of high order in the northern zone of India which is creating a number of environmental problems, such as air and water pollution, slum development and congested housing etc.

## GROWTH AND DEVELOPMENT OF DELHI

Delhi has been evolving since the 10th century B.C. and 17 Delhi's have so far come and gone in various locations within an area covering six by twelve miles. Gerald Breese in his book "Urban and Regional for the Delhi - N. Delhi area, Capital for conquerors and Country" published in 1974, has given the chronological order of development of Delhi (**Table, 1**).

Simultaneously, the physical growth of Delhi from 1803 A.D. to 1969 A.D. has also been mapped by Breese and the location of the 17 Delhis' indicated thereon.

In 1912 the territorial boundaries of Delhi were fixed when a decision was taken by the Emperor to transfer the capital of British India from Calcutta to Delhi at Delhi Darbar in 1911. At the time Delhi became an independent province comprising the whole of Delhi tehsil and a portion of Ballabgarh tehsil. In 1915, some villages from Ghaziabad tehsil of the Meerut district and the Shahadara town across the Yamuna river were transferred to the Delhi province. There has not been any boundary change since 1915. After the adoption of the Constitution of India, Delhi was given the status of part-C state and the Legislative Assembly was set up in 1952. In 1956, the State Re-organization Act came into force, under this Act, Delhi became the Union Territory. The Metropolitan Council was set up in 1966 which continued until the New Legislative Assembly came into being under the National Capital Territory Act of 1993. In this chapter, the National Capital Territory of Delhi is referred to as Delhi.

### Geographical Features

Delhi lies between 28°25' and 28°53' N latitude and 76°50' and 77°22' E longitude. It is situated between the Himalaya and Aravali ranges in the heart of the Indian subcontinent. It is surrounded in the east across the Yamuna River by Uttar Pradesh and on the other sides by Haryana. The major part of the territory lies to the West of the Yamuna River. Delhi can be broadly divided into three geographical segments.

1. The Yamuna Flood Plains
2. The Ridge, and
3. The Plains

The Yamuna Flood Plains region is low lying, sandy and flood prone. The Ridge Segment originates from the Aravali Hills of Rajasthan and enters Delhi from the south in a north-eastern direction and encircles the Territory to the northwest and west. The highest point is 318.5 metres. The plains is a major portion of the total area on which Delhi, N. Delhi, the Cantonment, and a number of villages are located. Delhi's altitude ranges between 213 to 305 metres above sea level, descending from north to south.

*Environmental Problems of Delhi and Governmental Concern*

*Table 1. Chronological order of development of Delhi: seven-seventeen Delhis'*

| Order | | Date | Name of Settlement | Founders Site | Present Probable Site |
|---|---|---|---|---|---|
| I | 1 | 900 B.C. | Indraprastha | Yodhistra | Purana Quilla |
| | 2 | 1020 A.D. | Suraj Kund | Anang Pal | Near the road linking Mathura Road and Mehrauli Road by the same name. |
| | 3 | 1052 | Lal Kot | Prithviraj Chauhan | Near Qutab site |
| | 4 | 1180 | Quilla Rai Pithora | Prithviraj Chauhan | |
| | 5 | 1288 | Kilokheri | Muir-ud-din Kaiquabad | |
| II | 6 | 1301 | Sri | Alauddin Khilji (1295-1315) | Near Hauz Khas |
| III | 7 | 1321-1323 | Tughlaqabad | Gayasudin Tughlaq (1321-1325) | On the link road connecting Mathura Road and Mehrauli Road near QutabMinar |
| | 8 | 1325 | Adilabad | Mohd.Tughlaq (1325-1351) | Near Tughlaquabad |
| IV | 9 | 1327 | Adilabad | Mohd.Tughlaq | Between Siri and Raipithora |
| V | 10 | 1354 | Ferozabad | FerozshahTughlaq (1351-1388) | Near Feroz Shah Kotla Stadium |
| | 11 | 1415 | Khirabad | Khira Khan | No trace |
| | 12 | 1425 | Mubarakabad | Mubarakshah | No trace |
| VI | 13 | 1530 (1533) | Dinpanah and Sher Garh | Humayun (1530,1538, 1555-1556), Left incomplete; completed by Shershah Suri (1538-1545) | Purana Quila |
| VII | 14 | 1638 | Shahjahanabad (1638-1649) | Shahjahan (1628-1658) | Old Delhi (walled city) |
| | 15 | 1912 (1911) | Delhi | British capital | North of Walled City Shajahanabad; Old (Civil lines) Secretariat etc. |
| | 16 | 1931 opened | New Delhi | British capital Designed by Lutyenes and Baker | Central vista, Connaught Place and near about area |
| | 17 | Aug. 15, 1947-date | New Delhi | Capital of free India subsequently designed by T.P.O. T.C.P.O. and DDA | Delhi Urban Area |

**Source**: Gerald Breese (1974) "Urban and Regional for the Delhi - N. Delhi area, Capital for conquerors and Country"

## Climate, Rainfall, and Seasons

The climate conditions in Delhi are typical of those prevailing in the western part of the tropic of cancer. Here the summer is very hot and winter is very cold with little moisture in the air. The mean daily temperature is at its highest in June at 38° to 45° centigrade. The percentage relative humidity varies between 29 in May to 77 in August. The southwest monsoon brings moderate to heavy rainfall in Delhi during months of June - September. Approximately 84 percent of the total rainfall in the year occurs during these four months. A total of 170-220 mm of rainfall occurs during July and August. Hot dry winds blow from the desert in the west and northwest during half of the summer days.

## Economy

Delhi is a hub of industry, trade, commerce and higher education. It has acquired a cosmopolitan character drawing people from all parts of the country and the world. As such it is also centre of international political, social and cultural interactions.

In Delhi, the majority (79%) of the population is engaged in the tertiary sector. Nearly 70% of the net domestic product was derived from the tertiary sector in 1990-91. 35% of the income in the tertiary sector comes from finance, insurance, real estate and business services, and another 26% comes from community, social, and personal services. Industry also occupies an important place in Delhi's economy. During the period 1981-87 to 1989-90, there was an increase of 62 per cent in the number of industrial units (NFHS, 1993).

## Demographic Profile

The population of National Capital Territory (NCT) has been growing at startling rate. Delhi started as a small town with a population of hardly 4.1 lakhs in 1911 and grew steadily to reach a population of 9.2 lakhs in 1941 with the decadal growth rate average approximately 30%. The sudden influx of migrants from other states raised the population from 9.2 lakhs in 1941 to 17.4 lakhs in 1951; registering a decade growth approximately 90%. Then onwards in the next four decades the decadal growth rate has constantly been above 50% and had resulted into a population of 167.87 lakhs in 2011. The **Table 2** depicts the population growth in Delhi from (1901-2011).

*Table 2. Population growth in Delhi (1901 to 2011)*

| Year | 1901 | 1911 | 1921 | 1931 | 1941 | 1951 | 1961 | 1971 | 1981 | 1991 | 2001 | 2011 |
|---|---|---|---|---|---|---|---|---|---|---|---|---|
| Population (in ooo's) | 405 | 413 | 488 | 636 | 917 | 1744 | 2658 | 4065 | 6220 | 9420 | 13850 | 16787 |

**Source:** Census of India

The main reasons for this high rate of growth in population in NCT of Delhi are:

1. It's becoming the capital of India in 1911.
2. The huge influx of displaced persons from Pakistan in 1947 onwards.
3. Multiplication and intensification of services during the post-independence era.
4. Expansion of commerce and trade.
5. Growing industrialization, mostly in the field of small scale industries during the last three decades, i.e. (1970-2000).
6. Availability of higher education prospects.

## MIGRATION TO DELHI

The population of Delhi is growing fast mainly because of migration; between 1981- 91 almost 50% of the population growth was contributed by migrants. Two-third of all migrants to Delhi were classified as literate. 90% of the total migrants were employed in different sectors. In 2001 alone, its population increased by 215,000 due to natural growth and 285,000 through migration. In year 1991 the total number of in migrants were 3723462, then in 2001 the number crossed to 6014458 and by latest census 2011 it has risen to 7663956 (Population Census: 1981, 2001 & 2011). By 2020, Delhi is expected to be the third largest metropolis after Tokyo and Mumbai (World Population Review, 2018).

Three reasons for migration to Delhi were reported as:

1. To find employment
2. Marriage
3. Family reasons
4. Others

**Figures 1** & **2** give the occupational profile of migration towards Delhi.

*Figure 1. Migration to Delhi (Census, 1981)*

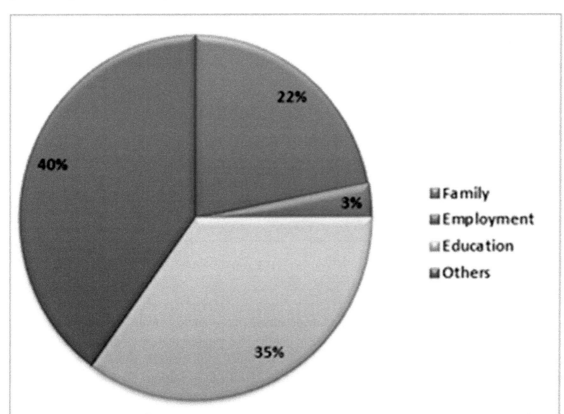

*Figure 2. Migration to Delhi (Census, 2011)*

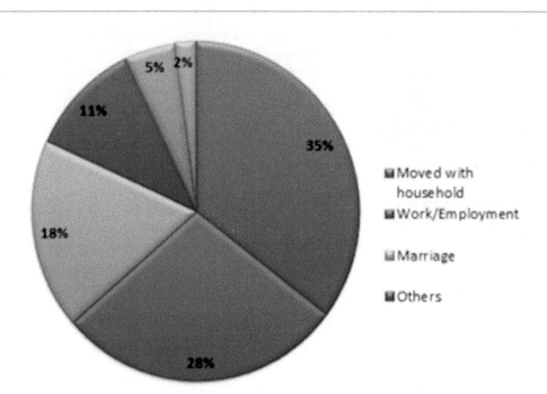

As per the census 2001, a good percentage of population in Delhi was migrant, majority of them from the neighboring states of Uttar Pradesh (43.13), Bihar (13.63) Haryana(10.43), Rajasthan(5.16), Punjab(4.81) etc. As shown in **Table 3**.

*Table 3. State-wise migration in Delhi-2001*

| S. No | States | Per cent | S. No | States | Per cent |
|---|---|---|---|---|---|
| 1. | Uttar Pradesh | 43.13 | 5 | Punjab | 4.81 |
| 2. | Bihar | 13.63 | 6 | West Bengal | 3.22 |
| 3 | Haryana | 10.43 | 7 | Madhya Pradesh | 1.90 |
| 4 | Rajasthan | 5.16 | 8 | Other states | 17.72 |

**Source:** Census of India 2001

Though, density of population is an important detriment of economic growth. There are many cities in the world having higher density than that of Delhi but the fact remains that given the stage of economic development and available technology, high density per square kilometre imposes great stress on the various public utilities provided by the government of Delhi and its various agencies. With the rapid increase in population without any relative increase in area, density of population in Delhi is the highest among all States and Union Territories in India (Central Statistical Organization, 1998). The density has increased 20 times in Delhi over a period of 70 years. The density of population in Delhi as per 1991 Census was 652 persons per square kilometres as against the national average of 273. By latest census of 2011, Delhi's density of population is 29 times higher than the density of the country (Census, 2011). The **Table 4** further clarifies variation in population density from 1921-2011.

*Table 4. Density of population in Delhi*

| Census Year | Persons per sq. Km | Census Year | Persons per sq. Km |
|---|---|---|---|
| **1921** | 318 | 1971 | 2738 |
| **1931** | 429 | 1981 | 5194 |
| **1941** | 617 | 1991 | 6352 |
| **1951** | 1165 | 2001 | 9340 |
| **1961** | 1792 | 2011 | 11320 |

**Source**: Census Handbook 2011, Department of Economics and Statistics

## GROWTH OF UNAUTHORIZED COLONIES: JHUGGIES AND JHOPRIES

Migration has been a major factor contributing to the rapid population growth of Delhi. It is estimated that migrants account for about 50% of the population increase every year. While a majority of the migrants come in search of better employment and higher income opportunities, due to the acute scarcity of land, shelter and infrastructure, many of them encroach public land and all types of vacant spaces to put jhuggies and jhopries. The number of jhuggies had noted a tremendous growth from a meagre 12749 in 1951 to over 4,80,000 in 1994 (M.N. Buch Committee Report). However by census 2011 their number has slightly declined to 367893. Latest report shows that there are still 1797 unauthorised colonies in Delhi (Hindustan Times, 2017). Around 18 million population of Delhi live in slums and unauthorized colonies without basic amenities (Times of India, 2012), which also poses quite reasonable challenge to environmental problems of Delhi as well as health implications to its inhabitants.

## DELHI'S ENVIRONMENTAL CONCERNS

In Delhi, pollution is one of the most critical problems facing the public and concerned authorities. According to the World Health Organisation (WHO), Delhi is the most polluted city in the world in terms of suspended particulate matter (SPM) (Hindustan Times, 2018). The growing pollution is responsible for increasing health problems. The deteriorating environment is the result of population pressure and haphazard growth. Industrial development has been haphazard and unplanned. Though very late in year 2002 metro rail service as safe and pollution free transport system was initiated and first phase got completed in year 2006, but literally it has not sufficed the growing transportation needs of the huge population.

## Vehicular Pollution

Road transport has remained the sole mode of public transport over the years; there has been a phenomenal increase in vehicle population, which has increased from 2 lakh in 1971 to 32 lakh in 1999 (Economic Survey of India, 1999-2000).

Data accessed from the transport department of Delhi government puts the total number of registered vehicles at 1,05,67,712 till May 25, 2017 among them 31,72,842 are registered cars, however motor cycles and scooters which make biggest chunk of the registered vehicles i.e. 66,48,730 are major air polluters due to poor emission standards. Other major categories of registered vehicles in the National Capital include goods carriers (2,25,438), motor cab (1,18,424), moped (1,16,092), passenger three wheelers (1,06,082), goods three wheelers (68,692), buses (35,332), e-rickshaws (31,555) and maxi cabs (30,207), the data reveals (Hindustan Times, 2018).

## Industrialization

Favourable infrastructural facilities, developmental scope, congenial tax structure, strategic location and somewhat non-uniform application or enforcement of municipal laws flourished industrial activities in NCT of Delhi. From the last couple of decades Delhi has rapidly grown into a major industrial area. At the time of the First Master Plan, Delhi had 17,000 industrial units. The Second Master Plan estimated that there were 46,000 units in 1981, a large majority of these outside their designated industrial zones (The Hindu Survey of Environment, 2001). And according to the Delhi Government's Fourth Economic Census in Delhi, in 1998, there were 126,175 manufacturing units which mean that over the last 20 years, on an average about 4,700 industrial units have sprouted in the city each year, most of them in residential, commercial and, other non-conforming areas. Moreover, the bulk of small-scale industries (SSIs) are in non-conforming areas. According to the 1996 survey, 98,000 units were non-conforming ones, operating outside the 28 industrial areas of Delhi (Frontline: 1-7). The averseness of the governments to discipline industrial sector created chaotic problem, which was again accentuated by the lack of environmental consciousness among Delhi's factory owners.

The first Master Plan of Delhi, MPD-62 for the period of 1961 and 1981 was sanctioned on September 1, 1962. The Delhi Development Authority (DDA) became the sole developer for the entire future extension of Delhi. However, the Plan remained largely on paper, amidst a totally chaotic growth of the city. The Master Plan 2001 was published in the Gazette of India on August 1, 1990, but was found unsatisfactory. Delhi Urban Art Commission, which was formed by Act of Parliament in 1974, was entrusted the job of reviewing the Master Plan. The Master Plan 2001 comprises a set of coordinated policies concerned with all aspects of development in the city. The concept underlying the plan was as follows:

1. Delhi to be planned as an integral part of its region;
2. Ecological balance to be maintained;
3. Central city area to be treated as special area;
4. Urban heritage of Delhi to be conserved;
5. The city centre to be decentralized;
6. Mass-transport system to be multimodal;
7. The urban development be low-rise high density;
8. Urban development to be hierarchical.

In ensuring composed regional development, migration to Delhi was to be checked through a restrictive policy on employment generation. The main guidelines for this policy were (i) only those new Central Government offices which directly served the ministries of the Government of India should be located in Delhi; (ii) existing offices of Public Sector Undertakings within Delhi should be encouraged to shift, while new offices to the extent possible within their operational areas be set up outside Delhi, (iii) industrial growth in Delhi should be restricted to small scale with stress on units which require skill, less manpower and energy and were clean and largely sub serve Delhi's economy; (iv) legal and fiscal measures should be adopted to restrict employment in industries and distributive trade.

Under the Delhi Master Plan 1962 and 2001, it was provided that

(a) Hazardous and noxious units were not to be permitted in Delhi;
(b) The existing industrial units of this type be shifted on priority within a maximum period of three years. Such a project report to execute shifting be prepared by the concerned units and submitted to the Authority within a maximum period of one year". The Master Plan also envisaged that heavy and noxious industries to be shifted to environmentally suitable locations in Delhi Metropolitan Area or National Capital Region (NCR)- participating states of NCR are Haryana, Uttar Pradesh and Rajasthan;
(c) The land which would thus become available would be used to make up the deficiency of the community services, especially to improve the lung space in Delhi.

However, the authorities did not take any concrete action to implement this plan and the situation continued to deteriorate. According to Master Plan, more than 34 per cent of the recreation area has been diverted to other uses. The Delhi Ridge, the only lung space left in Delhi, shrank due to encroachments, while the Government remained a spectator (The Hindu Survey of Environment, 2001).

## The Problem and Judicial Approach

As the apex court has viewed it over the years, there are two main aspects of the problem that M.C. Mehta an environmental lawyer and activist raised way back in 1985 (M.C. Mehta V. Union of India and Others, 1985). The first concern is air and water pollution in the city and the second is regarding non-compliance with the Master Plan i.e. the operation of the industrial units in the residential and non-conforming areas. Judgement delivered by division bench comprising Justice Kuldip Singh and Justice Faizanuddin, issued directions on July 8, 1996 and later on October 6 and 9, 1996 that:

1. 168 hazardous industries be moved from Delhi to suitable locations in the National Capital Region (NCR), or closed down by November 30, 1996;
2. 513, and 334 extensive industries in residential and non-conforming areas respectively (H-category in MPD-2001) be closed down by January 31, 1997 and relocated outside Delhi;
3. 46 hot mix plants be closed down by February 28, 1997;
4. 21 arc/induction furnaces be closed by March 31, 1997;
5. The Delhi administration ensures all (non-extensive F-category) units operating in non-conforming areas, except light and service industries as permitted by MPD-2001 (employing less than 10 workers and consuming less than 1 kw of power), be either relocated or closed down.

Undeniably, there is no record of the exact number of non-conforming units, the registration of their activities or their polluting status. Figures vary from source to source. In response to a court directive, a high-power committee was set up, which invited applications from non-conforming units for relocation. As of July 2000, 51,851 applications had been received by Delhi State Industrial Development Corporation (DSIDC), which had been entrusted in March 1998 by the high power committee to implement the relocation scheme. Of these, according to DSIDC, 27,915 applications were rejected on technical grounds; also it is unclear whether these 51,851applications represent the actual number of F-category units because of the existence of large number of unlicensed units, the lack of proper registration procedures and monitoring of these units. Even the 1996 survey, which is supposed to have generated the figure of 98,000, is apparently not reliable because it was done without fixing proper criteria or guidelines. It is learnt that school teachers were asked to collect the data for payment of an honorarium depending upon the number surveyed. The figure, particularly those of the non-conforming ones, could be unreliable, as per officials of Ministry of Environment and Forests (Ramachandran, 2000).

Furthermore, as the Steering Committee constituted to review MPD-2001 observed, several units obtained licenses declaring their operations to be non-polluting and within the allowed power-load. Subsequently, they changed their activities. With the implementation of municipal laws being ineffective and number of unauthorised power connections growing, the polluting units have multiplied in number. Given this situation, exact figures of the total number of industries and those operating in non-conforming areas simply do not exist (Ramachandran, 2000).

## Water Pollution

Although Delhi covers only 2 per cent of the length and basin area of the Yamuna River, it contributes 71% of the wastewater discharged into the river every day (The Citizens Fifth Report, 1997). Delhi poses the biggest challenge in the difficult task of cleaning the Yamuna. The largest polluter of the river, Delhi is also the largest consumer of water, with a production of 2,728 mld (million litres per day) of treated water. Though the length and basin area of the river in Delhi is only 2 per cent, it contributes 71 per cent of the wastewater and 55 percent of the total BOD (biochemical oxygen demand) load discharged into the river every day (The Citizens Fifth Report, 1997).

From the time the river enters Delhi at Wazirabad, the Yamuna is slowly asphyxiated. About 1,700 mld of untreated sewage finds its way through 18 notorious drains such as Najafgarh, Sen Nursing Home and Powerhouse Nullahs and ends up into the river at various points along its 22 Kilometre stretch in Delhi (The Citizens Fifth Report, 1997). The Najafgarh drain also contributed 60 per cent of the discharge and 40 per cent of BOD load into the Yamuna. Industrial wastes from 20 large, 25 medium and about 93,000 small scale industries located in Delhi also flow into the river through these drains. Though the large and medium industries form only about 0.05 per cent of the total industries located in Delhi, they contribute 50 per cent of the total 300 mld of industrial wastes generated every day. These are mostly engineering, textiles, chemicals, electronic and electrical goods factories (Agarwal, 1997).

The bulk of the pollution comes from untreated sewage, dumped into the river because the city administration lacks sufficient sewage treatment facilities, admits JC Kala, joint secretary official of Ministry of Environment and Forests (The Citizens Fifth Report, 1997). The effluents flowing into the river Yamuna comprise of municipal and industrial wastes. The Central Pollution Control Board (CPCB) has been monitoring the water quality of the Yamuna at the upstream of Wazirabad and at Okhla. Upstream of Wazirabad, the dissolved oxygen (DO) level is 7.5 mg/l (milligram per litre) and biochemical oxygen

## Environmental Problems of Delhi and Governmental Concern

demand (BOD) level is 2.3 mg/l, whereas, downstream at Okhla, the DO (dissolved oxygen) level declines to 1.3 mg/l with the BOD at 16 mg/l, indicating considerable deterioration in water quality in the stretch due to discharge of sewage and industrial effluents (White Paper on Pollution in Delhi, 1997).

The prescribed ambient water quality in terms of DO is 5 mg/l or above and 3 mg/l or and below in terms of BOD. The stretch between Wazirabad and Okhla is designated as bathing quality standard in terms of its water use. The coliform count at Wazirabad is 8,506/100 ml whereas at Okhla, it increases to 3,29,312/100 ml, as against the prescribed standard of 500/100 ml (White Paper on Pollution in Delhi, 1997).

The **Table 5** provides data on the water quality of the River Yamuna from 1988-97.

*Table 5. Water quality of River Yamuna in Delhi-Palla*

|     | 1988 | 1989 | 1990 | 1991 | 1992 | 1993 | 1994 | 1995 | 1996 | 1997 |
|---|---|---|---|---|---|---|---|---|---|---|
| **pH** | 8.2 | 8.28 | 8.41 | 8.1 | 8.21 | 8.0 | 8.1 | 8.4 | 8.0 | 8.31 |
| **DO** | 7.5 | 7.72 | 8.86 | 8.46 | 8.1 | 7.1 | 7.3 | 6.9 | 9.0 | 9.3 |
| **BOD** | 2.3 | 3.0 | 1.9 | 3.00 | 3.1 | 3.30 | 3.6 | 2.3 | 2.7 | 3.1 |
| **COD** | 9.5 | 10.5 | 16.7 | 11.5 | 12.6 | 10.2 | 10.9 | 8.3 | 18.2 | 49.2 |
| **TC** | 8506 |  |  |  |  |  |  |  | 5483 | 3645 |
| **FC** | 743 | 3300 | 35950 | 1505 | 3435 | 1580 | 795 | 193 | 3944 | 2901 |
| **TKN** |  | 0.45 | 0.74 | 0.63 | 0.6 |  | 3.1 | 1.3 | 1.1 |  |
| **WT** | 25.8 | 23.5 | 23.9 | 24.3 | 23.6 | 24.2 | 23.6 | 23 | 23.6 | 26.2 |
| **AMM** | 0.4 | 0.146 | 0.279 | 0.21 | 0.2 | 1.0 | 0.7 | 0.3 | 0.2 | 0.7 |

**Source:** Ministry of Environment and Forests (MOEF)
Note:
DO = Dissolved Oxygen, mg/l
FC = Fecal coliform, no./100 ml
BOD = Biochemical Oxygen Demand, mg/l
TKN = Total Kjeldahal Nitrogen, mg/l
COD = Chemical Oxygen Demand, mg/l
WT = Water Temperature, °C
TC = Total Coliform, no./100 ml
AMM = Ammonia, mg/l

The low self-purification capacity of the River Yamuna is due to want of minimum flow in the river and discharge of heavy municipal and industrial pollution load emanating from Delhi. Even though Delhi constitutes only 2% of the catchment of the Yamuna basin, yet the area contributes about 80% of the pollution load. There are 16 drains which discharge treated and untreated waste water/sewage of Delhi into the Yamuna. The municipal sector is the main source of water pollution in terms of its volume. Approximately 1,900 million litres per day (mld) of wastewater is discharged from municipal sector and 320 mld from industrial sector. The installed capacity for treatment is 1,270 mld. At the same time, the existing capacity for treatment is not up to the desired secondary treatment level. Thus, substantial quantity of untreated sewage and partially treated sewage is discharged into the Yamuna every day. The Najafgarh drain contributes 60% of total waste water and 45% of the total BOD load being discharged from Delhi into the Yamuna (White Paper on Pollution in Delhi, 1997). The municipal wastewater has increased from 960 mld in 1977 to 1,900 mld in 1997. The capacity for treatment has enhanced from 450 mld in 1977 to 1,270 mld in 1997 (White Paper on Pollution in Delhi, 1997). The pollution load being discharged into the Yamuna River from the drains, since 1982 to 1996 is presented in the **Table 6**.

Significant levels of pesticides such as dichloro-diphenyl trichloroethane (DDT), aldrin, dieldrin, heptachlor, benzene hexachloride (BHC) and endosulphan are present in the Yamuna waters **(Table, 7)**.

Whereas level of micro-pollutants due to heavy metals in Yamuna before it reaches Delhi are shown in **Table 8**.

*Table 6. Pollution load being discharged into The River Yamuna*

| Year | BOD Load, tonnes/day | Year | BOD Load, tonnes/day |
|---|---|---|---|
| 1982 | 117.3 | 1988 | 159.6 |
| 1983 | 132.3 | 1989 | 163.4 |
| 1984 | 119.4 | 1990 | 167.5 |
| 1985 | 123.2 | 1991 | 179.8 |
| 1986 | 165.1 | 1995 | 178.4 |
| 1987 | 148.5 | 1996 | 193.8 |

Source: MOEF

*Table 7. Levels of micro-pollutants (pesticides) in the Yamuna before it reaches Delhi*

| Micro-pollutant | July (ng/l) | March (ng/l) | Indian standards (ng/l) |
|---|---|---|---|
| T-BHC | 218.83 | 11.11 | Should be absent in drinking water |
| Aldrin | NT | NT | Should be absent in drinking water |
| T-Endosulphan | 51.30 | 90.23 | Should be absent in drinking water |
| Dieldrin | 30.44 | 20.42 | Should be absent in drinking water |
| T-DDT | 203.00 | 7.55 | Should be absent in drinking water |

NOTE: The monitoring station indicating is at Palla, 15 Kilometers upstream of Wazirabad barrage. July 1995 and March 1996 are indicative of monsoon and dry months, respectively.

**T-BHC**: All isomers of benzene hexachloride.
**T-Endosulphan**: All isomers of endosulphan.
**T-DDT**: All isomers of dichloro-diphenyl- trichloroethane.
**ng/l**: Nano gram per litre.
**NT**: Not traceable.
Source:(Anon 1996: CPCB, New Delhi)

The Indian standards clearly state that pesticide traces should be absent from drinking water. And yet, traces of certain pesticides exceeding 1,000 Nano grams per litre (a nano gram is one-billionth of a gram) were present in the Yamuna waters (Anon, 1996).

The CPCB's 16 month intensive study (from December 1994 to March 1996) of the 42 parameters of river quality at 19 locations on the Yamuna, done for Ministry of Environment and Forests (MoEF), acknowledged the presence of DDT and BHC at almost all monitoring stations. The Indo-Dutch study has also shown alarming levels of pesticides in the Yamuna. This study clearly states that "high concentra-

*Table 8. Levels of micro pollutants (heavy metals) in the Yamuna before it reaches Delhi*

| Micro-pollutant | July (mg/l) | March (mg/l) | Indian standards heavy metals (mg/l) |
|---|---|---|---|
| Cadmium | 0.01 | 0.01 | 0.01 |
| Chromium | 0.01 | NT | 0.05 |
| Copper | NT | 0.01 | 0.05 |
| Iron | 8.20 | 7.10 | 0.3 |
| Nickel | 0.03 | 0.05 | NS |
| Lead | NT | NT | 0.1 |
| Zinc | 0.60 | 0.06 | 5.0 |

**Note:** July 1995 and March 1996 are indicative of the monsoon and dry months respectively.
mg/l: Milligram per litre.
NS: Not specified.
NT: Not traceable.
Source: (Anon 1996: CPCB, New Delhi)

tions of pesticides are not detected by present monitoring systems and thus form a serious threat to the drinking water quality"(Agarwal, A 1997).Traces of chemicals and heavy metals in water are essentially due to industrial wastes and agricultural runoffs (State of India's Environment: 1997).

## Ground Water Quality

The city of Delhi has very little water sources of its own. The groundwater that is available is not adequate to serve the needs of the population of Delhi. Already 67% of the ground water resources has been utilised and the remaining water is of poor quality (National Capital Region Planning Board, 1995) and further exploitation may result in serious degradation. Results of extensive chemical tests indicate that except for isolated wells, the ground water of Delhi area is generally brackish and unfit for drinking or irrigation (National Capital Region Planning Board, 1995).

In the areas west of the ridge comprising blocks Najafgarh, Kanjhawala, Alipur and Mehrauli, the salinity of ground water generally increases towards south-west and north-west direction being fairly high in areas around Dhanse, Raota in south-west and Auchandi, Kanjhawala and Tikri Kalan in north-west (NCR Planning Board, 1999). Occurrences of high nitrate concentration are at several locations such as Saboli (Shahdara) and Kutabgarh (Kanjhawala) have nitrate contents above 100 mg/litre in ground water. (NCR Planning Board, 1999) Such high levels of nitrate concentration in shallow ground water could be due to leaching from solid wastes, discharge from sewage water etc. The ground water in the vicinity of the landfill in Yamuna flood plains has also high nitrate concentration (NCR Planning Board, 1999). Similarly fluoride and other harmful chemical concentrations beyond permissible limits are observed at several locations in Delhi.

## Effluent Treatment Plants

The order issued by Supreme court on September 13, 1999 relating to the pollution of river Yamuna and setting up of effluent treatment plants (ETPs), a Bench comprising Justice Kirpal, Justice S.P. Kurdukar and Justice V.N. Khare, after pursuing CPCB reports held, that "the quality of the river water was deteriorating at an alarming rate and it is perhaps dangerous to drink the said water .... The statistics also show that most of the pollution of the river takes place during its passage through Delhi (Frontline: 92).

On February 14, 1996 when the matter first came up, the Delhi Government had been ordered by the Apex Court to construct 15 common effluent treatment plants (CETPs). In September 13, 1999 order, the Bench observed that "three and half years had elapsed but till today no construction with regard to any CEPT has commenced" and that "the CEPTs were to be constructed with the Government contributing a major part of finance" (Frontline:92). The Delhi Government pleaded that the industrial units had not contributed in the required manner and to the required extent. However being lenient with the government, the order observed: "One can speculate that the industry has not chosen to contribute because non-construction of the CEPT has not resulted in any adverse action taken against them. Perhaps feeling may be that this state of affairs will be allowed to be continued so why pay". The court observed that at least two years were required for the completion of CEPTs. Also there was no authentic information on how many industries had even primary treatment plants. The Court said: "This being the state of affairs and there being no positive response with regard to the establishment of the treatment plants, there is no option left but to take appropriate steps in accordance with law (Frontline:92). The Delhi Government was directed to take measures to ensure that no industrial effluent was discharged into the river directly or indirectly from November 1, 1999. The Government was also asked to file a report within eight weeks. However it did not take appropriate steps to compel the units to confirm to effluent treatment norms. It also failed to file a report. Once again, on December 20, 1999, the Court observed that the Government had not taken any steps to ensure the non-pollution of the river and issued a show cause notice to the Chief Secretary regarding the non-compliance with its order. Following the 1996 order most of the large and medium units with water polluting potential, except the three thermal power plants, had been shut down while the small scale industries, which comprise nearly 99 per cent of the industrial units in Delhi, do not have ETPs, were still operating. The DPCC (Delhi Pollution Control Committee) identified 2,276 such water polluting ones. Despite the DPCC serving notices, ETPs were not being set up and when failing to get an extension of the six month deadline from the court, the committee began closure of units (Ramachandran, 2000).

CPCB (Central Pollution Control Board) officials also point out that this inventory has not included the 30 odd bus depots of the Delhi Transport Corporation (DTC), which discharge large quantities of effluents (Ramachandran, 2000). Interestingly, regarding the CETPs the World Bank which played a key role in popularizing the technology admitted to Greenpeace that: "CETPs generally fail to address toxic effluents" (The Hindu Survey of Environment, 2001).

A Greenpeace report adds, "End-of-pipe pollution control technologies such as CETPs are a waste of money. They do not destroy dangerous pollutants such as heavy metals and organic poisons. (The Hindu Survey of Environment, 2001) CETPs are themselves a source of high pollution loads, through the contaminated waste water and the large quantities of toxic sludge generated by their process, for instance in Vapi, the fisher folk of Kolak has complained for years that industrial effluents from industrial estate had choked the fish life in the Kolak and its confluence with the Gulf. In 1998, with the assistance and subsidies from World Bank, and the Central and State governments, the industry set up a common effluent treatment plant that would treat the effluent from 800 factories. It was found that CETP discharged red, foul-smelling, frothy effluent into the Damanganga (The Hindu Survey of Environment, 2001). An October 2000 analysis by Greenpeace of the treated wastewater revealed the presence of cadmium, chromium and even the deadly lead and mercury (The Hindu Survey of Environment, 2001).

## Atrophy of Political Will

Domestic and industrial sewage generated within the NCT of Delhi is the main source of pollution of the river Yamuna during its passage through the NCT. Despite over ten years of efforts and expenditure of Rupees 872 crore since 1994 on establishment of sewage treatment infrastructure for treatment of domestic and industrial sewage before its release into the river, the quality of water at the point where the river leaves Delhi has deteriorated drastically with large amounts of untreated sewage still falling into the river. The water quality of the river at the point of its entry into Delhi at Palla is adequate to sustain aquatic life and conforms to water quality of "bathing" standards in terms of Dissolved Oxygen (DO) and Bio-chemical Oxygen Demand (BOD).

However, at the point of its exit from Delhi at Okhla, the water quality of the river is unfit for any purpose with the BOD being 40 mg per litre against the norm of not more than 3 mg per litre while the DO deteriorates to almost nil against the norm of not less than 5 mg per litre. The coli form pollution which is already sub-standard at 217 times the norm when the river enters Delhi also deteriorates further to 1.39 lakh times the norm at the time of its exit from the NCT. Against the estimated domestic sewage generation of 719 Million Gallons per Day (MGD) in Delhi, the Government has created capacity for treatment of only 5122 MGD until March 2004. Even the created capacities of the STPs are not utilised optimally on account of construction of STPs in areas without adequate sewage load and non-synchronization of construction of trunk sewer lines and sewage pumping stations.

The progress of rehabilitation of the 28 trunk sewers of total length of 130 km which is crucial for conveyance of the sewage has been tardy. Only 335 MGD out of the estimated domestic sewage of 719 MGD is being treated before discharge into the river. The balance 384 MGD outfalls into the river untreated. Even the figures for actual treatment of sewage are arrived at through normative calculation on the basis of rated capacity of the STPs rather than through a robust system of measurement of the inflows and outflows. The quality of treated effluent did not meet the stipulated specifications implying that even the treated effluent is contributing to the deterioration in water quality. Of the 15 CETPs required for treatment of 42 MGD industrial sewage, only 10 are complete at an expenditure of Rupees 123 crore of which only four were commissioned as of March 2004. The utilization of even the commissioned plants is questionable. The construction and maintenance of the CETPs are afflicted by various constraints; viz. under-estimation of the cost leading to shortage of funds, cost escalation due to delay, delay in construction of conveyance systems, delay in formation of CETP societies, etc. Consequently, almost the entire industrial sewage was out falling into the river without treatment. The untreated sewage from the un-sewered areas continues to fall into the river through open drains and nallahs in the absence of trapping of the sewage from such areas for treatment. This further contributed to the pollution load in the river. While the physical progress of construction of sewage treatment and disposal works indicated under-performance, the Delhi Jal Board, which is entrusted with execution of such works, failed to utilize Rupees 159 crore out of Rupees 599 crore provided for the purpose by the Government during 1999-2004 (CAG Report, 2005).

Fortunately for the citizens, the Supreme Court has taken a strict view on the implementation of the Master Plan 2001. The authorities concerned have been ordered by the Court to clean up the mess that they have allowed to build up over a period of time. Subsequently once again in a significant judgement on Dec. 7, 2000, the Supreme Court has reiterated that access to drinking water is a fundamental right to life and "there is a duty on the State under Article 21 of the Constitution of India to provide clean drinking water to its citizens"(The Hindu, 2000) and ordered closure of all polluting industries operating out of the capitals non-conforming and residential area within a month (The Hindustan Times, 2000).

The Supreme Court's directions in brief are:

1. Shut all polluting industries in the capital within four weeks.
2. Nodal agency should ensure their closure.
3. Police should help in the task.
4. Submit compliance report within five weeks.
5. No order of any lower court can stop closures
6. MCD will prepare, within ten days, the list of industries in non-conforming areas.

Regarding the apex court's ruling to make Delhi largely pollution free, "the Court has not crossed its limits. It only asks the authorities concerned not to violate the law, but to enforce it. When the executive and legislature fail to give relief, a citizen has no recourse but to go to the Court, why did the Chief Secretary not hold an enquiry, as directed by court, against officials who erred in not enforcing the Master Plan? It is an unequal battle against those who have power and money. There is no political accountability" (Mehta, M.C, 2000). The unit owners, who defied all rules and regulations in connivance with bureaucrats to start units in non-conforming areas and the political parties both ruling and opposition who have contributed greatest to the mess by either giving licenses earlier (Daily Excelsior, 2000) and still espousing the cause of non-conforming units by altering the MPD, as a device to overcome the ruling of Apex Court in this way trying to put at stake once again the clean water and fresh air for over 10 million Delhi's residents.

The alteration involves reclassifying a residential area to an industrial area if 70 per cent of the area is industrial. Such a conversion of *de-facto* into *de-jure* would legalize a large number of units. This amounts to legalizing what is clearly and simply illegal. In fact, apart from being blatantly violation of the Apex Court's mandate this might lend a sort of legitimacy to the polluters and encroachers. The pressure has been handled effectively by the urban minister, Jagmohan, who pointed out that the interests of the polluting industries cannot be pitted against the health, well-being and peaceful living of the entire people in the NCT of Delhi. According to him "changes in the Master Plan or its norms to accommodate illegal activities would not only put premium on lawless activity, but would also cause injustice to law-abiding citizens who have continued to use their residential premises in accordance with the provisions of the Plan" Jagmohan as told to Rajya Sabha (The upper house of Indian parliament) in a written reply. He further stated that the figure of 70 per cent concentration of industries in these areas as a norm for reclassification to industrial area is not supported by any empirical evidence or data, nor has it been reconciled with documentary evidence of electoral rolls, ration cards, house tax, electricity and water connection bills etc. (Raja Sabha, 2000).

Another amendment that was suggested in the MPD is the definition of a non-conforming unit. i.e., any unit using one kilowatt of power load and employing 5 persons is treated as an industrial unit. The proposed suggestion was that definition of a non-conforming should be changed to any unit having 5 kilowatt and 20 persons due to the advanced technological and industrial inputs being introduced on a regular basis (The Hindustan Times, 2001).

Many non-conforming units function in residential areas. The exact number is not known as figures vary. For instance, in Delhi the estimates vary from 80,000 (according to Municipal Commissioner) to 100,000 (as per DPCC) to 270,000 (according to the Delhi Electric Supply Undertaking, DESU) (Benjamin and Mengani, 1996). Estimating the number of units is easier than finding out the exact number of workers, as there is gross under-reporting in order to escape the ambit of labour laws.

The workers, mostly migrant labourers work long hours in inhuman conditions without proper ventilation, lighting and sanitation. Employment of child labour cannot be ruled out. The situation of workers in industries like electroplating, anodising, PVC wire manufacture etc. is worse as there are inherently hazardous and workers are forced to inhale toxic fumes and dust and handle hazardous chemicals without any insurance against accidents (Kathuria, 2001). The PVC granules (locally referred as dana) manufacturing, leads to the emission of dioxins, some of which have been declared by the WHO as causing cancer. Besides, to give PVC flexibility and strength certain plasticising phthalates are used. Exposure to there over a sustained period can harm the reproductive system, and cause cancer of the liver and the kidneys. (Toxic Link, 2000). Besides workers in dingy units, those most exposed to toxic substances and chemicals are their families, as because of their low economic conditions can only afford to live in jhuggi-jhopris, many of which are in the vicinity of industrial areas. It is they who drink the ground water upon which, according to recent Federation of Indian Chambers and Commerce report entitled 'the resurgence of urban India' 45 per cent of slum dwellers in Delhi depend into which toxic pollutants seep.

In Wazirpur in North Delhi, a centre of steel pickling and electroplating industry, acids and chemicals used in steel processing collect in lethal puddles on the road. The air is heavy with the smell of acid. The drains overflow during the monsoon forcing people to walk through acid laden water. Over years of industrial activity, the acid has seeped into the ground and contaminated the ground water. Piles of solid waste can be seen on the edge of potholed roads.

Murty & Kumar (2002) assessed the cost of industrial water pollution abatement and found that these costs account for about 2.5 per cent of industrial GDP in India. (Parikh, 2004) shows that the cost of avoidance is much lower than damage costs. According to one estimate India lost about Rupees 366 billion, which account for about 3.95 per cent of the GDP, due to ill effects of water pollution and poor sanitation facilities in 1995 (Parikh, 2004). If India had made efforts for mitigating these affects in terms of providing better sanitation facilities and doing abatement of water pollution the required resources had ranged between 1.73 to 2.2 per cent of GDP. It may however, be emphasized that these damage costs do not fully reflect the loss in social welfare. These approximations only suggest that the abatement of pollution is socially desirable and economically justified.

It is quiet astonishing that water being elixir of life is facing colossal challenge from waters of river Yamuna as many Sewage Treatment Plants (STP) and Interceptor Sewage Projects (ISP) are yet to become operational or work on them has not been completed. After 22 years of Supreme Court monitoring and spending around rupees 4988 crore to clean Yamuna by governments of Delhi, Haryana and Utter Pradesh Yamuna continues to stink like a sewer in Delhi. The apex court like past has once again asked concerned agencies to file status report on above mentioned projects. Solicitor General Ranjit Kumar, as Amicus Curiae to court since 1994, told the bench that task of cleaning Yamuna has become difficult as the amount of untreated sewage flowing into river has increased exponentially and this has reached 1500-1600 million gallons per day (Economic Times, 2017). Agonized by unfruitful attempts by the governments of Delhi, Haryana and Utter Pradesh to clean a stretch of 22 kilometres of river which passes through Delhi has asked them to provide details that how much and where such huge amount was spent during the last two decades to clean river Yamuna (Economic Times, 2017).

## Air Pollution

It is estimated that about 3,000 metric tons (mt) of air pollutants are emitted every day in Delhi. The sources of air pollution in Delhi are: emissions from vehicles (67%), coal based thermal power plants (13%), industrial units (12%)) and domestic (8%). In 1991 the air pollutants emitted daily were 1,450 mt. There has been a rising trend, 1992: 1,700 mt, 1993: 2,010 mt, 1994: 2,400 mt, 1995: 2,890 mt. (White Paper on Pollution in Delhi, 1997).

In Delhi about 10,500 vehicles were added every month, consequently the number of motor vehicles in Delhi has increased from 5.73 lakhs in 1981 to 28.48 lakhs in 1998 and since then there has been an exponential growth of vehicular traffic on the roads of Delhi. On the other hand, the road space has not increased proportionately. The present number of motor vehicles in Delhi is more that of Mumbai, Calcutta, and Chennai put together. The result is extreme congestion on Delhi Roads, ever slowing speeds, increase in road accidents, fuel wastage and environment degradation.

## National Ambient Air Quality Standards

The **Table 9** shows national standards for ambient air quality notified under the Air (Prevention and Control of Pollution) Act, 1981 and in the Environment Protection Act, 1986.

The standards specify the maximum limit to which major air pollutants, such as Sulphur dioxide, oxides of nitrogen, suspended particulate matter, etc. are permitted in various zones which could be industrial, residential and sensitive zones. The monitoring of air quality is undertaken under the National Ambient Air Quality Monitoring Programme.

*Table 9. National ambient air quality standards*

| Air Pollutant | Time-weighted Average | Concentration in Ambient (micrograms per m$^3$) |||
|---|---|---|---|---|
| | | Industrial | Residential | Sensitive |
| Sulphur dioxide (SO$_2$) | Annual Average | 80 | 60 | 15 |
| | 24 Hours | 120 | 80 | 30 |
| Oxides of Nitrogen (NO$_2$) | Annual Average | 80 | 60 | 15 |
| | 24 Hours | 120 | 80 | 30 |
| Suspended Particulate | Annual Average | 360 | 140 | 70 |
| | 4 Hours | 500 | 200 | 100 |
| Respirable Particulate Matter (Size less than 10 microns) | Annual Average | 120 | 60 | 50 |
| | 24 Hours | 150 | 100 | 75 |
| Lead | Annual Average | 1.0 | 0.75 | 0.50 |
| | 24 Hours | 1.5 | 1.00 | 0.75 |
| Carbon Monoxide (mg/M$_3$) | 8 Hours | 5.0 | 2.0 | 1.0 |
| | 1 Hour | 10 | 4.0 | 2.0 |

**Source**: MOEF

**Monitoring:** The Central Pollution Control Board has been monitoring the ambient air quality regularly at various locations in Delhi, measuring Sulphur dioxide, oxides of nitrogen and particulates. The atmospheric concentrations of air pollutants have shown a rising trend. The monitoring stations are located at Ashok Vihar, Shahzadabagh, Siri Fort, Janakpuri, Nizamuddin, Shahdara. The ambient air quality data indicates high values of suspended particulate matter at all monitoring stations, namely, 367-452 µg/m3 (micrograms per cubic metre) on annual average basis as against the prescribed standards of 140-360 µg/m3. Though the annual mean value of Sulphur dioxide (15-26 µg/m3) and oxides of nitrogen (28-46 µg/m3) remain within the prescribed limits of 60-80 µg/m3, there is a rising trend (White Paper on Pollution in Delhi, 1997).

Compared to 1989, Sulphur dioxide atmospheric concentrations in 1996 have registered a 109 percent rise, and oxides of nitrogen 82 percent rise. The suspended particulate matter atmospheric concentration has shown only a nominal rise, because of installation of electrostatic precipitators by thermal power plants in Delhi.

To assess the impact of vehicular emissions on the ambient air, the Central Pollution Control Board monitored the air quality in terms of Sulphur dioxide, oxides of nitrogen and carbon monoxide at 10 major traffic intersections and at the IARI (Indian Agriculture Research Institute) campus, a relatively low pollution area. The study revealed marked differentials in air pollution between the IARI campus and the intersection points. It is notable that two thirds of the vehicles in Delhi are two-wheelers, operated on two stroke engines accounting for 70 percent of hydrocarbons and 50 percent of carbon monoxide emissions. The Central Pollution Control Board has been monitoring carbon monoxide at the ITO traffic intersections since 1989. The concentrations of carbon monoxide in 1996 show an increase of 92 percent) over the values observed in 1989. The annual average values are given in the **Table 10.**

**Lead:** The particulate lead concentrations appear to be getting in control which is attributable to the de-leading of petrol and restrictions on lead handling industrial units.

**Low Sulphur Content Diesel:** When we go by the definition of clean diesel in the developed countries like United States is one with Sulphur content of 0.001 per cent and also of much advanced engine technology and expensive particulate traps. Delhi Transport Corporation (DTC) has just about managed to get diesel quality with 0.05 per cent Sulphur about 33 times dirtier than California quality (Centre

*Table 10. Carbon monoxide concentration at ITO crossing, Delhi*

| Year | Annual Mean Concentration | % variation (Base 1989) |
|---|---|---|
| 1989 (Jan. – Dec.) | 2905 | – |
| 1990 ((Jan. – Dec.) | 2688 | (–) 7 |
| 1991 (Jan. – Dec.) | 3464 | (+) 19 |
| 1992 (Jan. – Dec.) | 3259 | (+) 12 |
| 1993 (Jan. – Dec.) | 4628 | (+) 59 |
| 1994 (Jan. – Dec.) | 3343 | (+) 15 |
| 1995 (Jan. – Dec.) | 3916 | (+) 35 |
| 1996 (Jan. – Dec.) | 5587 | (+) 92 |
| 1997 (Jan. – Aug.) | 4847 | (+) 103 |

**Source:** MOEF

for Science and Environment, 2000). Other countries are taking drastic steps to control diesel as a fuel when particulate levels in their countries hover around 60 to 70 microgram per cubic metre as opposed to a shocking 820 micrograms per cum in Delhi (Centre for Science and Environment, 2000).

The **Table 11** shows variation of Particulate Lead Concentration by comparing annual mean of 1989-90 with 1996-97.

*Table 11. Particulate lead concentration*

| Location | 1989 – 90 (microgrammes/m³) (Annual Mean) | 1996 – 97 (microgrammes/m³) (Annual Mean) | % Variation Base 1989 – 90 |
|---|---|---|---|
| Ashok Vihar | 0.200 | 0.179 | (–) 11 |
| Sirifort | 0.117 | 0.161 | (+) 27 |
| Janakpuri | 0.112 | 0.142 | (+) 21 |
| Nizamuddin | 0.218 | 0.092 | (–) 58 |
| B.S.Z. Marg | 0.251 | 0.210 | (–) 16 |
| Shahdara | 1.570 | 0.222 | (–) 86 |
| Shahzadabagh | 0.391 | 0.381 | (–) 3 |
| Average of Seven Stations | 0.408 | 0.198 | (–) 54 |

Source: MOEF

Air quality standards need to be made stricter if a lower level of pollutant affects health more than previously thought. However, in India, the standards were revised only once in 1994 to create a new category called respirable suspended particulate matter (RSPM) to account for small particulate emissions. Facilities were, however, not created to monitor RSPM separately (Aggarwal, 1993).

The then CPCB Chairperson D.K. Biswas said, "We have reviewed the NAAQS (national ambient air quality standards) in 1993 to include respirable particulate emissions. According to CPCB method, 40 per cent of the total suspended particulate matter is respirable. So any rise in total suspended particles would indicate the proportionate rise in respirable particulate emissions".

The annual mean concentration of $NO_2$ in Delhi rose steadily from 20μg/m3 in 1987 to 47μg/m3 in 1995. However, the rise in the maximum levels from 47μg/m3 in 1987 to 324μg/m3 in 1995- is even more worrying (The Citizens Fifth Report, 1997). Benzene is not monitored by CPCB. However, short-term measurements by research organizations indicate very high atmospheric concentrations in cities such as Delhi, and Calcutta. MP Keuken of Netherlands Institute for Applied Research conducted a study on benzene levels in Janpath. Air samples taken in October 1996 revealed concentrations of benzene 10-12 times higher than the European Union standards (Keuken, 1998).

Though the European Union has regulated standards for air borne benzene at10 μg/m3 but WHO considers there is no safe level for airborne benzene. The toxic effects of benzene in humans include damage to the central nervous system and the immune system (Chowdhury, 1998).

Keuken in his study Result of Preliminary Measurements and Segregations for a cost Effective Air Pollution Assessment mentions in the West of Janpath, the concentration measured was as high as 139μg/

## Environmental Problems of Delhi and Governmental Concern

m3 and in the east it was 115µg/m3. The average level of benzene measured in four sites in the heart of the capital exceeds 120µg/m3. This compares poorly with other cities in the world about 35µg/m3 in Cairo, 5µg/m3 in Berlin, and about 25µg/m3 in Rotterdam (Keuken, 1998).

Keuken stresses that his limited survey is only indicative of the alarming situation in Delhi and there is need for an elaborate survey. Keuken's findings have evoked strong responses from some Indian experts. H.B. Mathur, who chaired the committee to set mass emission standards for vehicles in 1992, feels benzene monitoring should be taken up on a priority basis, particularly as petrol with a high level of aromatics and benzene is consumed in India. "We have introduced unleaded petrol in great hurry, without lowering the levels of the benzene in it to a desirable level. It is more than 5 per cent in India, while the rest of the world is moving ahead to meet the limit of one percent", he says. Another study being conducted by C.K. Varshney of the School of Environmental Sciences, Jawaharlal Nehru University, shows that there has been a 50% rise in benzene levels after the unleaded petrol was made mandatory in Delhi for all categories of vehicles in September 1998 (Varshney, 1989). According to this study the levels of benzene and toluene at the two traffic intersection-All India Institute of Medical Science and Dhaula Kuan were 61-127µg/m3 and 125-252µg/m3, respectively (Varshney, 1989).

**Ozone Level:** According to report published in September 1997 by Central Road Research Institute (CRRI) - a government agency - the ozone level had already crossed the danger mark in 1993. In the Winter 1993, CRRI, monitored ozone levels at seven sites in Delhi. The average levels of ozone at Parliament Street, Daryaganj, Paharganj, Karol Bagh and Vasant Kunj measured over eight hours during the day exceeded the WHO mean standards of 100-120 micro grams/m3 by 10-40 per cent (Singh et al. 1997).

In areas of heavy traffic congestion such as Karol Bagh, Daiyaganj, and Parliament Street, the maximum ozone level was as high as 238 micro grams/m3, 251.6 micro grams/m3 and 247.8 micro grams/m3 respectively. The CRRI scientists point out that even short term exposure to high concentrations of ozone can be very lethal and therefore, talking generally about mean or average levels do not help to understand the damaging impact of ozone on health. Studies conducted in other countries have shown that short period of exposure to high ambient concentration of ozone can adversely affect lung function of even healthy people by causing inflammation of airways and attendant symptoms such as shortness of breath, wheezing and chest pain. Thus, the WHO has set two standards for ozone concentrations. The standard for an eight-hour average is 100-120 micro grams/m3 and for peak concentration for duration of one hour, the standard is 150-200 micro grams/m3 (Chowdhury, 1997).

## CNG as a Fuel: A Step Towards Abatement of Air Pollution

The increasing pollution levels became a matter of concern because of automobiles particularly those running on diesel and two wheelers and auto rickshaws powered by two-stroke petrol engines. Particulate and other pollution from centrally located electric power stations operating on poor quality of coal or by-passed pollution control equipment was ignored (The Hindu, 2001).

Way back on January 7, 1998, the Supreme Court directed the Ministry of Environmental and Forests (MoEF) to constitute an Environmental Protection (Prevention and Control) Authority (EPCA) to control the pollution in NCR. From this perspective EPCA is vested with powers to ensure compliance with regard to standards for the quality of environment in its various aspects, standards for emission or discharge of environmental pollutants from various sources and so on (Frontline, 2000).

Vehicular transport in Delhi is a major source of pollution and the powers of the EPCA specifically refer to this effect. To control vehicular pollution, the Authority has the powers "to take all necessary steps to ensure compliance of specified emission standards by vehicles including proper calibration of the equipment for testing vehicular pollution, ensuring compliance of fuel quality standards, monitoring and coordinating action for traffic planning and management".

With a view to check rapid deterioration of air quality in Delhi, which has become a health hazard certain directions have been issued by the Supreme Court from time to time (M.C. Mehta V. Union of India, 1985). On 28th July, 1998, some further directions were issued fixing a time schedule after taking note of the recommendations made by the Bhure Lal Committee. The Bhure Lal Committee stressed the importance of the use of CNG (Compressed Natural Gas) as a fuel and noted that it was imperative to have increased use of CNG as a fuel in Delhi and made various recommendations in this regard which were adopted in the form of directions by the Supreme Court.

The Supreme Court, on the basis of annexure VI recommendations of Bhure Lal Committee, has directed in its order dated 28th July 1998 with respect to Delhi:

1. That all pre-1990 autos and taxis be replaced with new vehicles running on clean fuels by 31st March 2000.
2. That financial incentive is made available for replacement of all post-1990 autos and taxis with new vehicles on clean fuels by 31st March 2001.
3. That by 1st April 2000, no 8 year buses are to ply except on CNG or other clean fuels [direction (f)].
4. That the entire city bus fleet be converted to single fuel mode of CNG By 31st March 2001 [direction (g)].
5. That Gas Authority of India Limited should expedite and expand, from 9 to 80 CNG outlets in Delhi by 31st March 2000.

Unfortunately, neither the Government authorities nor private bus operators acted seriously or diligently in taking steps for the purposes of complying with the aforesaid directions [direction (f) and direction (g)] the court said in its order on 26th March 2001 (M.C. Mehta V. Union of India, 1985). A number of applications have been filed and requests made at the bar seeking extension of deadline to convert the entire city bus fleet into a single fuel mode of CNG beyond 31st March 2001. In the applications filed for extension of time, difficulties being faced by the transporters because of non-availability of CNG conversion kits free from all defects; conversion of CNG at reasonable prices; lack of stabilization of CNG technology in respect of public transport as also the non-availability of CNG and CNG cylinders have been pointed out. There is, however, no satisfactory explanation offered either by the administration or the private transporters as to why they were sleeping over all this time and did not point out the difficulties earlier (M.C. Mehta V. Union of India, 1985). Keeping in view the interest of the health of citizens, the Court declined to give any blanket extension of its directions (g) and (f) as contained in the order dated 28th July 1998. However, in public interest and with a view to mitigate the sufferings of the commuter public in general and the school children, in particular made some relaxation or exemptions, such as:

1. Those schools which have as on 31st March, 2001, placed firm orders for replacement or conversion of school buses owned by them to CNG mode, but who have not so far obtained such buses running on CNG mode, are permitted to run their existing buses, equal to the number of buses for which conversion orders have been placed, provided such buses are not more than eight year old, upto 30th September, 2001.

### Environmental Problems of Delhi and Governmental Concern

2.  It was represented on the behalf of contract carriage operators of inter-state and tourist buses that the applicants were under the bonafide impression that the expression "city bus fleet" in the direction (g) of the order dated 28th July 1998 was not meant to take within its ambit buses owned by such tour operators as they run mostly on inter-state routes as Luxury Coaches. Even if that be so, their cases would certainly be covered by conditions (f) of the order dated 28th July 1998. Even if, it was bonafide believed that these buses were not to be converted to single fuel mode of CNG, they could not in any case ply except on CNG or other clean fuel, such buses which were not more than 8 year old. Diesel fuel particularly of the type available in India, is not regarded as a clean fuel whereas unleaded petrol with low benzene is considered as clean fuel, these bus operators definitely need to comply with the directions given by the Apex Court on 28 July, 1998 and it is for them to switch over to CNG or other clean fuel.
3.  The bus operators, including stage carriage permit holders, owners of other commercial vehicles, including autos, who have placed orders by 31st March 2001 for CNG buses or for conversion to CNG mode, were permitted to operate equal number of their existing buses which are not more than 8 year old till 30th September 2001. The various governmental bodies that were to ensure implementation of the court's orders either sat quiet (hoping that the "problem" would somehow disappear) or actively attempted to sabotage the implementation. Some of these actions were born of ignorance or at the behest of the vested interests (The Hindu, 2001). For example 32 months, and more was more than adequate for ensuring that enough CNG filling stations were in place to fill all vehicles that needed the gas and for setting up a proper distribution system to supply these filling stations in turn with CNG. Yet little was done to ensure either, clearly a failure on the part of both the Ministry of Petroleum and Natural Gas and the Delhi Administration in spite of the fact that Delhi is close to (Hajira-Bijaypur-Jagdishpur) HBJ gas pipeline and the CNG has been available in a limited way for automobile use since the early 1990s.

In its hearing on August 17, 2001, the Supreme Court bench comprising Chief Justice A.S. Anand had said: "We had been informed time and again that there is adequate availability of CNG to meet the requirements of entire city bus fleet, even today we are informed that there is no shortage of CNG to meet the present or future demand" (The Times of India, 2001).

The Supreme Court's Bench in its order on 26th March 2001 said: "It was contended before us that low Sulphur diesel should be regarded as a clean fuel and buses be permitted to run on that. It was submitted that in some other countries ultra-low Sulphur diesel (ULSD), which has Sulphur content of not more than 0.001 per cent, is now available. We direct the Bhure Lal Committee to examine this question and permit the parties submit their written representations to the committee in that behalf". The committee was asked to submit a report within a month as to which fuel can be regarded as clean (The Hindu, 2001).

In July 2001, the Bhure Lal Committee submitted its report on clean fuels to the Supreme Court and said clearly that there is no shortage of CNG. "There is no shortage of the gas as such but an enhanced and adequate allocation for Delhi's transport sector is needed... the allocation should keep pace with the demand".

On August 21, 2001, Union Petroleum Minister Ram Naik said: "we can provide enough gas for vehicles that have booked or ordered for CNG conversion till March 31, 2001" (The Times of India, 2001). While on August 30, 2001 Naik gives a contrasting statement in the parliament, he said: It would not be possible for the Union Government (read petroleum ministry) to cater to the growing demand of CNG in the Capital" (The Times of India, 2001). The Bhure Lal Panel had said: "The contention of

Petroleum and Natural Gas Ministry on the availability of CNG is not convincing. The HBJ pipeline has a capacity of 33.4 million standard cubic metres per day (mmscmd). Delhi has been given an allocation of 3.08 mmscmd. The power sector get 2.60 mmscmd and other sectors get 0.48 mmscmd (0.15 mmscmd for transport and 0.33 mmscmd for cooking gas in the households) (The Times of India, 2001). Since a paltry 0.15 mmscmd is allocated from the total capacity of 33.4 mmscmd, experts have argued that a mere enhancement allocation of CNG would be enough to meet the growing demands. Environmentalist Anil Aggarwal, who was member of the Bhure Lal Committee says, "The allocation can be easily enhanced by the gas linkages committee of the petroleum ministry. A mere 2 mmscmd would be enough to tide over the crisis" (The Times of India, 2001).

The Sheila Dixit the then chief minister of Delhi admits that CNG mother and online stations for buses and outlets for autos and taxis were grossly inadequate (Frontline, 2001). According to Jaswant Singh Arora, President of the Federation of the Delhi Transport Unions congress, both the governments were to blame for crisis. The Central government was not able to provide gas even for those vehicles in the CNG. Each taxi and auto rickshaw driver had to spend two to three hours to get his tank filled (Frontline, 2001). The government now appears to be seized of the recent problems caused by the CNG crisis as adequate number of CNG filling stations has come up.

At a phase when there were many excuses for CNG fuel in Delhi the developing country like Pakistan was running a staggering 1,60,000 vehicles on CNG and without a hitch as stated by Central Pollution Control Board chairman Mr. Dilip Biswas. He said Pakistan has been safely running these buses on CNG. Their conversion process has been on far the past fifty-six years (CleantechIndia).

## THE POLLUTANTS ARGUMENT

It would be desirable at this stage to briefly describe the nature of automobile pollution that is being argued about. It has been known for some time that particulates in diesel exhaust are often "coated" with the probably carcinogenic polycyclic aromatic hydrocarbons (PAH). Modern diesels emit very small quantities of particulates especially when they run on high quality low Sulphur fuel. However these particles are very small and can even penetrate the deep lung. Such as EPCA in its Fifth Report reveals that (1) Suspended Particulate Matter (RSPM) are very harmful pollutants with RSPM being carcinogenic. RSPM is deadly because they are breathed deep into the lungs and lodged there. (2) The modern diesel engine with advanced emission control technology, while reducing the total mass of particulate emission, would increase the number of ultra-fine particles, less than 2.5 micrometre in diameter. In support of this EPCA has relied upon the 1998 Report of the California Air Resource Board which has listed diesel particulate as a Toxic Air Contaminant (Frontline, 2000: 82).

Since petrol too has high fraction of Sulphur (0.2 per cent) in India, the problem of Sulphate particle emissions would be there for petrol engines too, on the other hand benzene and the other aromatic compounds in petrol (gasoline) are known carcinogens, so replacement of diesel vehicles with petrol ones is not necessarily a good idea on environmental grounds.

Compared to both diesel and petrol engines, emissions from vehicles running on CNG and LPG are very low. In addition, the effects of these emissions are also very low regardless of the weather (The Hindu, 2001).

The following chart shows the 'pollution performance' of widely representative engines operating on diesel, petrol, CNG and LPG.

## Exhaust Emissions and Effects With Gasoline, LPG, CNG, and Diesel

The **Figure 3** shows empirical values taken from five spark-ignition and five diesel-engine vehicles, status 1993 in Europe, including indirect emissions (manufacture and transport).

*Figure 3. Exhaust emissions from various fuels (a) Pollutant emissions (b) pollutant effect ("greenhouse gas" assessment, status 1998)*

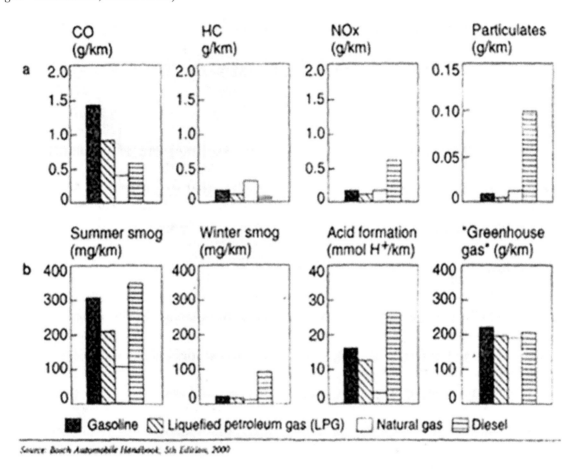

The figure clearly shows that superiority of the latter gaseous fuels. The engine designs are representative of current Indian practice (Robert Bosch of Germany is the world's leading manufacturer of diesel and petrol fuel injection equipment and is unlikely to publicise information potentially contrary to its own interest unless it is incontestably true).

Both CNG (which is 80 to 99 per cent methane) and LPG (mainly propane and butane) also do not contain Sulphur that can cause havoc in the combustion chamber. Most important these fuels cannot be adulterated in the many ways common in India.

## Safety

CNG is stored at high pressures in special tanks both on automobiles and at filling stations. These need to be specially manufactured and carefully installed. The current manufacturers and installers seem to be up to the job, but it is desirable (if not essential) that independent inspection agencies examine their work initially and at five yearly intervals. The gas itself is lighter than air and quickly and safely disperses to non-combustible levels if it leaks. In this respect it is much safer than diesel or petrol. Since the Supreme Court's orders that all Delhi's buses would run only on CNG along with other measures aimed at curbing vehicular pollution, there have been a plethora of statements by various authorities, and numerous articles and commentaries, furiously debating the pros and cons of CNG, especially as compared to other environment-friendly fuels, with a chorus of voices being raised against the singling out of CNG as the only option. Prominent among them is the Tata Energy Research Institute which is of the view that Ultra Law Sulphur Diesel (ULSD) is better than CNG. The institute's study rests its opposition to CNG on the results of one set of measurements conducted on one bus of the London Transport Buses in 1996/1997. The study claims to have found that a Euro II complaint diesel engine running on ULSD (with 0.005 per cent Sulphur) and fitted with continuously regenerating traps (CRT, which control particulate emissions) achieves emission results better than the CNG buses (Down to Earth, 2001).

Since its publication the study has come under serious scrutiny by several agencies that find it flawed in terms of the methods used. A 1998 study by the Expert Reference Group, commissioned by the Australian government, has been presented in the court by Fali Nariman, counsel for Tata Engineering and Locomotive Company. This, too, is based on London study. TERI conveniently forgot to inform the people of a 2000 report by the Australian Council for Scientific and Industrial Research Organization, which debunks the London study and states that CNG and LPG are the best options for combating air pollution and global warming (Down to Earth, 2001). While the balance of the evidence at present appears to be in favour of CNG, this does not mean that Supreme Court or the Bhure Lal Committee assisting it arrived at the correct decision in ordering that Delhi switch over completely to CNG. In fact this has been a grave error in that emissions should have been made the role criterion and yard-stick rather than the type of fuel used. For instance, CNG-fueled vehicles may today give the best emission figures but some better-fuel engine technology may give even better results tomorrow. In the USA two federal laws - the 1990 Clean Air Act (CAA) and the Energy Policy Act of 1992 (EPACT) - required certain fleets to operate on alternatives to petroleum fuels. The CAA requires individual States to implement clean-fuel fleet programmes and the EPACT requires the department of energy to implement an alternative-fuel fleet programme. The Law requires that 10 per cent of all states and 30 per cent of all fuel-provider new vehicle purchases should be alternative fuel vehicles. The requirements increase each year, and were in 1999 extended to municipal and private fleets. By 2006, 75 per cent of all municipal and private fuel providers and 70 per cent of all municipal and private new vehicle purchases must be alternative fuel vehicles. As per these laws, alternative fuels could be CNG/LNG, ethanol, propane, various bio-fuels or vehicles could be powered by fuel-cells, batteries etc., so long as they all conformed to common emission standards prescribed by the federal EPA and by respective state governments (Peoples Democracy, 2001).

At present, the State of California has the USA's most stringent emission standards, being even stricter than the federal standards. These emission standards have been prescribed for low-emission vehicles (LEV), Ultra Low Emission Vehicles (ULEV), Super Ultra-low Emission Vehicle (SULEV) or Zero Emission Vehicles (ZEV), the aim being to bring all vehicles at least to ULEV levels by 2007-2010. The State offers subsides for new purchases of such vehicles or conversions of older vehicles to these standards, provided that the operating fleet or manufacturing company attains prescribed average emission standards for the entire fleet, leaving to the individual fleet the relative proportion of LEV, ULEV, SULEV or ZEV that it acquires or manufactures (Peoples Democracy, 2001).

Once again, as we see, the aim and the criterion used to monitor progress towards it is not the fuel used or the type of vehicle but the level of emission set as the standard to be achieved within a stipulated time frame. The other noteworthy feature of the US system is that the process has been legislated, with clear executive and monitoring roles for the environment and energy/fuel departments or the ministries.

## RECENT GOVERNMENT INITIATIVES

The sectoral composition of Delhi's economy into Primary sector covering of agriculture, livestock, forestry, fishing, mining & quarrying, secondary sector comprising of manufacturing, electricity, gas, water supply and construction is decreasing and tertiary sector also called service sector comprising of trade, hotels and restaurants, transport, storage, communication, financing & insurance, real estate, business services, public administration, is a major contributor in the economy of Delhi and getting boosted regularly. The involvement of primary sector which was 3.85% during 1993-94 has come down to 0.97% in 2004-05 at 1993-94 constant prices. Similarly the contribution of secondary sector noted at 25.20% in 1993-94, has also declined to 19.92% in 2004-05. On the other hand, the contribution of tertiary sector worked out to 70.95% in 1993-94 has improved to 79.11% in 2004-05. This could be attributed to the rapid urbanization and consequential reduction in agricultural and allied activities on one hand and substantial increase in activities relating to the services sector on the other. The other aspect is, regular monitoring of environmental degradation by different government agencies on the directives of Supreme Court and consequent closing of polluting industrial units in and around Delhi (Confederation of Indian Industry, 2017).

The government and judiciary have proactively led the battle against pollution. India is to switch to Bharat Stage VI vehicle emission norms in April 2020, skipping an intermediate stage (Hindustan Times, 2018).

### New Industrial Policy 2010-2021

Industry department has formulated Industrial Policy1 for Delhi 2010-2021 that foresees devolvement of hi-tech, sophisticated knowledge based, service sector and IT industries in Delhi. The vision is to make Delhi a hub of clean, high-technology and skilled economic activities by 2021 by policy shift essentially to change industrial profile from low skilled to high tech and high-tech and high- skilled The objectives of the industries policies are to:

1. Promote high-technology and skilled industries in Delhi to keep in-migration of unskilled labour to a minimum
2. Develop world class infrastructure within planned industrial estates and regularized industrial clusters
3. Promote cluster approach and walk to work concept wherever possible
4. facilitate business through procedural simplification and e-governance measures

5. Promote transparent and business friendly environment The Industrial Policy envisages the strategies like: Infrastructure Development through better O & M of industrial assets, support skill development and other promotional measures like allowing knowledge based industries in industrial areas, decongesting industrial areas through redevelopment process, discouraging polluting industries through higher infrastructure development fee.

## Eastern Peripheral Expressway

This 135 Km long express way located in Haryana was inaugurated by the Prime Minister of India on May 28, 2018. It is expected to bring down the pollution level in Delhi by 27 per cent. The expressway is likely to bring down vehicular pollution by diverting at least 50,000 vehicles from Delhi going towards Jammu and Kashmir, Punjab, Haryana, Uttar Pradesh, Rajasthan and Uttarakhand. The Eastern Peripheral Expressway is India's first highway to be lit by solar power. It also has provisions of rainwater harvesting on every 500 metres on either side (India Today, 2018)

## Learning From China

China has realized that key solution to air pollution lies in restructuring energy consumption and eliminating production of highly polluting industries. In line with these initiatives, Green Peace East Asia has found that PM 2.5 smallest air pollution particles were 54% lesser in the Chinese capital during the fourth quarter of 2017 than during the same period of 2016. Concentrations of PM 2.5 in 26 cities across northern China, the province-sized metropoles of Beijing and Tianjin, were one third lower. (J.P., 2018)

The country since 2013 introduced tough anti-pollution measures such as the national action plan on air pollution. This imposed a nationwide cap on coal use, divided up among provinces, so that Beijing (for instance) had to reduce its coal consumption by 50% between 2013 and 2018. The plan banned new coal-burning capacity (though plants already in the works were allowed) and sped up the use of filters and scrubbers. These measures cut PM-2.5 levels in Beijing by more than a quarter between years 2012-13.State Council of China released another Three Year Action Plan 2018-2020, on air pollution control and roadmap for improving air quality. The plan mentions by 2020, emissions of sulfur dioxide and nitrogen oxide should drop by more than 15 percent compared with 2015 levels, while cities which fail to meet the requirement of PM2.5 density should see their density of PM 2.5, a key indicator of air pollution, fall by more than 18 percent from 2015 level.

Chinese created Environmental Protection Agency, with tough enforcement powers, in Beijing and its surroundings. In 2015 the Clean Air Alliance of China, an advisory group, reckoned that the investment cost of the 2013-18 National Plan in Beijing, Tianjin and the surrounding province of Hebei would be 250bn yuan ($38bn) (J.P., 2018).

On the other hand in Beijing the government made 28 million dollars as fines in 2015 (The Guardian, 2017).To combat vehicle exhaust smoke, which is responsible for one-third of Beijing's emissions, an annual quota of 150,000 new cars was established for 2017, with 60,000 allotted only to fuel efficient cars(The Wire, 2017). The government's efforts range widely such as an alternate to coal, China is rolling out the world's biggest investment in wind and solar power (Gardiner, 2017). In March 2017 the national government announced the closure or cancellation of 103 coal-fired power plants, capable of generating a total of more than 50 gigawatts of power. It said it would also cut steel production capacity by another 50 million tons (Gardiner, 2017).

Another positive thing which lies in controlling pollution is that the Chinese system is much decentralized; the provincial and local city authorities have a lot of power. All the city environmental protection bureaus are in charge of enforcing pollution laws, and one can envisage there is a wide range of enforcement standards across the country.

## RECOMMENDATIONS AND CONCLUSION

### Recommendations

### NCR Planning

Delhi is not an ideal location for industrial development. Yet, for variety of socio-political reasons, it has become a key industrial centre. This not only causes pollution, it also acts as a magnet for migration. Relocation of industries from Delhi should be one of the major objectives and this can be done in close collaboration with National Capital Region Planning Board (NCRPB) and adjacent States.

The physical environmental resources; land, water and air are in critical condition in Delhi and the capacity of Delhi to support an ever increasing population base is doubtful. The solution for containing the further growth of Delhi lies in framing the issue in a regional setting. There is need for an integrated planning and development of the National Capital Region (NCR). The success of NCR depends on the ability to transfer activities, and thereby population from Delhi. The emphasis therefore should have been on encouraging relocation of major activity centres to Priority NCR towns. Decongestion of Delhi, including dispersal of industries, to the NCR may become viable through adoption of a better working conditions and high standard infrastructural facilities. The Government of India should explore the possibility of shifting the headquarters of the public sector units from Delhi in a phased and time bound manner.

### Squatter Settlements and Sanitation

The problem posed by squatter clusters need to be addressed besides development of low cost sanitation and identifying appropriate reforms and investments. The role of improved household domestic practices likes disposal and segregation of waste and self-help schemes need to be encouraged at local level.

### Environmental Statistics

The status of pollution in Delhi will require constant updating through quantitative information relating to environmental quality and related parameters. Environmental epidemiological studies would be needed to assess the impacts of pollution on human health. A system will have to be institutionalized for collecting and analysing data on environmental health of the poorest, e.g. data on upper respiratory and gastro-intestinal illness. The costs and benefits of existing municipal and waste collection and management systems needs to be reviewed, including equipment installation, their utilization and maintenance, using the existing data and field observations.

## Conservation of Environmental Resources

*River Yamuna*

River Yamuna has been polluted to such an extent that quality of water is not even fit for animal consumption due to discharge of untreated effluent carrying drains. It is essential to intercept all drains discharging pollutants into the river. This needs a complete review of existing system of dealing with waste water, and suitable land use revision to allocate treatment plants.

*Ridge*

The ridge of Delhi is a unique environmental resource, and thus needs a careful conservation like notifying the ridge as 'protected zone'. Plantation of more trees in waste land areas will serve as carbon sink.

## Waste Disposal

Proper disposal of storm-water and sewerage is necessary to the health of citizens of the city. To meet this objective, urgent action is required in respect of (i) trapping of storm water drains, (ii) provision for sewer lines and treatment plants and (iii) construction of wastewater treatment plants in industrial estates as well as in individual large industrial units. There is also need for proper drainage of wastewater from slum areas or the areas where sewerage lines has not yet been constructed.

Industrial water effluents, in general, are highly polluting and primary treatment should be the unit level itself and only effluents conforming to the standards of discharge should be allowed to discharge into the public sewer system. The latter should be finally treated before its disposal into water bodies.

Government of Delhi should pay prompt attention to trapping and treatment of all sewage presently flowing through the drains/nallahs in line with its assurances given to the Supreme Court in December 1993 and prepare a comprehensive plan to complete the works in a time bound manner.

## Management of Solid Waste

The disposal of solid waste is done at sanitary landfill sites and the method of collection of municipal solid waste and its disposal is not satisfactory as the collection is done by traditional means. There is need for mechanization of the collection of municipal waste. The municipal waste needs to be managed in a more scientific way by means of incineration/ composting etc. for which Municipal Corporation will need additional funds. Moreover, the solid wastes from hospitals and slaughter houses being hazardous in nature should be disposed of at separate sites. Proper sites have not been developed for disposal of the hazardous waste generated from different types of industrial activities. The development of the proper sites for disposal is primary requirement.

## Transport

Vehicular transport is a major polluter of air so it will be advisable that metro rail transport system be given greater importance and expansion. Without any doubt the metro will bring immense benefits to the city in terms of fast, comfortable, safe and non-polluting means of transport for large numbers of people. Whatever metro train facility is available with government, need is to increase their frequency and spread the network to far-flung areas of National Capital Region so that people will prefer to travel by metro rather their own personal vehicles, hence a means to relieve vehicular pollution burden on Delhi's environment. Synchronization of traffic signaling system should be encouraged for steady and regulated movement of vehicles which minimized chances of air pollution at the road intersections.

## Use of Advanced Technology in Automobiles

The government should impress upon the automobile industries for the use of catalytic converters in different types of vehicles which is already in vogue in developed countries. These devices convert the harmful toxic gases into non-harmful exhausts. Although they will increase the cost of vehicles and which may slow down the demand from consumer side. In this context government should provide subsidies on these vehicles to a limited phase of time period till the technology becomes cheaper in developing societies. Other possibility is that government shall reduce registration fee as well as road tax on purchase of such vehicles which will encourage customers to go for these vehicles.

## Stubble Burning

Stubble burning in neighboring state of Punjab and Haryana is major air pollutant contributor to Delhi's air pollution. The farmers of Haryana feel it as easy route in burning their stubble rather than taking scientific route for its disposal. Ploughing stubble of leguminous family will work as nitrogen fixers; while as non-leguminous crops can be composted which requires government intervention. The constitution of India puts certain obligations on state as well citizens for preservation of environment. Article 48A, a Directive Principle of State Policy, provides that: 'The State shall endeavour to protect and improve the environment and safeguard the forests and wildlife of the country'. Moreover, article 51A(g) imposes a similar responsibility on every citizen to protect and improve the natural environment including forests, lakes, rivers and wildlife, and to have a compassion for living creatures....'Together, these articles, highlight the national consensus on the importance of the protection and improvement of the environment.

# Future Strategies

The modalities of incorporating NCR and Delhi into continuous spatial frame have to be evolved. This may require various sacrifices from Delhi. The guidelines, therefore, are divided into time-based frame of (1) short term, (2) medium term and (3) long term.

## Short Term

1. All urban projects by both public and private agencies should be subjected to Environmental Impact Assessment (EIA), and be implemented only when the EIA is favourable.
2. A land use and development monitoring cell for each zone should be set up in the Delhi Development Authority (DDA), with power to stop misuse of land and initiate prosecution for offenders.
3. The State Pollution Control Board and Central Pollution Control Board (CPCB) should independently monitor pollution and advice DDA for initiating action and if necessarily prosecute themselves.

## Medium Term

1. Formulate employment location and housing strategies in consultation with National Capital Region Planning Board (NCRPB). A task force may be constituted by the NCRPB to formulate action oriented strategy for the purpose.
2. Establish administrative set up to encourage effective dispersal of offices, commercial establishments and industries from Delhi to other towns of NCR.
3. Revise the Plan to designate open spaces and pollution abettor open spaces and city forests so that the quality of ambient air can be improved.
4. Establish administrative machinery to continuously monitor the development of Delhi and set up monitoring stations for recording air pollution and water pollution.

## Long Term

1. A rapid transit network to integrate Delhi with NCR towns should be developed.
2. A population and activity distribution policy for Delhi and NCR, within the objectives of National Urbanization Policy need to be evolved.
3. An environmental fund should be created by instituting cess on various developments, and on various modes of transportation.
4. Continuous monitoring of NCR plan should be done, so that effective interventions can take place.
5. The private sector can be made more socially responsive for a cleaner environment and there needs modifications in corporate strategies compelling greater investment in anti-pollution efforts.
6. Public participation is main requirement for any plan to control pollution; this includes the element of enhancing the general level of awareness about the effects of rising pollution and developing community spirit.

## CONCLUSION

If a good life is the aim of man, then its pursuit and achievement involves the fulfilment of certain conditions. Among them at very first place is the enjoyment of right to life. Right to life in its full content cannot be enjoyed if natural environment is put at stake by various pollutants beyond a permissible level. A number of environmental problems, such as air and water pollution, slum development and congested housing etc. are creating gastrointestinal diseases, respiratory ailments, heart and various microbial infections abortions, miscarriage, stillborn, primary and secondary sterility, psychosomatic disorders, mental retardation, schizophrenia epilepsy and learning disabilities among school going children as observed in advanced societies.

In case of Delhi Apex Court seems to be playing predominant role to control pollution rather executive and legislature should have played proactive role in making Delhi a model city. Delhi needs an Action Plan for a particular phase of time to introduce tough anti-pollution measures. It is quite imperative that an effective policy decision is needed for appropriate allocation of land and control over unsustainable developmental process, that will result in meaningful impact on the existing environment and it is also vital for sustaining the desirable environmental quality in Delhi.

## REFERENCES

Agarwal, A. (1997). *Homicide by Pesticides*. New Delhi: Centre for Science and Environment.

Aggarwal, A. L. (1993). *Air Quality Management Goals in India*. Paper presented at International Workshop on Urban Air Quality Management, Mumbai, India.

Anon. (1996). *Report on Water Quality Monitoring of Yamuna River*. New Delhi: Central Pollution Control Board, mimeo.

Biswas, A. K., & Hartly, K. (2017). Asist K Biswas & Kris Hartly. Delhi should follow Beijing example in tackling air pollution. *The Wire.*

Breese, G. (1974). *Urban and Regional for the Delhi-N. Delhi area, Capital for conquerors and Country*. Academic Press.

Census. (2011). *Govt. of India.*

Choudhary, A. A. (2017). Amit Anand Choudhary. Yamuna pollution: Supreme Court seeks report on sewage treatments plants. Economic Times.

Chowdhury, A. R. (1997). Linking Threat. Down to Earth, Society for Environmental Communications, 16(14), 20-24.

Chowdhury, A. R. (1998). Breathing Benzene. Down to Earth, Society for Environmental Communications, 7(1), 15.

Confederation of Indian Industry. (2017). https://cii.in/WebCMS/Upload/Economic_Overview_Delhi.pdf

Daily Excelsior. (2000). *As many as 15,000 units were given licences even after the Supreme Court's 1996 Order.* Author.

Economic Survey of India. (1999-2000). Planning Department, Govt. of India.

Frontline. (2000a, Feb. 4). Article. *Frontline, 82.*

Frontline. (2000b). Article. *Frontline, 17*(25), 1-7.

Frontline. (2000c). Article. *Frontline, 17*(4), 92.

Frontline. (2000d). Article. *Frontline, 17*(2).

Gardiner, B. (2017). China's Surprising Solutions to Clear Killer Air. *National Geographic.*

Hindustan Times. (2018). *Delhi world's most polluted city, Mumbai worse than Beijing: WHO.* Author.

India Today. (2018). *Eastern Peripheral Expressway inaugurated by PM Modi is likely to decrease Delhi pollution by 27 per cent.* Author.

J.P. (2018, Jan. 25). How China cut its air pollution. *The Economist.*

Kathuria, V. (2001). *Relocation, Ban or Boon*. Chennai: Hindu Survey of Environment.

Keuken, M.P. (1998). *Result of Preliminary Measurements and Segregations for a cost Effective Air Pollution Assessment.* Paper presented at workshop on Integrated Approach to Vehicular Pollution Control in Delhi.

Mahapatra, D. (2012). Half of Delhi's Population lives in slums. *The Times of India.*

M.C. Mehta V. Union of India, (1985). *Writ Petition (C) No. 13029.*

M.C. Mehta V. Union of India and Others, (1985). *Case No. LA. No. 1254, Writ Petition (C), No. 4677.*

M.C. Mehta V. Union of India and Others, (1985). *Writ Petition 4677/1985 (1996. 07.08)*

Mehta, M.C. (2000). Mashesh Chander Mehta in an interview with V. Venkatesan. *Frontline.*

MOEF. (1997). *White Paper on Pollution in Delhi.* Author.

Murty, M. N., & Kumar, S. (2002). Measuring Cost of Environmentally Sustainable Industrial Development in India: A Distance Function Approach. *Environment and Development Economics,* 7(3), 467–486. doi:10.1017/S1355770X02000281

National Capital Region Planning Board. (1995). *Status of the Environment: NCT.* Author.

NCR Planning Board. (1999). *Delhi 1999, A Fact Sheet.* New Delhi: NCR Planning Board, India Habitat Centre.

NFHS. (1993). *Population Research Centre.* Delhi: Institute of Economic Growth.

Parikh, J. (2004). *Environmentally Sustainable Development in India.* Available at http://scid.stanford.edu/events/ India2004/JParikh.pdf

Peoples Democracy. (2001). Weekly Organ of the Communist Party of India (Marxist), 25(18).

Raja Sabha. (2000). *Short duration discussion, shifting of industries from residential areas of Delhi.* Retrieved at www.rsdebate.nic.in

Rajalakshmi, T. K., & Venkatesan, V. (2001). Commuters Crisis. *Frontline,* 18(8), 14-27.

Rajput, A. (2017). MCD polls: Delhi slum clusters wait for promised development. *Hindustan Times.*

Ramachandran, R. (2000). The Lethal Zones. *Frontline,* 17(25).

CAG Report. (2005). *CAG (The Comptroller & Auditor General of India) Report on Government of Delhi.* Author.

Singh, A. (1997). Ozone Distribution in Urban Environment of Delhi during Winter Months. In *Atmospheric Environment.* Elsevier Science Ltd. doi:10.1016/S1352-2310(97)00138-6

The Citizens Fifth Report. (1997). *State of India's Environment.* New Delhi: CSE.

The Guardian. (2017). Jamie Fullerton. Beijing hit by dirty smog but observers say air is getting better. *The Guardian.*

The Hindu Survey of Environment. (2001). Chennai: Academic Press.

The Times of India. (2001). *Naik takes a U-turn, says not enough CNG*. Author.

Toxic Link. (2000). Cloning Bhopal: Exposing the Dangers in Delhi's Environment. Author.

Varshney, C. K. (1989). *VOC and Ozone Pollution: Health Implications*. New Delhi: Jawaharlal Nehru University, School of Environmental Sciences.

World Population Review. (2018). https://worldpopulationreview.com/world-cities/delhi-population

# Chapter 9
# Hydrogen:
## An Environmental Remediation

**Athule Ngqalakwezi**
https://orcid.org/0000-0003-2024-0389
*University of the Witwatersrand, South Africa*

**Diakanua Bevon Nkazi**
*University of the Witwatersrand, South Africa*

**Siwela Jeffrey Baloyi**
*Mintek, South Africa*

**Thabang Abraham Ntho**
*Mintek, South Africa*

## ABSTRACT

*Global warming is a pertinent issue and is quintessential of the environmental issues that the world is facing, and thereby, remedial actions and technologies that aim to alleviate this issue are of paramount importance. In this chapter, hydrogen has been discussed as an alternative energy that can potentially replace traditional fuels such as diesel and gasoline. The storage of hydrogen as a gas, liquid, and solid was discussed. The key issues in hydrogen storage were also highlighted. Furthermore, regulations and legislations concerning the emission of greenhouse gases from fossil fuels-based sources were discussed.*

## INTRODUCTION

During the 1700s, industrialization and modern civilizations improved the standard of living for many people around the world. However, this had adverse effects on the environment, resulting in environmental issues like global warming. In 2015, the United States Environmental Protection Agency (EPA) reported that approximately 6,587 billion metric tons of carbon dioxide ($CO_2$), a major contributor to global warming, were emitted into the atmosphere (Liu et al., 2018). Its significant environmental impact

DOI: 10.4018/978-1-7998-3576-9.ch009

## Hydrogen

can no longer be ignored. Therefore, the world has experienced a paradigm shift in which clean systems and technology must be implemented.

Hydrogen energy has received attention as an alternative energy source to replace traditional sources like diesel, gasoline, and coal. Hydrogen, which has a zero-emission characteristic, is renewable. In addition, its energy content is higher than current traditional energy sources. Table 1 shows the variation in energy content between traditional energy sources and hydrogen.

*Table 1. Energy density and specific energy of different energy sources*

| Energy Source | Specific Energy (MJ\Kg) | Energy Density (MJ\L) |
|---|---|---|
| Diesel | 48 | 35.8 |
| LPG | 46.4 | 26 |
| Gasoline | 46.4 | 34.2 |
| Hydrogen | 142 | 10 |
| Coal | ~30 | ~38 |

Hydrogen contains a higher chemical energy per mass compared to current traditional fuels. It also contains higher energy content by weight compared to gasoline. It combusts more rapidly than gasoline and is significantly less dense than gasoline (AECC, n.d.). For this reason, hydrogen has been considered for storage of on-board applications.

Apart from the introduction of green technology, legislation and goals have been instituted to mitigate current $CO_2$ emissions. This is due to the global population's dependency on fossil fuels, which implies that the environment continues to be polluted with greenhouse gases. The legislation is set to manage emissions while research explores environmentally friendly technologies.

## Legislation on Emissions and the Environment

### Euro Standards

Both remedial actions and legislation to regulate emissions from fossil fuels have been implemented. The European emission standards are set to regulate exhaust emission limits for new cars sold in European Environment Agency (EEA) member states and the European Union (EU). Euro standards govern and limit the emission of nitrogen oxides (NOx), volatile organic compounds (VOCs), carbon monoxide

(CO), and other particulate matter into the atmosphere (Prabhu, Nayak, Kapilan, & Hindasagen, 2017). These standards are updated every four years to reduce the permitted minimum emissions (see Table 2).

*Table 2. Euro standards with minimum allowable pollutants*

| Euro Stand. | NOx | PM | CO | HC &NOx | Year |
|---|---|---|---|---|---|
| Euro 1 | — | 0.14 | 2.72 | 0.97 | 1992 |
| Euro 2 | — | 0.08 | 1.00 | 0.70 | 1996 |
| Euro 3 | 0.50 | 0.05 | 0.64 | 0.56 | 2000 |
| Euro 4 | 0.25 | 0.025 | 0.50 | 0.30 | 2005 |
| Euro 5 | 0.18 | 0.005 | 0.50 | 0.23 | 2009 |
| Euro 6 | 0.08 | 0.005 | 0.50 | 0.17 | 2014 |

The Euro standards have significantly reduced emissions into the environment since the inception of the first set of standards in 1992. Each year, the standards are more stringent. For example, car manufacturers are not allowed to put vehicles on the road without meeting all requirements. The Euro standards regulate the emission of NOx, particulate matter (PM), CO, and hydrocarbons (HCs) into the atmosphere. NOx has been the most difficult to reduce among these pollutants. Technologies, like selective catalytic reduction systems, lean NOx trap, and lean NOx catalyst have been implemented for the abatement of NOx on the environment to meet the Euro standards.

The selective catalyst reduction system injects a liquid reductant agent into the exhaust stream of a diesel engine through a special catalyst. Its liquid reductant, diesel exhaust fluid (DEF), is usually auto-

*Figure 1. Selective catalytic reduction system (Chemical Transfer Solutions, 2015)*

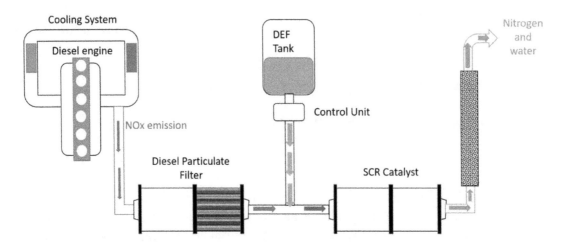

*Hydrogen*

*Figure 2. Sustainable development goals (United Nations, 2019)*

motive grade urea. The reductant initiates the chemical reaction that converts NOx into nitrogen, small quantities of $CO_2$, water, and constituents of the air (Diesel Technology Forum, n.d.).

The lean NOx trap or adsorbents adsorb and store NOx under lean conditions. This technology accelerates the conversion of nitric oxide to nitrogen dioxide utilizing an oxidizing catalyst to increase the storage of nitrogen dioxide as a nitrate on alkaline earth oxides (AECC, n.d.). This process is very sensitive because small changes in the stoichiometric ratios may result in the desorption of the stored NOx (AECC, n.d.). These technologies have been employed to reduce the concentration of NOx into the environment.

## Sustainable Development Goals

Sustainable development goals are set by members of the United Nations (UN) to protect the environment, end poverty, and ensure that all people enjoy wealth and live in serenity by 2030 (Sustainable Development Goals, n.d.). These 17 goals, which represent a blueprint to attain a more sustainable future for all, ensure balance between environmental, economic, and social sustainability.

Several goals (i.e., 7, 11, and 13) address environmental issues. Goal 13, for instance, explores climate action. Progress has been made to ensure that this goal is put into action. Since May 2019, 28 countries have been granted access to green climate fund financing for the development of national adaptation (Green Climate Fund, 2019).

These efforts aim to save the environment and sustain the livelihood of future generations by researching sources of alternative energy. Our dependence on energy derived from fossil fuels. Therefore, the magnitude of fossil fuels cannot be overlooked. Replacing these sources, as well as correcting their damage, will take decades (if not centuries). This has pushed the world into a paradigm shift where green systems and processes must be implemented to secure the health of the environment.

## $CO_2$ Emission Regulations

$CO_2$ is a greenhouse gas that contributes to global environmental issues. Concentrations of $CO_2$ have reached new highs, specifically increasing between 2015 and 2017. The global concentration of $CO_2$ averaged 400.1 ppm mole fractions in 2015 to 405.5 ppm in 2017 (Sustainable Development Goals, 2019). Many states introduced legislation to curb $CO_2$ emissions. In February 2019, South Africa's parliament announced that it will introduce a carbon tax in which companies will be charged for carbon emissions. The law includes a R120 (approximately $10) fine per tonne of carbon emitted for primary greenhouse gas emitters (South African Government, 2019). The law will be implemented in phases between June 2019 and December 2021. Fines are subject to a 2% annual increase (South African Government, 2019). The South African government hopes to lower their level of $CO_2$ emissions through the institution of this legislation.

The U.S. EPA (n.d.), established in 1970, seeks to protect both environmental and human health through legislation and standards. The EPA was formulated to respond to rising environmental issues faced in the U.S. during the 1950s and 1960s. It regulates the use of chemicals and other pollutants, as well as their manufacturing, processing, and distribution (EPA, n.d.). Regulations are enforced through sanctions, fines, and other procedures.

The EPA's recent regulations under the Trump administration curb carbon emissions from automobiles, power plants, and other major carbon emitters that contribute to climate change (Milman, 2018). In 2017, the EPA saved $37 billion through their voluntary partnership programs and avoided emitting about 433 million tons of carbon into the atmosphere. These programs are all effective in reducing the concentration of greenhouse gases in the atmosphere (EPA, n.d.).

Other countries have also joined the fight against climate change by regulating major carbon emitters and developing green technologies. Countries like Peru, Singapore, Canada, Bulgaria, Tunisia, Romania, and Belarus have taken a stand in efforts to reduce and restore the environment for future generations.

## HYDROGEN AS THE ENVIRONMENTAL REMEDIATION

Hydrogen has garnered a lot of attention due to its environmentally friendly characteristics. Hydrogen-related studies fall under the hydrogen economy, a system that is proposed to produce hydrogen and utilize it as an energy carrier. Success in the development and implementation of the hydrogen economy has many benefits for the economy, environment, and energy security (O'Malley, 2015). The hydrogen economy has three legs: (1) hydrogen generation; (2) hydrogen storage; and (3) hydrogen distribution

*Hydrogen*

(Anderson & Gronkvist, 2019). This chapter explores the economy's storage of hydrogen for automotive applications in which hydrogen is a potential source to replace diesel and gasoline.

Hydrogen, as a molecule, has an extremely low density, making its storage difficult and impractical for practical on-board applications. The three-stage storage of hydrogen has been researched and will be discussed in this chapter. The shift from a carbon-based to economy-based economy on renewable, green sources necessitates trustworthy, easy energy storage mechanisms (Abe, Popoola, Ajenifuja, & Popoola, 2019). As such, various methods have been explored for storing hydrogen as a gas, liquid, and solid. These efforts aim to develop robust hydrogen storage materials that will store hydrogen and be equally efficient in terms of kinetics and thermodynamics requirements set by the U.S. Department of Energy (DOE) for practical on-board applications.

The development of hydrogen storage systems must be safe, cost effective, and practical for on-board applications. Acceptable characteristics for a material deemed suitable for hydrogen storage include (Trygve, Sandrock, Ulleberg, & Vie, 2005):

- Cyclability
- Minimized loss of energy during the charge and discharge of hydrogen
- Decreased dissociation temperature
- Rapid kinetics
- Lowered costs in recycling and charging infrastructures
- Longer stability strength against oxygen and moisture or dampness of air
- Moderate dissociation temperature
- Decreased heat dissipation during the exothermic formation of the hydride
- Reversibility and decreased heat of formation to reduce the energy required for release

To date, no material has met all the requirements for practical on-board applications. Current technologies in hydrogen storage include composite tanks for the gaseous storage of hydrogen, liquid storage of hydrogen, and other materials that store hydrogen as a solid like lanthanum penta-nickel.

## Gaseous Storage of Hydrogen

The storage of hydrogen in gaseous form is the most established form of storage (Abe et al., 2019). Hydrogen tends to have a low density at 0.089 kg/m$^3$. Due to this, it requires tremendously low temperatures and very high pressures (Hwang & Varma, 2014; Pesonen & Alakunnas, 2017; Prabhukhot, Wagh, & Gangal, 2016; Salameh, 2014; Sheriff, Yogi, Stefanakos, & Steinfield, 2014). Gaseous storage of hydrogen and liquid storage are two techniques to store hydrogen in its purest form.

Currently, hydrogen fuel cell applications require that hydrogen be pressurized at high pressures (ranging from 35 MPa and 70 MPa). This pressure range is excessively high because hydrogen is extremely light (Abe et al., 2019; Pesonen & Alakunnas, 2017). Due to high pressure, there are safety risks and high possibilities for leaking tanks. As such, a significant amount of money must be invested on materials to construct high-quality storage tanks in fuel cells.

Construction materials for the hydrogen storage tanks include steel, aluminum (Al), and carbon fiber (Abe et al., 2019). Carbon fiber has been a preferred material because of its strength and high resistance to impact (Hirscher, 2009; O'Malley et al., 2015; Salameh, 2014; Sheriff et al., 2014). Gaseous hydrogen

*Figure 3. Glass microsphere (James Shelby and Matthew Hall, Alfred University)*

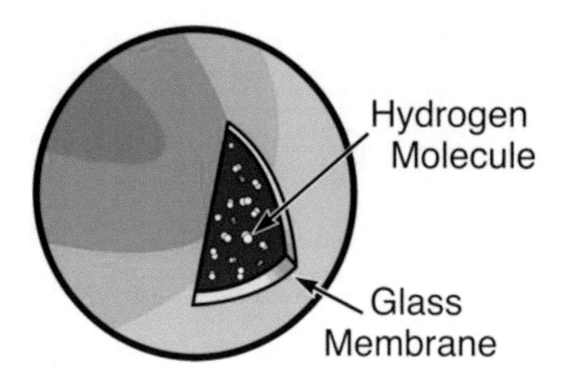

can be stored in composite tanks, underground, or in microglass spheres at elevated temperatures. The storage of hydrogen in these materials will be discussed in this chapter.

## Microglass Spheres

The technique of using glass microspheres for hydrogen storage has been practised for several years due to their light weight, low cost, reusability, environmentally friendly characteristics, mechanical strength, and chemical durability. In this technique, the hydrogen gas diffuses at a high pressure and temperature through the reedy lining of the glass microsphere. The hydrogen is trapped as it cools to an ambient temperature.

The process of hydrogen uptake in microglass spheres can be summed up in three processes: (1) charging; (2) filling; and (3) discharging. The main disadvantage of glass microspheres is their safe storage of compressed hydrogen gas relies on the glass microspheres' absence of leaks. The important technical advantage of glass microspheres is the illustrated storage density of 5.4 wt.% $H_2$. Another disadvantage to this technique is poor dehydrogenation of the gas due to poor thermal conductivity of a packed bed of glass microspheres. Other issues associated with the use of glass microspheres include increased pressures for filling, breakage of the glass microspheres during the cycling process, and achievement of low volumetric density. Issues that have already been addressed include cost, time to refuel, durability, weight, and volume. Issues that need to be addressed include the optimization of hydrogen permeability.

## Composite Tanks

The composite tanks technique also stores gaseous hydrogen. Composite tanks are generally divided into the storage part and the compressor part (Anderson & Gronkvist, 2019). These tanks are manufactured to withstand elevated pressures. Therefore, the material of construction is of high importance. Traditional composite tanks are made up of corrosion resistant, lightweight polymer lining re-enforced with corrosion resistant carbon fiber. The tanks are also covered with an impact resistant dome and fitted with a re-enforced external protective shell.

Commercialization of tanks for hydrogen storage is governed by a set of regulations, codes, and standards for manufacturers. These regulations, codes, and standards look at the scope, nominal working pressure, service life and filling cycles, and tank temperature (Wang et al., 2019). Over the past decade the UN, International Organization for Standardization (ISO), the EU, Japan, Canada, U.S., and China have focused on the development of regulations, codes, and standards for composite tanks for hydrogen storage on-board applications (Wang et al., 2019). Many researchers have discussed the regulations around the construction of composite tanks for hydrogen storage. The development on standards and codes for the storage of hydrogen at high pressure was discussed by Zheng et al. (2012). Mair, Scherer, and Hoffman (2015) also analyzed the minimum burst pressure of composite gas in cylinders in current standards. Regulations, codes, and standards like ISO 19881, SAE J2579, and ANSI HGV 2 govern the technical requirements of these tanks (CSA America, Inc., n.d.; ISO, 2018; SAE International, 2018).

The main advantage of composite tanks is their light weight. This meets the standards of the DOE for on-board applications. In addition, they are commercially available, do not require internal heat exchange, and can safely store hydrogen in pressures ranging from 350-700 bars. The first fuel cell vehicles utilized an on-board pressure of 700 bars. At this very high pressure, a significant amount of money must be invested to ensure the safety of the tank (O'Malley et al., 2015).

There are several disadvantages associated with this technology. First, penalties are given for compressing gas at very high pressures. Second, the large physical volume does not meet DOE targets. Third, there is a significant cost associated with ensuring that the tank's integrity is not compromised at high pressures.

## Liquid Storage of Hydrogen

Liquid storage of hydrogen is another form of its physical storage. Storage as a liquid is often through liquefying it at cryogenic temperatures. This form of storage provides a higher density of about 71 g/L at a cryogenic temperature of about 20 K (Abe et al., 2019). The density of hydrogen as a liquid is 1.8 times higher than hydrogen pressurized up to 70 MPa at 288 K (Midilli, Ay, Dincer, & Rosen, 2005; Webb, 2015). Liquid storage has an advantage in that its high volumetric and gravimetric hydrogen densities meet the standards of the DOE for practical on-board applications. These materials have valuable safety characteristics for practical on-board applications. Liquid storage has been recognized for hydrogen distribution because of the high density (Cardella, Decker, & Klein, 2017).

A key concern with the liquid storage of hydrogen is that the liquefaction process is energy intensive (Anderson & Gronkvist, 2019). According to Anderson and Gronkvist (2019), it requires high energy because hydrogen does not cool down during the throttling process at temperatures above -73°C and hydrogen boils at cryogenic temperatures (-253°C and 1 bar). Furthermore, the throttling process requires a precooling step during liquefaction through the evaporation of liquid nitrogen (Valenti, 2015). This energy intensive process results in a 30% to 40% loss of energy.

Another main disadvantage is that a lot of money must be invested into storage tanks to ensure quality. Specialized double-walled tanks are lined with an efficient insulation system to prevent leakages in the tanks (Abe et al., 2019). Liquid storage faces restrictions because of the required amounts of energy and the persistent boiling off of hydrogen.

There are other ways that hydrogen can be stored as liquid. These include sodium borohydride solutions, rechargeable organic liquids, liquid organic hydrogen carriers, ammonia borane, formic acid, and hydrazine borane. These liquids have many advantages like control and safe on-board generation of hydrogen (Trygve et al., 2005).

However, due to the toxic levels of many of these chemicals, special handling and infrastructures are required. This would add to the cost. Thus, storage of hydrogen in the chemicals will not be conducive for practical on-board applications (Trygve et al., 2005). The liquefaction of hydrogen is well established in the aerospace industry (for example, the National Aeronautics and Space Administration [NASA]), with an installed capacity of 355 tons per day and a large plant that produces 34 tons per day. For on-board application, this would prove to be a challenge.

## Solid State Storage of Hydrogen

Solid storage of hydrogen for on-board application has received a great deal of attention due to its safety and economic advantages compared to liquid and gaseous storage. Much advancement has been made in developing materials in solid storage for on-board applications. Yet none has met the standard instituted by the DOE, which governs and releases hydrogen storage system targets every five years.

The targets include cycle life, volumetric and gravimetric capacities, costs, and refilling time (Durbin & Malardier-Jugroot, 2013). Hua et al. (2011) defined volumetric capacity as the amount of gas in a given volume and gravimetric capacity as the amount of electricity given off by a specified weight of fuel. These two requirements are considered with cost and on-board system efficiency.

Figure 4 shows updated 2020 hydrogen energy targets. Several materials have attracted attention for hydrogen storage applications. However, to date, no material has met DOE requirements and standards for on-board applications.

Various materials have been considered for hydrogen storage. However, lightweight, metal-based materials have been very attractive for solid state storage of hydrogen due to high hydrogen capacities that meet DOE requirement (Khafidz, Yaakob, & Lim, 2016). Table 1 depicts the 2017 revised DOE standard for light duty vehicles.

The kinetics of these materials have hindered their application as an alternative energy. Lightweight metals (i.e., lithium, magnesium [Mg], calcium, potassium, and sodium) have been researched for this application. Lithium has the ability to store energy with high density as hydrogen storage material through the formation of hydride (Wang, Quadir, & Aguey-Zinsou, 2016). The thermodynamic stability of this material has precluded it as an applicable material for practical hydrogen storage. The desorption temperature of this material goes up to 700°C.

Like lithium, Mg is a choice material for hydrogen storage because of its proved high gravimetric capacity of 7.6 wt.% with volumetric density that doubles to that of liquid hydrogen. Mg's other in-

*Hydrogen*

*Figure 4. DOE system standards target for 2020 (US Office of Energy efficiency and renewable energy, 2019)*

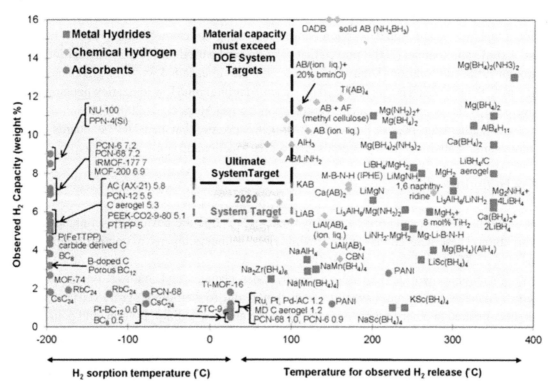

*Table 3. 2017 DOE standards for light duty vehicles*

| H₂ Storage Material Requirements | 2017 | Units |
|---|---|---|
| Gravimetric capacity | 1.5 | Kg H₂/kg system |
| Volumetric capacity | 1.8 | Kg H₂/L systems |
| Cycle life | 1500 | Cycles |
| On-board efficiency | 90 | Percentage % |
| Storage system delivery pressure | 5Fuel Cell/35 Internal Combustion Engine | Bars |
| Charging and discharging rates |  |  |
| System fill time | 3.3/1.5 | Kg H2/min |
| Start time to flow | 5 | S |
| Transient response | 0.75 | S |
| Minimum full flow rate | 0.02 | (g/s)/KW |

teresting properties include recyclability, reversibility, resistance to heat, and absorption of vibrations (Sakintuna, Lamari-Darkrim, & Hirscher, 2007; Selvam, Viswanathan, Swamy, & Srinivasan, 1986). Limitations with this material include slow absorption, increased reactivity with oxygen and air, and desorption temperatures and high desorption temperatures ranging from 350°C to 400°C.

Ataca, Akturk, and Ustunel (2009) proved that calcium can be used to functionalize carbon nanomaterial through surface decoration. Their work predicted a gravimetric 8.4 wt.% for graphene decorated with calcium. Calcium, which is characterized by vacant 3d orbitals and low ionization potential, permits easy movement of charges between calcium and carbon atoms (Ataca et al., 2009).

Many lightweight metals possess interesting hydrogen capacities that meet DOE standards. However, the thermodynamic stability of these materials affects their kinetics and application for hydrogen storage.

## Complex Hydrides

Complex metal hydrides are molecules with hydrogen atoms that are covalently semibound within a polyatomic anion. These materials have high gravimetric capacities compared to standard simple hydrides (Satyapal et al., 2007; Jain, Jain, & Jain, 2010).

Low weight complex hydrides like imides, borohydrides, alanates, and amides have garnered significant interest. Alanates and borates have been studied more due to their higher gravimetric capacity and light weight compared to other groups. Alanates from metals (i.e., sodium, lithium, and potassium) are reported to have high gravimetric capacities. However, alanates have been reported to have decomposition issues during desorption of hydrogen (Sakintuna et al., 2007).

Bogdanovic et al. (2006) proved that the kinetics of sodium alanate can be improved by adding titanium chloride. After his work, more approaches (i.e., nanoscaling, use of catalysts, and destabilization) were employed to develop and improve complex hydrides.

Infiltration into highly porous silica is a reported technique that improves the desorption reaction. Kim et al. (2015) reported on improved dehydrogenation of ammonia borane using this technique. The infiltration technique includes solvent mediated infiltration, ball milling, and melt infiltration (Xia, 2015). However, individually, these methods have been reported to have drawbacks like requirements for solvents that can dissolve metal hydrides, severe collision results in the destruction of porous material, and a decomposition temperature of porous material greater than the melting point of hydrides.

Lithium borohydride is reported to destabilize through nanoconfinement (Fang, Kang, & Wang, 2010). Fang et al. (2010) showed that ball milled lithium borohydride can discharge 10 wt.% hydrogen at 450°C within one hour. Christian and Aguey-Zinsou (2010) confined lithium borohydride with porous carbon materials of pore size 2 nm through the melt infiltration method. This resulted in a significant decrease of the dehydrogenation temperature (460°C to 220°C) and the elimination of both the melting and structural phase transitions (Christian & Aguey-Zinsou, 2010).

Due to doping, complex hydrides are reported to have higher gravimetric capacities than normal hydrides. Figure 5 shows complex hydrides and simple hydrides considered for hydrogen storage.

## Carbon-Based Materials

Porous carbon materials like fullerenes, graphite nanofibers, and carbon nanotubes have been the subject of research as potential materials for hydrogen storage (Chambers, Park, Baker, & Rodriguez, 1998; Trygve et al., 2005; Yoon, Yang, & Wang, 2007). Properties of carbon materials (i.e., large pores,

*Hydrogen*

*Figure 5. Metal hydrides for hydrogen storage*

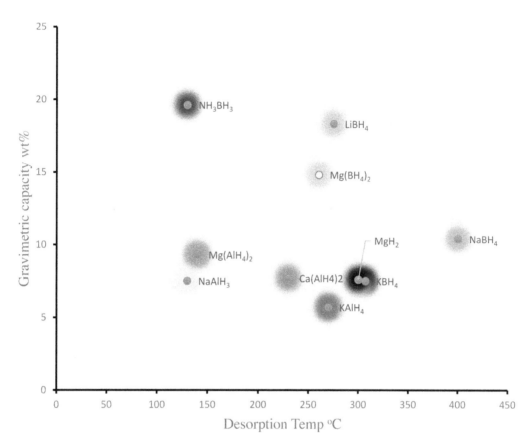

extensive diversity in structural forms, and low densities) have made them interesting, particularly for hydrogen storage (Iijima, 1991).

Dillon et al. (1997) first reported that single walled carbon nanotubes permit hydrogen gas to condense to a gravimetric density between 5 wt.% and 10 wt.%, which results in an increased potential of hydrogen absorption by these materials. Carbon nanotubes are reported to be the best allotropes of carbonaceous material for gas absorption due to their high surface area (Lee et al., 2014).

Hydrogen adsorption in pure carbon nanotubes transpires on the surface of these materials due to weak Van der Waals forces between carbon and hydrogen atoms. Carbon nanotubes possess good reversibility qualities with kinetics that are much faster than metal hydrides. Seemita et al. (2015), after studying the variation of hydrogen storage capacities reported for these materials, found that the discrepancies are due to the sensitivity of hydrogen storage apparatus, existence of metallic impurities in carbon nanotubes, and existence of amorphous carbon (Lee et al., 2014).

Asalatha et al. (n.d.) proved that the presence of nitrogen and boron in carbonaceous materials greatly improves the hydrogen storage capacities and the catalytic activity of these materials. In addition, they reported on the complications and complexities encountered while doping carbon materials with nitrogen and boron, thus making the process obstinate.

Lee et al. (2014) and Tahir et al. (2013) reported graphitic carbon nitride (g-C$_3$N$_4$) as a promising material for hydrogen storage. This material, characterized by several mesopores, has a high nitrogen content and low surface area. Koh, Zhang, and Pan (2012) studied predicted that the high storage capacity of g-C$_3$N$_4$ is due to its different nitrogen structure.

It has been proven that the decoration of carbonaceous material with metals improves the catalytic activity and increases their hydrogen capacities. Decorated carbon materials are said to be the future of hydrogen storage due to faster kinetics and positive reversibility. Prolonged research to find a resolution for slow kinetics and issues associated with properties of metal hydrides make these materials more attractive.

## Hybrid Materials

Hybrid materials are momentarily gaining popularity over metal hydrides due to the endless efforts that have been employed on metal hydrides. Yet they have yielded neither positive nor constructive directions. Carbonaceous materials like carbon nanotube, graphene, and fullerenes are being investigated for practical application in the hydrogen storage economy. According to recent studies, new classes of materials, including H4-alkanes and vanadium- based materials, have been explored for hydrogen storage.

Hybrid materials are slowly gathering research due to their properties. The materials' properties are enhanced by the decoration of their surrounding metal nanocrystals. Carbon nanotube, carbon black, fullerenes, graphene, and carbon nanofibers have been studied in this regard (Antolini, 2003, 2009; Huang & Wang, 2012). According to Antolini (2003, 2009), metal nanoparticles are scattered on highly stable carbon material with high electrical conductivity.

Transition metals, alkali earth metals, and alkali metals have often been used to decorate carbonaceous material to increase their hydrogen capacities (Chandrakumar & Ghosh, 2008; Liu, Ren, He, & Cheng, 2010; Yoon et al., 2007). Yoon et al. (2007) suggested that calcium is a superior metal to functionalize carbon nanomaterial for hydrogen storage due to the 8.4 wt.% capacity obtained on graphene decorated with calcium. The charge transfer between the calcium atom and adsorbed carbon atoms transpires due to the empty three-dimensional (3-D) orbitals and lower ionization potential (Yoon et al., 2007). The discovery of graphene led to other two-dimensional (2-D) carbon allotropes (Yadav, Zhu, & Singh, 2014). The four 2-D carbon allotropes are C$_{65}$, C$_{64}$, C$_{63}$, and C$_{41}$, respectively (Yadav et al., 2014). Seemita et al. (2015) used wet impregnation to show the successful dispersion of Pd nanoparticles on the surface of activated carbon.

## H4-Alkanes

H4M is a methane-based material with four physisorbed hydrogen molecules (H$_2$)$_4$CH$_4$. These materials have very high gravimetric and volumetric capacities of 50.2 wt% and 0.15 kg H$_2$/L, respectively (Ahluwalia, Peng, & Hua, 2015; Li & Thonhauser, 2012; Mao, Koh, & Sloan, 2007). These values far exceed the DOE standards. According to Somayazulu, Finger, Hemley, and Mao (1996), H4M was discovered two decades ago.

Struzhkin et al. (2007) reported that these materials are only stable at extreme conditions (5-6 GPa at ambient temperature and 65K at room pressure). Efforts to stabilize these materials have been dismissed. However, they are utilized as inspiration for novel class of materials (Harrison, Welchman, & Thonhauser, 2017).

H4-alkanes were derived using H4M as an inspiration. Harrison et al. (2017) showed that the stability of H4M materials can be enhanced by increasing the chain length of the alkanes, deriving H4-alkanes using this formula $(H_2)_{4n} C_n H_{2n+2}$ based on the alkane's structural formula $C_n H_{2n+2 \, [140]}$.

## Vanadium-Based Storage Materials

Vanadium has come under the analytical eye of many researchers and scientists due to its easy absorption and desorption kinetics at ambient temperatures and relatively high hydrogen capacity at 4 wt.% (Kumar & Krishnamurthy, 2011; Lototsky, Yartys, & Zavaliy, 2005; Peterson & Nelson, 1985; Schober, 1996; Veleckis & Edwards, 1969). However, the β-phase of this material is thermally stable, limiting the cyclic hydrogen storage capacity at ambient temperature (Kumar et al., 2017). Furthermore, this material has weak pulverization quality due to a big crystal misfit between the metal hydride phase and metal.

Several researchers have reported on the influence of alloyed metals on the kinetics, thermodynamics, and stability of vanadium-based materials (Kagawa, Ono, Kusakabe, & Sakamoto, 1991; Kumar et al., 2012). Kumar, Taxak, and Krishnamurthy (2013b) reported an increase in the rate of hydrogen uptake in titanium (Ti) alloyed vanadium materials. Ti was also reported to increase the terminal solid solubility of hydrogen in vanadium.

## Zeolites

Zeolites are abundantly available due to their natural occurrence. The general formula of these materials is given by $Mx/m \, [ \, (AlO_2)x \, (SiO_2)y \, ] \, zH_2O$, where z is the number of $H_2O$ molecules, M is the cation with valence m, and x and y are integers such that y/x is more than or equal to 1 (Langmi, 2005).. These crystal alumino-silicate materials are characterized by constant pore size and orderly structures.

Zeolites have been tested and proved to be very good adsorbents for gases like $CO_2$, hydrogen, and moisture (Ataca et al., 2009; Bogdanovic et al., 2006; Jain et al., 2010; Sakintuna et al., 2007; Selvam et al., 1986). For this reason, these materials have been studied for hydrogen storage.

It has been reported that a typical zeolite (at 77K and 1 bar) can adsorb 0.7 wt.%, a value far less compared to highly porous activated carbons, which at a similar condition adsorb 2.1 wt.% (Anderson, 2005). Langmi (2003, 2005) reported that NaX zeolites at 77K and 40 bars yield an adsorption capacity of 2.55 wt.%.

The Langmuir adsorption model has been utilized to explain the physisorption capacities of zeolites. This model simulates that a monolayer of gas forms on the sites available for storage. Kleperis et al. (2013) proved that depositing palladium (Pd) on zeolites improves the hydrogen uptake of these materials at room temperature. Furthermore, Li and Yang (2006) showed that the spill over mechanism and bridging technique can be used to improve the hydrogen adsorption in these materials.

Another technique of hydrogen adsorption on zeolites was proposed due to cryogenic temperatures required for hydrogen uptake. In this regard, encapsulation was tested for hydrogen uptake. Anderson (2008) reported that at 350°C and 900 bar hydrogen can be enforced into the structure of zeolite where it remains intact in the structure even at ambient conditions. This technique, referred to as "encapsulation," is proportional to the type of cations available in the zeolite framework, temperature, and pressure. However, when using this technique, it was observed that hydrogen is less adsorbed with reported gravimetric capacities ranging from 0.1 wt.% to 0.6 wt.% (Anderson, 2008). With porous material, it has been noted that the hydrogen capacity lessens with greater pore size. Thus, surface area is an important criterion for hydrogen uptake.

## EFFORTS TO SOLVE KINETIC ISSUES IN HYDROGEN STORAGE

Kinetics of hydrogen storage materials play a vital role in its absorption and desorption reactions. The catalyst concentration, morphology, surface composition, size, and dimensions of particles play a crucial role in how fast or slow the intake and release of hydrogen transpires (Dornheim et al., 2007; Graetz, 2012). Using kinetic models to study reactions and properties gives a better understanding of their kinetic mechanism (Pang & Li, 2016). Many steps have been taken to improve the kinetics of solid-state materials, including metal hydrides and complex hydrides.

### Alloying

The following categories of composites have been identified based on various stoichiometry's of alloys: (1) AB; (2) $AB_2$; (3) $AB_3$; and (4) $A_2B$. These classes are also known as "intermetallic compounds." Lightweight metal-based hydrogen storage materials were initially alloyed to improve sorption kinetics by decreasing the bond strength of the metal to hydrogen (Chen & Zhu, 2008; Dinca et al., 2006).

These alloys can be prepared mechanically or through arc melting. Mechanical alloying is a dry powder method with at least two metal-metal powders or metal-nonmetal powders. This method is broadly used to enhance kinetics and increase hydrogen storage capacities. It introduces properties like a new surface that speeds up hydrogen sorption and diffusion, increases solid interface volume, and defects the concentration to provide a pathway for diffusion.

Chen and Lin (2007) and Simchi, Kaflou, and Simchi (2009) investigated hydrogen kinetics of $Mg_2Ni$ with Ti using mechanical alloying. They showed that hydrogen absorption by the Ti alloy can be entirely achieved in 10 minutes at a temperature of 300°C. In addition, Chen and Lin (2007) showed improvement of Mg desorption kinetics through mechanical alloying.

The arc melting method of dissolving pure components of the alloy in the required stoichiometry uses an electric arc under unreactive conditions (Ma & Cheng, 2012). This method is used in small-scale applications or laboratories. Staszewski, Osadnik, Czepelak, and Swoboda (2011) reported on successfully synthesizing the $AB_2$ and $AB_5$ alloys. Balcerzak, Nowak, and Jurczyk (2012) concluded that alloys made via arc melting possess higher hydrogen capacities than alloys made through mechanical alloying.

### Ball Milling

An important part of pure lightweight metals is its surface, which manages the dissociation of hydrogen molecules and permits effortless diffusion of hydrogen into the bulk (Sakintuna et al., 2007). The ball milling technique is reported to improve kinetics by diminishing the thickness of the penetrable hydride layer that is developed during hydrogen absorption (Zahiri, Amirkhiz, & Mitlin, 2010). Ball milling disintegrates the surface of pure metals, inducing new properties like amplified surface areas, the establishment of defects on the surface and internally, and the creation of nanostructures. A defected surface helps with the diffusion of hydrogen by providing several sites with low activation energy of diffusion (Guoxian, Erde, & Shoushi, 2000).

Structural and performance discrepancies between unmilled and milled $MgH_2$ were investigated. Milled $MgH_2$ showed an improvement in hydrogen desorption kinetics and reduction in activation energy (Huot et al., 1999). Zaluska, Zaluski, and Strom-Olsen (2000) reported decreased desorption temperatures for ball milled $MgNiH_4$. Ball milling under unreactive conditions is an effective method to enhance the hydrogen absorption rate (Aoyagi, Aoki, & Masumoto, 1995). This creates a decline in the size of the powder and establishment of novel surfaces. Ball milling under an inert condition is effective in reducing the oxidation issue faced by lightweight metals considered for hydrogen storage (Aguey-Zinsou & Ares-Fernandez, 2010; Aguey-Zinsou, Ares-Fernandez, Klaasen, & Bormann, 2007).

Milling time is also critical as it directly affects hydrogen absorption (Aoyagi et al., 1995). This process has been deemed a necessary step when synthesizing and processing solids with high hydrogen capacities (Beattie, Setthanan, & McGrady, 2011). Li, Li, Ma, and Chen (2007) reported that reduction in particle size resulted in an increased time to acquire full hydrogen desorption paralleled to bigger particle sizes.

It has been suggested that particles of grain size less than 1.3 nm decrease the desorption temperature (Aguey-Zinsou & Ares-Fernandez, 2010). A different approach is being implemented where materials are milled under a hydrogen environment. This step diffuses hydrogen into the material (Huot, Akiba, & Takasa, 1995). Milling under inert conditions has proven to play a key role in the hydriding reactions. It was concluded that nanostructured materials are effective in the storage of hydrogen because their properties have been proven to improve kinetics in high hydrogen storage materials (Jain, Lal, & Jain, 2010).

## Catalysis

A number of catalytic additives have been doped into lightweight metal-based systems, complex hydrides, and others to improve their surface kinetics (Dufour & Huot, 2007; Hanada, Ichikwa, & Fujii, 2005; Yermakov et al., 2006). The role of these additives is to expedite the electron exchange between the hydrogen molecules and material (Barkhordarian, Klaasen, & Bormann, 2006; Paunovic, Popovski, & Dimitrov, 2011). Received metals like nickel (Ni) and Ti have garnered attention due to their positive effects on kinetic properties of hydrogen storage materials (Bogdanovic et al., 2006; Ma & Cheng, 2012; Daryani et al., 2014; Jeon et al., 2011).

Ichikawa et al. (2005) used Ti in $LiH/LiNH_2$ systems, reporting an improvement in absorption and desorption kinetics and improved purity in the desorbed hydrogen (Ichikwa et al., 2005). Other work claimed that adding Ti-based catalysts to Mg hydride may improve the hydrogen kinetics by lessening the bond energy between Mg and $H_2$. It also improves the diffusion of hydrogen to particle surfaces through the accessibility of grain margins and large surface area of Mg hydride particles (Daryani et al., 2014). Rafi-ud-din, Zhang, Ping, and Xuanhui (2010) showed that lithium alanate ($LiAlH_4$) doped with TiC showed an enhancement on the dehydrogenation kinetics with an optimum desorption rate of 7 to 8 rapid compared to that of pure $LiAlH_4$.

Shahi, Tiwari, Shaz, and Srivastava (2013), after studying the effects of nanocrystalline $MgH_2$ doped with Ni, iron (Fe), and Ti, found that hydrogen sorption rate was improved compared to undoped/pure ball milled Mg hydride $MgH_2$. Shimada, Tamaki, Higuchi, and Inoue (2014) illustrated that the $MgH_2$-$Ni$-$Ni_2Si$, $MgH_2$-$Ni$, and $MgH_2$-$Ni$-$Si$ composites took up 5 wt% to 6 wt.% of hydrogen within half a minute. The same amount of hydrogen took 15 minutes in ball milled $MgH_2$.

Metal oxides have also been reported as good catalytic agents for hydrogen absorption and desorption of Mg Hydride. Barkhodarian, Klaasen, and Bormann (2006) discovered that $Nb_2O_5$ is an excellent catalyst for absorption and desorption reactions. According to their study, $Fe_3O_4$ had the best effects on desorption kinetics compared to other metal oxides. Bhat et al. (2008) reported that high surface area nanocrystalline niobium (Nb) oxide was enhanced compared to traditional $Nb_2O_5$. Another improvement was observed by Patah, Takasaki, and Szmyd (2009) in the sorption properties of $Nb_2O_5$ doped with $MgH_2$.

Emerging efforts include the synthesis of composites of $MgH_2$ with carbonaceous material (Lu, 1996; Reda, 2009). Their features, including dispersive nature and separation of carbon at grain boundary, have prominent effects in hydrogen storage by increasing the hydrogen storage capacity while decreasing the desorption temperatures (XiangDong & GaoQing, 2008).

## Thin Films

An emerging unconventional method used to enhance the kinetics of hydrogen storage materials is the use of thin films. These films allow for the creation of novel structures on the surface where hydrogen sorption takes place. The film, which is characterized by a large surface area, permits rapid hydrogenation and dehydrogenation (Khafidz et al., 2016).

Thin films offer a better approach for the synthesis of nanostructured materials with improved power over stoichiometry, contamination, and morphology (Jain et al., 2010). These films have the same abilities as mechanically milled materials. In these materials, a tinny layer of catalyst is deposited to avoid reactions with oxygen and to propagate better division of hydrogen molecules into singular hydrogen atoms during desorption of hydrogen (Baldi et al., 2015).

Mooij (2013), upon studying the model system of thin films, found that Mg can be manipulated by changing the thickness and interfacing materials of the Mg layer (Mooij, 2013). This eradicates the need for a high desorption temperature for $MgH_2$. Furthermore, the effect of energy on the surface layer can destabilize nanocrystalline $MgH_2$ (Ham et al., 2014). Baldi et al. (2015) studied thin Mg films doped in multilayer arrangements with metals like Ni, Pd, and Al. In using this process, they were able to create firm Mg alloys and escalate the equilibrium pressure of hydrogen compared to pure Mg in bulk (Baldi et al., 2015).

It was discovered that Mg films doped with Ni and Pd had higher equilibrium pressures than films deposited with Vanadium, Ti, and Nb. This is due to the immiscibility of these materials in Mg (Khafidz et al., 2016). Wang et al. (2004) studied the hydrogenation characteristics of $Mg/MmNi_{3.5}$ $(CoAlMn)_{1.5}$ film on a silicon substrate. They proved that the uptake of hydrogen occurs at 200°C and the release of hydrogen at 250°C. Stoltz and Popovic (2007) showed that adding Ni to Mg films enhances their hydrogenation properties (Stoltz and Popovic, 2007). Song, Guo, and Yang (2004) proved that adding Al to Mg films enhances the dehydrogenation reaction conditions.

Other studies have looked at the effect of Al, MgAl, and Mg on the behaviour of hydrogen. These studies proved that adding Al is beneficial on the hydrogenation properties of Mg films (Ferrer et al., 2007; Pranevicius, Milcius, Pranevicius, & Thomas, 2004).

## CONCLUSION

Undoubtedly, fossil fuels have caused a lot of damage to the environment. However, managing and restricting the use of these materials has and will continue to reduce environmental impacts. Legislation and regulations regarding the use of these materials is a remedial action for recorded improvements. Other remedial actions include seeking alternative green energy sources like hydrogen to replace fossil fuels.

The hydrogen economy and its storage for on-board applications remain in the development phase. However, great progress has been made in this field. The collective energy of the government, corporate businesses, and research and development spheres across the globe has made a valuable impact in the fight on climate change.

## REFERENCES

Abe, J. O., Popoola, A. P. I., Ajenifuja, E., & Popoola, O. M. (2019). Hydrogen energy, economy and storage: Review and recommendation. *International Journal of Hydrogen Energy, 44*(29), 15072–15086. doi:10.1016/j.ijhydene.2019.04.068

Aguey-Zinsou, K.-F., & Ares-Fernandez, J.-R. (2010). Hydrogen in magnesium: New perspective toward functional stores. *Energy & Environmental Science, 3*(5), 252. doi:10.1039/b921645f

Aguey-Zinsou, K.-F., Ares-Fernandez, J.-R., Klaasen, T., & Bormann, R. (2007). Effect of Nb2O5 on MgH2 properties during mechanical milling. *International Journal of Hydrogen Energy, 32*(13), 2400–2407. doi:10.1016/j.ijhydene.2006.10.068

Ahluwalia, R. K., Peng, J -K., & Hua, T. Q. (2015). Sorbent material property requirements for on board hydrogen storage for automotive fuel cell systems. *International Journal of Hydrogen Energy, 40*, 673.

CSA America, Inc. (n.d.). *ANSI HGV 2-2014.* ANSI Webstore.

Anderson, J., & Gronkvist, S. (2019). Large scale storage of hydrogen. *International of Hydrogen Energy, 44*(23), 11901–11919. doi:10.1016/j.ijhydene.2019.03.063

Anderson, P. A. (2008). Storage of hydrogen in zeolites. In G. Walker (Ed.), *Solid-state hydrogen storage: Materials and chemistry* (pp. 233–260). Woodhead Publishing.

Antolini, E. (2003). Formation of carbon supported PtM alloys for low temperature fuel cells: A review. *Materials Chemistry and Physics, 78*(3), 563–573. doi:10.1016/S0254-0584(02)00389-9

Antolini, E. (2009). Carbon supports for low temperature fuel cell catalysts. *Applied Catalysis B: Environmental, 88*(1-2), 1–24. doi:10.1016/j.apcatb.2008.09.030

Aoyagi, H., Aoki, K., & Masumoto, T. (1995). Effect of ball milling on hydrogen absorption properties of FeTi, Mg2Ni and LaNi5. *Journal of Alloys and Compounds, 231*(1-2), 804–809. doi:10.1016/0925-8388(95)01721-6

Asalatha (n.d.).

Association for Emissions Control by Catalyst (AECC). (n.d.). *Adsorbers*. Retrieved from https://www.aecc.eu/technology/adsorbers/

Ataca, C., Akturk, E., & Ustunel, H. (2009). High-capacity hydrogen storage by metallized graphene. *Applied Physics Letters*, *93*(4), 043123. doi:10.1063/1.2963976

Balcerzak, M., Nowak, M., & Jurczyk, M. (2012). Nanocrystalline TiNi, Ti2Ni alloys for hydrogen storage. *Materials Science and Engineering*, *33*, 370–373.

Baldi, A., Palsson, G. K., Gonzalez-Silveira, M., Schreuders, H., Slaman, M., Rector, J. H., ... Griessen, R. (2015). Mg/Ti multilayers: Structural, optical and hydrogen absorption properties. *Physical Review. B*, *81*(22), 224203. doi:10.1103/PhysRevB.81.224203

Barkhordarian, G., Klaasen, T., & Bormann, R. (2006). Catalytic mechanism of transition-metal compounds on Mg hydrogen sorption reaction. *The Journal of Physical Chemistry B*, *110*(22), 11020–11024. doi:10.1021/jp0541563 PMID:16771356

Beattie, S. D., Setthanan, U., & McGrady, G. S. (2011). Thermal desorption of hydrogen from magnesium hydride (MgH2): An in situ microscopy study by environmental SEM and TEM. *International Journal of Hydrogen Energy*, *36*(10), 6014–6021. doi:10.1016/j.ijhydene.2011.02.026

Bhat, V. V., Rougier, A., Aymard, L., Nazri, G. A., & Tarascon, J. M. (2008). High surface area niobium oxides as catalysts for improved hydrogen sorption properties of ball milled MgH2. *Journal of Alloys and Compounds*, *406*(1-2), 507–512. doi:10.1016/j.jallcom.2007.05.084

Bogdanovic, B., Felderhoff, M., Pommerin, A., Schuth, F., & Spielkamp, N. (2006). Advanced hydrogen-storage materials based on Sc-, Ce-, and Pr-doped NaAlH4. *Advanced Materials*, *18*(9), 1198–1201. doi:10.1002/adma.200501367

Cardella, U., Decker, L., & Klein, H. (2017). Roadmap to economically viable hydrogen liquefaction. *International Journal of Hydrogen Energy*, *42*(19), 13329–13338. doi:10.1016/j.ijhydene.2017.01.068

Chambers, A., Park, C., Baker, R. T. K., & Rodriguez, N. M. (1998). Hydrogen storage in graphite nanofibers. *Journal of Physical Chemistry*, *102*(22), 4253–4256. doi:10.1021/jp980114l

Chandrakumar, K. R. S., & Ghosh, S. K. (2008). Hydrogen adsorption on Na-SWCNT systems. *Nano Letters*, *8*, 13–19. doi:10.1021/nl071456i PMID:18085807

Chen, C.-Y., & Lin, S.-K. (2007). Improvement of hydrogen absorption performance of mechanically alloyed Mg2Ni powders by silver doping. *Materials Transactions*, *48*(5), 1113–1118. doi:10.2320/matertrans.48.1113

Chen, P., & Zhu, M. (2008). Recent progress in hydrogen storage. *Materials Today*, *11*(12), 36–43. doi:10.1016/S1369-7021(08)70251-7

Christian, M., & Aguey-Zinsou, K. (2010). Destabilization of complex hydrides through size effects. *Nanoscale*, *2*(12), 2587–2590. doi:10.1039/c0nr00418a PMID:20886168

Daryani, M., Simchi, A., Sadati, M., Hosseini, H. M., Targholizadeh, H., & Khakbiz, M. (2014). Effects of Ti-based catalysts on hydrogen desorption kinetics of nanostructured magnesium hydride. *International Journal of Hydrogen Energy*, *39*(36), 21007–21014. doi:10.1016/j.ijhydene.2014.10.078

Diesel Technology Forum. (n.d.). *About clean diesel: What is SCR?* Retrieved from https://www.dieselforum.org/about-clean-diesel/what-is-scr

Dillon, A. C., Jones, K. M., Bekkedahl, T. A., Kiang, C. H., Bethune, D. S., & Heben, M. J. (1997). Storage of hydrogen in single walled carbon nanotubes. *Nature*, *386*(6623), 377–379. doi:10.1038/386377a0

Dinca, M., Dailly, A., Liu, Y., Brown, C. M., Neumann, D. A., & Long, J. R. (2006). Hydrogen storage in a microporous metal organic framework with exposed Mn2+ coordination sites supporting information. *Journal of the American Chemical Society*, *128*(51), 16876–16883. doi:10.1021/ja0656853 PMID:17177438

Dornheim, M., Doppiu, S., Barkhordarian, G., Boesenberg, U., Klaasen, T., & Gutfleisch, O. (2007). Hydrogen storage in magnesium-based hydrides and hydride composites. *Scripta Materialia*, *56*(10), 841–846. doi:10.1016/j.scriptamat.2007.01.003

Dufour, J., & Huot, J. (2007). Rapid activation, enhanced hydrogen sorption kinetics and air resistance in laminated Mg-Pd 2.5 at%. *Journal of Alloys and Compounds*, *439*(1-2), L5–L7. doi:10.1016/j.jallcom.2006.08.264

Durbin, D. J., & Malardier-Jugroot, C. (2013). Review of hydrogen storage techniques for on board vehicle applications. *International Journal of Hydrogen Energy*, *38*(34), 14595–14617. doi:10.1016/j.ijhydene.2013.07.058

Fang, Z.-Z., Kang, X.-D., & Wang, P. (2010). Improved hydrogen storage properties of LiBH4 by mechanical milling with various carbon additives. *International Journal of Hydrogen Energy*, *35*(15), 8247–8252. doi:10.1016/j.ijhydene.2009.12.037

Ferrer, R. D., Sridharan, M. G., Garcia, G., Pi, F., & Viejo, J. R. (2007). Hydrogenation properties of pure magnesium and magnesium-aluminum thin films. *Journal of Power Sources*, *169*(1), 117–122. doi:10.1016/j.jpowsour.2007.01.049

Graetz, J. (2012). Metastable metal hydrides for hydrogen storage. *ISRN Materials Science*, 18.

Green Climate Fund. (2019). *Process for readiness support*. Retrieved from https://www.greenclimate.fund/gcf101/empowering-countries/readiness-support

Guoxian, L., Erde, W., & Shoushi, F. (2000). Hydrogen absorption and desorption characteristics of mechanically milled Mg-35 wt% FeTi1.2 powders. *Journal of Alloys and Compounds*, *297*, 222–231.

Ham, B., Junkaew, A., Arroyave, R., Park, J., Zhou, H. C., Foley, D., ... Zhang, X. (2014). Size and stress dependent hydrogen desorption in metastable Mg hydride films. *International Journal of Hydrogen Energy*, *39*(6), 2597–2607. doi:10.1016/j.ijhydene.2013.12.017

Hanada, N., Ichikwa, T., & Fujii, H. (2005). Catalytic effect of nanoparticles 3D-transition metals on hydrogen storage properties in magnesium hydride MgH2 prepared by mechanical milling. *Journal of Physical Chemistry*, *109*(15), 7188–7194. doi:10.1021/jp044576c PMID:16851820

Harrison, D., Welchman, E., & Thonhauser, T. (2017). H-4 alkanes: A new class of hydrogen storage material? *International Journal of Hydrogen Energy, 42*(4), 1–6. doi:10.1016/j.ijhydene.2016.12.144

Hirscher, M. (2009). *Handbook of hydrogen storage: New materials for future energy storage.* Weinheim, Germany: Wiley-VCH Verlag GmbH & Co. KGaA.

Hua, T. Q., Ahluwalia, R. K., Peng, J.-K., Kromer, M., Lasher, S., McKenney, K., ... Sinha, J. (2011). Technical assessment of compressed hydrogen storage tank systems for automotive application. *International Journal of Hydrogen Energy, 36*(4), 3037–3049. doi:10.1016/j.ijhydene.2010.11.090

Huang, H. J., & Wang, X. (2012). Pd nanoparticles supported on low-defect graphene sheets for use as high performance electrocatalysts for formic acid and methanol oxidation. *Journal of Materials Chemistry. A, Materials for Energy and Sustainability, 22*, 22533–22541.

Huot, J., Akiba, E., & Takasa, T. (1995). Mechanical alloying of Mg-Ni compounds under hydrogen and inert atmosphere. *Journal of Alloys and Compounds, 231*(1-2), 815–819. doi:10.1016/0925-8388(95)01764-X

Huot, J., Liang, G., Boily, S., Neste, A. V., & Schulz, R. (1999). Structural study and hydrogen sorption kinetics of ball milled magnesium hydride. *Journal of Alloys and Compounds, 293-295*, 495–500. doi:10.1016/S0925-8388(99)00474-0

Hwang, H. T., & Varma, A. (2014). Hydrogen storage for fuel cell vehicles. *Current Opinion in Chemical Engineering, 5*, 42–48. doi:10.1016/j.coche.2014.04.004

Ichikwa, T., Hanada, N., Isobe, S., Leng, H. Y., & Fujii, H. (2005). Hydrogen storage properties in Ti catalyzed Li-N-H system. *Journal of Alloys and Compounds, 404-406*, 435–438. doi:10.1016/j.jallcom.2004.11.110

Iijima, S. (1991). Helical microtubes of graphitic carbon. *Nature, 354*(6348), 56–58. doi:10.1038/354056a0

SAE International. (2018). *Standard for fuel systems in fuel cell and other hydrogen vehicles.* SAE International.

International Organisation of Standardization (ISO). (2018). *ISO 19881: Gaseuos hydrogen – Land vehicle fuel containers.* ISO.

Jain, I. P., Jain, P., & Jain, A. (2010). Novel hydrogen storage materials: A review of lightweight complex hydrides. *Journal of Alloys and Compounds, 503*(2), 303–339. doi:10.1016/j.jallcom.2010.04.250

Jain, I. P., Lal, C., & Jain, A. (2010). Hydrogen storage in Mg: A most promising material. *International Journal of Hydrogen Energy, 35*(10), 5133–5144. doi:10.1016/j.ijhydene.2009.08.088

Jeon, K.-J., Moon, H. R., Ruminski, A. M., Jiang, B., Kisielowski, C., Bardhan, R., & Urban, J. J. (2011). Air-stable magnesium nanocomposites provide rapid and high capacity hydrogen storage without using heavy-metal catalysts. *Nature Materials, 97*(4), 083106–083103. doi:10.1038/nmat2978 PMID:21399630

Kagawa, A., Ono, E., Kusakabe, T., & Sakamoto, Y. (1991). Absorption of hydrogen by vanadium-rich V-Ti based alloys. *Journal of the Less Common Metals, 172-174*, 64–70. doi:10.1016/0022-5088(91)90433-5

Khafidz, N. Z. A., Yaakob, Z., & Lim, K. L. (2016). The kinetics of lightweight solid-state hydrogen storage materials: A review. *International Journal of Hydrogen Energy, 41*, 13131–13151. doi:10.1016/j.ijhydene.2016.05.169

Kim, S. K., Hong, S. A., Son, H. J., Han, W. S., Michalak, A., Hwang, S. J., & Kang, S. O. (2015). Dehydrogenation of ammonia borane by cation Pd(II) and Ni(II) complexes in a nitromethane medium: Hydrogen release and spent fuel characterization. *Dalton Transactions (Cambridge, England), 44*(16), 7373–7381. doi:10.1039/C5DT00599J PMID:25799252

Kleperis, J., Lesnicenoks, P., Grinberga, L., Chikvaidze, G., & Klavins, J. (2013). Zeolite as material for hydrogen storage in transport applications. *Latvian Journal of Physics and Technical Sciences, 50*(3), 59–64. doi:10.2478/lpts-2013-0020

Koh, G., Zhang, Y.-W., & Pan, H. (2012). First-principles study on hydrogen storage by graphitic carbon nitride nanotubes. *International Journal of Hydrogen Energy, 37*(5), 4170–4178. doi:10.1016/j.ijhydene.2011.11.109

Kumar, S., Jain, A., Ichikawa, Y., Kojima, Y., & Dey, G. K. (2017). Development of vanadium based hydrogen storage material: A review. *Renewable & Sustainable Energy Reviews, 72*, 791–800. doi:10.1016/j.rser.2017.01.063

Kumar, S., & Krishnamurthy, N. (2011). Synthesis of V-Ti-Cr alloys by aluminothermy co-reduction of its oxide. *International Journal of Applied Ceramic Technology, 5*, 181–186.

Kumar, S., Taxak, M., & Krishnamurthy, N. (2013a). Hydrogen absorption kinetics of V-Al alloy. *Journal of Thermal Analysis and Calorimetry, 112*(1), 5–10. doi:10.100710973-012-2558-1

Kumar, S., Taxak, M., & Krishnamurthy, N. (2013b). Synthesis and hydrogen absorption in V-Ti-Cr alloy. *Journal of Thermal Analysis and Calorimetry, 112*(1), 51–57. doi:10.100710973-012-2643-5

Kumar, S., Taxak, M., Krishnamurthy, N., Suri, A. K., & Tiwari, G. P. (2012). Terminal solid solubility of hydrogen in V-Al solid solutions. *International Journal of Refractory Metals & Hard Materials, 31*, 76–81. doi:10.1016/j.ijrmhm.2011.09.009

Langmi, H., Book, D., Walton, A., Johnson, S. R., Al-Mamouri, M. M., Speight, J. D., ... Anderson, P. A. (2005). Hydrogen storage in ion-exchanged zeolites. *Journal of Alloys and Compounds, 404-406*, 637–642. doi:10.1016/j.jallcom.2004.12.193

Langmi, H., Walton, A., Al-Mamouri, M. M., Johnson, S. R., Book, D., Speight, J. D., ... Harris, I. R. (2003). Hydrogen adsorption in zeolites A,X,Y and RHO. *Journal of Alloys and Compounds, 356-357*, 256–357, 710–715. doi:10.1016/S0925-8388(03)00368-2

Lee, J. H., Ryu, J., Kim, J. Y., Nam, S.-W., Han, J. H., Lim, T.-H., ... Yoon, C. W. (2014). Carbon dioxide mediated, reversible chemical hydrogen storage using Pd nanocatalyst supported on mesoporous graphitic carbon nitride. *Journal of Materials Chemistry. A, Materials for Energy and Sustainability, 2*(25), 9490. doi:10.1039/c4ta01133c

Li, Q., & Thonhauser, T. (2012). A theoretical study of the hydrogen storage potential of (H2)4 CH4 in metal organic framework materials and carbon nanotubes. *Journal of Physics Condensed Matter*, *24*(42), 424204. doi:10.1088/0953-8984/24/42/424204 PMID:23032298

Li, W., Li, C., Ma, H., & Chen, J. (2007). Supporting information-magnesium nanowires: Enhanced kinetics for hydrogen absorption and desorption. *Journal of the American Chemical Society*, *129*(21), 6710–6711. doi:10.1021/ja071323z PMID:17488082

Li, Y., & Yang, R. T. (2006). Hydrogen storage in low silica type X zeolites. *The Journal of Physical Chemistry B*, *110*(34), 17175–17181. doi:10.1021/jp0634508 PMID:16928014

Liu, W., Setijadi, E., Crema, L., Bartali, R., Laidani, N., Aguey-Zinsou, K. F., & Speranza, G. (2018). Carbon nanostructures/Mg hybrid materials for hydrogen storage. *Diamond and Related Materials*, *82*, 19–24. doi:10.1016/j.diamond.2017.12.003

Liu, Y., Ren, L., He, Y., & Cheng, H. P. (2010). Titanium-decorated graphene for high capacity hydrogen storage studied by density functional simulations. *Journal of Physics Condensed Matter*, *22*(44), 445301. doi:10.1088/0953-8984/22/44/445301 PMID:21403342

Lototsky, M. V., Yartys, V. A., & Zavaliy, I. Y. (2005). Vanadium based BCC alloys phase structural characteristics and hydrogen sorption properties. *Journal of Alloys and Compounds*, *404-406*, 421–426. doi:10.1016/j.jallcom.2005.01.139

Lu, K. (1996). Nanocrystalline metals crystallized from amorphous solids: Nanocrystallization, structure and properties. *Materials Science and Engineering*, *16*(4), 161–221. doi:10.1016/0927-796X(95)00187-5

Ma, H., & Cheng, F. (2012). Nickel-metal hydride (Ni-MH) rechargeable batteries. In J. Zhang, L. Zhang, H. Liu, A. Sun, & R.-S. Liu (Eds.), *Electrochemical technologies for energy storage and conversion* (pp. 175–237). Wiley-VCH Verlag. doi:10.1002/9783527639496.ch5

Mair, G. W., Scherer, F., & Hoffman, M. (2015). Type approval of composite gas cylindres-probabilistic analysis and standards requirements concerning minimum burst pressure. *International Journal of Hydrogen Energy*, *40*(15), 5359–5366. doi:10.1016/j.ijhydene.2015.01.161

Mao, W. L., Koh, C. A., & Sloan, E. D. (2007). Clathrate hydrates under pressure. *Physics Today*, *60*(10), 42–47. doi:10.1063/1.2800096

Midilli, A., Ay, M., Dincer, I., & Rosen, M. (2005). On hydrogen and hydrogen energy strategies I: Current status and needs. *Renewable & Sustainable Energy Reviews*, *9*(3), 255–271. doi:10.1016/j.rser.2004.05.003

Milman, O. (2018, January 1). Vehicles are now America's biggest CO2 source but EPA is tearing up regulations. *The Guardian*. Retrieved from https://www.theguardian.com/environment/2018/jan/01/vehicles-climate-change-emissions-trump-administration

Mooij, L. (2013). Destabilization of magnesium hydride through interface engineering. *LPA Mooij*. Retrieved at https://d1rkab7tlqy5f1.cloudfront.net/TNW/Afdelingen/ChemE/PIs/Dam%2C%20Bernard/Supervised%20theses/PhD_Thesis_L.P.A._Mooij_1_October_2013_small.pdf

O'Malley, K., Ordaz, J., Adams, J., Randolph, K., Ahn, C., & Stetson, N. T. (2015). Applied hydrogen storage research and development a perspective from the US department of energy. *Journal of Alloys and Compounds, 645*(1), 419–422. doi:10.1016/j.jallcom.2014.12.090

Pang, Y., & Li, Q. (2016). A review on kinetic models and corresponding analysis methods for hydrogen storage materials. *International Journal of Hydrogen Energy, 41*(40), 18072–18087. doi:10.1016/j.ijhydene.2016.08.018

Patah, A., Takasaki, A., & Szmyd, J. S. (2009). Influence of multiple oxide (Cr2O3/Nb2O5) addition on the sorption kinetics of MgH2. *International Journal of Hydrogen Energy, 34*(7), 3032–3037. doi:10.1016/j.ijhydene.2009.01.086

Paunovic, P., Popovski, O., & Dimitrov, A. T. (2011). Hydrogen economy: The role of nano-scaled support materials for electrocatalysts aimed for water electrolysis. In J. P. Reithmaier, P. Paunovic, W. Kulisch, C. Popov, & P. Petkov (Eds.), NATO science for peace and security series B, Physics and biophysics (pp. 545-563). Academic Press.

Pesonen, O., & Alakunnas, T. (2017). *Energy storage: A missing piece of the puzzle for the self-sufficient living*. Lapland University of Applied Sciences.

Peterson, D. T., & Nelson, S. O. (1985). The isopiestic solubility of hydrogen in vanadium alloys at low temperatures. *Metallurgical and Materials Transactions. A, Physical Metallurgy and Materials Science, 16A*, 36–374.

Prabhu, S. S., Nayak, S. N., Kapilan, N., & Hindasagen, V. (2017). An experiment and numeric study on the effects of exhaust gas temperature and flowrate on deposit formation in Urea Selective catalytic reduction (SCR) system of modern automobiles. *Applied Thermal Engineering, 111*, 1211–1231. doi:10.1016/j.applthermaleng.2016.09.134

Prabhukhot, P. R., Wagh, M. M., & Gangal, A. C. (2016). A review on solid storage of hydrogen material. *Advances in Energy and Power, 4*(2), 11–22.

Pranevicius, L., Milcius, D., Pranevicius, L. L., & Thomas, G. (2004). Plasma hydrogenation of Al, Mg and MgAl films under high-flux ion irradiation at elevated temperatures. *Journal of Alloys and Compounds, 373*(1-2), 9–15. doi:10.1016/j.jallcom.2003.10.029

Rafi-ud-din, Zhang, L., Ping, L., & Xuanhui, Q. (2010). Catalytic effects of nano-sized TiC additions on the hydrogen storage properties of LiAlH4. *Journal of Alloys and Compounds, 508*(1), 119–128. doi:10.1016/j.jallcom.2010.08.008

Reda, M. R. (2009). The effect of organic additive in Mg/Graphite composite as hydrogen storage materials. *Journal of Alloys and Compounds, 480*(2), 238–240. doi:10.1016/j.jallcom.2009.02.021

Sakintuna, B., Lamari-Darkrim, F., & Hirscher, M. (2007). Metal hydride materials for solid hydrogen storage: A review. *International Journal of Hydrogen Energy, 32*(9), 1121–1140. doi:10.1016/j.ijhydene.2006.11.022

Salameh, C. M. (2014). *Synthesis of boron or aluminum based functional nitrides for energy applications (hydrogen production and storage)*. Material chemistry. Université Montpellier II - Sciences et Techniques du Languedoc.

Satyapal, S., Petrovic, J., Read, C., Thomas, G., & Ordaz, G. (2007). The US Department of Energy's national hydrogen storage project: Progress towards meeting hydrogen-powered vehicle requirements. *Catalysis Today, 120*(3-4), 246–256. doi:10.1016/j.cattod.2006.09.022

Schober, T. (1996). Vanadium, Niobium and tantalum-hydrogen. *Diffusion and Defect Data, Solid State Data. Part B, Solid State Phenomena, 49-50*, 357–422. doi:10.4028/www.scientific.net/SSP.49-50.357

Seemita, B., Dasgupta, K., Kumar, A., Ruz, P., Vishwanadh, B., Joshi, J. B., & Sudarsan, V. (2015). Comparative evaluation of hydrogen storage behavior of Pd doped carbon nanotubes prepared by wet impregnation and polyol methods. *International Journal of Hydrogen Energy, 40*(8), 3268–3276. doi:10.1016/j.ijhydene.2015.01.048

Selvam, P., Viswanathan, B., Swamy, C. S., & Srinivasan, V. (1986). Magnesium and magnesium alloy hydrides. *International Journal of Hydrogen Energy, 11*(3), 169–192. doi:10.1016/0360-3199(86)90082-0

Shahi, R. R., Tiwari, A. P., Shaz, M. A., & Srivastava, O. N. (2013). Studies on de/rehydrogenation characteristics of nanocrystalline MgH2 co-catalyzed with Ti, Fe, and Ni. *International Journal of Hydrogen Energy, 38*(6), 2778–2784. doi:10.1016/j.ijhydene.2012.11.073

Sheriff, S. A., Yogi, G. D., Stefanakos, E., & Steinfield, A. (2014). *A handbook of hydrogen energy*. CRC Press. doi:10.1201/b17226

Shimada, M., Tamaki, H., Higuchi, E., & Inoue, H. (2014). Kinetic analysis for hydrogen absorption and desorption of MgH2-based composites. *Journal of Materials and Chemical Engineering, 2*, 64–71.

Simchi, H., Kaflou, A., & Simchi, A. (2009). Synergetic effect Ni and Nb2O5 on dehydrogenation properties of nanostructured MgH2 synthesized by high-energy mechanical alloying. *International Journal of Hydrogen Energy, 34*(18), 7724–7730. doi:10.1016/j.ijhydene.2009.07.038

Somayazulu, M. S., Finger, L. W., Hemley, R. J., & Mao, H. K. (1996). High pressure compounds in methane hydrogen mixtures. *Science, 271*(5254), 1400–1402. doi:10.1126cience.271.5254.1400

Song, Y., Guo, Z. X., & Yang, R. (2004). Influence of selected alloying elements on the stability of magnesium dihydride for hydrogen storage applications: A first principles investigation. *Physical Review. B, 69*(9), 094205. doi:10.1103/PhysRevB.69.094205

South African Government. (2019, May 26). *President Cyril Ramaphosa signs 2019 Carbon Tax Act into law* [media statement]. Retrieved from https://www.gov.za/speeches/publication-2019-carbon-tax-act-26-may-2019-0000

Staszewski, M., Osadnik, M., Czepelak, M., & Swoboda, P. (2011). Hydrogen storage alloys prepared by high energy milling. *Journal of Achievements in Materials and Manufacturing, 44*, 154–160.

Stoltz, S. E., & Popovic, D. (2007). A high-resolution core-level study of Ni-catalyzed absorption and desorption of hydrogen in Mg-films. *Surface Science, 601*(6), 1507–1512. doi:10.1016/j.susc.2007.01.016

Struzhkin, V. V., Militzer, B., Mao, W. L., Mao, H.-K., & Hemley, R. J. (2007). Hydrogen storage in molecular clathrates. *Chemical Reviews*, *107*(10), 4133–4151. doi:10.1021/cr050183d PMID:17850164

Sustainable Development Goals. (2019). *Progress of goal 13 in 2019*. Retrieved from https://sustainabledevelopment.un.org/sdg13

Sustainable Development Goals. (n.d.). *Sustainable development goals*. United Nations Department of Information. Retrieved from https://sustainabledevelopment.un.org/?menu=1300

Tahir, M., Cao, C., Butt, F. K., Idrees, F., Mahmood, N., Ali, Z., ... Mahmood, T. (2013). Tubular graphitic-C3N4: A prospective material for energy storage and green photocatalysis. *Journal of Materials Chemistry. A, Materials for Energy and Sustainability*, *1*(44), 1. doi:10.1039/c3ta13291a

Trygve, R., Sandrock, G., Ulleberg, O., & Vie, P. J. S. (2005). Hydrogen storage - Gaps and priorities. *HIA HCG*, 1-13.

United States Environment Protection Agency (EPA). (n.d.). *EPA history*. Retrieved from https://www.epa.gov/history

Valenti, G. (2015). Hydrogen liquefaction and liquid hydrogen storage. *Compendium of Hydrogen Energy*, 27-51.

Veleckis, E., & Edwards, R. K. (1969). Thermodynamic properties on the system vanadium hydrogen, Niobium-hydrogen and tantalum-hydrogen. *Journal of Physical Chemistry*, *73*(3), 683–692. doi:10.1021/j100723a033

Wang, D., Liao, B., Zheng, J., Huang, G., Hua, Z., Gu, C., & Xu, P. (2019). Development of regulations, codes and standards on composite tanks for on-board gaseous hydrogen storage. *International Journal of Hydrogen Energy*, *44*(40), 22643–22653. doi:10.1016/j.ijhydene.2019.04.133

Wang, H., Ouyang, L. Z., Peng, C. H., Zeng, M. Q., Chung, C. Y., & Zhu, M. (2004). MmM5/Mg multilayer hydrogen storage thin films prepared by dc magnetron sputtering. *Journal of Alloys and Compounds*, *370*(1-2), L4–L6. doi:10.1016/j.jallcom.2003.09.019

Wang, L., Quadir, M. Z., & Aguey-Zinsou, K.-F. (2016). Ni coated LiH nanoparticles for reversible hydrogen storage. *International Journal of Hydrogen Energy*, *41*(15), 6376–6386. doi:10.1016/j.ijhydene.2016.01.173

Webb, C. J. (2015). A review of catalyst-enhanced magnesium hydride as a hydrogen storage material. *Journal of Physics and Chemistry of Solids*, *84*, 96–106. doi:10.1016/j.jpcs.2014.06.014

Xia, G. (2015). *Light metal hydrides for reversible hydrogen storage applications*. University of Wollongong Research Online.

XiangDong, Y., & GaoQing, L. (2008). Magnesium-based materials for hydrogen storage: Recent advances and future perspectives. *Chinese Science Bulletin*, *53*, 2421–2440.

Yadav, S., Zhu, Z., & Singh, C. V. (2014). Defect engineering of graphene for effective hydrogen storage. *International Journal of Hydrogen Energy*, *39*(10), 4981–4995. doi:10.1016/j.ijhydene.2014.01.051

Yermakov, A. Y., Mushnikov, N. V., Uimin, M. A., Gaviko, V. S., Tankeev, A. P., Skripov, A. V., ... Buzlukov, A. L. (2006). Hydrogen reaction kinetics of Mg-based alloys synthesized by mechanical milling. *Journal of Alloys and Compounds, 425*(1-2), 367–372. doi:10.1016/j.jallcom.2006.01.039

Yoon, M., Yang, S., Wang, E. Z. Z., & Zhang, Z. (2007). Charged fullerenes as high capacity storage media. *Nano Letters, 7*(9), 2578–2258. doi:10.1021/nl070809a PMID:17718530

Zahiri, B., Amirkhiz, B. S., & Mitlin, D. (2010). Hydrogen storage cycling of MgH2 thin film nanocomposites catalyzed by bimetallic CrTi. *Applied Physics Letters, 97*(8), 083106. doi:10.1063/1.3479914

Zaluska, A., Zaluski, L., & Strom-Olsen, J. O. (2000). Lithium-berylium hydrides: The lightest reversible metal hydride. *Journal of Alloys and Compounds, 307*(1-2), 157–166. doi:10.1016/S0925-8388(00)00883-5

Zheng, L., Liu, X., Xu, P., Liu, P., Zhao, Y., & Yang, J. (2012). Development of high pressure gaseous hydrogen storage technologies. *International Journal of Hydrogen Energy, 37*(1), 1048–1057. doi:10.1016/j.ijhydene.2011.02.125

# Chapter 10
# Healthcare in India:
## Challenges and Innovative Solutions

**Reenu Kumari**
*MIET, India*

## ABSTRACT

*India facing some complex challenges in healthcare sectors like adaptation of hi tech (digital health), financing, approachability, and adaptability. Therefore, this chapter explores an overview of health and healthcare, source of financing in Indian healthcare, challenges and innovative solutions for the problems. This chapter also discussed the healthcare industry (global and India perspective) and found that total healthcare industry size will be increased US$ 160 Billion (2017) and US$ 372 billion (2022). Health financing is by a number of sources: (1) the tax-based public sector that comprises local, state, and central governments, in addition to numerous autonomous public sector bodies and the private sector. There are major challenges in Indian healthcare: population, non-commutation disease, new technology, inaccessibility to primary healthcare, and medical education. This chapter concluded by providing appropriate knowledge to the multidisciplinary groups such as researchers, policymakers, administrators, and politician.*

## INTRODUCTION

### Concept of Health and Healthcare

Health sector has a play vital role in Indian economy because contribution of healthcare in Indian GDP is 4.2% in 2012. At the time of independence Indian ministries developed one more ministry, which was known as the Ministry of Health. Basically, this ministry gives priority to health area and they have tried to consider healthcare in five year plans, each of which determinants state spending priorities for the coming five years. In 1983, Parliament countersigned to The National Health Policy and this policy was aimed global healthcare coverage 2000, and this program was renewed in 2002. On the basis of "World Health Organization (WHO, 1999)" global health report India spends on health only 20% of that of China's. Healthcare sector has many social impacts. Among these the healthcare sector drives

DOI: 10.4018/978-1-7998-3576-9.ch010

directly the GDP growth, because a strong and efficient healthcare system will augment GDP growth via an employment, productivity, export and entrepreneurship. Better healthcare in India will be possible when healthcare system becomes strong. The health-care system in India currently is multi-layered and complex, which makes it difficult to understand its true potential and appropriate quality of services. In India healthcare system is very approachable, but there remains many differences in quality of healthcare system between rural and urban areas. Most of the private hospitals are located in urban areas while public hospitals predominantly exist in rural areas. This is the main reason behind the unequal access to healthcare system in India. Private hospitals try to maintain their reputation in market through providing low cost medicine and high quality of services. On the other hand, the public hospitals provide low cost medicine, but are often short of basic hygiene and cleanness as well as such basic facilities and services as servant services, compounders, qualified doctors, appropriate infrastructure and equipment. Furthermore, a large number of private hospitals are not able to provide modern healthcare, because they are not fully aware and trained in modern medicine and, instead, rely heavily on traditional medicine and such practices as the *ayurveda, siddha, unani* and homœopathy.

According to FICCI, (2015) A robust healthcare system drives GDP growth in the presence of adequate investments and a conducive environment by not only acting as a productivity and employment generator, but also as a magnet to attract foreign exchange earnings and provide opportunities for innovation and entrepreneurship. Atun *et.al*, (2007 a) explored how policies help to access new medical technologies (like machines) and manage healthcare costs. Government polices tend to seek reductions in pharmaceutical budgets without having detrimental effects on health system.

Srinivisan (2012) mentions that positive scenario of Indian health system is largely premised on an average 8% rate of economic growth. Srinivisan further explained three sets of consequences of present shortcomings, the fist of which was related to low investment in water saving measures and low priority given to basic public health in terms of ignoring the role of personal hygiene as the basis of good health. Second consequences were related to inadequate levels of funding dedicated to facilities as public expenditures remain a sensitive issue. Lastly, third set of consequences is related to the availability of medicines. On the basis of the WHO report, Indian healthcare industry is not free from all kinds of challenges and presence of diseases (physical as well as mental ones). There are various indicators of health such as life expectancy at birth; the percentage of children being underweight; infant mortality rates (IMRs); the percentage of women with body mass index (BMI) below 18.5; quality-adjusted life years (QALYs); and disability-adjusted life years (DALYs). Similarly, in empirical studies (Kumari and Sharma, 2018; Woodward, 1992; Cheng and Kwan, 2000; Bouoiyour, 2007; Alsan et al. 2006; Kurtishi-Kastrati et al. 2016) health of population is measured by different indicators such as the Life expectancy rate; the Gender parity index (GPI) and the School enrollment rate, but here we have considered healthcare only, but it is important to remember that issues of health are closely related to other social issues and, therefore, it is important to pay attention to a broad range of indicators.

Balarajan, (2015) tried to find out how to best improve in healthcare and also make progress with equality in terms of both geography and socioeconomic status. He mentioned key challenges for achievement equity being especially in service provision, equity in financing and financial risk protection in India. These challenges included inequality with resource allocation, accessibility of high quality health services and high expenditure related to health equipment. Duggal and Gangolli (2014) discussed healthcare as a main factor, which affects the status of whole Indian population. If healthcare is not provided in an appropriate manner, it will have unsatisfactory consequences for the health whole population. NHA report "2004-2005" identified the financing of healthcare as being the most critical determinant of healthcare system in contemporary India. Problems with financing can push the whole population to danger zone. On the account of this type of situation health system in India needs quick improvement especially in decision-making related to health policy. India needs to adopt new strategies that will ensure the protection of health of all individuals. Similarly, Kumar *et, al.* (2011) mentioned that due to health financing problem in India there is marked health inequity, unequal accessibility to proper equipment and poor quality in terms of hospital services. In rural areas expenses of private hospitals are out of pocket and, on the other hand, in urban areas people can better afford such services. Moreover, rural citizens are not even aware about facilities and different types of medical insurances. So, every year government of India has tried to increase the amount of money allocated to healthcare (expenditure on healthcare) but these policies are not implemented in systematic and effective enough way. If the government increases public expenditure (including health) it should also be able to contribute to batter quality in the public and private healthcare sector. The healthcare also includes many other aspects such as the clinical trials, medical devices, telemedicine, health insurance, medical tourism and medical equipment. The Indian government faces a real challenge in bringing quality and good governance to all areas of public health.

## Importance of the Healthcare in Indian Economy

Healthcare consumption is expected to increase progressively in the future, in line with economic growth. The emergence of disruptive technologies is likely to aid care delivery and lead to consumers who are more informed, engaged, discerning and value conscious (see Figure 2). Significant growth and refinement of health infrastructure are anticipated. Investments by financial investors (PE/VCs)[1] in healthcare have surged recently, and the government has announced a sharp focus on transforming India's health system and an emerging paradigm emphasizes access to quality healthcare as a basic human right. Investment in healthcare will create a virtuous cycle of productivity, employment and consumption, resulting in overall economic growth.

## Healthcare Industry Overview (Global & Indian Prospective)

In terms of revenue and employment, India has become largest country who investment in Health sector. In this section chapter discussed industry based different segments like:

- **Hospital:** This category covers extraction, manufacturing, processing and other services.
- **Diagnostics:** This category covers business and laboratories that offer analysis services, including body fluid analysis.
- **Medical Equipment and Supplies:** This category covers equipment related to the diseases for example orthopedic instrument and surgical dental.
- **Medical Insurance:** This category covers health insurance and medical reimbursement. It helps to recover all or part of the individual's hospitalization or other medical expenses incurred due to sickness.
- **Telemedicine:** This category bridges the gap between rural and remote areas and provides applications for training, management and education in the field of healthcare.

The total healthcare industry size is predicted as US$ 160 billion (2017) and US$ 372 billion (2022). The Indian hospital industry currently stands at US$ 61.79 billion (2017) and it is expected to rise at a

*Figure 1. Trend of healthcare sector growth (US$ billions)*
**Note:** *E- Estimated, F-Forecast*
**Source:** *Frost & Sullivan (n.d.)*

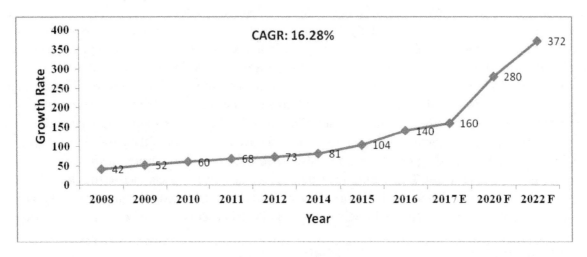

Compound Annual Growth Rate (CAGR) of 16-17%, which means that Indian hospital industry will reach US$ 132.84 billion in 2022. In some areas India is already the most competitive country when compared to other Asian and Western countries. However, it is difficult to compare Indian surgery costs with the American or Western European costs, because the overall cost structure as well as the provided facilities are so different explaining the high costs charged in many places outside India. Healthcare industry is growing due to many different reasons such as providing managerial skills to employees, better coverage, better services, strengthening and escalating expenditures by various players (public and private).

Healthcare industry growth can be analyzed on the basis of market size and investment. Market size of healthcare is set to increase three-fold to US$ 133.44 billion by year 2022. India is experiencing a 22-25% growth in medical tourism and this industry is expected to reach the level of US$ 9 billion by 2020. On the other hand, Department of Industrial Promotion Policy (DIPP, n.d.) report reveals that investment in diagnostic centers and hospitals engrossed foreign direct investment (FDI) worth of US$6.09 billion between April 2000 and March 2019. Among the most recent investment in the healthcare industry in India is according to the Ministry of Health And Family Welfare (2005), Government of India that India and Cuba signed a Memorandum of Understanding (MoU) to increase cooperation in the areas of health and medicine. Moreover, Indian healthcare sector witnessed 23 deals, worth US$ 679 million in H1 2018. Latest news related to healthcare is that Fortis healthcare and Manipal Hospital enterprise have received an approval of de-merge. Dr. Ranjan and TGP could invest US$ 602.41 million in Manipal Hospital Enterprise. If we compare between two countries, the United States and India, there are still marked differences. As reported in Dartmouth Atlas, the United States spends over 17% GDP on healthcare (across geographic regions) while India spends only 4% of GDP on healthcare. This tells about the societal inequalities in resources allocation and access to healthcare. Compared to the United States government, which spends to much on national health, central and state governments in India spend only one quarter of national budget on healthcare. In the United States most of the population is nowadays covered by the health insurance and as a result, these people are free or in a better position to deal with physicians and hospitals. But in India, only a small part of population is covered by insurance

*Healthcare in India*

and, therefore, most people have to worry far more than in the United States about which physicians and hospital they can use. Finally, in the United States the situation of hospitals and medical insurance has been socially and politically contested for the past decade; but in India these two sectors are booming and new solutions are spreading dynamically, although the government is not always in full control of the situation, as noted before.

## Notable Trends in the Indian Healthcare Sector

There are notable changes in Indian healthcare sector, which are summed up below:

- In India most of the communicable disease (high blood pressure, diabetes, and heart disease) fall under the category of life style diseases.
- Vaatsalya healthcare is one of the first hospitals in India who provide facilities to Tier II and Tier III cities.
- In India telemedicine[2] is the fastest emerging medical sector in India. Telemedicine facilities are adopted by Apollo, AIIMS and Narayana Hrudayalaya. In India telemedicine market is expected to have risen at a CAGR rate of 20% during financial year 2016-2017. Telemedicine will at least help to partially fill the gap between rural and urban communities.
- In India engagement and adaptation of AI-Based applications helps to directly talk with doctors and take expertise for the best treatment.
- Investment and development in IT sectors and integration with medical electronics have provided high quality medicine care to everyone. The home healthcare market is estimated to reach US$ 4.46 billion by end of 2018 to 2020.
- Health insurance is gaining year to year. Gross direct premium income is grew 20.06% in FY 2019 (Statista, n.d.).
- Till 2020, strong mobile technology infrastructure provide in India. (e.g. Cycle Tel Humsafer[3]).
- Indian medical technology[4] sector is forecasted to US$ 9.60 billion by 2020.
- In India provide luxurious services (e.g. Pick and drop patient by helicopter)

## FINANCING IN HEALTHCARE

In India healthcare financing will reach a level of appropriate state when there will be adequate funding for healthcare and when equitable access will become available to all population groups (higher, middle & lower). Finally, there is an urgent need to reduce financial barrier to make it possible to utilize health services. Basically, healthcare financing deals with the issues of generating funds and allocation how we can use of the financial resources in healthcare sector. Healthcare financing will directly contribute to improvement of health and to better access to health services. Health financing depends on a number of sources. Most notable is the tax-based public sector that comprises local, State and Central Governments. In addition, there are numerous autonomous public sector bodies and the private sector.

The total healthcare expenditure is classified into two categories first public expenditure and private expenditure. All public expenditure is taken care by the central and state governments. Central government includes the Ministry of Health and Family Welfare and other central ministries/departments. The

*Figure 2. Trends in healthcare sector in India*

**Source**: *IRDA, CII, Grant Thornton, Gartner, Technopak, Aranca Research, Business Standard, Ministry of External Affairs, Ministry of External Affairs (Investment and Technology Promotion Division)*

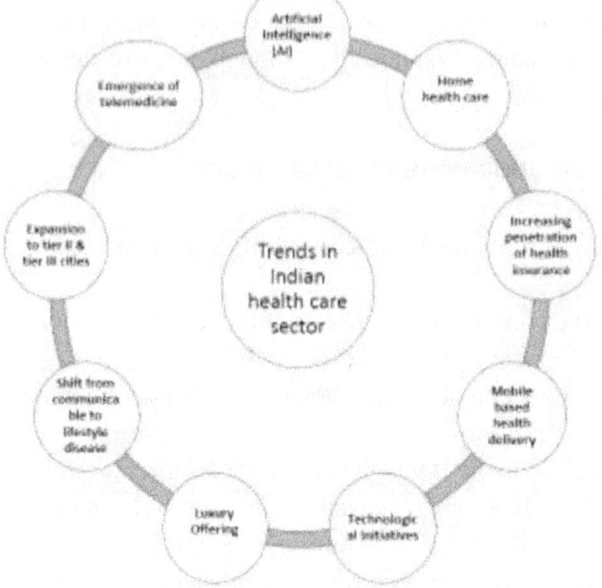

*Figure 3. Sources of financing in Indian healthcare*
**Source:** *Ministry of Health and Family Welfare, Government of India (2005)*

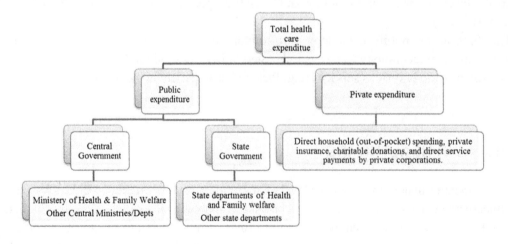

*Healthcare in India*

state governments include state departments of health and family welfare and other state departments. On other hand, the private expenditure includes direct household (out of pocket) spending, private insurance, charitable donations, and direct service payments by private corporations. Although, taxation is work as a financing resources to the middle class population and this is the mobilizing resources form richer sections to needy. Moreover, taxation also full fills the requirement of poor people who suffered health problem due to poor nutrition, unhealthy leaving conditions etc.

## MAJOR CHALLENGES IN INDIAN HEALTHCARE

There are so many challenges facing the Indian society and the issue of economic growth is closely related to the future of Indian healthcare system. Healthcare in India is still characterized by notable coverage gaps, in particular for urban, rural, poor and marginalized groups. On the basis of WHO research (2010), half the world population is still unable to access needed health services, while 100 million people are pushed into extreme poverty each year due to rapidly increasing health expenses. Moreover, 800 million people spend their own income on healthcare expenses. These high expenses also disturb the mental health of many people. If healthcare services will better ensure the quality of services to all income levels, the life struggle would ease for many people and that would have major improvement also for mental health. There are main challenges discussed below:

- **Population:** Indian daily life is disturbed because of rapidly aging population and their growing needs. Normal salaried persons can not afford the cost of new equipment and services, doctors fees, other charges, etc.
- **Non-Nicable Diseases (NCDs):** This includes various type of disease (cardiovascular, cancer, diabetes& mental health), which are not communicable easily. Globally, 70% deaths happen due to non- communicable diseases and the majority of deaths take place in low- and -middle-income countries. Simultaneously, communicable diseases (maternal mortality, unmet need for sexual and reproductive health services, and malnutrition) also affect to the global economy.
- **New Technologies:** New technology basically means that something new is discovered and people can make their lives better with the help of new technology, but in Indian context the technology tends to be in hands of educated people and elites. Urban population can easily access new technology in comparison to the rural population. Thus, the access to new technology is still a major challenge (in addition to other related challenges such as the domination of private health sector, expensive technology, poor quality of service, monopolies and limited resources, because the way that new technology is utilized and understood in India.
- **Inaccessibility to Primary Healthcare:** The access to primary healthcare is a universal goal. The healthcare in India is a vast system and as such is much like the country herself: full of complexity and paradoxes. Although researchers and policymakers are constantly studying why challenges remain and how we could remove these challenges, the goals remain elusive and even the objectives and priorities can not always be listed in simple ways.
- **Medical Education:** In medical education still two major questions remain. The first one is how to improve quality and actual learning in medical education? The second is how to ensure that the education and changes in educational system will fit the social, cultural and economical contexts of very different countries (E.g. North American or European or African or Asian procedures

*Figure 4. Major challenges in healthcare*
**Source:** *Frost and Sullivan, LSI Financial Services, Deloitte, Aranca Research*

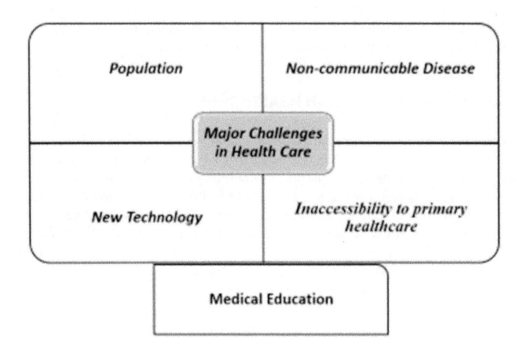

and quality rules will produce different results in other countries). Another important question is 'how we can adopt universal standards to contribute to the improvement in access to healthcare' (Segouin, Hodges & Brechat, 2005).

## INNOVATIVE SOLUTION AND GOVERNMENT POLICIES

There are many initiatives that have been adopted by the Government of India but here I include only those which are most important and characterize the Indian healthcare industry best:

## OPPORTUNITIES FOR HEALTHCARE INDUSTRY

India provides full opportunities for the investors or players in the medical industry. Healthcare sector is a one of the fastest growing sector in India and it is expected to reach $280 billion by year 2020. Healthcare sector in India is very large in every segment and measurement such as medical technology, tax-payers, medical facilities providers, etc. As a result of an increase in the competition in medical sector many companies and individuals want to start new facilities and are looking into exploring for the

*Table 1. Steps taken by government of India*

| Date | Steps Taken/Mission | Amount/Percentage |
| --- | --- | --- |
| 23 Sept, 2018 | *Launched* Pradhan Mantri Jan Arogya Yojana (PMJAY) | Health insurance worth Rs. 5,00,000 to 100 Million/Pa. |
| Aug, 2018 | *Ayushman Bharat-National Health Protection Mission (BNHPM)* | 60:40 (central and state government sponsored) for all states. 90:10 (Hilly North Eastern States) 60:40 for Union Territories with legislature. Central government 100% contributes without legislature. |
| Dec, 2018 | *Indradhanush Mission* | This mission 90% covers unvaccinated and partially vaccinated children in rural and urban areas of India. |
| **Achievements (2017)** | | |
| 2017 | *National Nutrition Mission (NNM)* | This mission was approved by two government bodies (Ministry of Health and Family Welfare (MoHFW) and the Ministry of Women and Child development (WCD). This mission attempts to interrupt inter-generational cycle of under-nutrition |
| *November 15, 2017* | *Medicines and Reasonable Implants for Treatment (AMRIT) Pharmacies.* | 4.45 Million People have made use of this facility in India. |
| *December 15, 2017* | *National Medical Commission Bill 2017* | This bill was passed by the government of India and serves as the basis of medical education reforms. |

**Source:** Compiled by Author (on the basis of study of *Frost & Sullivan, n.d.*)

latest trends, which may have a positive impact on business. In India, hospital industry is estimated to increase to US$ 132.84 billion (FY22) to US$ 61.79 billion (FY17) at a CAGR 16-17% (India Services Medical Value Travel Services, 2018). Similarly, Indian medical sector or medical industries scenario is developing rapidly and is reflected in the new scheme of Abbreviated New Drug Application (ANDA)[5]. While promoting business the government of India has also focused on investment in healthcare infrastructure in both urban as well rural areas. In particular, India provides new offers to the investors in R&D departments and medical tourism as these segments provide good business opportunities and profits.

## CONCLUSIONS AND FUTURE SCOPE

What is needed in India is a major restructuring and strengthening of the health system. This involves two major ingredients: popular mobilization for operationalizing the right to healthcare and the political will to implement policy changes necessary to transform the health system. Healthcare in India is a vast system and can be much like the rest of the country: full of complexity and paradoxes. Although researchers and policymakers are constantly studying why challenges remain and how we could remove these challenges, the goals remain elusive and even the objectives and priorities can not always be listed in simple ways.

This chapter has discusses a great variety of interrelated issues, which are at the core of healthcare system in India. The objective of the chapter is to identify what kind of challenges the Indian healthcare system faces and how we can solve these challenges and problems. Moreover, this chapter has also discussed the sources of financing healthcare. It is hoped that this chapter has developed ideas that prove to be useful for multidisciplinary analysis and for various interested groups such as the researchers, policymakers, administrators and politicians.

## REFERENCES

Alsan, M., Bloom, D. E., & Canning, D. (2006). The effect of population health on foreign direct investment inflows to low-and middle-income countries. *World Development, 34*(4), 613–630. doi:10.1016/j.worlddev.2005.09.006

Atun, Gurol-Urganci & Sheridan. (2007a). Uptake and diffusion of pharmaceutical innovations in health systems. *International Journal of Innovation Management, 11*(2), 299-321.

Atun, R. A., & Sheridan, D. (2007). Editorial Innovation In Healthcare: The Engine Of Technological Advances. *International Journal of Innovation Management, 11*(2), v–x. doi:10.1142/S1363919607001692

Bouoiyour, J. (2007). The determining factors of foreign direct investment in Morocco. *Savings and Development*, 91-106.

Cheng, L. K., & Kwan, Y. K. (2000). What are the determinants of the location of foreign direct investment? The Chinese experience. *Journal of International Economics, 51*(2), 379–400. doi:10.1016/S0022-1996(99)00032-X

Chokshi, Patil, Khanna, Neogi, Sharma, Paul, & Zodpey. (2016). Health Systems in India. *J Perinatol., 36*(3), S9–S12. doi:10.1038/jp.2016.184

DIPP. (n.d.). *Government of India*. Retrieved from https://dipp.gov.in/

FICCI. (2015). *Healthcare: The neglected GDP driver Need for a paradigm shift*. FICCI.

Frost & Sullivan. (n.d.). *Healthcare sector growth trends (US $Billion)*. Retrieved from https://www.researchgate.net/figure/Healthcare-sector-growth-trend-Source-Frost-and-Sullivan-LSI-Financial-Services_fig1_328540551

Government of India. (2009). *National Health Accounts – India (2004–05) –with provisional estimates from 2005-06 to 2008-09). National Health Accounts Cell*. New Delhi: Ministry of Health and Family Welfare.

India Services Medical Value Travel Services. (2018). *Sector Overview*. Retrieved from https://www.indiaservices.in/medical

Kumar, A.K., Chen, L.C., Choudhury, M., Ganju, S., Mahajan, V., Sinha, A., & Sen, A. (2011). *Financing healthcare for all: challenges and opportunities*. Doi:10.1016/S0140-6736(10)61884-3

Kurtishi-Kastrati, S., Ramadani, V., Dana, L. P., & Ratten, V. (2016). Do foreign direct investments accelerate economic growth? The case of the Republic of Macedonia. *International Journal of Competitiveness, 1*(1), 71–98. doi:10.1504/IJC.2016.075903

Ministry of Health and Family Welfare Government of India. (2005). *National Family Health Survey (NFHS-3)*. Retrieved from https://dhsprogram.com/pubs/pdf/frind3/frind3-vol1andvol2.pdf

Segouin, C., Hodges, B., & Brechat, P.-H. (2005). Globalization in health care: Is international standardization of quality a step toward outsourcing? *International Journal for Quality in Health Care, 17*(4), 277–279. doi:10.1093/intqhc/mzi059 PMID:16033805

Srinivasan, R. (2012). *Healthcare in India – 2025 Issues and prospects*. Retrieved from http://planningcommission.gov.in/reports/genrep/bkpap2020/26_bg2020.pdf

Statista. (n.d.). *Value of gross direct premiums from general insurance industry across India from FY 2002 to FY 2019*. Retrieved from https://www.statista.com/statistics/1075200/india-insurance-gross-direct-premium-value/

WHO. (2008). *Commission on social determinants of health. Closing the gap in a generation: health equity through action on the social determinants of health: final report of the commission on social determinants of health*. Geneva: World Health Organization, 2008. Retrieved from https://www.who.int/social_determinants/thecommission/finalreport/en/

WHO. (2010). *World health statistics 2010*. Retrieved from https://www.who.int/whosis/whostat/2010/en/

Woodward, D. P. (1992). Locational determinants of Japanese manufacturing start-ups in the United States. *Southern Economic Journal, 58*(3), 690–708. doi:10.2307/1059836

World Health Organization. (1999). *The World Health Report 1999*. Geneva: WHO. Retrieved from https://apps.who.int/iris/handle/10665/42167

## ENDNOTES

[1] PE/VC represented investment by private companies/ VC represents capital contribution made by different investors.

[2] Telemedicines provide the facilities of consultation, medicines and health assessments by telecommunication.

[3] Cycle Tel Humsafer is a SMS based mobile service, which is intends to help women to organize their family lives in better ways.

[4] Medical technology includes electronic records related to medical docs, mobile healthcare record, PRACTO, Hospital information system, etc.

[5] Abbreviated New Drug Application (ANDA).

# Chapter 11
# Medical Tourism Patient Mortality:
## Considerations From a 10-Year Review of Global News Media Representations

**Alicia Mason**
*Pittsburg State University, USA*

**Sakshi Bhati**
https://orcid.org/0000-0002-6305-7704
*Pittsburg State University, USA*

**Ran Jiang**
https://orcid.org/0000-0002-1542-3124
*Soochow University, China*

**Elizabeth A. Spencer**
*University of Kentucky, USA*

## ABSTRACT

*Medical tourism is a process in which a consumer travels from one's place of residence and receives medical treatment, thus becoming a patient. Patients Beyond Borders (PBB) forecasts some 1.9 million Americans will travel outside the United States for medical care in 2019. This chapter explores media representations of patient mortality associated with medical tourism within the global news media occurring between 2009-2019. A qualitative content analysis of 50 patient mortality cases found that (1) a majority of media representations of medical tourism patient death are of middle-class, minority females between 25-55 years of age who seek cosmetic surgery internationally; (2) sudden death, grief, and bereavement counseling is noticeably absent from medical tourism providers (MTPs); and (3) risk information from authority figures within the media reports is often vague and abstract. A detailed list of health communication recommendations and considerations for future medical tourists and their social support systems are provided.*

DOI: 10.4018/978-1-7998-3576-9.ch011

## INTRODUCTION

As an industry medical tourism involves both the treatment of illness and the facilitation of wellness, with travel. Broadly, medical tourism involves a process through which a consumer travels from his/her place of residency (usually to another country), receives medical treatment or care thus becoming a patient, and typically involves at least one overnight stay. Those who engage in the process of medical tourism are called 'medical tourists' and they do so for a variety of reasons. Many seek access to advanced medical technology, higher quality of care, and quicker access to novel or restricted medical treatments and procedures in their home countries.

Medical tourists are motivated by several factors. Lower-cost procedures and discretionary cosmetic operations represent only a small segment of the global patient base. Data from the Medical Tourism Association's (MTA) *2016-2017 Global Buyers Guide* report is founded on 1,110 prequalified global buyers of healthcare services. Global buyers are defined as "individuals or companies that either refer patients to healthcare destinations or are involved in the selection of healthcare providers for the networks offered to traveling patients," (p. 3). The report highlights that 72% of respondents seek orthopedic/spine and oncology/cancer care, 60% are those seeking cosmetic/plastic surgery, 54% are seeking cardiovascular treatment, 52% neurology, while 40% are seeking IVF/fertility treatments (MTA, 2017). Medical tourism patients enter the marketplace in a variety of ways with 54% using the assistance of a medical tourism facilitator (MTF), insurance company or government program, 16% are referred by local physicians, 12% receive interpersonal word of mouth (WOMC) information, 10% rely on internet research, and 8% use other means (MTA, 2017).

Proponents of medical tourism argue the marketplace is an organic and emergent bi-product of the globalized healthcare systems, digital communication technology, and international transportation that provides opportunities for affordable access to high quality healthcare in premier facilities, and expedited access to novel or emerging treatments (e.g., stem cell treatments; Hopkins, Labonté, Runnels, & Packer, 2010). Critics maintain medical tourism is a poorly regulated healthcare industry (Turner, 2011), that increases patient risk due to weak pre-operative counseling (Crooks, Turner, Snyder, Johnston, & Kingsbury, 2011), and poorly coordinated post-operative treatment plans (Mason & Wright, 2011). Critics also argue that in some destinations the process creates preferential treatment centers, or dual delivery healthcare systems, that exacerbate current health disparities in local populations (Mason, 2014; Snyder, Johnston, Crooks, Morgan, & Adams, 2017).

Affordability is a driver. For example, when quantifying the medical costs, cross-border and local transportation fees, inpatient stay, and post-operative accommodation expenses, Patients Beyond Borders (PBB) estimates the global media tourism market size is $65-87.5 billion USD, with approximately 20-24 million cross-border patients spending an average of $3,410 USD per visit. PBB also forecasts some 1.9 million Americans will travel outside the United States for medical care in 2019 (PBB, 2019).

## HISTORY OF MEDICAL TOURISM IN WESTERN MEDIA

Medical tourism, sometimes referred to as 'healing holidays,' 'medical voyages,' and 'therapeutic journeys' have appeared in U.S. print news publications as far back as the 1870's. These publications often use travel narratives and promotional advertising placements to attract 'health seekers.' Health seekers in this era were often invalids, those with incipient consumption, and TB patients. As a result of westward

U.S. expansion and improved railway transportation, newspaper advertising in the early 1890's attracted health seekers with the allure of 'palatial day coaches' and 'salubrious and health-giving environments,' (St. John's, 1907).

During this era, medical tourism media promotion often appeared in the form of persuasive advertising appeals targeting the environmental-consciousness of patients. These health messages drew upon alluring aspects of elevation, temperature, and variability of year-round climate in key destination locations. Other persuasive health appeals drew upon the history and knowledge of indigenous U.S. populations to establish credibility. For example, Rhea Springs, Tennessee promoted the healing properties of the local land and water by asserting: "Even before the white man stepped his foot on American soil the healing properties of this famous water were appreciated by the medicine men of the Cherokee Indian Tribe," (Presbyterian, 1911, p. 21).

By 1914 the Denver Health Seekers Bureau (DHSB) was established to offer advice to current and newly arriving health seekers. Working in coordination with the Denver Visiting Nurses Association these patients were provided information on the importance of proper hygiene, and instructed not to solely rely on the benefits of the Colorado climate for recovery. Serving as precursor to today's *medical tourism facilitators* (MTF), intermediaries or liaisons between patients and medical tourism providers (MTP), members of the DHSB would contact health seekers upon arrival and assist in finding sanitary quarters (Bell, 1914).

These early media representations of medical tourism did not frequently report stories of patient mortality resulting from complications of travel, quality of accommodation, or substandard healthcare services and treatments. Contrary to these early forms of medical tourism promotion, today's media frequently report high-profile deaths resulting from medical tourism. Merlock, Jackson, Payne, and Stolley (2013) argue that American media "often scrutinizes high-profile medical tourists, they frame narratives of desperation, hope, caution, technology, choice - even a bit of adventure - and these stories resonate with the public," (p. 126). Popular celebrity examples include the death of Steve McQueen, who was at one point was the highest paid movie star in the world, died at the age of 50 in Mexico while seeking unorthodox cancer treatments unavailable in the United States. Others include Farrah Fawcett, an American actress, who died after receiving failed cancer treatments in Germany not available to U.S. patients at the time. Tameka Foster, ex-wife to international R&B singer Usher, died following cosmetic surgery complications in Brazil. Media representations of medical tourists have appeared in a wide variety of media outlets and across multiple channels and platforms over an extended period of time. The next section provides a review of recent scholarly inquiries into medical tourism and patient mortality.

## MEDICAL TOURISM AND HEALTH COMMUNICATION

Global healthcare is in a transformative period. Networks of globalized healthcare systems are becoming more centralized and dense, and economic development is accelerating through the advancement of transportation and technology in special economic zones (SEZs) spurring the continued expansion of the global health marketplace. Over the past decade health communication scholars have employed a variety of qualitative and quantitative methodologies to better understand the impact this process has on patients.

Some lines of academic inquiry have been *patient-focused* examining the communication needs of medical tourists (Ngamvichaikit & Beise-Zee, 2014), while others focus on the attitudes, motivations, and justifications expressed by medical tourists for engaging in the process

(Cameron et al., 2014; Khan, Chelliah, Haron, & Ahmed, 2017). Still, other work has been *message-focused* involving inquiries that examine specific forms of persuasion, including the elements of emotion and risk in medical tourism advertising appeals (Kemp, Williams, & Porter, 2015), and first-person patient testimonials (Hohm & Snyder, 2015).

The promotional health messages associated with the medical tourism industry have been studied across *media channels* including websites (Lee et al., 2014) and newspapers (Jun & Oh, 2015). Additional research efforts to more fully understand the digital promotion techniques have employed a *regional-focus* such as examining the Korean medical tourism industry (Jun, 2016) and Canadian medical tourism brokers (Penny et al., 2011); a *treatment-focus* analyzing specific medical treatments such as weight loss surgery (Glenn, McGannon, & Spence, 2013) and transplant tourism (McKay, 2016); and a *socio-cultural-focus* including studies that explore differences between East–West medical tourism facilitators' websites (Frederick & Gan, 2015), differences in medical tourism news representations between low- and middle-income countries (Imison & Schweinsberg, 2013), and also how specific treatments are presented to diverse audience bases (e.g., reproductive tourism representations between Israel and Germany; Bassan & Michaelsen, 2013).

Extant literature has helped to inform our current understandings of many patient-centered concepts associated with the medical tourism including: risk, disclosure and informed consent, patient decision-making, organizational and physician credibility, in addition to calling attention to the ethical, financial, and legal considerations. This field of study has successfully provided many pragmatic considerations necessary for the continued growth and advancement of the global medical tourism industry. In an attempt to add to this growing body of scholarship, the present chapter explores media representations of patient mortality in the global news media occurring between 2009-2019, in order to better understand how the deaths of medical tourism patients are framed to global audiences.

Turner (2013) previously analyzed news representations of international medical tourism patient deaths ranging from 1993-2011 who exclusively received either cosmetic and/or bariatric surgery. Findings from this work argued: (1) targeted public safety messages about risks of cosmetic surgery procedures to low and middle income women, (2) communicating the importance of regulation and licensing of healthcare professionals and ambulatory environments in medical tourism, (3) increasing coordination of public awareness through U.S. consulate offices and other services such as Yellow Book©, (4) improving advertising regulation, (5) critically reflecting about the "routinization and normalization" of popular, commercialized procedures such as liposuction, breast augmentation, and lap band surgery, and finally, (6) advocated the need for better tracking strategies for assessing patient flow, service, and treatment outcomes (Turner, 2013).

Our work seeks to determine if prior results will be replicated in a data set that includes both domestic and international medical tourists, in a sample that is not treatment-specific, and aims to extend upon prior work in three key ways. First, our analysis offers an enhanced focus on medical tourism providers' response characteristics presented in the media accounts of patient death. Second, this analysis explores how post-mortem patient advocacy by friends and family of victims are integrated into the reporting process. Finally, we provide new considerations for planning and preparation prior to seeking elective or urgent healthcare abroad.

This chapter is segmented into two main sections. The *Approach* section details the sampling procedures used to create the data set, and details the database and digital search strategies used to identify cases of medical tourism patient mortality. A comprehensive list of cases included in this study are presented with the support materials. The *Analysis & Interpretation* section is presented as a qualitative content analysis and includes several health communication recommendations and considerations for patients and their families, applied health practitioners, and members of the academic community.

## APPROACH

Historically some scholars have chosen to limit the definition of 'medical tourism' as the act of traveling from one country to another, specifically for the purpose of seeking medical treatment, procedures, or services. This analysis does not predicate the conceptual definition as contingent upon 'international' travel. Medical tourism exists as an option for domestic and international patients globally. Patients traveling from California, U.S. seeking treatment in Florida, U.S. would travel significantly farther than an individual leaving Florida, U.S. traveling to Cuba, or the Dominican Republic for healthcare purposes. In this chapter we define medical tourism as a consumer transaction through which one travels, either domestically or internationally, for the specific purposes of obtaining medical or healthcare treatment and services, thus becoming a patient.

To create a sample for analysis, three researchers identified patient mortality cases using a variety of digital and database search strategies. The strategic focus was media representations of patient deaths and our research efforts concentrated on identifying cases presented in U.S., India, and Chinese media outlets. Cases were identified using several databases including: *Newswire, Nexis Uni,* and *WebNews*. Online search engines were also used to locate cases including: *Google, Yahoo!,* and *Bing*. Chinese media representations were identified using *Baidu*, the largest search engine in China, and CNKI.[1] Cases were also identified through regional news networks including Chinese news websites (e.g., *Sing Pao Daily News, Souhu News,* and *Sina News*) and news websites in India (e.g., *India Times*). Publicly available hospital reports and prior cases in medical studies were considered. If a news report referenced a case not previously identified, additional research efforts were made in order to identify the patient and further search and code the case. When available, social media content about the case was analyzed and integrated into the analysis. A variety of non-treatment specific search terms were used including: *medical tourism, medical tourist, patient death, medical tourists, patient's death, fatal, death, medical, hospital,* and *abroad*. Our analysis excludes travelers who received care on an emergency basis (i.e., ordinary tourists who become sick on a cruise), wellness tourists (i.e., those seeking non-medical massages or acupuncture), and expatriates seeking care in their country of residence.

Cases were omitted if the medical tourist was specifically traveling for assisted suicide (e.g., suicide tourism, death tourism), if a patient was injured but not deceased, or if the case involved a foreign tourist who died in a foreign hospital, but did not travel initially for medical purposes.

Data for each case were coded into four categories including: *patient background*: home country, sex, age, cause of death, identity descriptors (i.e., lawyer, mom, business owner, etc.); *process characteristics:* location of treatment/surgery, place of treatment surgery, place of convalescence/recovery (i.e., hospital, hotel, apartment), type of treatment received, physician qualifications, facility accreditation/licensing, use of medical tourism facilitators, reputational and performance history of physicians, facilitators and facilities; *case outcomes* such as patient advocacy, fundraising, the launch of formal, criminal or administrative investigations; a physician suspended, arrested, charged; a treatment facility temporarily or permanently shut down; and also the *educational or instructional health information* including statements telling audiences what to do to avoid negative outcomes (i.e. be aware, do more research, etc.)

A total of 57 cases were identified for consideration, each case representing one, individual medical tourist. Upon secondary review, seven cases were disqualified as some were found to be children under the age of 18 which were not decision-making consumers at the time services were received, individuals receiving treatment in one's hometown, and cases pre-dating the sample time frame 2009-2019. [2]

The final dataset included 50 cases, of these 80% are females ($n=40$) and 20% are males ($n=10$) who died as a result of medical tourism during the January 2009 to March 2019 time period. Twenty percent of the cases ($n=10$) resulted from domestic medical tourism, 80% ($n=40$) were international medical tourists. For the purposes of this analysis and consistent with extant literature, the patients' names will not be disclosed. The sample consists of a variety of medical treatments including: *cosmetic procedures* (i.e., tummy tucks, breast augmentation, liposuction, eye lid surgery), *bariatric procedures* (i.e., lab-band surgery, gastric bypass), as well as *addiction treatments, dental treatments, liberation therapy,* and *organ transplants*. A total of 72 treatments/procedures were noted for these 50 cases, with some patients obtaining 2-5 surgeries in a single visit. Only 10% of the cases ($n=5$) would be considered high-profile involving models, actors/actresses, beauty queens.

The next section of the chapter details the qualitative analysis and interpretation of the cases involved in this study. Findings from this thematic analysis will be drawn upon and used to frame recommendations and strategies for consideration in future practice. These areas of recommendation include: *informational and educational health messages, medical tourism patient and family recommendations,* and the *MTP organizational response considerations.*

## INTERPRETATION AND ANALYSIS

### Information and Educational Health Messages

Upon review of the public advice and procedural recommendations included within the media reports included in our sample, it was evident these statements are often vague and abstract. The advisory statements following patient deaths are often delivered by highly credible sources such as attorneys, government officials, professional and law enforcement agencies, officers, and directors. The public statements tell lay audiences that hospitals should be 'authentic and qualified' (P8), medical tourists should 'watch out' what hospital is selected (P9), medical tourists should understand local regulations to 'protect themselves' (P3), and that medical tourists should have 'a basic ability to identify false propaganda and minimize risk' (P6). Audiences are further instructed to 'do their homework', and 'go to a place that has the right facilities', and '"properly trained" doctors'.

While sound advice, these generalized claims fail to address important message design aspects of risk communication processes. The recommendations noted in our sample offer little detail, and lack 'how to' messages of efficacy that audiences would be able to behaviorally follow. Medical tourists, like other patients, have specific health information seeking behaviors and differing degrees of health literacy. In many of these cases, the referring source to the medical tourist's treatment clinic or physician came through interpersonal relations, word of mouth (WOM) recommendations from friends and family. The interpersonal influence of trusted friends, who are known associates of prior medical tourists, may undermine additional health information seeking behaviors commonly advocated by authorities within these reports. *Health information seeking* (HIS) is the intentional, active effort to obtain specific information above and beyond the normal patterns of media exposure and use of interpersonal sources

(Atkin, 1973; Griffin, Dunwoody, & Neuwirth, 1999). These HIS behaviors include non-routine media use or interpersonal conversation about a specific health topic and includes behaviors such as "viewing a special program about a health-related treatment, using a search engine to find information about a particular health topic, and/or posing specific health-related questions to a friend, family member, or medical practitioner outside the normal flow of conversation," (Niederdeppe et al., 2007, p. 155).

Several cases reported learning about medical tourism opportunities through online searching, receiving digital promotional advertising, and coordinating with a MTF. Although authority figures in the media reports advocated for medical tourists to discern between 'good' and 'bad' physicians, and 'real' and 'fake' propaganda, the methods or techniques for doing so were not provided. Health information seeking (HIS), specifically online, poses challenges for medical tourists to 'do their homework.' Medical tourists may not be able to access information about a hospital or physician's performance record in their native language, or may find the information is presented in lingo, jargon, or highly technical language. Online HIS efforts may yield false claims about health services, products, and treatments without providing supportive evidence of the promotional health claims (Dow et al., 1996). Digital health claims may be strategically ambiguous, presented at an advanced reading level, and may overuse textual formats than can exacerbate language barriers (McGrath, 1997). Assessing credibility of the health claims is also challenging as the authorship may be misleading or outdated. The webpage may appear savvy and look "official", leading consumers to believe the health claims (Pereira & Bruera, 1998). A family member of P47 stated, 'I saw the website. It looks great. You would think this was the place, but it's not. It's a death palace.'

It is difficult to glean what information is omitted from online advertising content strategically framed to present physicians, treatments, and/or facilities in a positive light and distributed to medical tourists prior to travel. It is not enough for patients to research a treatment facility and use the organization's accreditation as the sole indicator of high-quality healthcare expectations. In several cases, at the time of death the medical facility/clinic had the proper licensing and certification, it was the operating physician/ surgeon who did not. While many healthcare organizations can easily demonstrate strong organizational credibility through international accreditation (i.e., JCI, AAAAF); physician expertise and performance history are equally as paramount.

Even if a medical tourist possessed all of the necessary, pertinent information prior to electing treatment abroad, it doesn't mean they will comprehend the nuanced details. Understanding the pre-travel health literacy of medical tourism patients is critical, yet today relatively little is known about the general knowledge and health literacy of this patient population. *Health literacy* is the degree to which individuals can obtain, process, understand, and communicate about health-related information needed to make informed health decisions (Berkman, Davis, & McCormack, 2010; Berkman et al., 2011; McCormack et al., 2010; Ratzan & Parker, 2000). Patients with high health literacy may be more likely to understand the discrete differences between a cosmetic physician and plastic surgeon, and thereby proactively engage in health information seeking behaviors to confirm specific physician qualifications and skills training. For those with low health literacy this may result in negative outcomes. The Agency for Healthcare Research and Quality (AHRQ) found low health literacy is associated with more emergency department visits and hospital readmissions, less preventive care, and poor medication administration skills (AHRQ, 2017).

Health literacy as a concept, is multifaceted and includes printed literature, oral communication, and numeracy. Printed literature requires reading and writing ability. This may affect a patient's ability to understand the promotional, medical, travel, legal/contractual, and financial documents prior to travel. Oral literacy requires the ability to listen and speak. Patients in medical tourism scenarios may need to address intercultural communication challenges with providers and staff, and communicate post-operative messages to convalescent teams or caretakers in off-site locations (e.g., hotels, apartments). Numerical literacy, or numeracy, is the skill and ability to understand and use numbers which is essential with respect to knowing financial documentation, medical paperwork, and medication dosages. Statements from authority figures in these cases often lack specific language that directs the public to factual and reliable information resources and fails to deliver messages of efficacy explaining how to do so in a way that lay audiences are able to comply.

## Medical Tourism Patient Recommendations

From the experiences of patient deaths included in this sample, several lessons can be learned to reduce risk. Six categories for health awareness and education of medical tourists are described below including: (1) understand there are 'bad' actors, (2) arrange post-operative and convalescent care, (3) comprehend and understand signed documentation, (4) avoid bypassing planned procedures, (4) disclosure of honest and accurate information, and (6) focus on short- and long-term care planning.

### Understand There Are Bad Actors

Several cases in this sample can be traced to a select group of physicians, some unlicensed, and poorly performing procedures on unwitting medical tourists. In 2009 Dr. Johan Tapia Bueno owner and operator of *Clinica Brazil* was arrested for malpractice. The clinic, operating out of an apartment building, was closed for lack of licensing and noncompliance with poor sanitary requirements. Bueno was not registered with the Dominican Society of Plastic Surgeons at the time (Actualidad, 2009). Dr. Hector Cabral owned and operated the *International Center for Advanced Plastic Surgery* and is connected to P21, P48, and P47 deaths. In 2011 Cabral was arrested and charged with 10 counts of unauthorized practice of medicine in the United States. Cabral later pleaded guilty to one charge, received no jail time, paid a fine, and completed 250 hours of community service in the Dominican Republic to continue practice. The Dominican authorities would later close the clinic following the deaths of three international patients (Vincent & Klein, 2013; Allred, 2015). A third doctor connected to P29 and P39, Dr. Victor Hector Ramirez Hernandez, was arrested, charged, temporarily suspended, and eventually found innocent in the death of P29. Following the trial, Ramirez quickly asserted credibility by noting membership in the Mexican Plastic Surgery Council and the International Association of Plastic Surgeons. His response did not address the closure of his medical facility due to (a) stockpiles of expired medications, (b) poorly maintained and obsolete equipment, (c) improper management of biological waste, and (d) having undivided recovery areas (Lopez, 2015).

Other cases were connected to specific surgical centers, but because the names of the business were changed several times "patients were often unable to connect the deaths to the business," (Sallah & Perez, 2019, n.p.). *Jolie Plastic Surgery*, in Miami, FL, U.S., was previously known as *Eres Plastic Surgery*, previously known as *Encore,* previously known as *Vanity*, and were all operated and directed by owner Dr. Ismael Labrador. Of the 39 physicians promoted on the website, 24 were determined to be not board-certified in plastic surgery, a specialty which is earned after six years of residency and advanced safety training.

### Arrange Post-Operative and Convalescent Care

Twenty percent of these cases involved treatment and recovery outside of an ambulatory facility including locations of private apartments, hotel suites, and houses ($n=10$). In some cases patients received operations or treatments in hotels, houses, penthouse apartment spaces, and recovered in off-site locations such as other hotels, or apartments. These locations would not meet basic accreditation standards from the American Association for Accreditation of Ambulatory Surgery Facilities, Inc. (AAAASF) for either treatment or postoperative care and recovery. Staff at off-site convalescent locations, often marketed as 'recovery suites' may only be authorized to deliver first aid treatment, if needed. Medical tourists seeking high-risk procedures at high-volume facilities need to understand that distance adds delay to urgent post-operative medical needs.

### Comprehend and Understand What You Sign

Prior to surgery, medical tourists often sign documentation they do not read, with terms and conditions they do not understand. Some documentation has been found to contain legal clauses restricting negative comments from being shared online or through social media, under penalty of a defamation lawsuit from the MTP. In some cases, patients are asked to sign medical documents in a non-native language, reducing their ability to understand what they are or are not agreeing to. By not comprehending or understanding the medical forms, medical tourists may be restricted from publicly discussing negative experiences, forgo patient legal rights, or unintentionally agree to procedures they are not fully aware of at the time.

### Avoid Bypassing Procedures

Several patient death cases resulted from the bypassing of standardized procedures. In the case of P21, the patient's inbound flight was delayed, causing a late-night arrival for an operation the following morning. Media reports indicated the prescreening and medical questionnaire screening protocols were bypassed. P21 later died due to a pulmonary embolism. On the day of surgery, P38 was directed to an off-site location to obtain a pre-operation ultrasound. The location was closed, no ultrasound was obtained, and the staff reportedly proceeded with multiple cosmetic surgeries. In the case of P24, an investigation concluded that the preoperative consultation did not meet an international standard. Medical tourists need to be educated on the importance of pre-screening protocols and educated on the expectations of pre-operative counseling.

### Disclose Honest and Accurate Information

Medical tourists need to disclose honest and accurate information to both physicians and family throughout the medical tourism process. In several cases, the families had no prior knowledge of the patient's choice to seek treatment abroad. P27 relatives only discovered the patient had a procedure when they arrived to identify the body at Rio's Institute of Legal Medicine (IML). P23's family stated, "We had no idea what [P23] was going to do. If we had known we would have done everything to stop her." Medical tourists should be encouraged to disclose accurate information to physicians and MTF/MTPs. In some cases, patients were found to withhold relevant personal health histories, others were engaging in unauthorized illegal drug use prior to surgery, and still others allegedly used sleeping medication post-operatively, without physician or caregiver knowledge. Honest, accurate, and transparent communication is needed throughout the medical tourism process.

## Focus on Short- and Long-Term Care Planning

Post-operative care planning needs to be established before engaging in high-risk medical tourism procedures. Post-operative care planning should include identifying and meeting with healthcare specialists in one's home country who agree to provide follow-up care and treatment should complications occur after returning home. P33 returned from Costa Rica following therapy and demonstrating vast improvements in health to friends and family. Within months, P33 developed a blood clot in a stent placed by an MTP. A local hospital reportedly refused treatment because the operation was performed internationally. With no other choice, P33 returned to Costa Rica. After a month- long process and receiving medicine to dissolve the clot, P33 died. Multiple patients died several weeks following the procedures for which they traveled. Throughout the convalescent period, medical tourists remain vulnerable, and not all medical tourists have the financial means for return travel or corrective procedures, should they be denied care in their home countries.

## Medical Tourism Family and Social Support Recommendations

Of the 50 cases included in this study, 40% involved posthumous engagement by friends and family ($n=20$). Twenty-six cases reported investigations into the operating physician, staff, and/or surgical facilities (e.g., clinic, hospital). Of these cases, 24 resulted in physician arrest, physician conviction, revocation of licensure, temporary or permanent closure of the clinic, facility, or hospital, and/or civil legal actions. The strong association between post-death patient advocacy by friends and family and civil and criminal processes was notable. In the unlikely event a death was to occur, we offer three recommendations for friends, family, and care providers of medical tourists.

## Obtain and Review Documentation of Procedures and Processes

Patient advocates need to understand the importance of requesting, reviewing, and recording correspondence and documentation with MTP/MTFs. When a family member of P36 requested the medical chart, she noticed there was no EKG readout attached. She reportedly asked the doctor where it was, because it is customary to monitor the heart during surgeries. He replied that the EKG machine had run out of paper. When the family of P29 arrived in Mexico, they discovered the paperwork the patient was supposed to have read and signed before the surgery was incomplete. These types of discrepancies are not unusual. In one case the initial autopsy reports listed the cause of death as either natural causes or a heart attack; however, a second, independent autopsy from a coroner revealed significant puncture wounds to many of the patient's organs.

## Prepare to Experience Barriers

One of the most challenging barriers immediately experienced following a medical tourism patient death is the return of the body. Several reports noted the struggle of families to gain access the bodies of dead medical tourists. Days after learning of P25's death, the family were still waiting for explanations and also trying to arrange return transportation of the body. P39's family reported that it took a week to get the body out of Mexico, with little assistance from locals, the Mexican consulate, or the American consulate; while families were also reportedly receiving contradictory and conflicting information to many questions. Additional financial resources may be needed to overcome these barriers. For example, because the cosmetic procedures for P49 weren't approved by her U.S.-based healthcare provider, the company would not transport back the body from the Dominican Republic.

## Fundraising and Patient Advocacy

Western patient mortality cases most frequently involved grassroots, public fundraising and outreach efforts requesting assistance and financial support from others through social media (e.g., Go Fund Me). With the exception of one case from Australia, post-mortem fundraising was a feature exclusively presented in Western media reports. The advocacy actions of friends and family members appeared in many forms including: speaking in front of international congressional assemblies, protesting outside healthcare facilities, creating fundraising campaigns, giving media interviews, developing websites, creating awareness, and offering social support to others (e.g., Evitaa.com).

## Medical Tourism Providers

For MTPs the organizational response to a medical tourist patient's death in local and international media is a defining moment, in terms of reputation management. Prior scholarship has found when patients feel they have been mistreated or their concerns are not addressed, the likelihood for litigation increases (Hickson et al., 2001). Common reasons for pursuing legal recourse include communication problems, inadequate explanations of procedures and treatments, feeling hurried by a physician, a lack of interpersonal warmth, and non-empathy (Hickson et al., 2001). After evaluating the organizational statements from physicians and facilities involved in the cases included in our sample, two improvements for applied practitioners were noted, including the need for: (1) responding with honestly and empathy, and (2) improving death and grief counseling services.

## Responding With Honesty and Empathy

Following the death of a medical tourist patient, it is often traveling companions and/or distant family members who make important decisions such as requesting a second autopsy (if needed), requesting formal investigations, and making transportation arrangements for the return of the patient's body. In several cases, MTPs did not immediately contact family members or the emergency contacts listed on medical contact forms (P1). Some families received different accounts of the stories about the patient's death (P12); and there were discrepancies about the cause of death between hospitals, police, and mortuaries (P1). Brown and colleagues (2001) argue malpractice is reduced when providers can demonstrate clear evidence the patient was fully informed about the risks and benefits of the treatments and was provided the opportunity to make an informed choice. Still in some MT mortality cases, representatives in healthcare facilities of MTPs denied the patient was ever a client, and avoided family and friends phone calls.

## Improved Death and Grief Counseling

Organizational empathetic responses can range from making referrals and recommendations to social support groups to assisting with grief resulting from sudden death. Grief counseling was not present in any media reports following a medical tourism patient's death. Bereavement counseling is assistance and support to people with emotional and psychological stress following the death of a loved one. The emotional loss was expressed by many family members of medical tourists. P47's mother exclaimed "This is so devastating. We can't believe it. It's still like a nightmare. My daughter is now coming home in a box." Sudden death bereavement refers to a situation in which there is no expectation of death prior to a person's passing. Death is emotionally costly, legally complex; a practical challenge in social and physical terms. Those who experience a sudden death are "unprepared for bereavement and suffer more than people who have expectancy of death prior to a person's passing," (Frost, Honeycutt, & Heath, 2017, p. 327). Bereaved individuals, whose loved ones pass unexpectedly, are more likely to grieve more intensely (Parkes, 1975), endure longer lasting grief (Parkes & Weiss, 1983; Tsai et al., 2016), and may develop psychological and physical health issues themselves (Lundin, 1984). Extant literature indicates that people who are unprepared for bereavement have a stronger need for social support from their community and from health professionals (Tsai et al., 2016). No evidence of grief counseling was found within this dataset. This is a noted weakness throughout the medical tourism industry, including practitioners, facilitators, and navigators. Additional focus on the development of health information resources and intervention materials is needed to help extended family members and friends transition and cope with unexpected, sudden death resulting from surgical or anesthesia complications in medical tourism encounters.

## DISCUSSION AND LIMITATIONS

Our analysis found that the instructional and educational information contained within media reports of medical tourism patient deaths is often vague and abstract, holding little informational value for audiences. Further refinement of key messages and detailed strategies to produce positive outcomes is needed. We argue that patients need to realize there are bad doctors and adapt their health information seeking behaviors to accurately research the physician's background and performance record, including the skills and qualifications to perform the specific procedure(s) sought. We also advocate for careful consideration of post-operative convalescent care for high-risk operations. Patients need to be aware of what they sign, and should avoid bypassing steps in the process, including necessary pre-operative medical screenings and protocols. We feel these cases highlight the importance of trust in the physician-patient dyad and reveal the need for medical tourists, prior to travel, to disclose honest and accurate information to MTFs, MTPs, and family and friends. Finally, these cases call attention to the short-term and long-term needs of medical tourists, and highlight the importance of proper financial, legal, will, and estate and healthcare planning. No media representations included in this study indicated if the patients had advanced care directives or powers of attorney in place prior to seeking and receiving treatment abroad. Appropriate legal preparation and planning is needed to ensure the patients' wishes are clear. Because complications may occur several weeks or months following surgery, long-term planning considerations are needed to prepare for complications while anticipating positive health outcomes, relative to the scope and complexity of the procedure(s).

Learning from the experiences of others can provide a useful guide for those who may one day face similar circumstances. This review of patient mortality cases and outcomes revealed that families, friends, and/or patient advocates need to be prepared to request, obtain, and review important patient documentation, and to psychologically and behaviorally overcome obstacles related to intercultural differences and international policies. The experiences of those who have lost loved ones due to medical tourism provides valuable insight into successful advocacy, outreach, and fundraising highlights the legal complexity and ethical challenges of the global healthcare marketplace. This case review of patient mortality in medical tourism also supports the need for individual MTPs to better provide empathetic responses to the family and friends of those who died seeking care, and exposes the need for grief and bereavement transitional counseling services throughout the medical tourism industry, both providers and facilitators.

We acknowledge this analysis has several weaknesses and limitations. To begin, once a case was identified, extensive efforts were made to create a robust picture of the patient experience documented within the media reports. We cannot be certain in each case that specific facts and details are accurate, as we relied on the sources within the media reports. This sample is not a generalizable representation of the medical tourism industry. Based on the 2017 MTA Global Buyers Report, several popular treatments obtained though medical tourism are not represented in the sample (e.g., oncology, neurology, etc.). We are currently seeing rising rates of patient deaths in male populations seeking hair transplant surgery in India (Doley, 2016; Narayan & Debroy, 2019), and these cases are not included in this data set. Our sample was primarily female who sought international medical treatments, and this may have resulted from the search strategy and search terms used to identify cases. Some sample characteristics are explainable by travel policies which limit or restrict patient access to international healthcare options. For example, most countries require Chinese patients to enter with a valid visa, which involves a complex and troublesome application process. As a result, the process and policies may restrict Chinese patients from becoming eligible for outbound medical tourism options and help to explain why so many of those cases resulted from domestic medical tourism within the media sources analyzed. Many of these patient mortality cases emerged from the cottage industry of small, privately-owned clinics as opposed to large healthcare systems that are more likely to hold JCI or AAAAF accreditation (i.e., All India Institute of Medical Sciences).

Future research that explores the treatment-specific barriers and challenges will offer new insight for shaping the risk communication such as health advisories and patient warnings for those seeking specific medical treatment abroad. Improved warning and advisory systems are needed to alert patients in high-volume, high-risk areas of cosmetic medical tourism destinations. Death is the ultimate price, and many seekers do not die, but are instead left with scars, disfigurements, or remain in need of long-term care. Patient morbidity, in contrast to mortality, is a domain of exploration that may also yield novel considerations and recommendations for those considering international healthcare treatments. Patient morbidity requires unique long-term planning and treatment processes. Currently, clusters of patients are returning to the United States with superbugs that are difficult, if not impossible, to treat, keeping medical tourism patients reliant on healthcare teams and pharmaceuticals in their home locations at great costs for significant lengths of time (Lamotte, 2019).

What is clear is that the most dominant global media representations of medical tourism patient mortality occurring between 2009 and 2019 are middle-class, minority females between 25-55 years of age who seek cosmetic surgery internationally. There are no records or international databases that can be used to determine the actual numbers of patient mortality cases either domestically or internationally. In lieu of this, our analysis sought to glean insight to better formalize recommendations to those choosing to engage in the medical tourism marketplace.

## CONCLUSION

This chapter details a qualitative content analysis of 50 cases of patient mortality resulting from the process of domestic and international medical tourism. Cases were identified using worldwide news databases, and regional news networks in India, China, and the West. Findings show the most dominant global media representation of medical tourism patient mortality occurring between 2009 and 2019 is framed as minority females between 25-55 years of age who seek cosmetic surgery internationally. This work extends Turner's (2013) findings and offers a several pragmatic considerations for medical tourism patients, the social support systems for these patients, applied medical tourism practitioners MTF/MTPs, and the academic community.

## REFERENCES

Actualidad.com. (2009, November 25). *Clinic closed and surgeon arrested for malpractice.* Accessed March 3, 2019 retrieved from https://www.diariolibre.com/actualidad/clinic-closed-andsurgeon-arrested-for-malpractice-MIDL224822

AHRQ. (2017). *Agency for Healthcare Research and Quality. National healthcare disparities report.* Washington, DC: Government Printing Office.

Allred, A. (2015, July 30). *5 On Your Side Investigates: Going under the knife, "on the cheap"* Accessed on March 6, 2019 from https://www.usatoday.com/story/news/health/2015/07/30/5-onyour-side-investigates-going-under-on-cheap/30868471/

Atkin, C. K. (1973). Instrumental utilities and information seeking. In P. Clarke (Ed.), *New models for communication research* (pp. 205–242). Beverly Hills, CA: Sage.

Bassan, S., & Michaelsen, M. A. (2013). Honeymoon, medical treatment or big business? an analysis of the meanings of the term "reproductive tourism" in German and Israeli public media discourses. *Philosophy, Ethics, and Humanities in Medicine; PEHM, 8*(1), 1–8. doi:10.1186/1747-5341-8-9 PMID:23962355

Bell, B. (1914, Oct. 2). Health Seekers Bureau. *Summit City Journal.*

Berkman, N. D., Davis, T. C., & McCormack, L. (2010). Health literacy: What is it? *Journal of Health Communication, 15*(S2), 9–19. doi:10.1080/10810730.2010.499985 PMID:20845189

Berkman, N.D., Sheridan, S.L., Donahue, K.E., Halpern, D.J., Viera, A., Crotty, K., … Viswanathan, M., (2011, March). *Health literacy interventions and outcomes: An updated systematic review.* Evidence Report/Technology Assessment No. 199.

Cameron, K., Crooks, V. A., Chouinard, V., Snyder, J., Johnston, R., & Casey, V. (2014). Motivation, justification, normalization: Talk strategies used by Canadian medical tourists regarding their choices to go abroad for hip and knee surgeries. *Social Science & Medicine*, *106*, 93–100. doi:10.1016/j.socscimed.2014.01.047 PMID:24556288

Crooks, V. A., Turner, L., Snyder, J., Johnston, R., & Kingsbury, P. (2011). Promoting medical tourism to India: Messages, images, and the marketing of international patient travel. *Social Science & Medicine*, *72*(5), 726–732. doi:10.1016/j.socscimed.2010.12.022 PMID:21310519

Doley, K. (2016). *How a botched hair transplant put rs 1.25 lakh cr beauty biz in india under scanner*. Financial Express.

Dow, M. G., Kearns, W., & Thornton, D. H. (1996). The internet II: Future effects on cognitive behavioral practice. *Cognitive and Behavioral Practice*, *3*(1), 137–157. doi:10.1016/S1077-7229(96)80035-6

Frost, J. K., Honeycutt, J. M., & Heath, S. K. (2017). Relational maintenance and social support in the aftermath of sudden and expected death. *Communication Research Reports*, *34*(4), 326–334. doi:10.1080/08824096.2017.1350573

Glenn, N. M., McGannon, K. R., & Spence, J. C. (2013). Exploring media representations of weight-loss surgery. *Qualitative Health Research*, *23*(5), 631–644. doi:10.1177/1049732312471731 PMID:23282795

Griffin, R. J., Dunwoody, S., & Neuwirth, K. (1999). Proposed model of the relationship of risk information seeking and processing to the development of preventive behaviors. *Environmental Research*, *80*(2), S230–S245. doi:10.1006/enrs.1998.3940 PMID:10092438

Hickson, G. B., Federspiel, C. F., Pichert, J. W., Miller, C. S., Gauld-Jaeger, J., & Bost, P. (2002). Patient complaints and malpractice risk. *Journal of the American Medical Association*, *287*(22), 2951–2957. doi:10.1001/jama.287.22.2951 PMID:12052124

Hohm, C., & Snyder, J. (2015). "It was the best decision of my life": A thematic content analysis of former medical tourists' patient testimonials. *BMC Medical Ethics*, *16*(1), 1–7. doi:10.1186/1472-6939-16-8 PMID:25614083

Hopkins, L., Labonté, R., Runnels, V., & Packer, C. (2010). Medical tourism today: What is the state of existing knowledge? *Journal of Public Health Policy*, *31*(2), 185–198. doi:10.1057/jphp.2010.10 PMID:20535101

Imison, M., & Schweinsberg, S. (2013). Australian news media framing of medical tourism in low- and middle-income countries: A content review. *BMC Public Health*, *13*(1), 1–12. doi:10.1186/1471-2458-13-109 PMID:23384294

Jun, J. (2016). Framing service, benefit, and credibility through images and texts: A content analysis of online promotional messages of Korean medical tourism industry. *Health Communication*, *31*(7), 845–852. doi:10.1080/10410236.2015.1007553 PMID:26644259

Jun, J., & Oh, K. M. (2015). Framing risks and benefits of medical tourism: A content analysis of medical tourism coverage in Korean American community newspapers. *Journal of Health Communication*, *20*(6), 720–727. doi:10.1080/10810730.2015.1018574 PMID:25942506

Kemp, E., Williams, K. H., & Porter, M. III. (2015). Hope across the seas: The role of emotions and risk propensity in medical tourism advertising. *International Journal of Advertising, 34*(4), 621–640. doi:10.1080/02650487.2015.1024385

Khan, M. J., Chelliah, S., Haron, M. S., & Ahmed, S. (2017). Role of travel motivations, perceived risks and travel constraints on destination image and visit intention in medical tourism: Theoretical model. *Sultan Qaboos University Medical Journal, 17*(1), e11–e17. doi:10.18295qumj.2016.17.01.003 PMID:28417022

Klein, H. J., Simic, D., Fuchs, N., Schweizer, R., Mehra, T., Giovanoli, P., & Plock, J. A. (2017). Complications after cosmetic surgery tourism. *Aesthetic Surgery Journal, 37*(4), .474-482.

Labonté, R., Crooks, V. A., Valdés, A. C., Runnels, V., & Snyder, J. (2018). Government roles in regulating medical tourism: Evidence from Guatemala. *International Journal for Equity in Health, 17*(1), 1–10. doi:10.118612939-018-0866-1 PMID:30236120

LaMotte, S. (2019, January 10). *Surgeries in Mexico linked to antibiotic-resistant infections in US, CDC says.* Retrieved March 3, 2019 from https://www.cnn.com/2019/01/10/health/mexicosurgery-antibiotic-resistant-infection-cdc/index.html

Lee, H., Wright, K. B., O'Connor, M., & Wombacher, K. (2014). Framing medical tourism: An analysis of persuasive appeals, risks and benefits, and new media features of medical tourism broker websites. *Health Communication, 29*(7), 637–645. doi:10.1080/10410236.2013.794412 PMID:24138286

Mason, A. (2014). Overcoming the dual-delivery stigma: A review of patient-centeredness within the Costa Rican medical tourism industry. *International Journal of Communication and Health, 4,* 1–9.

Mason, A., & Wright, K. B. (2011). Framing medical tourism: An examination of appeal, risk, convalescence, accreditation, and interactivity in medical tourism web sites. *Journal of Health Communication, 16*(2), 163–177. doi:10.1080/10810730.2010.535105 PMID:21161812

McGrath, I. (1997). Information superhighway or information traffic jam for healthcare consumers? *Clinical Performance and Quality Health Care, 5*(2), 90–93. PMID:10167219

McKay, L. (2016). Generating ambivalence: Media representations of Canadian transplant tourism. *Studies in Social Justice, 10*(2), 322–341. doi:10.26522sj.v10i2.1421

Merlock Jackson, K., Lyon Payne, L., & Shepherd Stolley, K. (2013). Celebrity treatment: The intersection of star culture and medical tourism in American society. *Journal of American Culture, 36*(2), 124–134. doi:10.1111/jacc.12019

MTA. (2017). *Medical Tourism Association 2016-2017 Global Buyers Guide.* Retrieved online March 12, 2019 from https://www.medicaltourismassociation.com/en/prod40_global-buyersreport-2016-2017.html

Narayan, V., & Debroy, S. (2019). Hair transplant death: Procedure went on for 12 hours till man complained of pain. *The Times of India.*

Ngamvichaikit, A., & Beise-Zee, R. (2014). Communication needs of medical tourists: An exploratory study in Thailand. *International Journal of Pharmaceutical and Healthcare Marketing, 8*(1), 98–117. doi:10.1108/IJPHM-10-2012-0010

Niederdeppe, J., Hornik, R., Kelly, B., Frosch, D., Romantan, A., Stevens, R., ... Schwarz, S. (2007). Examining the dimensions of cancer-related information scanning and seeking behavior. *Health Communication, 22*(2), 153–167. doi:10.1080/10410230701454189 PMID:17668995

Parkes, C. M. (1975). Determinants of outcome following bereavement. *Journal of Death and Dying, 6*(4), 303–323. doi:10.2190/PR0R-GLPD-5FPB-422L

Parkes, C. M., & Weiss, R. S. (1983). *Recovery from bereavement.* New York, NY: Basic Book.

PBB. (2017). *Patients Beyond Borders. Medical tourism statistics and facts.* Accessed March 1, 2019 from www.patientsbeyondborders.com/medical-tourism-statistics-facts

Penney, K., Snyder, J., Crooks, V. A., & Johnston, R. (2011). Risk communication and informed consent in the medical tourism industry: A thematic content analysis of Canadian broker websites. *BMC Medical Ethics, 12*(1), 1–9. doi:10.1186/1472-6939-12-17 PMID:21943392

Pereira, J., & Bruera, E. (1998). The Internet as a resource for palliative care and hospice: A review and proposals. *Journal of Pain and Symptom Management, 16*(1), 59–68. doi:10.1016/S0885-3924(98)00022-0 PMID:9707658

Ratzan, S. C., & Parker, R. M. (2000). *Health literacy.* National Library of Medicine website. www.nlm.nih.gov/archive/20061214/pubs/cbm/hliteracy.html#15

Sallah, M., & Perez, M. (2019, January 30). This business helped transform Miami into a national plastic surgery destination. Eight women died. *USA Today.* Retrieved February 20, 2019 from https://www.usatoday.com/in-depth/news/investigations/2019/01/31/miami-doctors-plasticsurgery-empire-becomes-floridas-deadliest-clinics/2729802002/?fbclid=IwAR2imyhjTKSLbXWGfTgPGg_KX7uBRAb9eEbqt-dxZoDnl7RLhrw-qFt--iW0

Snyder, J., Johnston, R., Crooks, V. A., Morgan, J., & Adams, K. (2017). How medical tourism enables preferential access to care: Four patterns from the Canadian context. *Healthcare Analysis: HCA: Journal of Health Philosophy and Policy, 25*(2), 138–150. doi:10.100710728-015-0312-0 PMID:26724280

St. Johns herald and Apache news. *[volume]* (St. Johns, Apache Co., Ariz.), 11 April 1907. (n.d.). *Chronicling America: Historic American Newspapers.* Lib. of Congress. Retrieved from: https://chroniclingamerica.loc.gov/lccn/sn95060582/1907-04-11/ed-1/seq-8/

Tsai, W., Prigerson, H. G., Li, C., Chou, W., Kuo, S., & Tang, S. T. (2016). Longitudinal changes and predictors of prolonged grief for bereaved family caregivers over the first 2 years after the terminally ill cancer patient's death. *Palliative Medicine, 30*(5), 495–503. doi:10.1177/0269216315603261 PMID:26311571

Turner, L. (2013). *Patient mortality in medical tourism.* Oxford University Press; doi:10.1093/acprof:oso/9780199917907.003.0001

Turner, L. G. (2011). Quality in healthcare and globalization of health services: Accreditation and regulatory oversight of medical tourism companies. *International Journal for Quality in Healthcare: Journal of the International Society for Quality in Healthcare, 23*(1), 1–7. doi:10.1093/intqhc/mzq078 PMID:21148210

## ENDNOTES

[1] Due to regulatory oversight and public information restrictions in China, Google and other popular western search engines were determined to be insufficient in locating stories of patient mortality cases in this region.

[2] The case of P9 was marked as DNC "Death Not Confirmed," but death was likely, as continued reporting on this case was not available to confirm death.

# APPENDIX

*Table 1. Descriptions of patient mortality cases included in analysis*

| | | |
|---|---|---|
| P1: A 29-year-old woman from Jiangxi, China travelled to Guangzhou in May, 2015 for subcutaneous mastectomy and Nipple areola reduction surgery. The patient experienced abnormal breathing and heart pulse, heart arrest and died 2.5 hours following surgery. Family members doubted the cause of death. The family is still negotiated with the hospital and the case was under investigation. | P7: A 22-year-old model traveled from Beijing to Harbin for liposuction and fat fill in Sept. 2013. The patient was assured about the safety before operation. The patient suffered from breathing problems, lost consciousness lost and died. Nurses withheld the developments from the patient's friend during surgery. An investigation followed. | P13: A 18-year-old female, traveled to India from Africa for an emergency surgery after being diagnosed for Posterior fossa medulloblastoma. She was declared dead due to Ventriculoperitoneal shunt sepsis. |
| P2: A 25-year-old model from Dongguan travelled to Shenzhen in Feb. 2018 for double eyelid surgery and breast augmentation. The patient suffered and died from anesthetic allergy. An investigation was launched. | P8: A 45-year-old man travelled to Shandong from Beijing for his kidney transplant in June, 2015. The patient was taken to a house in the countryside. He died from the illegal surgery. The criminal, individual broker Cui was arrested after the patient's parents called the police. The parents sued Cui and Cui was criminally sentenced by the court. | P14: A 70-year-old male, traveled to India from Africa for a nonemergency surgery after being diagnosed for Nonfictional Pituitary adenoma. He was declared dead due to Pulmonary embolism. |
| P3: A 52-year-old business woman work in jewelry industry, travelled from Hongkong to Taiwan to a surgery center for breast augmentation and liposuction. She died in Jan 2018 due to heart and respiratory arrest. Her son arrived in Taiwan one day later and sought charges against the clinic. Four doctors and nurses were arrested. | P9: A 25-year-old Chinese woman, travelled to South Korea in May 2016 for breast augmentation. She suffered from heart and respiratory arrest and fell into deep coma for days. Family members requested an ambulance flight and transferred the patient to a Chinese hospital. She still had to rely on the booster to maintain blood pressure, and had been unable to get out of the ventilator, and had been in a deep coma. Family members had been accompanying the patient since transferred to China. DNC. | P15: A 41-year-old male, traveled to India from Africa for an emergency surgery after being diagnosed for Traumatic Complete Cervical 5 myelopathy. He was declared dead due to Pulmonary embolism. |
| P4: A 38-year-old woman from Yueyang, China travelled to Changshan, China for her first Orthodontic Surgery in April, 2017. The surgery was postponed due to patient hesitation. The patient suffered from anesthesia accident, transferred to another hospital and died the same day. An investigation occurred. Results were either pending or unable to be disclosed. The family first rejected compensatory damages from the hospital but later accepted. The case was resolved privately. | P10: A 50-year-old Chinese woman traveled to Souel, South Korea in January, 2015 for cosmetic surgery. The patient suffered from brain death and cardiac arrest due to anesthetic overdose. The Association of Plastic Surgeons requested an investigation. The family intended to sue the hospital. The clinic was later shut down. | P16: A 65-year-old male, traveled to India from Africa for an emergency surgery after being diagnosed for glioblastoma multiformes. He was declared dead due to status epilepticus. |
| P5: A 30 years old female, travelled from Gansu to Xian in April, 2018 for breast augmentation and liposuction. The patient was taken to a hotel for the operation, and later found dead. The cause of death was under investigation. The physician fled and the anesthesiologist arrested. Policeman were looking for other suspects. The anesthetist was suspended. | P11: A Japanese woman travelled to South Korea for cosmetic surgery in November, 2015. The patient experienced post-operative complications. Returned to the hospital 3 days after surgery, and died 5 days after in a hotel. Sleeping pills were found in the room too. The patient suffered from insomnia. An investigation sought to determine if the patient died due to an embolism from the surgery or drug poisoning. | P17: A 35-year-old female from Singapore traveled to Chennai, India in November, 2017 for liposuction surgery. Six hours after surgery, the patient experienced cardiac arrest and died in the intensive care unit of the same hospital. Her husband requested to know the cause for death. He was denied information, pressed charges against the hospital for inadequate information, and negligence. The case remains unresolved. |
| P6: A 36-year-old woman from Malaysia received cosmetic surgery in Hunan in June, 2018. The patient experienced post-operative complications and died 7 days after surgery from cerebral embolism. Outcomes are pending. The provider's beauty department was suspended temporarily. | P12: A 25-year-old mother traveled from the United States to Columbia for liposuction surgery in July, 2018 against her mother's will. She suffered a heart attack right before an incision was made and died on the table. Her husband has been advocating to find out the reason of the death but the CECILIP website and the surgeon have been unresponsive. | P18: A 56-year-old realtor from Bangladesh traveled to Kolkata, India in April, 2014 for liposuction surgery. Within few hours of the surgery, the patient was return to his hotel, suffered a heart attack and died. His family members accused the provider. The provider has stated the patient's friends have undergone the same procedure before. No further updates on allegations were found. |
| P19: A 31-year-old America-born-Indian actress traveled to the U.S. in June, 2015 for liposuction surgery. During the surgery she suffered complications from a respiratory disorder and died on the table. She was a Tollywood actress and had worked in U.S before working in India. No official complaint was lodged against the hospital or the clinic. | P25: A 46 yr. old female, mother travelled from the UK to Turkey for a Brazilian Butt Lift (BBL) in Aug. 2018. The patient died due to embolism. | P31: A 33 yr. old female, mother travelled from the U.S. to Mexico in May, 2011 for stomach banding surgery. The patient died from hypervolemic shock and cardiac arrest. An investigation confirmed the physician and hospital, where the operation was performed, both had proper licensing and followed the correct protocols. |
| P20: A 37-year-old Egyptian woman traveled to Mumbai for bariatric weight loss surgery in September, 2017. The surgery went well and the patient flew to Abu Dhabi within 24 hours of the surgery. The patient began experiencing intestinal shock and cardiac arrest after 2 days. Family member has accused the surgeons of lying about the weight loss risks. No charges were pressed against the hospital or the doctors. | P26: A 46 yr. old female, mother travelled within Brazil for a Brazilian Butt Lift (BBL) in July, 2018 at an apartment of the physician. The patient received an unsafe dose of PMMA and died of cardiac arrest six hours after the procedure. The Dr. was later arrested and charged with murder. | P32: A 24 yr. old, female travelled from the U.K. to Thailand for liposuction in Oct., 2014. The clinic had proper licensing, the physician did not. The patient contracted an infection 10 days after the initial operation and retuned to the facility. The patient died after receiving an intravenous anaestthic. The clinic was licensed and closed for 60 days. The physician was not certified to perform cosmetic surgery. |
| P21: A 28 yr. old female mother travelled from the U.S. to the Dominican Republic in Feb., 2014 for cosmetic surgery including: liposuction, tummy tuck and breast augmentation. The patient suffered an embolism and died. | P27: A 24 yr. old female, model travelled from Denmark to Brazil in July, 2018 for a Brazilian Butt Lift (BBL). The patient died of a heart attack in a hotel following the procedure. An investigation occurred and the operating physician was arrested and tried for manslaughter. | P33: A 35 yr. old, male travelled from Canada to Costa Rica for liberation therapy in Oct., 2010. The patient had a mesh stent inserted into a neck vein to counter the alleged effects of CCSVI. The patient was operated on three times for the stent to be inserted successfully. He returned home, experienced complications, specialists in Canada refused treatment because of concerns about his out-of-country procedures. He returned to Costa Rica, and died after doctors were unsuccessful in dissolving a blood clot. |
| P22: A 22 yr. old female, mother travelled from West Virginia to Florida in May, 2016 for a Brazilian Butt Lift (BBL). Death resulted from fat particles which clogged arteries in her lungs and heart. The physician's license was revoked. | P28: A 46 yr. old male, restaurant owner travelled from the U.S. to Brazil in Sept., 2011 for weight loss surgery. The patient died from pulmonary embolism and cardiac arrest within two weeks following the procedure. | P34: A 38 yr. old female, mother, model travelled from Argentina to Buenos Aires in Nov., 2009 for Brazilian Butt Lift (BBL). The patient died due to an embolism following the procedure. |
| P23: A 20 yr. old female, actress travelled from the UK to USA in Feb., 2011 for a Brazilian Butt Lift (BBL). The patient died due to embolism and liver failure. The physician was not certified and arrested for the death. | P29: A 29 yr. old female travelled from Australia to Mexico for Brazilian Butt Lift (BBL) in March, 2015. The patient suffered a heart attack and embolism during the procedure. An autopsy revealed the patient's lungs had been punctured four times. The physician was suspended, charged for homicide and operating without a license. The physician is connected to the deaths of at least 3 other patients. | P35: A 36 yr. old female, mother travelled from the U.S. to Mexico for rhinoplasty and breast implant replacement in Oct., 2018. Patient died due to complications of the anesthesia. The anesthesia was inserted into the wrong place in the patient's spine, her brain swelled, kidneys failed and she went into cardiac arrest. |
| P24: A 31 yr. old male travelled from Australia to Malaysia for liposuction, thigh lift, chest sculpting, chin tuck and upper eye lift in April, 2014. Patient suffered a pulmonary thromboembolism one day after returning home for treatment. | P30: A 51 yr. old, female travelled from the U.S. to Mexico in Oct., 2017 for liposuction after learning about the clinic online through Facebook. Following the operation, the patient was unconscious and in a coma prior to being retuned home. The patient died approximately two weeks later. | P36: A 21 yr. old, female travelled from the U.S. to the Dominican Republic for breast augmentation and liposuction in Feb., 2009. The patient died due to a heart attack. The family sued three physicians. The case was settled out of court. |

## Table 1. Continued

| | | |
|---|---|---|
| P37: A 42-yr old, female travelled from Norway to Costa Rica in 2014 for addiction treatment. The patient suffered a heart attack. The provider had a prior history of patient deaths and was shut down. | P43: A 35 yr. old, female, business owner travelled from China to Australia for breast augmentation in Sept., 2017. The physician was arrested and charged. | P49: A 38 yr. old female, mother, nurse travelled from Texas to Dominican Republic for cosmetic surgery in May, 2016. The patient died due to an allergic reaction. Dominican doctors were able to revive the patient, she was placed into a medically-induced coma until she could be brought back to the U.S. |
| P38: A 44 yr. old female, mother travelled from San Diego, U.S. to Miami, U.S. for breast augmentation, liposuction, and a tummy tuck. The day after the operation the patient fell unconscious in her friend's home and died hours later at Jackson Memorial Hospital. This was the eighth patient to die as a result of treatment at the provider's facility in the past 2 years. | P44: A 30 yr. old female, mother travelled from Illinois, U.S. to Florida, U.S. for plastic surgery in June, 2016. The patient died from an embolism. An investigation occurred. The clinic was shut down temporarily. The physician was not board certified in plastic surgery. | P50: A 35 yr. old female, wife travelled from NYC to Florida, U.S. in Jan., 2011. The patient died of cardiac arrest after undergoing liposuction and a buttocks enhancement surgery. The operating physician who was previously disciplined by the state in connection with two additional plastic surgery deaths in 2004. |
| P39: A 52 yr. old female travelled from the U.S. to Mexico to receive a tummy tuck in 2014. After being counseled by the physician the procedure was changed before surgery. She died the next day from complications of a buttock augmentation. The physician had a prior history of patient deaths. | P45: A 50 yr. old male business man travelled from Mexico to Thailand for a kidney transplant. The organ was received in November, complication occurred over several months. The kidney was rejected, the patient returned home and passed in Jan. A criminal suit was filed against the physician. The death has led to a series of investigations into the practice of illegal transplants at private clinics. Authorities declared it an illegal transplant at an unauthorized health facility. | |
| P40: A 30 yr. old female travelled from Texas to Panama in May, 2009 for liposuction. Patient died due to respitory failure. Provider maintained they conducted all required medical evaluations, briefed the patient on the risks of surgery, and the next day conducted the surgery. | P46: A 41 yr. old female travelled for BBL surgery in July, 2017. The patient died days later due to post-operative complications. | |
| P41: A 25 yr. old female, mother travelled from the U.S. to the Dominican Republic in June, 2017 for breast augmentation, tummy tuck and butt augmentation. The patient died a few weeks later from a blood clot. | P47: A female, mother travelled from the U.S. to the Dominican Republic. The patient died from complications after receiving liposuction and a tummy tuck. the New York Attorney General's office previously fined the provider more than $20,000, and charged him with unauthorized practice of medicine, because he recruited patients in the U.S. by examining them in beauty salons. | |
| P42: A 26 yr. old, female, mother travelled from the U.S. to Columbia for cosmetic surgery in Dec., 2016. She was persuaded to add a breast operation to her planned liposuction The patient died from cardiac arrest. A federal investigation occurred. | P48: A 31-yr old female travelled from U.S. to the Dominican Republic for liposuction, tummy tuck and breast augmentation. The patient died 17 days later. The Dr. was not licensed in the U.S. but was previously found to recruit patients through nail salons. | |

# Chapter 12
# Magnetocaloric as Solid-State Cooling Technique for Energy Saving

**Ciro Aprea**
*University of Salerno, Italy*

**Adriana Greco**
*University of Naples Federico II, Italy*

**Angelo Maiorino**
*University of Salerno, Italy*

**Claudia Masselli**
https://orcid.org/0000-0002-6869-6724
*University of Salerno, Italy*

## ABSTRACT

*Magnetocaloric is an emerging cooling technology arisen as alternative to vapor compression. The main novelty introduced is the employment of solid-state materials as refrigerants that experiment magnetocaloric effect, an intrinsic property of changing their temperature because of the application of an external magnetic field under adiabatic conditions. The reference thermodynamic cycle is called active magnetocaloric regenerative refrigeration cycle, and it is Brayton-based with active regeneration. In this chapter, this cooling technology is introduced from the fundamental principles up to a description of the state of the art and the goals achieved by researches and investigations.*

DOI: 10.4018/978-1-7998-3576-9.ch012

## INTRODUCTION

Nowadays, the refrigeration is responsible of more than 20% of the overall energy consumption all over the world and most modern refrigeration units are based on Vapor Compression Plants (VCP). The traditional refrigerant fluids employing VCP, i.e. ChloroFluoroCarbons (CFCs) and HydroChloroFluoroCarbons (HCFCs), have been banned by the Montreal Protocol in 1987 (Montreal Protocol, 1987), because of their contribution to the disruption of the stratospheric ozone layer (Ozone-Depleting Potential substances - ODPs). Over the time, periodical meetings among the Parties to the Montreal Protocol have been succeeded. Since 2000, the usage of HCFCs in new refrigerating systems is forbidden, letting HydroFluoroCarbons (HFCs) the only fluorinated refrigerants allowed because of their zero ODP characteristic. Since 2009, each meeting related to Montreal protocol, initially dedicated to the phase-out of the substances depleting the stratospheric ozone layer, namely CFCs and HCFCs, had been leading to conflicting exchanges on high Global Warming Potential HFCs which replace CFCs and HCFCs most of the time. Year after year, human activities have been increasing the concentration of greenhouse gases in the atmosphere, thus resulting in a substantial warming of both earth surface and atmosphere. The impact of greenhouse gases on global warming is quantified by their GWP (Global Warming Potential). As a result, over the year measures to reduce global warming have been taken, beginning with the Kyoto Protocol (Kyoto Protocol, 1997) and consequently with the EU regulations applying the prescriptions derived from it, like EU regulation 517/2014 (EU No 517/2014) on fluorinated greenhouse gases.

The above described general frameworks led scientific community to studying and applying solutions with environmentally friendly gases, with small GWP and zero ODP (Aprea & Greco, 1998; Greco & Vanoli, 2005): one of the most focused classes of new generation refrigerants is hydrofluoroolefins (HFO) (Aprea et al., 2016a; 2016b; 2018a; Mota-Babiloni et al., 2016), descending of olefins rather than alkanes (paraffins) and they are known as unsaturated HFCs, with environmentally friendly behavior, quite low costs. Despite all these advances, it is essential to underline that a vapor compression plant produces both a direct and an indirect contribution to global warming. The former depends on the GWP of refrigerant fluids and on the fraction of refrigerant charge which is either directly released in the atmosphere during operation and maintenance or is not recovered when the system is scrapped. The indirect contribution is related to energy-consumption of the plant. In fact, a vapor compression refrigerator requires electrical energy produced by a power plant that typically burns a fossil fuel, thus releasing $CO_2$ into the atmosphere. The amount of $CO_2$ emitted is a strong function of the COP of the vapor compression plant.

For all these reason, in the last decades the interest of scientific community has oriented itself in studying and developing new refrigeration technologies of low impact in our ecosystem: a class of them is composed by solid-state cooling (Kitanovski et al., 2015; Aprea et al., 2018b), which are gaining more and more attention, due to their potential in being performing and ecological methodologies. Recent discoveries of giant caloric effects (Pecharsky and Gschneidner, 1997; Lu et al., 2010; Liu et al., 2014) in some ferric materials opened the door to the use of solid-state materials for caloric cooling as an alternative to gases for conventional and cryogenic refrigeration.

Currently, the most mature, eco-friendly, solid-state caloric technique is Magnetocaloric Refrigeration (MR) (Kumaran et al., 2018), whose main innovation consists in employing solid materials with magnetic properties as refrigerants, able to increase or decrease their temperature when interacting with a magnetic field. Instead of the fluid refrigerants, proper of vapour compression, a magnetic refrigerant is a solid and therefore it has essentially zero vapour pressure, which means no direct ODP and zero GWP.

Specifically, magnetic refrigeration is related to MagnetoCaloric Effect (MCE) (Warburg, 1881), a physical phenomenon observed in some material with magnetic properties and it consists in an increment of temperature in the MCE material because of the magnetization of the material under adiabatic conditions. Magnetocaloric refrigeration in founded on a regenerative Brayton based thermodynamical cycle, called Active Magnetic Regenerative refrigeration (AMR) cycle. The Active Magnetic Regenerator is the core of a magnetic refrigerator system. It is a special kind of thermal regenerator made of material with magnetic properties, which works both as a refrigerant and as a heat-regenerating medium. The performance of an AMR system strongly depends on the magnetocaloric effect of the magnetic material used to build the regenerator, on the geometry of the regenerator and on the operating conditions of the system. The cooling efficiency in magnetic refrigerators can reach 60% of the theoretical limit, as compared to the 40% achieved by the best vapor compression plant.

Recent studies identified the weakness (Trevizoli et al., 2017; 2018) points of magnetic refrigeration and tried to improve them (Silva et al., 2014; Teyber et al., 2018): as a matter of fact, most of the criticism lays in the high costs in magnetic field generating and the expensiveness of the most promising magnetocaloric refrigerants (Coey, 2012). In this chapter magnetocaloric cooling is introduced from the fundamental principles up to the state of the art and the goals achieved by the research and investigations.

## BACKGROUND

Magnetic refrigeration, based on magnetocaloric materials exhibiting a caloric effect under a variation of an external magnetic field, has been the pioneer of caloric cooling since it has been well-studied and investigated for room temperature application over the past thirty years. The interest in magnetic refrigeration began with the discovery of the magnetocaloric effect attributed to Weiss and Piccard (Weiss and Piccard, 1918) in 1918. In truth, MCE was discovered casually in 1881 by Emil Gabriel Warburg (Warburg, 1881), a German professor of Physics at University of Strasbourg, during some experiments on ferromagnetic materials. Cryogenic was the first branch of refrigeration where MCE found application, since 1920s (Debye, 1926; Giauque, 1927) and nowadays it is maturely used in liquefaction of hydrogen and helium (Luo et al., 2007).

The research on room temperature magnetocaloric cooling began in 1976 with the first prototype developed by Brown at the NASA Lewis Research Centre (Brown, 1976). In 1982, the employment of a reciprocating thermal regenerator coupled with magnetocaloric cycle, well known as Active Magnetocaloric Regenerator (AMR), was introduced by Barclay (Barclay, 1982).

Since then an appreciable number of experimental prototypes was developed, with different configurations of regenerators, different magnetic materials and different magnetic sources (Yu et al., 2010; Greco et al., 2019). The real potential shown, both from environmentally friendly and energy point of view, pushed towards the birth of a community of scientists and researchers devoted to a common goal: make magnetocaloric refrigeration the technology of tomorrow.

# MAGNETOCALORIC EFFECT

## Thermodynamics and Phase Order Transitions

Magnetocaloric effect is a physical phenomenon, related to solid-state materials with magnetic properties. MCE consists in a coupling between the entropy of the Magnetocaloric Material (MM) and the variation of an external magnetic field applied to the material, which causes magnetic ordering in the MM structure. If the magnetization is done isothermally, it will decrease the magnetic entropy of the material, by the isothermal entropy change $\Delta S_M$ and, therefore, a change in the total entropy of the magnetic material is registered.

On the other side, if the magnetization occurs in an adiabatic process, without any interaction with the external environment, the total entropy of MM remains constant and the consequent decreasing in magnetic entropy is countered by an increase in the lattice and electron entropy contributions. This causes a heating of the material and a temperature increase given by the adiabatic temperature change ($\Delta T_{ad}$). In Figure 1 are clearly visible the quantities $\Delta S_M$ and $\Delta T_{ad}$ during an adiabatic magnetization of a ferromagnetic material.

Dually, under adiabatic demagnetization the magnetic entropy increases, causing a reduction in lattice vibrations and therefore a temperature decrease. At its Curie temperature, where is located its own magnetic phase transition, a MM shows the peak of MCE, in terms of $\Delta T_{ad}$ and $\Delta S_M$. This occurs because the two opposite forces (the ordering force due to exchange interaction of the magnetic moments, and the disordering force of the lattice thermal vibrations) are approximately balanced near the Curie point $T_{Curie}$.

Hence, the isothermal application of a magnetic field produces a much greater increase in the magnetization (i.e., an increase of the magnetic order and, consequently, a decrease in magnetic entropy, $\Delta S_M$) near the Curie point, rather than far away from $T_{Curie}$. The effect of magnetic field above and below $T_c$ is significantly reduced because only the paramagnetic response of the magnetic lattice can be achieved for $T \gg T_{Curie}$, and for $T \ll T_{Curie}$ the spontaneous magnetization is already close to saturation and cannot be increased much more. Similarly, the adiabatic application of a magnetic field leads to an increase in the magnetic material temperature, $\Delta T_{ad}$, which is also sharply peaked near the $T_{Curie}$.

The two possible magnetic phase changes observable at the Curie point are First Order Magnetic Transition (FOMT) and Second Order Magnetic Transition (SOMT). At the Curie point a magnetic transition has FOMT characteristics when the material exhibits a discontinuity in the first derivative of the Gibbs free energy (G.f.e.), magnetization function, whereas SOMT behavior is registered when the gap is detected in the second derivative of G.f.e., magnetic susceptibility, while its first derivative is a continuous function. Most of the magnetic materials order with a SOMT from a paramagnet to a ferromagnet, ferrimagnet or antiferromagnet.

Considering an internally reversible process, the MCE for SOMT materials can be evaluated as:

$$\left(\Delta S_M\right)_T = \mu_0 \int_{H_i}^{H_f} \left(\frac{\partial M}{\partial T}\right)_H dH \tag{1}$$

*Figure 1. The magnetocaloric effect detected as $\Delta S_M$ and $\Delta T_{ad}$ in a ferromagnetic material where it is magnetized by magnetic field going from $H_0$ to $H_1$*

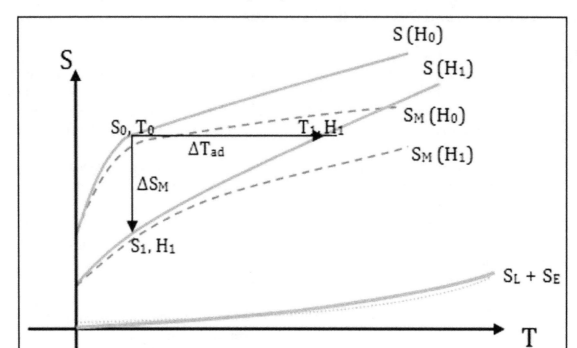

$$\left(\Delta T_{ad}\right)_s = -\mu_0 \int_{H_i}^{H_f} \frac{T}{C_H}\left(\frac{\partial M}{\partial T}\right)_H dH \qquad (2)$$

Indeed, in a SOMT material, while $\left|(dM/dT)_H\right|$ exhibits its maximum value, one can register the peak of MCE, located at the Curie temperature or near absolute zero, respectively if the material is a ferromagnetic or paramagnetic. Instead in a FOMT material, the phase transition should ideally take place at constant temperature for an infinite value of $\left|(dM/dT)_H\right|$ resulting in a giant magnetocaloric effect. The FOMT materials exhibit both an instantaneous orientation of magnetic dipoles and a latent heat correlated with the transition. Moreover, some FOMT materials show a coupled magnetic and crystallographic phase transition. As a result, the application of a magnetic field to these FOMT materials can produce both the magnetic state transition from a paramagnet or an antiferromagnet to a ferromagnet, simultaneously with a structural variation or a significant phase volume discontinuity but without showing a clear crystallographic alteration. Since that partial first derivatives of G.f.e. under T or H present a discontinuity at the first order phase transition, the bulk magnetization varies ($\Delta M$) and $C_H$ has ideally an infinite value at the Curie temperature. Therefore, the isothermal magnetic entropy change ($\Delta S_M$) in a FOMT material can be estimated by (3) coming from the Clasius-Clapeyron equation:

$$(\Delta S_M)_T = \left(\frac{dH}{dT_c}\right)_{eq} (\Delta M)_T \qquad (3)$$

where the derivative is kept at equilibrium.

## FOMT and SOMT Materials

The discovery of giant magnetocaloric effect in $Gd_5(Si_{4-x}Ge_x)$, reported by Pecharsky and Gschneidner in 1997 (Pecharsky and Gschneidner Jr., 1997), stimulated a considerable growth of researches in finding new materials whose MCE would be potent; great interest rose also to study the role that magneto-structural transitions play in this phenomenon. Giant MCE in FOMT materials, already observed, comes out since it is the result of two phenomena: the conventional magnetic entropy-driven process (magnetic entropy variation $\Delta S_M$) and the difference in the entropies of the two crystallographic alterations (structural entropy variation $\Delta S_{ST}$). The second terms consider the greater contribution due to entropy variation in FOMT with respect to SOMT materials:

$$(\Delta S)_T = (\Delta S_M)_T + (\Delta S_{ST})_T \qquad (4)$$

Assuming $T/C_H$ constant, in a FOMT material $\Delta T_{ad}$ could be estimated as:

$$(\Delta T_{ad})_s = -\mu_0 \left(\frac{T}{C_H}\right)\left(\frac{dH}{dT_c}\right)(\Delta M)_T \qquad (5)$$

The peak of adiabatic temperature variation is greater in most of the FOMT than SOMT materials. There are fundamental differences in the behavior of FOMT and SOMT materials:

- in a first order transition the MCE is concentrated in a narrow temperature range, whereas in a second order the transition spread over a broad temperature range.
- In a SOMT materials the adiabatic temperature variation is nearly immediate whereas it doesn't happen in FOMT materials because of the consequent alteration in their crystallographic structure, which produces a displacement of the atoms, needing thus of an amount of time required many orders of magnitude larger. The non-instantaneous response to the magnetic field of FOMT material, could become a significant problem since the operation frequencies usually adopted by magnetic refrigerators are in [0.5÷10] Hz range, where the delay in materials' answer could assume significant values.
- A substantial hysteresis has generally been detected (Aprea et al, 2013) in FOMT materials, whereas it is very low in SOMTs. The consequences of hysteresis when analyzing the MCE are: history dependence and energy dissipation. The former is related to parameters such as magnetization and heat capacity that become non-single valued functions of temperature and field; the latter accounts the heat dissipated during the magnetization process.

- The large volume variation that one can observe in FOMT materials is another defect that should be taken into account when deciding to employ these materials.

## THERMODYNAMIC CYCLES FOR MAGNETOCALORIC REFRIGERATION

A general thermodynamic magnetocaloric refrigeration cycle consists of magnetization and demagnetization in which heat is expelled and absorbed respectively, and two other benign middle processes.

The basic cycles for magnetocaloric refrigeration are: Carnot, Stirling, Ericsson and Brayton cycles (Yu et al., 2003; Kitanovski and Egolf, 2006,). Among them, the magnetic Ericsson and Brayton cycles are applicable for room temperature magnetic refrigeration. The employment of a regenerator in the Ericsson and Brayton cycles lets the achievement of a large temperature span and easy operating.

### Ericsson Cycle

The Ericsson cycle for magnetocaloric refrigeration consists of two isothermal processes/ stages and two isofield processes as illustrated in Figures 2(a) and 2(b).

1. In the isothermal magnetization process I (A®B in Figure 2(a)), when magnetic field increases from $H_0$ to $H_1$ the heat transferred from magnetic refrigerant to the upper regenerator fluid:

$$Q_{ab} = T_1\left(S_a - S_b\right) \qquad (6)$$

makes the latter increasing its temperature.

2. In the isofield cooling process II (B®C in Figure 2(a)), while magnetic field is kept constant at the maximum value $H_1$, both the refrigerant and the magnet which generates the field move downward to bottom and hence the heat:

$$Q_{bc} = \int_{S_c}^{S_b} TdS \qquad (7)$$

is transferred from magnetic refrigerant to regenerator fluid. Then a temperature gradient is set up in the regenerator.

3. In the isothermal demagnetization process III (C®D in Figure 2(a)), when magnetic field decreases from $H_1$ to $H_0$, the magnetic refrigerant absorbs heat:

*Figure 2. (a) The four processes of magnetocaloric Ericsson cycle in T-S diagram (b) Principle of magnetic Ericsson cycle refrigerator*

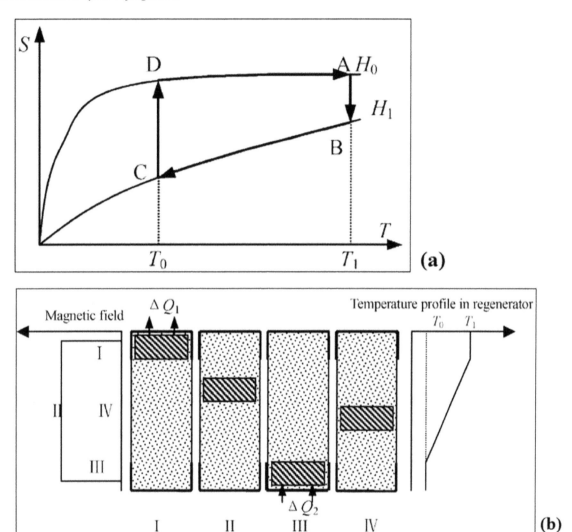

$$Q_{ab} = T_0 \left( S_d - S_c \right) \quad (8)$$

from the lower regenerator fluid. After that the fluid decreases its temperature.

4. 4. In the isofield heating process IV (D®A in Figure 2(a)), while the magnetic field has kept constant at the minimum $H_0$, both the magnetic refrigerant and magnet move upward to the top and the regenerator fluid absorbs heat

$$Q_{da} = \int_{S_d}^{S_a} TdS \qquad (9)$$

To make the Ericsson cycle possess the efficiency of magnetic Carnot cycle, it is required that the heat transferred in two isofield processes $Q_{bc}$, $Q_{da}$ are equal. For an ideal Ericsson cycle, 'parallel' T–S curves are optimal, that is, $\Delta S_M$ keeps constant in the cooling temperature range.

## Brayton Cycle

The Brayton cycle for magnetocaloric refrigeration consists of adiabatic processes and two isofield processes as shown in Figure 3.

*Figure 3. The four processes of magnetocaloric Brayton cycle in T-S diagram*

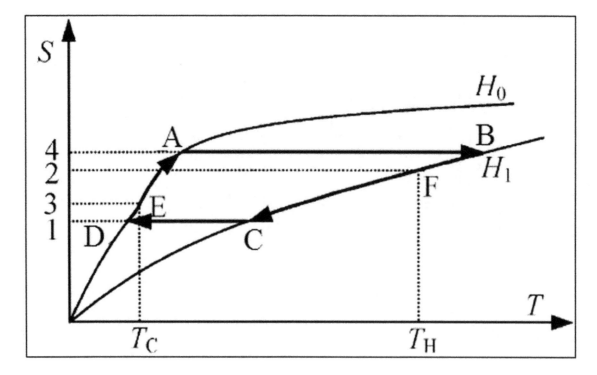

The first process (A®B in Figure 3) consists in and adiabatic magnetization, due to magnetic field increasing from $H_0$ to $H_1$, that leads to a temperature increment of $\Delta T_{ad'AB}$ in the solid refrigerant. With the field kept constant at $H_1$, in the isofield process (B®C in Figure 3) the heat of the area BC14 is transferred from the refrigerant to the hot heat exchanger at temperature $T_H$. In the adiabatic demagnetization (C®D in Figure 3) process the applied magnetic field falls from $H_1$ to $H_0$ and the solid refrigerant reduces its temperature of a quantity $\Delta T_{ad'CD}$ as a result of magnetocaloric effect. In the isofield process at $H_0$ the heat of the area AD14 is subtracted from the cold heat exchanger at $T_C$. No heat flows from and out of the magnetic refrigerant during the adiabatic magnetization and demagnetization processes.

The Brayton cycle can exhibit optimal performance as well with magnetic refrigerants having parallel T–S curves.

## Carnot Cycle

The magnetic Carnot cycle, consists of two adiabatic and two isothermal processes, applied to a magnetic refrigerant, as shown in Figure 4.

*Figure 4. The four processes of the magnetocaloric Carnot cycle in T-S diagram*

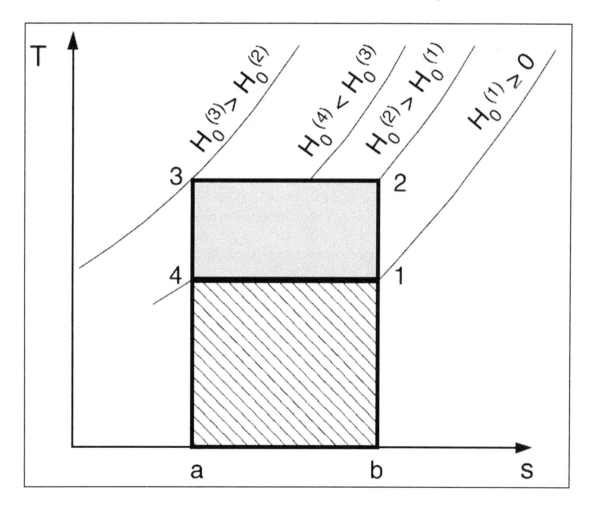

An adiabatic magnetization occurs in process 1⊗2 (Figure 4) and it continues with a further magnetization in stage 2⊗3, which is now an isothermal magnetization. During this process the generated heat is extracted from the system. The next step, named 3⊛4 sees an adiabatic demagnetization process. Connecting the system with a heat source isothermal demagnetization occurs, resulting in process 4⊛1.

It becomes clear that the Carnot cycle can only be run, if a minimum of four different magnetic field intensities occurs, through which the magnetocaloric material is moved. In the vertical process 1⊛2 the alteration of the magnetic field must be applied quickly, not allowing heat to diffuse away or be transported out by convection. In 2⊛3 the isothermal magnetization requires an alteration of the magnetic field and simultaneous rejection of heat. This process will therefore be slower. The area between (1–2–3–4) represents the work required and the area (1–4–a–b) is related to the thermal cooling energy.

## CASCADE SYSTEM REGENERATORS

All the cycles previously discussed are ideal cycles. Currently the existing magnetocaloric materials could not show temperature differences sufficiently wide for some frequently occurring refrigeration and heat pump applications. A solution to this problem is to build magnetic refrigerators and heat pumps, which take advantage of cascades. However, both—the regeneration and the cascade systems—show additional irreversibility in their cycles. These lead to lower coefficients of performance.

Cascade systems are well known in conventional refrigeration technology. A cascade system is a serial connection of some refrigeration apparatuses. They may be packed into one housing to give the impression of having only a single unit. Each of these apparatuses has a different working domain and temperature range of operation. This can be seen in Figure 5(a) by the decreasing temperature domains of stages I– III. In this figure the cooling energy of stage I (surface: ef14) is applied for the heat rejection of stage II (surface cd23). Analogously, the cooling energy of stage II (surface cd14) is responsible for the heat rejection in stage III. The cooling energy of the entire cascade system is represented by the surface ab14 of the last stage (white domain). The total work performed in the total cascade system is given by the sum of the areas 1234 of all present stages I, II, and III.

*Figure 5. Two cascade systems based on the Brayton cycle. In case (a) all stages (I, II and III) are designed to have a different optimally adapted material, whereas in case (b) they are produced with the same material.*

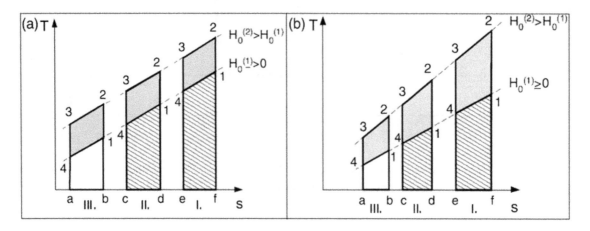

The magnetocaloric effect is maximal at the Curie temperature. It is large only in the temperature interval around this temperature, with decreasing effect in the case of greater (temperature) differences. It is advantageous that the operating point of the cooling system and this temperature interval of optimal magnetocaloric effect would coincide. If the temperature span of the refrigeration process is too wide, a decrease in efficiency occurs. A solution to this problem is to work with a cascade system, where each internal unit has its own optimally adapted working temperature. Each stage of a cascade system contains a different magnetocaloric material (Figure 5a) or it contains the same (Figure 5b).

The disadvantage of a cascade system is that the cycles of each stage must be designed to avoid overlaps. An overlap leads to a narrower temperature span and reduced efficiency.

A major advantage of a magnetic refrigeration cascade system over a conventional one is that in the magnetic refrigeration machine no external heat exchangers are required between the cooling process of the higher stage and the heat rejection process of the lower stage. This is because the magnetocaloric material is solid and a single fluid may be transferred to both stages

## THE REGENERATORS

The introduction of regenerators in refrigeration systems, with reference to magnetocaloric refrigeration, has the purpose to recover heat fluxes involved in Brayton cycle, that otherwise would have gone lost. The approach is particularly needed in room temperature magnetic refrigeration, where the vibrational entropy contribution becomes too high to be neglected. Therefore, a part of the magnetocaloric refrigeration cooling capacity is spent to cool the thermal load offered by the crystalline lattice, thus reducing the useful effect of the cycle. As a matter of fact, if a regenerator is added to the system, the heat rejected by the crystalline structure in one of the process of the cycle, is stored and then given back in the other process. By the way, the energy spent to cool the crystalline structure could be utilized effectively to increase of effective entropy change and temperature span (Yu et al., 2003).

Figure 6 shows the effect of the regeneration in Brayton cycle: thanks to the thermal interchange is possible to reach a temperature range $T_A$-$T_C$ 'greater than $T_A$-$T_C$, that would occur without regeneration, and a higher entropy change.

It is useful underlining the characteristics proper of the different typologies of regenerator. There are three types of regenerators employable in magnetic refrigerators:

- external;
- internal;
- active.

Heat transferring through external regenerator between the regenerator material (which generally is solid-state) and the MM is completed by an auxiliary heat transfer fluid. With internal regenerator solution, the magnetic refrigerant is placed in regenerator together with the regenerator material and heat is transferred directly between themselves so that the regenerator should be in the magnetic field area. If the regenerator is active, the magnetic material plays the double role of refrigerant and regenerator at the same time. The latter solution leads two advantages: both the irreversibility losses due to the auxiliary fluid and the regenerating fluid mixing at different temperature are reduced. The irreversible

*Figure 6. The regeneration effect in the Brayton cycle*

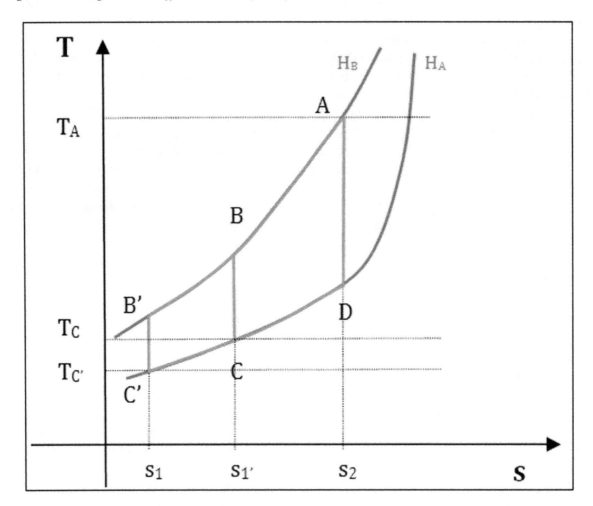

losses yielded by the second heat transferring in external regenerator or by mixing of the regenerator fluid with different temperature in internal regenerator, can be reduced.

The relative heat capacity between auxiliary fluid and magnetic refrigerant, plays a key role in choosing the best-fitting configuration of regenerator. If the heat capacity of the regenerator's fluid is higher than refrigerant magnetic material, it is better adopting an internal regenerator, where the temperature span in the regenerator is easily influenced by the fluid mixing. On the contrary case the best choice falls on Active Magnetocaloric Regenerator (AMR), where a high temperature span could be triggered and the requested fluid flow rate, on equal cooling power, is lower than the previous case.

There are several desirable characteristics for the perfect regenerator (Barclay, 1994):

- infinite thermal mass compared to the working material being cooled or heated;
- infinite heat transfer (a product of thermal conductance multiplied by the contact area) between working material and regenerator mass;
- zero void volume;

- zero pressure drop for convection of fluid through the regenerator;
- zero longitudinal conduction along the regenerator;
- uniform, linear temperature gradient from the hot end to cold end of the unit.
- The irreversible heat losses of the regenerator have a great influence on the performance of the whole magnetic refrigeration system. The effect of thermal resistances and regenerative losses on the performance of magnetic Ericsson cycle was analyzed qualitatively (Chen and Yan, 1998). Main irreversible heat losses are:
- loss of finite heat transfer between regenerator material and heat transfer fluid;
- loss of pressure drops yielded by flow resistance;
- thermal conduction along the magnetic material.
- loss of mixing of regenerator fluid in the internal regenerator;
- loss of heat leakage;
- losses of magnetic hysteresis and eddy currents;
- loss caused by viscous dissipation in the regenerator fluid;
- loss of "dead volume".

In an AMR, some amount of the heat transfer fluid is always in the connecting lines between the beds and the heat exchangers and never cycles both through the beds and the heat exchangers. This trapped heat transfer fluid, commonly referred to as the "dead volume," is a significant source of inefficiency in active magnetic regenerators (Lawton et al., 1999).

Some typical structural configurations of AMR regenerators are:

1. parallel plate;
2. perforated plate;
3. grid;
4. packed bed.

## ACTIVE MAGNETOCALORIC REGENERATIVE REFRIGERATION CYCLE

Barclay firstly introduced the Active Magnetic Regenerative (AMR) refrigeration cycle, well known as AMR, in 1982 (Barclay, 1982). The innovative idea leads to a new magnetocaloric cycle, different from the previous ones (Carnot, Ericsson, Brayton, or Stirling). It is founded on a magnetic Brayton cycle as shown in Figure 7.

The main innovation consists of introducing the AMR regenerator concept, i.e. the employment of the magnetic material both as refrigerant and regenerator. A secondary fluid is used to transfer heat from the cold to the hot end of the regenerator. Substantially every section of the regenerator experiments its own AMR cycle, according to the proper working temperature. Through an AMR one can appreciate a larger temperature span between the magnetocaloric regenerator and the auxiliary fluid.

During the entire cycle the magnetic field induction varies between a minimum value, $B_0$, and a maximum, $B_1$. The regenerator works between a warm reservoir at $T_H$ and a cold one at $T_C$. The AMR cycle consists of four processes. During the magnetization (1®2), the magnetic field is increased adiabatically with no fluid flow, causing the increase of the temperature of the magnetic material due to MCE, as displayed in Figure 8(a). In the cold-to-hot flow process (2®3), visible in Figure 8(b) the secondary fluid

*Figure 7. The four processes of AMR cycle in T-s diagram*

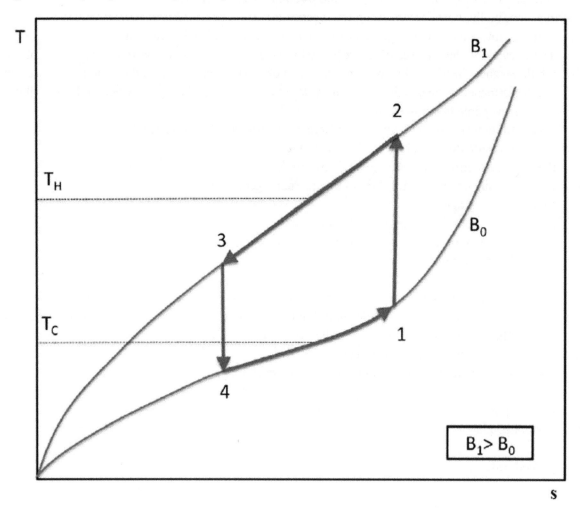

blows from the cold to the hot end of the regenerator when the field has kept constant at the maximum value. The fluid absorbs heat from the bed, reaching a temperature above $T_H$ and rejects it through the warm heat exchanger.

The next process, shown in figure 8(c), is the adiabatic demagnetization (3®4) where, without any fluid flowing, the magnetic field is removed, and the magnetic material temperature decreases because of the MCE. Finally, the heat transfer fluid flows through the bed from the hot to the cold end (4®1), as Figure 8(d) exhibits, with the applied magnetic field kept fixed at the minimum. The resulting hotter fluid cools the bed and reach a temperature lower than $T_C$. At this stage the secondary fluid absorbs heat from the cold heat exchanger producing a cooling load.

The AMR cycle described in this section has the peculiarity to employ a heat transfer fluid which has alternative motion. Other project solutions could take to slightly modifications of the cycle implementation.

*Figure 8. a) First process of AMR cycle: adiabatic magnetization; b) Second process of AMR cycle: cold-to-hot side fluid flowing; c) Third process of AMR cycle: adiabatic demagnetization; d) Fourth process of AMR cycle: hot-to-cold end fluid flowing.*

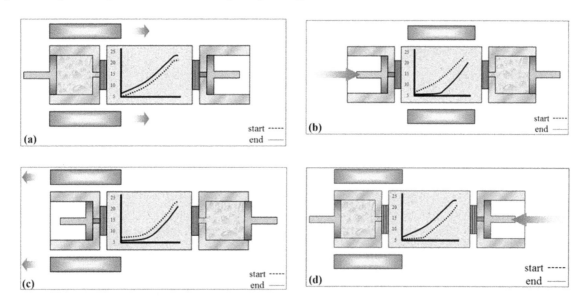

## MAGNETOCALORIC MATERIALS

Since in 1976 the ferromagnetic material gadolinium (Gd) was employed for the first time in a room temperature magnetic refrigerator by Brown (Brown, 1976), the research field of the magnetocaloric materials devoted to refrigeration has greatly expanded. Firstly, some ferromagnets concerning the second order transition were investigated for the large MCE manifesting. Recently the MM undergoing a first-order magnetic transition became more interesting, since the giant MCE was discovered in GdSiGe alloys. Nowadays a huge part of researchers is orienting its attention on some new alloys of magnetocaloric materials: as a result, many new materials with large MCEs (and many with lesser values) have been discovered, and a much better understanding of this magneto-thermal property resulted. Several new magnetocaloric substances with substantial cooling capacity were discovered: the lanthanide aluminides; the La manganites, some of which having MCEs comparable to Gd between 220 and 290 K; the tunable giant MCE for room temperature range.

### The Criteria for Selecting Magnetocaloric Refrigerant

In terms of the theoretical analyses and the magnetocaloric nature of existing materials, the criteria for selecting magnetic refrigerants for active magnetic refrigerators are given as follows (Phan and Yub, 2007; Aprea et al. 2015):

- the large magnetic entropy change and the large adiabatic temperature change (i.e., the large MCE) at moderate magnetic field;

*Figure 9. An example of the evaluation of relative cooling power based on the temperature dependence of magnetic entropy change, RCP(S)*

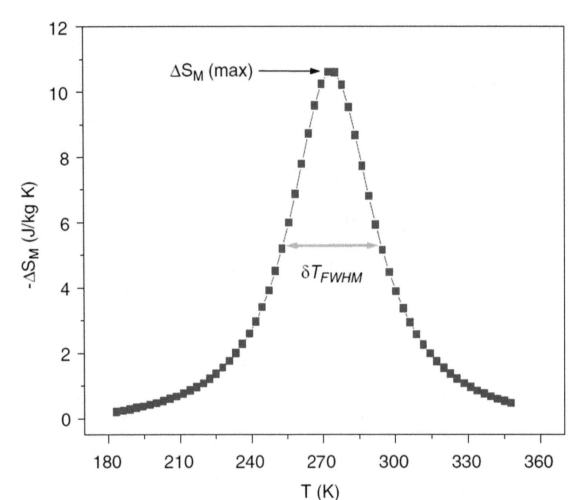

- the large density of magnetic entropy (it is an important factor contributing to the working efficiency of materials);
- the small lattice entropy;
- nearly zero magnetic hysteresis (it is related to the working efficiency of a magnetic refrigerant material);
- very small thermal hysteresis (this is related to the reversibility of the MCE of a magnetic refrigerant material);
- small specific heat and large thermal conductivity (these ensure remarkable temperature change and rapid heat exchange);
- large electric resistance (i.e., the lowering eddy current heating or the small eddy current loss);
- high chemical stability and simple sample synthesis route are also required for magnetic refrigerant materials;
- no toxic or too expensive elements;

- corrosion resistance material;
- limited number of elements in the compound in order to facilitate the control of reproducibility;
- synthesis process compatible with large-scale production;
- instantaneous MCE corresponding to a magnetic field variation;
- small volume variation.

## Magnetocaloric Cooling Efficiency

The magnetic cooling efficiency of a magnetocaloric material can be, in simple cases, evaluated by considering the magnitude of $\Delta S_M$ or $\Delta T_{ad}$ and its full width at half-maximum ($\delta T_{FWHM}$) (Phan and Yub, 2007).

It is possible to define the product of the $\Delta S_M$ maximum and the full-width at half-maximum:

$$\delta T_{FWHM} = T_2 - T_1 \tag{10}$$

as

$$RCP(S) = -"S_M(T,H) \times \delta T_{FWHM} \tag{11}$$

which stands for the so-called Relative Cooling Power (RCP) based on the magnetic entropy change. An example is reported in Figure 9.

Similarly, the product of the maximum adiabatic temperature change $\Delta T_{ad}$ and the full-width at half-maximum $\delta T_{FWHM}$ is expressed by:

$$RCP(T) = "T_{ad}(T,H) \times \delta T_{FWHM} \tag{12}$$

which stands for the so-called RCP based on the adiabatic temperature change.

## Classification of Magnetocaloric Materials

### Crystalline Materials Containing Rare Earth Metals: Gd and Its Alloys

The benchmark material for the refrigeration at room temperature is gadolinium, a rare-earth metal which exhibits a second order paramagnetic to ferromagnetic transition at its Curie temperature ($T_{Curie} = 294$ K) which exhibits excellent magnetocaloric properties. Gadolinium (Gd) is the only rare earth which orders magnetically near room temperature and is often considered to be a simple Heisenberg ferromagnet, i.e., a representative classical ferromagnet. The paramagnetic to ferromagnetic phase transition is a second-order phase transformation. There are several reasons why Gd is considered a benchmark materials for magnetic cooling applications: since Gd is considered a classical ferromagnet and it orders near room temperature, it is quite easy to carry out experiments to gain a better understanding of the nature of the paramagnetic to ferromagnetic phase transformation. Second, this transformation has a potentially large practical importance, specifically regarding magnetic cooling and heating near room temperature.

The heat capacity is a useful tool for studying magnetic phase transitions in the lanthanide metals and their compounds. Usually the heat capacity at constant pressure $C_p(T)$ behaves abnormally near the temperature where magnetic phase transitions occur. Hence, the behavior of $C_p(H,T)$ as a function of temperature and magnetic field can be used to examine the nature of the magnetic phase transition.

In a study of the magnetic properties of Gd-based materials with Curie temperatures between 250 and 350 K, Pecharsky and Gschneidner, Jr. (Pecharsky and Gschneidner Jr., 1997) in 1997 have discovered an extraordinarily large magnetocaloric effect in $Gd_5(Si_2Ge_2)$.

$Gd_5Si_4$ orders ferromagnetically at 335 K. The related phase $Gd_5Ge_4$, orders antiferromagnetically at much lower temperature. When Ge substitutes for Si in $Gd_5(SiGe)_4$, the Curie temperature is lowered from 335 to 295 K at the $Gd_5(Si_2Ge_2)$ composition. Further substitution of Ge for Si induces the appearance of a then unidentified intermediate phase and the beginning of a sharp decrease in the ordering temperature when the Si:Ge ratio is further reduced (Holtzberg et al., 1967). Experimental studies (Pecharsky and Gschneidner Jr., 1997) confirm that the extended solid solution $Gd_5(Si_xGe_{1-x})_4$ exists for 1£x<0.5 and that the Curie temperature of $Gd_5Si_4$ is gradually lowered from 335 to 300 K near the $Gd_5(Si_2Ge_2)$ composition. $Gd_5(Si_xGe_{1-x})_4$ compounds (where1£x<0.5) are among the most promising compounds for magnetic refrigeration. Alloys of gadolinium, silicon and germanium, $Gd_5(Si_xGe_{1-x})_4$ show a First Order Magnetic Transition (FOMT) characterized by a peak of $\Delta T_{ad}$ and $\Delta S_M$ much greater than gadolinium ones, but the whole function is quite sharper in $Gd_5(Si_xGe_{1-x})_4$ alloys. In particular $Gd_5(Si_2Ge_2)$, which exhibits the larger MCE among $Gd_5(Si_xGe_{1-x})_4$ compounds and which is named GIANT magnetocaloric effect, presents two different phase transitions. At 276 K one can observe in $Gd_5(Si_2Ge_2)$ a FOMT which constitute the MCE highest temperature peak, whereas at 299 K it is possible to appreciate a Second Order Magnetic Transition (SOMT) where, according to it, the material orders paramagnetically. Nevertheless Gd5($Si_2Ge_2$) shows a moderate hysteresis of 2 K (von Moos et al., 2015).

The adiabatic temperature rises for $Gd_5(Si_2Ge_2)$ is quite sharp and extends over a narrow temperature range, just as the $\Delta S_M$. The $\Delta T_{ad}$ values of $Gd_5(Si_2Ge_2)$ are larger than the corresponding $\Delta T_{ad}$ values for Gd ³ 30%, by comparing the peak values, regardless of the temperature.

The magnetic field dependence of the heat capacities $C_p(H,T)$ of $Gd_5(Si_2Ge_2)$ compared with Gd are significantly different. For $Gd_5(Si_2Ge_2)$ the low temperature heat capacity peak is quite sharp, and it essentially retains the same shape, whereas the peak height falls off and the peak temperature increases more-or-less in a linear fashion with increasing field. In contrast, for Gd the peak is reduced in height by low magnetic fields, becomes rounded off, and is spread out over a wide temperature with increasing field.

This difference is because for $Gd_5(Si_2Ge_2)$ the lower temperature and dominant magnetic ordering [ferromagnetic (I) ® ferromagnetic (II)] is a first order phase transformation, while for Gd it is a second order transformation. The field dependence of $Gd_5(Si_2Ge_2)$ entropy, indicates that the S(T) discontinuity is shifted to higher temperature by increasing magnetic field and that the transition remains thermodynamically a first order phase transformation in magnetic fields as high as 7.5 T (and, maybe, even 10 T).

## Crystalline Materials Containing Rare Earth Metals: La[Fe(Si,Al)]$_{13}$ Compounds

The La-Fe binary phase diagram shows an immiscible system, in which no intermetallic compounds form. The addition of small amount of Si or Al allows the formation of a ferromagnetic face-centred-cubic structure; this intermetallic compound exhibits very interesting magnetic behaviour with GMCE for $LaFe_{11.4}Si_{1.6}$ ($\Delta S_M$ close to 20 J $kg^{-1}K^{-1}$ at 210 K, for H = 5 T) (Hu et al., 2001). This was ascribed to a first-order itinerant electron metamagnetic transition, where a volume change of approximately 1%

does not change the crystal symmetry. The itinerant electron metamagnetic transition can be lost due to compositional effects, leading to SOMT, which reduces MCE but implies negligible hysteresis losses.

Tuning the temperature transition either by changing the Fe/Si ratio or substituting a different Rare Earth metal for La yielded an almost linear relationship between $S_M$ and $T_C$. Moreover, the FOMT changed to a SOMT as temperature increased. A more efficient way to shift Curie temperature to higher temperatures, however, is by partial Co substitution for Fe, as the magnetic moment of Fe is not seriously affected by Co substitution, but for $LaFe_{13-x}Si_x$ the average magnetic moment of Fe atoms decreases 0.286 μB per Si atom. Hydrides (H⁻) are also known to shift the temperature of the FOMT to higher values, without diminishing $|\Delta S_M|$ values, and with an enhancement of $\Delta T_{ad}$ as the heat capacity remains almost unchanged. However, hydrides in this compound are unstable above 423 K.

Single-phase $La(Fe,Si)_{13}$ alloys are very difficult to obtain and generally contain impurities, even after prolonged annealing at high temperature. These impurities must be factored into any comparison of different results, as composition can be inhomogeneous, and thus multiphase, rather than single-phase, behaviour should be expected. The use of Al instead of Si to stabilize the face-centred-cubic structure is less effective with respect to MCE (Franco et al., 2012).

Two interesting compounds among $La(Fe,Si)_{13}$ alloys for magnetic refrigeration at room temperature are: $LaFe_{11.384}Mn_{0.356}Si_{1.26}H_{1.52}$ (Morrison et al., 2012), which has a first order magnetic phase transition at 290 K, and $LaFe_{13-x-y}Co_xSi_y$ (Bjørk et al., 2010a), where the chemical composition of these are $LaFe_{11.06}Co_{0.86}Si_{1.08}$, $LaFe_{11.05}Co_{0.94}Si_{1.01}$ and $LaFe_{10.96}Co_{0.97}Si_{1.07}$, in order to varying the Curie temperature in 275.8÷289.8 K.

The properties of the LaFeCoSi samples are compared with the properties of commercial grade gadolinium, here simply termed Gd. This gadolinium is much cheaper than pure gadolinium, but the purity is also lower. The commercial grade gadolinium contains 99.5% rare earth metal, of which 99.94% are gadolinium. In LaFeCoSi samples, under a magnetic field whose induction varies in the 0÷1 T range, the peak temperature increases together with x, whereas the highest absolute value of $\Delta S_M(K)$ is registered for $LaFe_{11.06}Co_{0.86}Si_{1.08}$ at its Curie temperature (275.8 K). Considering the trends of $\Delta T_{ad}(H,T)$ of LaFeCoSi alloys, under 1T as magnetic field induction, which starts from a zero value, the transition temperatures are 277.1 K, 287.1 K, 289.6 K, respectively for $LaFe_{11.06}Co_{0.86}Si_{1.08}$, $LaFe_{11.05}Co_{0.94}Si_{1.01}$ and $LaFe_{10.96}Co_{0.97}Si_{1.07}$. No hysteresis is observed in the LaFeCoSi samples. The position of the peak of both ΔS, Cp and ΔTad changes between the three different composites of LaFeCoSi alloy. Thus, it is clearly that the peak position of the materials belonging to this alloy is tuneable in room temperature range.

## Rare Earth-Free Crystalline Materials: MnAs Alloys

MnAs alloys also assume the role of candidates for magnetocaloric refrigeration at room temperature because those compounds have a giant MCE exhibiting at first order magnetic transition; by varying the composition of the alloys, the Curie temperature could be switched in the range 220÷318 K. MnAs shows also a thermal hysteresis that is bad for practical application. Among Mn-based compounds, $MnFeP_{1-x}As_x$ compounds that are stable for $0.15 < x < 0.66$ and exhibit interesting magnetic properties associated with a first order metamagnetic transition. The Curie temperature of the alloy increases linearly with the As contents. $MnFeP_{1-x}As_x$ shows a large magnetic entropy change with the same magnitude as $Gd_5Si_2Ge_2$.

$MnFeP_{0.45}As_{0.55}$, is an interesting MnAs alloys: it undergoes a FOMT from paramagnetic to ferromagnetic at 307 K (on heating) according to a rapid decrease of the material parameters which changes its Debye temperature and its electronic structure (de Oliveira and von Ranke, 2005). The first order

*Table 1. Physic and MCE characteristics of MM in the room temperature range*

| Materials | $T_{Curie}$ [K] | $\Delta T_{ad}$ [K] | $\Delta S_M$ [J/kgK] | $\rho$ [kg/m$^3$] | k [W/mK] |
|---|---|---|---|---|---|
| Gd | 294 | 6 | 5 | 7900 | 10.9 |
| $Gd_5(Si_2Ge_2)$ | 276 | 7.8 | 14 | 7205 | 5.8 |
| $LaFe_{11.384}Mn_{0.356}Si_{1.26}H_{1.52}$ | 290 | 5 | 10.5 | 7100 | 9 |
| $LaFe_{11.05}Co_{0.94}Si_{1.01}$ | 287 | 3.17 | 5.5 | 7290 | 8.9 |
| $MnFeP_{0.45}As_{0.55}$ | 307 | 4 | 12 | 7300 | 2.5 |
| $Pr_{0.65}Sr_{0.35}MnO_3$ | 295 | 1.5 | 2.5 | 5800 | 1.8 |

transition occurs at 302.8 K on cooling, and at 306.6 K on heating. This indicates a thermal hysteresis of 3.8 K. The change in volume at the transition temperature is about 2.2%. The adiabatic temperature change for these compounds is relatively low and the thermal conductivity is significantly lower than that of gadolinium and other magnetic materials.

## Oxide Materials: Manganese Perovskites $Pr_{1-x}Sr_xMnO_3$

The best oxides for MCE around room temperature are manganese perovskites: among them, the series $Pr_{1-x}Sr_xMnO_3$ is noteworthy, since it has the advantage to contain fewer elements, so to have an easier synthesis and a better control of Curie temperature. Since the $T_C$ in $Pr_{1-x}Sr_xMnO_3$ was found to be very sensitive to the substitution level (displacement of about 4 K per % of Sr, in this range of x), several x values could be investigated.

$Pr_{0.65}Sr_{0.35}MnO_3$, investigated by Guillou et al (Guillou et al., 2012), is a SOMT manganese perovskite whose Curie temperature is located at 294 K, where a modest MCE has been observed. Figure I.41 reports the isothermal entropy change $\Delta S(T,H)$ evaluated for $Pr_{0.65}Sr_{0.35}MnO_3$ under a magnetic field induction which rises from 0 to 1 T. The intensity of the peak, located at 295 K, is about -2.3 J/kg K.

## General Summary of Magnetocaloric Materials

Table 1 summarizes the physical and MCE characteristics of the presented magnetocaloric materials in the room temperature range: $\Delta T_{ad}$ and $\Delta S_M$ reported are the peak value evaluated under a magnetic field induction $\Delta B$ of 1.5 T. The FOMT materials have peak values of $\Delta Tad$ and $\Delta S_M$ always higher than Gd ones. $Gd_5(Si_2Ge_2)$ shows the higher peak value, but at lower temperature with respect to the ones proper of the other materials considered. Gadolinium has the greatest thermal conductivity value, but it is comparable with the ones of $La(FeSi)_{13}$ alloys.

Table 2 reports the main parameters of the different materials in order to compare them in terms of RCP vs. costs. In terms of RCP(T) and RCP(S), Gd is the best material for magnetic refrigeration; $Gd_5(Si_2Ge_2)$ also shows acceptable values. From a commercial point of view, $Gd_5(Si_2Ge_2)$ and Gd are very expensive and the magnetic transition metals are more adequate than the rare earths for industrial production of magnetic cooling engines scopes. The price of $MnFeP_{0.45}As_{0.55}$ is quite low, but processing of As is complicated due to its toxicity.

Table 2. A comparison among magnetocaloric materials in terms of relative cooling powers and costs

| Materials | RCP(T) [K²] | RCP(S) [J/kg] | Price [€/kg] |
|---|---|---|---|
| Gd | 187 | 240 | 3000 |
| $Gd_5(Si_2Ge_2)$ | 117 | 154 | 9000 |
| $LaFe_{11.384}Mn_{0.356}Si_{1.26}H_{1.52}$ | 50 | 115 | 1200 |
| $LaFe_{11.05}Co_{0.94}Si_{1.01}$ | 30 | 83 | 1200 |
| $MnFeP_{0.45}As_{0.55}$ | 40 | 180 | 1500 |
| $Pr_{0.65}Sr_{0.35}MnO_3$ | 27 | 38 | 1050 |

## STATE OF THE ART AND FUTURE PERSPECTIVES

The introduction of AMR concept marked a turning point, since most of the subsequent projects on magnetocaloric devices based on it. Over the last 20 years, magnetocaloric has assumed the role of the reference Not-In-Kind cooling technology to constitute an alternative to vapor compression. The birth of the community of scientists and researchers focused on magnetocaloric refrigeration moved in the direction of building an interactive working group, that rose in 2004 through International Institute of Refrigeration (IIR/IIF) as "Working Party on Magnetic Refrigeration at Room Temperature". In 2005 with "Thermag", the first international conference on magnetic refrigeration at room temperature, took place in Montreux (Switzerland). Since then, eight Thermag conferences have been organized, approximately every two-years with the purpose to give a periodic overview of the current developments on magnetocaloric materials and prototypes. During such period, a remarkable number of prototypes of magnetic refrigerators or heat pumps was built, as well as a relevant number of numerical studies focused on materials and cycles was accounted (Aprea et al., 2018c; 2019).

In their review Yu et al. (Yu et al., 2010) classified the magnetocaloric devices introduced to community up to 2010 dividing them in a first and a second generation of room temperature magnetic refrigerators and heat pumps. The first generation identified all the room-temperature-operating devices that presented the common points of employing superconducting magnets and working at low-frequency of operation. The transition to the second generation protypes occurred when the Astronautic Corporation of America presented the first rotating magnetocaloric refrigeration, mounting permanent magnets (Zimm et al., 2003). The application of rotating permanent magnets allowed reaching higher frequencies of operation. It gives the birth to the second generation of magnetocaloric prototypes, showing the common points of being rotary and operating with higher frequencies. Greco et al. (Greco et al., 2019) classified all the magnetocaloric prototypes built from 2010 to 2019, configuring their review as a consecutive prosecution of the work started by Yu et al., in 2010. The last decade was characterized a huge increment of developed prototypes was registered. The number of research groups working on magnetocaloric, and then more generally on caloric, is growing day after day. Nowadays the number of developed magnetocaloric prototypes amounts to 75 machines; half of them was realized in the last decade. Among them, the most remarkable is the one introduced by Jacobs et al. (Jacobs et al., 2014) for Astronautics Corporation of America: it is a rotary magnetic refrigeration system, on a permanent magnets magnetic system able to achieve up to 3 kW as cooling power during no load tests. This is the best result achieved ever by the research community of magnetocaloric refrigeration.

The bottleneck of magnetocaloric refrigeration is constituted by the magnetocaloric effect shown by MM that is not enough satisfactory to allow the achievement of good performances in temperature ranges as wide as real commercial cooling applications require. Indeed, today, magnetocaloric technology is for sure not competitive with the other commercial techniques but it is not excluded that it will penetrate different markets in the future. Brown and Domanski (Brown and Domanski, 2014) and Goetzler et al. (Goetzler et al., 2014) proposed by two technology-foresight studies, in which they concluded that that magnetocaloric energy conversion systems (i.e., refrigerators, air conditioners, freezers, heat pumps or power generators) will certainly penetrate particular market niches; however, they will not represent, in the short term, a serious alternative for the full replacement of vapor compression. However, in the medium or a long term, this may happen. There is still a lot of very strong research effort needed, in both material science and engineering. Kitanovski et al. (Kitanovski et al., 2015) observed that only strong, international and interdisciplinary collaborations between different experts can bring these technologies towards the first market applications. In order to attract strategic industrial partners, good, operating, prototype devices are needed, which could operate under conditions that fulfill the requirements of some market applications.

## CONCLUSION

Downstream of all this, what we can say is that magnetocaloric refrigeration and heat pumping technology has made great strides in the last decade, and we hope that it will soon be able to make the decisive one for carrying all the efforts done toward a large-scale implementation as well as the competitiveness of the solid-state prototypes based on magnetocaloric effect.

## REFERENCES

Aprea, C., & Greco, A. (1998). An experimental evaluation of the greenhouse effect in R22 substitution. *Energy Conversion and Management, 39*(9), 877–887. doi:10.1016/S0196-8904(97)10058-9

Aprea, C., Greco, A., & Maiorino, A. (2013). The use of the first and of the second order phase magnetic transition alloys for an AMR refrigerator at room temperature: A numerical analysis of the energy performances. *Energy Conversion and Management, 70*, 40–55. doi:10.1016/j.enconman.2013.02.006

Aprea, C., Greco, A., Maiorino, A., & Masselli, C. (2015). A comparison between rare earth and transition metals working as magnetic materials in an AMR refrigerator in the room temperature range. *Applied Thermal Engineering, 91*, 767–777. doi:10.1016/j.applthermaleng.2015.08.083

Aprea, C., Greco, A., Maiorino, A., & Masselli, C. (2018a). The drop-in of HFC134a with HFO1234ze in a household refrigerator. *International Journal of Thermal Sciences, 127*, 117–125. doi:10.1016/j.ijthermalsci.2018.01.026

Aprea, C., Greco, A., Maiorino, A., & Masselli, C. (2018b). Solid-state refrigeration: A comparison of the energy performances of caloric materials operating in an active caloric regenerator. *Energy, 165*, 439–455. doi:10.1016/j.energy.2018.09.114

Aprea, C., Greco, A., Maiorino, A., & Masselli, C. (2018c). Energy performances and numerical investigation of solid-state magnetocaloric materials used as refrigerant in an active magnetic regenerator. *Thermal Science and Engineering Progress*, *6*, 370–379. doi:10.1016/j.tsep.2018.01.006

Aprea, C., Greco, A., Maiorino, A., & Masselli, C. (2019). The Employment of Caloric-Effect Materials for Solid-State Heat Pumping. *International Journal of Refrigeration*. doi:10.1016/j.ijrefrig.2019.09.011

Aprea, C., Greco, A., Maiorino, A., Masselli, C., & Metallo, A. (2016a). HFO1234yf as a drop-in replacement for R134a in domestic refrigerators: A life cycle climate performance analysis. *International Journal of Heat and Technology*, *34*(2), S212–S218. doi:10.18280/ijht.34S204

Aprea, C., Greco, A., Maiorino, A., Masselli, C., & Metallo, A. (2016b). HFO1234ze as drop-in replacement for R134a in domestic refrigerators: An environmental impact analysis. *Energy Procedia*, *101*, 964–971. doi:10.1016/j.egypro.2016.11.122

Barclay, J. A. (1982). Theory of an active magnetic regenerative refrigerator (No. LA-UR-83-1251; CONF-821237-1). Los Alamos National Lab.

Barclay, J. A. (1994). Active and passive magnetic regenerators in gas/magnetic refrigerators. *Journal of Alloys and Compounds*, *207*, 355–361. doi:10.1016/0925-8388(94)90239-9

Bjørk, R., Bahl, C. R. H., & Katter, M. (2010a). Magnetocaloric properties of LaFe$_{13-x-y}$Co$_x$Si$_y$ and commercial grade Gd. *Journal of Magnetism and Magnetic Materials*, *322*(24), 3882–3888. doi:10.1016/j.jmmm.2010.08.013

Brown, G. V. (1976). Magnetic heat pumping near room temperature. *Journal of Applied Physics*, *47*(8), 3673–3680. doi:10.1063/1.323176

Brown, J. S., & Domanski, P. A. (2014). Review of alternative cooling technologies. *Applied Thermal Engineering*, *64*(1-2), 252–262. doi:10.1016/j.applthermaleng.2013.12.014

Chen, J., & Yan, Z. (1998). The effect of thermal resistances and regenerative losses on the performance characteristics of a magnetic Ericsson refrigeration cycle. *Journal of Applied Physics*, *84*(4), 1791–1795. doi:10.1063/1.368349

Coey, J. M. D. (2012). Permanent magnets: Plugging the gap. *Scripta Materialia*, *67*(6), 524–529. doi:10.1016/j.scriptamat.2012.04.036

Dan'Kov, S. Y., Tishin, A. M., Pecharsky, V. K., & Gschneidner, K. A. (1998). Magnetic phase transitions and the magnetothermal properties of gadolinium. *Physical Review. B*, *57*(6), 3478–3490. doi:10.1103/PhysRevB.57.3478

de Oliveira, N. A., & von Ranke, P. J. (2005). Theoretical calculations of the magnetocaloric effect in MnFeP0. 45As0. 55: A model of itinerant electrons. *Journal of Physics Condensed Matter*, *17*(21), 3325–3332. doi:10.1088/0953-8984/17/21/025

Debye, P. (1926). Einige bemerkungen zur magnetisierung bei tiefer temperatur. *Annalen der Physik*, *386*(25), 1154–1160. doi:10.1002/andp.19263862517

Franco, V., Blázquez, J. S., Ingale, B., & Conde, A. (2012). The magnetocaloric effect and magnetic refrigeration near room temperature: Materials and models. *Annual Review of Materials Research*, *42*(1), 305–342. doi:10.1146/annurev-matsci-062910-100356

Giauque, W. F. (1927). A thermodynamic treatment of certain magnetic effects. A proposed method of producing temperatures considerably below 1 absolute. *Journal of the American Chemical Society*, *49*(8), 1864–1870. doi:10.1021/ja01407a003

Goetzler, W., Zogg, R., Young, J., & Johnson, C. (2014). Alternatives to vapor-compression HVAC technology. *ASHRAE Journal*, *56*(10), 12.

Greco, A., Aprea, C., Maiorino, A., & Masselli, C. (2019). A review of the state of the art of solid-state caloric cooling processes at room-temperature before 2019. *International Journal of Refrigeration*, *106*, 66–88. doi:10.1016/j.ijrefrig.2019.06.034

Greco, A., & Vanoli, G. P. (2005). Flow boiling heat transfer with HFC mixtures in a smooth horizontal tube. Part II: Assessment of predictive methods. *Experimental Thermal and Fluid Science*, *29*(2), 199–208. doi:10.1016/j.expthermflusci.2004.03.004

Griffel, M., Skochdopole, R. E., & Spedding, F. H. (1954). The heat capacity of gadolinium from 15 to 355 K. *Physical Review*, *93*(4), 657–661. doi:10.1103/PhysRev.93.657

Guillou, F., Legait, U., Kedous-Lebouc, A., & Hardy, V. (2012). Development of a new magnetocaloric material used in a magnetic refrigeration device. In *EPJ Web of Conferences* (*Vol. 29*, p. 21). EDP Sciences. 10.1051/epjconf/20122900021

Holtzberg, F., Gambino, R. J., & McGuire, T. R. (1967). New ferromagnetic 5: 4 compounds in the rare earth silicon and germanium systems. *Journal of Physics and Chemistry of Solids*, *28*(11), 2283–2289. doi:10.1016/0022-3697(67)90253-3

Hu, F. X., Shen, B. G., Sun, J. R., Cheng, Z. H., Rao, G. H., & Zhang, X. X. (2001). Influence of negative lattice expansion and metamagnetic transition on magnetic entropy change in the compound LaFe 11.4 Si 1.6. *Applied Physics Letters*, *78*(23), 3675–3677. doi:10.1063/1.1375836

Kitanovski, A., & Egolf, P. W. (2006). Thermodynamics of magnetic refrigeration. *International Journal of Refrigeration*, *29*(1), 3–21. doi:10.1016/j.ijrefrig.2005.04.007

Kitanovski, A., Plaznik, U., Tomc, U., & Poredoš, A. (2015). Present and future caloric refrigeration and heat-pump technologies. *International Journal of Refrigeration*, *57*, 288–298. doi:10.1016/j.ijrefrig.2015.06.008

Kumaran, P., Giridharan, V., Gokulakrishnan, R., & Hariharan, M. (2018). A Review of room temperature magnetocaloric materials for home appliance. *International Journal of Pure and Applied Mathematics*, *119*(16), 2053–2059.

Kyoto Protocol to the United nation Framework Convention on climate change. (1997). United Nation Environment Program (UN).

Lawton, L. M., Jr., Zimm, C. B., & Jastrab, A. G. (1999). *U.S. Patent No. 5,934,078*. Washington, DC: U.S. Patent and Trademark Office.

Liu, Y., Infante, I. C., Lou, X., Bellaiche, L., Scott, J. F., & Dkhil, B. (2014). Giant room-temperature elastocaloric effect in ferroelectric ultrathin films. *Advanced Materials*, *26*(35), 6132–6137. doi:10.1002/adma.201401935 PMID:25042767

Lu, S. G., Rožič, B., Zhang, Q. M., Kutnjak, Z., Li, X., Furman, E., ... Blinc, R. (2010). Organic and inorganic relaxor ferroelectrics with giant electrocaloric effect. *Applied Physics Letters*, *97*(16), 162904. doi:10.1063/1.3501975

Luo, Q., Zhao, D. Q., Pan, M. X., & Wang, W. H. (2007). Magnetocaloric effect of Ho-, Dy-, and Er-based bulk metallic glasses in helium and hydrogen liquefaction temperature range. *Applied Physics Letters*, *90*(21), 211903. doi:10.1063/1.2741120

Montreal Protocol on substances that deplete the ozone layer. (1987). United Nation Environment Program (UN).

Morrison, K., Sandeman, K. G., Cohen, L. F., Sasso, C. P., Basso, V., Barcza, A., ... Gutfleisch, O. (2012). Evaluation of the reliability of the measurement of key magnetocaloric properties: A round robin study of La (Fe, Si, Mn) Hδ conducted by the SSEEC consortium of European laboratories. *International Journal of Refrigeration*, *35*(6), 1528–1536. doi:10.1016/j.ijrefrig.2012.04.001

Mota-Babiloni, A., Navarro-Esbrí, J., Molés, F., Cervera, Á. B., Peris, B., & Verdú, G. (2016). A review of refrigerant R1234ze (E) recent investigations. *Applied Thermal Engineering*, *95*, 211–222. doi:10.1016/j.applthermaleng.2015.09.055

EU No 517/2014, European Regulation No 517/2014 on fluorinated greenhouse gases and repealing Regulation (EC) No 842/2006, 2014, *Off. J. Eur. Union*.

Pecharsky, V. K., & Gschneidner, K. A. Jr. (1997). Giant magnetocaloric effect in Gd 5 (Si 2 Ge 2). *Physical Review Letters*, *78*(23), 4494–4497. doi:10.1103/PhysRevLett.78.4494

Phan, M. H., & Yu, S. C. (2007). Review of the magnetocaloric effect in manganite materials. *Journal of Magnetism and Magnetic Materials*, *308*(2), 325–340. doi:10.1016/j.jmmm.2006.07.025

Silva, D. J., Ventura, J., Araújo, J. P., & Pereira, A. M. (2014). Maximizing the temperature span of a solid state active magnetic regenerative refrigerator. *Applied Energy*, *113*, 1149–1154. doi:10.1016/j.apenergy.2013.08.070

Teyber, R., Trevizoli, P. V., Christiaanse, T. V., Govindappa, P., Niknia, I., & Rowe, A. (2018). Semi-analytic AMR element model. *Applied Thermal Engineering*, *128*, 1022–1029. doi:10.1016/j.applthermaleng.2017.09.082

Trevizoli, P. V., Nakashima, A. T., Peixer, G. F., & Barbosa, J. R. Jr. (2017). Performance assessment of different porous matrix geometries for active magnetic regenerators. *Applied Energy*, *187*, 847–861. doi:10.1016/j.apenergy.2016.11.031

Trevizoli, P. V., Peixer, G. F., Nakashima, A. T., Capovilla, M. S., Lozano, J. A., & Barbosa, J. R. Jr. (2018). Influence of inlet flow maldistribution and carryover losses on the performance of thermal regenerators. *Applied Thermal Engineering*, *133*, 472–482. doi:10.1016/j.applthermaleng.2018.01.055

von Moos, L., Bahl, C. R. H., Nielsen, K. K., & Engelbrecht, K. (2014). The influence of hysteresis on the determination of the magnetocaloric effect in Gd5Si2Ge2. *Journal of Physics. D, Applied Physics*, *48*(2), 025005. doi:10.1088/0022-3727/48/2/025005

Warburg, E. (1881). Über einige Wirkungen der Coërcitivkraft. *Annalen der Physik*, *13*, 141. doi:10.1002/andp.18812490510

Weiss, P., & Piccard, A. (1918). Sur un nouveau phénomène magnétocalorique. *CR (East Lansing, Mich.)*, *166*, 352–354.

Yu, B., Liu, M., Egolf, P. W., & Kitanovski, A. (2010). A review of magnetic refrigerator and heat pump prototypes built before the year 2010. *International Journal of Refrigeration*, *33*(6), 1029–1060. doi:10.1016/j.ijrefrig.2010.04.002

Yu, B. F., Gao, Q., Zhang, B., Meng, X. Z., & Chen, Z. (2003). Review on research of room temperature magnetic refrigeration. *International Journal of Refrigeration*, *26*(6), 622–636. doi:10.1016/S0140-7007(03)00048-3

Zimm, C. B., Sternberg, A., Jastrab, A. G., Boeder, A. M., Lawton, L. M., & Chell, J. J. (2003). *U.S. Patent No. 6,526,759*. Washington, DC: U.S. Patent and Trademark Office.

# Chapter 13
# Global Racial, Ethnic, and Religious Subculture and Its Impact on Healthcare Organizations

**Rohit Singh Tomar**
*Amity University Madhya Pradesh, India*

**Meenal Kulkarni**
*Symbiosis Institute of Health Science, India*

## ABSTRACT

*This chapter deals with the racial, ethnic, and religious subculture and its impact on the global healthcare organizations and their practices. This study is exploratory in nature where secondary sources have been analyzed to find out the answers of the selected objectives. A discussion-based approach has been used to compare and contrast various information regarding healthcare practice and the role of ethnicity and religion in affecting it. Healthcare seems to be affected by race, ethnicity, and religion, but there is a huge scope of quantitative analysis to get a detailed and comprehensive result.*

## INTRODUCTION

This chapter deals with the racial, ethnic and religious subculture and its impact on the global health care organizations and their practices. This study is exploratory in nature where a variety of sources covering different disciplines and approaches have been analyzed to find out the answers of the following questions-

1. How ethnic and racial subculture put a negative effect on global health practices?
2. What are ethnic and racial differences in healthcare practices of developed and developing countries/societies?
3. Why, how and where diseases discriminate among races and ethnicity?

DOI: 10.4018/978-1-7998-3576-9.ch013

4. What are positive innovative practices adopted globally to handle ethnic and racial issues in health care industries?
5. What is the historical perspective of health care industry based on Religion?
6. What are the religious taboos in treatment of diseases?
7. What is required by health care industry to handle religious sensitivities?
8. What is the role of spirituality in health care?
9. What are the positive impacts of religion on health care?

A discussion-based approach has been used to compare and contrast various information regarding health care practice and role of ethnicity and religion in affecting it. Results show that, some races have poor health as compared to others because of multiple reasons. Major reasons are socio-economic and biological (or genetic) in nature. Some of the poor health practices prevalent among major races dwelling across the globe have been tabulated in the chapter. Race based differences in health care system can be observed on the bases of developed and developing societies. Developed countries have more lifestyle-based diseases while developing societies suffer from lifestyle as well as vector born diseases. It has been found that diseases discriminate among races where some races are more prone to certain disease; details are tabulated in the chapter. It has been observer that health institutions across the world intentionally or unintentionally practice racism. However, some healthcare institutes and public health organizations are using innovative and technology driven practices to fight against racism. Diversification in staff ethnicity and race along with their social mobility are useful techniques which has reduced racism in hospital settings. Chapter provides examples and cases where innovative practices improved the behavior of staff towards patients. In many tribes and religion certain diseases are considered as taboo and treated with the help of traditional practices. Spiritual and religious practices have played an important role in the treatment of the psychological and mental illness. Much evidence is presented in research to support the role of spirituality and religion in the patient care and recovery from the diseases. However some studies do not support that religion and spirituality plays an important role in reducing human health related problems. Some studies don't even find any correlation in health care and religion. Health care seems to be affected by race, ethnicity and religion but there is a huge scope of quantitative analysis to get a detailed and comprehensive result. Study needs to be conducted across the world based upon primary data where it could compare various culture, sub-culture, ethnicity, races and religion and its impact on healthcare industry and patients.

## ETHNIC, RACIAL, AND RELIGIOUS IMPACT ON GLOBAL HEALTH PRACTICES

Culture is a complex term it is stable, yet dynamic, and sensitive, yet enduring. Culture is an integration of shared values, norms, belief, traditions, regional rules, myths, rituals, ethnicity, races, religion and many more variables existing in a society. This cultural composite produces variety of results as per situational demand. Cultural practices admired by one society can be a shame or sin for others. Race, ethnicity and religions bring myths and rituals simultaneously, where certain rituals are the outcome of established myths. No part of world and profession is untouched by the popular myths prevalent in humans. Most scientific field of professions like healthcare, pharmacy, and wellness are directly or indirectly influenced by these myths. Ancient Egyptians used bloodletting to cure hypertension and to maintain body fluids. Tobacco toothpaste is an outcome of strong belief of Indians that tobacco can alleviate tooth-ache and help in removing cavity. In 1900s tapeworms were used for the weight loss purpose. Mental illnesses like epilepsy are caused by evil spirits is a popular myth in India and other parts of the world, many religious house are treating such mental trauma by their own absurd ways. Full moon is harmful for pregnant women and it could break the water bag is one of the myth in Britain. No country, developed or underdeveloped, is protected from myths prevalent in health and allied industry. Is this negative phenomenon observed because of religion, race or ethnic intervention?

*Myths are defined as popular stories or saying having symbolic influence on society. It has dual dimensions one is opposed by the rationalist and the other is favored by the orthodox followers*

*Rituals are the set practices applied on occasions and based upon cultural demands of the societies*

*Religion is a way of living developed through faith in a sect; usually it is inherited by individual from his family*

On the other hand a strong positive impact of religion could be observed on the health of the society, where religious practices helped individuals and groups to win over diseases at early stages. This could be because of better food and living habits promoted by usually all religions. Recent research has indicated that religious identity and practice can impact health outcomes at the population level as well as individual clinical decisions of patients (Koenig, 2015; VanderWeele, Balboni, & Koh, 2017). Health professionals need to create a way to accommodate belief and myths. It is expected from a health professional to apply scientific knowledge to explore possibility of treatment with due care of religious and ethnic sensitivity. Patients too display multiple dimensions of their belief about religion during his or her treatment. Some strongly believe that God will cure the disease and others blame God for their sufferings. For fast recovery patients needs to learn de-ethnicization, and have to detach themselves from existing ethical norms. Same should be practice by the health professionals, as religious belief too varies from one ethnicity to another. For example, American 'blacks' tend to be more religious as compared to 'white' Americans. Races and religion have been important markers of discrimination and socio-economic conditions of the population. Meanwhile, the socio-economic structure of society or country has played a significant role in the healthcare sector. Better socio-economic conditions are responsible for better health and overall well beings of the individuals. In India caste system is derived from the races and religion, which is responsible for the poor economic and social condition of some natives of the country. Similarly Black Americans are still relatively poorer than white Americans. Diseases are also selective, they pick race of its choice. For example, Sickle Cell Anemia is quite common in Black Africans, perhaps because of similar DNA sequences of people of common race. Global differences and similarities among race, ethnicity and religion immensely affect health care and wellness industry, government policies and programs. Acculturation of the health professionals is must to handle patients of different cultural background. In next section some harmful ethnic and racial practices adopted at global level are discussed. Some of these are monocultural practiced adopted by all major societies of the world while some are regional specific practices.

## NEGATIVE IMPACT OF ETHNIC AND RACIAL SUBCULTURE ON GLOBAL HEALTH PRACTICES

*Race is a socially constructed concept and does not have clear biological definition. The modern genetics and DNA studies have exposed the limitations of earlier beliefs on 'race' and the concept is not used in modern natural sciences. However, races as a social construction are still often distinguished in different parts of the world on the basis of visible differences among groups and individuals and in some parts of the world people continue to use rather carelessly 'race' and racial categories, which, as such, can be regarded as an act of racism.*

In the 19th century science, largely for the purposes of colonialism and empire-building, four primary races were distinguished in the world – Caucasoid, Mongoloid, Negroid and Australoid. All these were supposed to have different physical and biological attributes. It was difficult to divide these races further as more careful analysis would immediately produce a long list of sub-categories and question the whole idea of four primary races. In this chapter the focus is on the negative impact created within different ethnic groups on global health practices and it seems like ethnicities are still often understood or demarcated by rather arbitrary racial terms and identifications. Before we put forward our discussion regarding races and malpractices in health we must reason to find out whether scientific development happening in the field of health and well-being should be based upon ethnicities/ 'races' or not. Some patients don't want to be labeled or treated on the basis of their race or ethnicity. To understand documented differences, one must come to understand covert as well as overt racism and the multifaceted dimensions of institutional racism in medical and health institutions (King, 1996). Physicians too have different views regarding patient treatment on the basis of different races. Number of researchers believes that races and treatments are correlated phenomenon while a set of experts finds no connection between races and treatment. Above all it is difficult for physicians to identify the race of an individual on the basis of physical appearance. On the other hand it is also difficult to create a tool for patients to self-identify his or her race. In a single race sub races are immerged which are considerably different from each other. Identity of race further depends upon country of origin or social status in the adopted nation. An African American could have any kind of mixture of ancestors from East Africa, Central Africa or West Africa – and more often than not also ancestors from other parts of the world. Furthermore, on the basis of modern understanding of human evolution it is most likely that we all evolved largely in Africa. Ethnicity is always better explained by culture than race. But for sure (socially constructed) races and ethnicity are often related to their own malpractices, taboos and limitations which brings plethora of problems for the health care professionals.

*Ethnic Subculture is defined as group of people who share common cultural or genetic ties, where both its members and others recognize it as distinct category*

Some ethnic groups have created practices that have negative impact on human health because of poor economic and social conditions that make it difficult to fully understand the consequences of these practices. Most people in Africa have relatively poor economic conditions. Serious economic problems have given birth to social problems and make quick improvement difficult. Ancestral cultural norms are often hostile to certain sections of society and the largest group that often falls victim to such outdated norms are the females. However some ethnic groups have developed particular health problems because of cultural orthodoxy transferred from one generation to the next generation (Table 1 shows some of the poor health practices among different ethnic groups). This means that social problems could be a cause of public health problems, but these problems are clearly different from general and individual health problems. The social action or unintended consequence of purposive action theory states that all social programs have both intentional and unintentional outputs and society needs to be vigilant about any unintentional outcome of the health program (Robert K. Merton). One such unintentional output of practicing racism is negative impact on health, which means because of racial discrimination an individual can suffer from disease. Perceived ethnic discrimination in African Americans is strongly associated with increased risk for hypertension (Brondolo, Rieppi, Kelly, & Gerin, 2003).

## DEVELOPED VS DEVELOPING COUNTRIES/ SOCIETIES: ETHNIC AND RACIAL DIFFERENCES IN HEALTHCARE PRACTICES

Developed countries can be differentiated from underdeveloped countries on the economic basis. Developed countries have better infrastructure, Gross Domestic Product, Per capita Income, economic wealth, stable currency and industries as compared to underdeveloped countries. Economic parameter affects social issues like education, religious and ethnic tolerance, culture and overall aspects of human development including 'health'. Developed countries have better health infrastructure and health policies for their people while underdeveloped countries are lacking in the same. But when we try and analyze the social and behavioral aspects of people involved in health care practice while treating patients of different race and ethnicity we find an enduring imprints of racism which do not differentiate between developed and underdeveloped countries. Only the way this discrimination is practiced across the world is different. People of developed countries feel and express this bigotry quite often while people of underdeveloped nations feel it but don't express or at times don't want to feel this discrimination. In this section we will discuss the ethnic and racial differences in healthcare practices with respect to developed and underdeveloped countries.

Developed countries have modern and latest healthcare technology with a subconscious, unnoticed and dormant feeling of racism in some of if not most healthcare professionals. Implicit biased or subconscious prejudice can affect the way clinicians' treats patients (King & Redwood, 2016). It is well-established that blacks and other minority groups in the U.S. experience more illness, worse outcomes, and premature death compared with whites (Health, United States, 2015 with Special Feature on Racial and Ethnic Health Disparities, 2015). This racism is not a result of explicit behavior while it is implicit and inherited by the racist system (Rachel, Eduardo, & Katy, 2016). This kind of racist behavior is not one way, patients and their family members too practice racial discrimination against doctors and health professionals of Black, Indian and Jewish origin (Saadi, 2016).

In underdeveloped country healthcare sector is facing different problems regarding medical complications and wellbeing of different races and ethnicity. Many races in minorities suffer while undergoing treatment at health centers and hospitals over majority ethnic group in underdeveloped countries. Problems are more associated with the economic conditions of the races. Because of poor education they are unable to express discriminating behavior displayed by the health providers. Moreover malnutrition prevailing in most of underdeveloped countries is a primary factor behind retarded mental and physical growth, which has restricted them to even to identify discriminating behavior. Internalized racism is so much rooted in the individual of underdeveloped country that they hold negative ideas about their own

*Table 1. Shows some traditional cultural practices that have contributed to poor health among different groups of people*

| Country Groups/ Nations/ Ethnicities | Poor health practices |
|---|---|
| Turkish, Jewish, Arab | · Don't take medicine during Ramadan |
| Arabian Peninsula | · Believes in traditional remedies and evil spirits<br>· Mental illness are not disclosed<br>· Disclosure of any illness in family reduces chances of marriage of young adults<br>· Visiting patients in hospitals with gift is quite common, they don't like following visiting rules of hospitals<br>· Patients are usually communicated only about good news of diagnosis |
| India | · Child Marriage<br>· Mental illness caused by evil spirits<br>· Traditional treatment of diseases through rituals and prayers<br>· Popularity of alternative treatments like home remedies<br>· Talking about sex is a taboo<br>· Poor economic and social condition of deprived section of society like tribes, and schedule caste<br>· Water contamination of rivers because of rituals causing water born diseases<br>· Poor living standard and illiteracy<br>· Female feticide in certain section of society<br>· Poor knowledge of vaccination |
| China, Nepal, North-east of India | · Patients believe in home remedies and it is used before western medicines<br>· Most illness are caused by loss of vital energy like qi, yin and yang<br>· Cause of mental illness is emotional unbalance and evil spirits<br>· Avoid surgery as they want to keep their soul intact inside their body<br>· Avoid blood transfusion as it brings weakness |
| North and South Korea | · Use Han yak similar to Chinese Yin and Yang, as well as the balance of earth, water, metal, and earth to treat patients<br>· Herbs, acupuncture and cupping are traditional way of treatment used with western medication<br>· Illness is associated with failure to pray<br>· Talking about disease and death makes disease difficult to treat |
| Japan | · Mental illness is because of evil spirits and not considered as illness |
| Nigeria | · Sex is a private subject and not to be discussed with young girls, resulting high rate spread of HIV and STD<br>· Way the houses are build in Africa are one of the causes behind spread of vector born diseases<br>· Some food are ban for consumption in Africa because of cultural taboos<br>· Early marriages of girls and big age gap between married couple<br>· Poor socio economic status of Negroid groups as compared to other ethnic groups |

Source: Compiled by Author

cultural beliefs. In India many tribes belonging to Australoid race are economically poor and hence could not avail best of medical facility. They depend upon government policies and health centers where they directly or indirectly face discrimination. In African continent where almost all nations are underdeveloped, along with economic problem, cultural norms among sub races are different and hence bring natural discrimination. Child marriages, polygamy, age difference in marriages are major cause behind various psychological, physical and sexual disorders. Racial and ethnic preferences along with religious and cast based preferential treatments of patients could be seen in underdeveloped countries.

Poor economic conditions of African, South American and Asian countries have indiscriminately increased the spread of Malaria, diarrhea, Hepatitis, HIV, COPD, leprosy, and skin diseases. General well being is also poor among these countries because of impoverished diet, poor medical facilities, less competent alternative health facilities, improper vaccination of children. However lifestyle born diseases are common in rich and poor countries but causes are different. Rich countries' have fast lifestyle, more dependency on technology, high alcohol abuse, smoking and low physical activities. While poor countries have problems like drugs, unprotected sex, malnutrition, poor awareness about health or education, pollution, poor working conditions which are responsible for the lifestyle related diseases. In major parts of the world whites earnings are more as compared to earnings of African Americans, Hispanic, Asians and Indians, this differentiates in their living condition and health. Major parts of Africa and Middle East are war prone resulting more unnatural deaths of people of these races.

## RACIAL AND ETHNIC DISCRIMINATION OF DISEASES

Genes are common link between ethnic groups and health, as common genes affects the ethnic group as well as health, and health issues are more biological in nature. Diseases create bigotry on the basis of races and ethnicity or human beings discriminates among races while treating any disease? Or pervasive environmental issues including socio-economic environment are responsible for the suffering of one race and well being of another. It can be genetic makeup of individual which is responsible for the spread of specific disease more in one type of gene and less in another. Researchers have differences in their views, some geneticists hold a strong view that races can be separated on the bases of their genetic makeup while some claimed that genetic differences are minimum and unstably distributed across the world (Risch, Burchard, Ziv, & Tang, 2007; Barbujani, Ghirotto, & Tassi, 2013). Death caused by major diseases like heart disease, brain stroke, kidney failure, liver cirrhosis, diabetes and hypertension have been found to be more prevalent among the African Americans than other major ethnic groups in the United States (Kung, Hoyert, Xu, & Murphy, 2008). Some researches note that often the 'race' remains more important indicator of health risk than socio economic position of the patients. African American female college degree holders carry a higher infant mortality rate than Latino, Asian and Pacific Islander women who have not completed their high schools (Pamuk, Makuk, Heck, & Reuben, 1998). African Americans fall under high risk of blood pressure, HDL cholesterol, inflammation risk even if they have better health behavior like low level of smoking, better diet and better physical activity. Studies have found poor health conditions of African American even after having better socio economic conditions than other races (Crimmins, Kim, Alley, Karlamangla, & Seeman, 2007). Differences have been found in rates of STDs among some racial minority or Hispanic groups when compared with rates among Whites (Newman & Berman, 2008; Hogben & Leichliter, 2008). Table 2 shows diseases/health condition /major cause of death prevalent in different races.

Some evidences proved that diseases discriminate between races and genes even after keeping other causes of disease as constant. But prevailing social and economic issues in some races are also major cause of spreading communicable diseases.

*Table 2. Major cause of death in different races and ethnicity*

| Ethnicity | Major Diseases among different ethnic groups | Major Diseases due to socio economic conditions | Remark |
|---|---|---|---|
| Japanese | Ischemic heart disease, Alzheimer's disease, Stroke, Lower respiratory infect, Lung cancer, Stomach cancer, Colorectal cancer Liver cancer, Chronic kidney disease,, COPD, Chronic kidney disease, Pancreatic Cancer | Self-harm like suicide | http://www.healthdata.org/japan |
| Chinese | Cardiovascular Diseases, Neoplasms, Chronic Respiratory Diseases, Diabetes/Urological/Blood/Endocrine Diseases, Transport Injuries, Neurological Disorders, Unintentional Injuries, Diarrheal Diseases, Cirrhosis, Self-harm & Violence | | https://www.cdc.gov/globalhealth/countries/china/default.htm |
| African | Diabetes, Blood Pressure, Cancer, Sickle cell anemia, Heart disease, Stroke | Vector born disease, HIV, COPD | http://www.webmd.com |
| Hispanic | Diabetes mellitus, Diabetes retinopathy, Glaucoma, Gestational diabetes, Lactose intolerance, Metabolic syndrome, Spina bifida, Obstructive sleep apnea, Sickle cell disease, Asthma and obesity in children | Depression and Alcohol Abuse | https://www.ranker.com/list/hispanic-diseases-with-this-risk-factor/reference |
| Indians | Ischemi heart disease, Chronic obstructive pulmonary disease, Stroke, Diarrhea, Lower respiratory infection, Tuberculosis, Asthma, Diabetes, Chronic kidney disease | | https://www.cdc.gov/globalhealth/countries/india/default.htm |
| European/ American Whites | Heart Disease, Cystic Fibrosis, Lactose intolerance, Cancer, Coronary heart disease, Stroke | | |

Source: Self Compiled

## POSITIVE INNOVATIVE PRACTICES ADOPTED GLOBALLY TO HANDLE ETHNIC AND RACIAL ISSUES IN HEALTH CARE INDUSTRIES

Unfortunately, people practice racism, and globally prevalent racial issues are difficult to handle in almost all professions, including healthcare. Racism is practiced at four levels- individual level, interpersonal level, institutional level and structural level (Intergroup Resources, 2012). The real challenge in healthcare sector is to fight against racism at every level. Organizations have come up with their own methods of battling with this social evil at all levels. Primarily racism germinates from self, and creates a perception in an individual about himself as a weak member of the society. This is an individual or internalizes racism, which is most difficult to handle. Almost all governments have made laws against practicing racism but still it is inherited by an individual, giving a self-feeling of being weaker race. This type of racism is widespread in commonwealth nations like India. But strategy to fight with this racism lies in eradicating structural racism. Structural racism is practiced across the world, where discrimination is inherited in the social and economical institutions of the world. Institutions for education, culture, health, and other public services are not untouched by this kind of racism. Story of suffering from racism begins at childhood from the first few institutions of socialization. Children of one race/ethnic group learn at the earliest about their standing in society, which is weaker than some of the privileged groups. This gives a sense of insecurity which creates a permanent block in their mind and they doubt on their capabilities. Individual or internalized racism is unnoticed by the governments and public welfare organizations. Making laws to eradicate this kind racism will not be enough; rather a holistic environmental shift is required. Education against any kind of racist approach is required to be a mandatory part of all nations' education policy; this will inculcate a sense of confidence in the underprivileged ethnic or social group. Structural change in education has been done by many nations to eradicate racism or social discrimination. Dôwa education system implementation in Japan helped in eradicating racism against *Burakumin* children, who are of same ethnic origin as the main ethnic group, *wajin*. Buraku is the name of living quarters of the poor people (called Burakumin), who became subject to discrimination largely due to their occupations, because of rigid class system during the Edo Period. In Japan, Burakumin can't be distinguished from other people on the basis of physical characteristics. However, the potential employers and marriage partners were/are known to categorize people on the basis of the information that the family registry *(koseki)* reveals.

Global education initiative launched by the United Nations Children's Funds (UNICEF) for Arab, Jordan, Oman, Syria and West bank taught antiracism culture, by asking students to analyze cultural, ethical and value based differences in the society. Racism No way! A campaign started by Australian government to teach school kids to fight against racism (Rights, 2000). These are some examples from the existing world, which suggests how at Individual and Interpersonal level fight against racism has eradicated institutional and structural racism.

Whereas it is also inherited to consider one as superior race has known as interpersonal racism, this is prevalent in western world Northern America, Europe and in Southern hemisphere. The Greensboro Health Disparities Collaborative partnered with Cone Health's Wesley Long Cancer Center and the University of Pittsburgh Medical Center's Hillman Cancer Center in a project known as ACCURE (Accountability for Cancer Care Through Undoing Racism and Equity) found that Black American women suffer less from breast cancer than white women but their mortality rate is high. Similarly, lung cancer patients' treatment completion rate in the Black Americans was less than the White Americans. Staff members of the both cancer institute's were trained to handle ethnic and racial training, they found that black women surviving from the breast cancer has experienced poor treatment and non-interpretation of diagnosis along with ill-treatment from front desk staff. Team also studied the treatment procedure and flow of treatment for cancer patients. A nurse navigator has been used to detail patients about their treatment and related aspects, which improved patients' satisfaction and provide transparency in the system. Therefore, technology helped patients to overcome distrust raised by racism. Study also focused up on customized support to cancer patients, which can completely eradicate treatment bias between black and white patients.

Diversification of the healthcare professionals are equally important, a balanced ratio of diversified staff members will reduce incidence of racial, ethnical and cast based discrimination. India government has made law of reservation to ensure equal and just participation of all caste members in the government workforce. Government is the major healthcare provider to the Indian population, especially poor tribes and *dalits* gets subsidized treatment in these healthcare settings. Unfortunately race and caste based discrimination has been confirmed by many studies done in India in various healthcare setups. Clinicians need more training to handle patients of different race, caste and ethnicity. Clinicians require learning acculturation; it is a process to understand the culture of others.

Social mobility is another strategy, which can be implemented and adopted at structural level to eradicate all kind of discrimination. Change in current social status of deprived race and movement of health professionals at all level of societies existing in a state or country is required to better understand health needs of individuals based upon cultural and ethnic norms.

## RELIGION AND HEALTHCARE PRACTICES

In this section will discuss about religious belief and its influence on healthcare practices across major religions of the world.

## HEALTH CARE INDUSTRY BASED ON RELIGION: A HISTORICAL PERSPECTIVE

Many aspects manage the delivery of healthcare services. Until this day, many countries around the world are coping with tensions between religious traditions, culture and medical science.

This brings us again to understand religion in healthcare perspective. However, before we draw parallels, the understanding of religion is important. A group of people understands religion as an essential and agreed beliefs and practices. These set of beliefs concern the cause, nature and purpose of the universe, and involve devotional and ritual observances. In some or the other way these have laid the basis of healthcare practices followed by many religious groups in the past and even today.

Indian languages like Sanskrit and Hindi have translated religion as 'dharma'. Which originates from the Sanskrit root ' *dhri* ', meaning 'to uphold, support or sustain,' thus in the connotation of the word, dharma is what upholds existence- life, and growth- being and becoming (Howard, 2017).

There are many examples from past and even present that signifies the role of religion in giving treatment and care, like in Egypt, where the physicians were the priests who delivered natural therapeutic measures using mainly herbal yields. The system of Ayurveda followed largely by Hindus has its roots in ancient religious texts of the *Atharvaveda*. Until today Hindus are staunch believers that knowledge of medicine possessed by sages and monks, is sacred, almighty being the definitive base of this knowledge of life. Thus, the association between religion, spirituality and health goes centuries back and there are many instances to support this association.

One such instance, which highlights the amalgamation of religion and curative care, dates back to Buddhism Era, where Buddha inculcated in his disciples the principle of caring for ailing persons. Emperor Asoka a firm believer and follower of Buddhism too followed the same principle owing to which many provisions were made in his kingdom for medical treatment for everyone.

There are evidences in the literature which points out at other religious groups like Christianity and Islam, played a pivotal role in establishing hospitals in initial stages. In earlier times, Hospital that was referred as 'Hospice' a home caring for sick and terminally ill. Gradually, these religious sects have contributed immensely in building care centers providing modern medical care with knowledge and technique.

This close relationship between religion and medicine being centuries old there were fewer evidence of understanding the nature and type of disease to provide curative care in earlier times. This limited the focus on advancing the medical knowledge during this phase (Chattopadhyay, 2007).

### Effects of Religion on Treatment

The further discussion delves in understanding further the contribution of any ethnic group in developing the healthcare services that we see today.

Each ethnic group brings its own perceptions and standards/ morals to the health care system, and many health care beliefs and practices vary. Whether these bring positive or negative effects in total care of population has been a debatable topic ever since. It is from patient point of view that these beliefs and faith exists.

Hospitals were non-existent prior to the Christian era. In Greek and Roman periods, the general population incapable of affording a private physician or treatment in an Asclepiad temple sent to their families or were left to die unattended. The first major hospital in western civilization was built in Asia Minor around 370 AD at the insistence of St. Basil, bishop of Caesarea.

A Buddhist prince, Hsiao Tzu-Liang, started the first hospital in China in 491 AD (Koenig H.G, 2000).

During the course of the Middle Ages and up through the French Revolution, doctors were often clergy. The religious organizations for hundreds of years were responsible for declaring physicians to practice medicine.

During similar time in the American colonies, many clergy practiced medicine as a second job that helped them to complement their scanty income from church work (Koenig H.G, 2000).

It is a notable fact, that many schools of medicine have religion as an elective course in their curriculum along with spirituality and medicine. This establishes the fact that even if religion is dissociated with medicine in modern times, the influence cannot be ignored and the interest in terms of spreading this knowledge and association continues signifying the major impact of religion on modern medical practices too (Dein et.al, 2010).

"The Hippocratic Corpus" was an initial effort to think about diseases, not as retribution from the gods, but as a disparity of human being with his surroundings. This aspect is considered a very important step forward.

With this step, also came the logic that if origin of disease in imbalance with one's environment or diet then the cure should also be according to it. This thought provided some sort of justification for accepting the underlying causes associated with health and disease.

During the same time, the Greek physicians recommended variations in diet or lifestyle and occasionally invented drugs or did surgery to give relief from suffering. Regardless of such contributions of the Corpus, the progress with respect to medicine and science in Europe was pretty slacked for several centuries.

Over the period, the population grew, and cities became thickly populated, leading to waste generation and disposal. Without any systematic plan in place for it. The growing pile and lack of sanitation gave rise to endemic disease and periodic epidemics. There are multiple evidences from past which gives us enough historical evidences of many epidemics leading to loss of life and absence of proper care and treatment to the ones who suffered.

Further, we should understand that the relationship established between determinants of health and disease was invented and progressed over centuries.

The below table highlights few of the discoveries that are quoted verbatim to retain the elements the table wants to portray in terms of evolution that took place and remains as a strong foundation for the medical sciences and breakthrough discoveries (*Public Health Module, Boston Medical Campus, n.d*):

*Table 3. Breakthrough discoveries in medical science*

| Expert Name & Duration | Work that led to breakthrough discoveries |
|---|---|
| Girolamo Fracastoro (1546) | Outlined his concept of epidemic diseases in "*De contagione et contagiosis morbis*" and proposed that disease could be transmitted by direct contact, through the air, or on contaminated clothing and linens. |
| John Graunt - The Bills of Mortality (1662) | Analysed the data regarding common causes of death, higher death rates in men, seasonal variation in death rates, and the fact that some diseases had relatively constant death rates, while others varied considerably. Estimated population size and rates of population growth, first one to construct life table. |
| Anton van Leeuwenhouk (1670s) | Credited with creation of microscope. Used it to see first time the structures of bacteria (1674), yeast, protozoa, sperm cells, and red blood cells. |
| John Pringle and "Jail Fever" (1740s) | Gave number of solutions aimed at improving the health of soldiers including betterment of hospital ventilation and camp sanitation, proper drainage, adequate latrines, and the avoidance of marshes. |
| James Lind and Scurvy (1754) | Conducted the world's first controlled clinical trial on 12 sailors with scurvy |
| Francois Broussais & Pierre Louis (1832) | Used bloodletting to treat many diseases, including cholera |

Source: Public Health Module, Boston University medical Campus, http://sphweb.bumc.bu.edu

## DISEASES AND RELIGIOUS TABOO IN TREATMENT

A taboo is something, which is considered as forbidden. Going by this explanation there are so many things in all religions across that are forbidden or have been associated with origin of certain diseases, patterns and beliefs which people have made a very strong part of their understanding and associating treatment of diseases with the same. The thing that diseases are form of curse from some supernatural being is still believed.

In many tribes and settlements, people resort to the measures where, diseases are treated either by magic or religious performances, which involves dealing with the soul of the patient or in urging or forcing the evil spirits, that have entered the body of the patient to leave to relieve the agony and suffering. It is further understood by the many tribal groups that if sickness is minor and physically tolerable then no specialist is consulted, as the temporary sickness will soon disappear. In certain cases, if physical pain increases, the patient himself, herself, or any relative on behalf of the patient consults a magic-religious specialist, also referred as a medicine man, or healer. This specialist healer diagnose the sickness from evident symptoms, or may resort to prophecy and other mystic-religious practices'[7] How far these things result into cure and long-term benefit is debatable.

A common belief that is shared amongst the various societies and cultures is that spirituality and religion have played an imperative role in the healing of diseases. The scientific works done on same area has also stressed that spirituality or religiousness can be helpful in recovery from illnesses including cancer, mental disorders, cardiovascular diseases, and rheumatoid arthritis and head injuries. Individuals and group of people have claimed to get relief from diseases, by following the acts of devotion (Romeo. N, *et.al* 2015).

Aboriginals frequently believed that medicos trained in west are not well informed to address their physical and spiritual apprehensions. Their beliefs and perceptions of ill health came from their socio-cultural context. According to them, the indigenous healers are capable of curing them (Marrone, 2007).

Skinner (1948, 1953) through his classic work discovered that superstitious behavior could arise through conditioning. During the growing up years, the individuals see what is practiced around them, these practices results in superstition and are developed into their belief system.

It can be argued that people carry out rituals in an unclear situation; in which the outcome is both uncertain and important. This situation is changing due to the awareness brought in by Education and spread of knowledge.

Despite this, a significant proportion of the society, still turn to these self-confessed god-men, especially when it comes to ailments- physical or mental giving rise to morbidity and mortality. It also brings discredit to the medical profession, when the doctors are unable to help such cases due to loss of crucial duration of treatment that could have saved one's life.

The dark side of healing process involves spiritual protections, sacrifices, spiritual cleansing, appeasing the Gods by following certain specific ritual etc.

The traditional tribes and early-established still practice prescription of herbs, application of herbs and clay and counseling by elders, the kind of food the person should or should not eat. This is mostly done when person has violated a taboo. Further, the role of diviner cannot be denied in certain communities. Mystics treat illnesses primarily through enabling the direct intervention of spiritual world. They play an intermediary role between the spirit and real world. In many societies, they are believed to be the custodian of the theories of healing and the hope of society (White, 2015).

## HEALTH CARE INDUSTRIES AND RELIGIOUS SENSITIVITY / HEALING THROUGH SPIRITUAL RELIGIOUS PRACTICES

In healthcare industry religion and health are used simultaneously although both have different meaning and applications (Miller & Thorensen, 2003). Religion is more institutional based and communal in nature. Religiousness involves faith, belief and collective worship for the god, saint, symbol of monument. Whereas spirituality in more individualistic in nature and harness inner strength of the practicing person (Pesut, Fowler, Taylor, Reimer-Kirkham, & Sawatzky, 2008). It is most vital issue of life with subjectivity at all level of human development including physical senses. Spirituality is obscure; it tries to find association with nature and god to explore meaning and goals of life. The main features of spirituality are based on, connecting, transcendence and value (Martsolf & Mickley, 1998) . Impact of religion on health is explained in the next section of this chapter. In this section impact of spirituality on the health has been discussed. Spirituality means connecting soul with the god, in practical term it is considered as relaxation technique. Meditation is a major practice of spirituality, which is connected with the physiological measures.

*Spirituality is recognized as a factor that contributes to health in many persons. The concept of spirituality is found in all cultures and societies. It is expressed in an individual's search for ultimate meaning – The Association of American Medical College*

Researches done earlier suggest that meditation has reduced blood pressure of the patients. Regular practice of meditation done for longer time duration, more than two months has reduced blood pressure considerably (Patel, Marmot, Terry, Carruthers, Hunt, & Patel, 1985). A study conducted on African American showed that meditation has reduced systolic and diastolic blood pressure more as compared to muscle relaxant and group using lifestyle improving classes (Schneider, et al., 1998). Dhammakaya Buddhist meditation program showed a reduction in systolic and diastolic pressure, compared to a control group of male college students that did not follow the program (Sudsuang et al. 1991).

People who are spiritual have lower level of Cortisol stress hormone in their blood. Some people believe that meditation is a separate practice, in fact it is a regular practice followed by many spiritual people. Way of practicing meditations may be different in different culture but end product is common. Meditation reduces cortisol level in body and hence reduces stress level in the human body. Stress level hormone cortisol in human body has reduced significantly in those who practice meditation for longer time duration as compared to those groups who do not practice any technique for reducing stress (Jevning, Wilson, & Davison, 1978). In one more study it has been found that those who practice meditation have shown lower level of aldosterone and norepinephrine along with cortisol (Walton, Pugh, Gerderloos, & Macrae, 1995).

Besides reducing stress level in human beings, meditation is also useful in decreasing heart rate, respiratory rate, metabolism and brain waves. It has improved chronic pain, insomnia, anxiety, and found to be supportive for the treatment of HIV (Benson, Beary, & Carol, The relaxation response, 1974).

If spirituality and religions have such a wonderful effects on the human health then why is it ignored by clinicians and scientists? Perhaps they don't find it more scientific in nature neither it has a scope for scientific study. There is a gorge between clinicians and patients regarding role of spirituality in healing diseases. Few training programs for the clinicians on the importance of spirituality and meditation on health and other religious issue (Shafranske & Malony, 1990). Weak connect between clinicians' and spirituality experts (Pirl, Roth, Cotton, Levin, & Fitzpatrick, 2000) with polarization of religious belief against science and vice- versa are majorly responsible for the distance between science and religious spirituality (Wuthnow, 1998).

## POSITIVE IMPACT OF RELIGION ON HEALTH

Religious practice brings positivity in human life by promoting good health practices. This positive energy improves physical and mental health. Visiting religious place and praying brings in happiness by reducing depression and improving self-esteem. Religious practice improves physical strength and increases longevity; it improves an individual's chances of recovering from illness, and lessens the incidence of many diseases (Stark, 2012).

Mental and physical well-being is received through religious worship and practices. Almost all religions bring people at a common praying point where they meet and share their issues related to life. In Christianity confession is a process where one can express his sins and wrong-doings. This process purge the internal thought process and therefore individual feels happy and relieved. In Hindu religion public gathering is quite common at temples where religious Gurus give lectures on handling day-to-day problems. Many more examples can be sighted from other religions where religion is a cause of mental peace in human beings. Religious festivals are celebrated in almost all religions across globe, giving a reason to its followers to enjoy holidays and remain stress free. Those who participated in community religious services had lower levels of depression than those who did not fellowship in a religious community but prayed alone (Ellison, 1995). Participation in public activity, team-work and feeling of contributing to society are major reasons which keeps an individual protected from depression and stress. Conversely, a lack of religious affiliation correlated with an increased risk of suicide (Tovato, 1990).

Perhaps religion connects an individual with god, which is nothing but a manifestation of faith in life. Most religion condemns self-destruction of any kind. Beside that other followers serves as a support network, which unites an individual with family and friends. All religions suggests regimented pattern of living life where due importance is attached to cleanliness, eating habits, nutrition, exercise or any physical activity, way to live life and to be abstemious. This results in reduced mortality risk from infectious diseases and diabetes (Hummer, Roger, Nam, & Ellison, 1999). Accidents because of drunken driving are less in religious people. Also, religious people are less likely to be suffered from liver cirrhosis, emphysema and cardiovascular diseases (Comstock & Patridge, 1972). Those who are involved in religion lives longer life as compared to those who do not follow any religious practice, regardless of their gender, caste, race, ethnicity and health history (Johnson, Tompkins, & Webb; Johnson, Tompkins, & Webb, 2017). Further religious institutes representing most religions organize medical camps in association with hospitals and health institutes.

Some researches done on the impact of religious preaching and its importance on healthcare concluded with no benefits of religious prayers on the health of individual. Even spirituality has shown no relevance with pain and distress caused by pain (Skevington, Carse, & Williams, 2001). While some studies says that religion and health are not associated at all. There are evidences among the devotees of religion who repent on their sufferings from diseases and blame god for their ill health.

## REFERENCES

Barbujani, G., Ghirotto, S., & Tassi, F. (2013). Nine things to remember about human genome diversity. *Tissue Antigens*, *82*(3), 155–164. doi:10.1111/tan.12165 PMID:24032721

Benson, H., Beary, J. F., & Carol, M. P. (1974). The relaxation response. *Psychiatry*, *37*(1), 37–46. doi:10.1080/00332747.1974.11023785 PMID:4810622

Benson, H., Beary, J. F., & Carol, M. P. (1974). The relaxation response. *Psychiatry*, *37*(1), 37–46. doi:10.1080/00332747.1974.11023785 PMID:4810622

Brondolo, E., Rieppi, R., Kelly, K. P., & Gerin, W. (2003). Perceived racism and blood pressure: A review of the literature and conceptual and methodological critique. *PubMed*, *25*(1), 55–56. doi:10.1207/S15324796ABM2501_08 PMID:12581937

Chattopadhyay, S. (2007). Religion, spirituality, health and medicine: Why should Indian physicians care? *Journal of Postgraduate Medicine*, *53*(4), 262. doi:10.4103/0022-3859.33967 PMID:18097118

Comstock, G. W., & Patridge, K. B. (1972). Church Attendance and Health. *Journal of Chronic Diseases*, *25*(12), 665–672. doi:10.1016/0021-9681(72)90002-1 PMID:4648512

Crimmins, E. M., Kim, J. K., Alley, D. E., Karlamangla, A., & Seeman, T. (2007). Hispanic paradox in biological risk profiles. *American Journal of Public Health*, *97*(7), 1305–1310. doi:10.2105/AJPH.2006.091892 PMID:17538054

Dein, S. (2010, January 10). *Religion, Spirituality, and Mental Health*. Retrieved from https://www.psychiatrictimes.com/schizophrenia/religion-spirituality-and-mental-health

Ellison, C. G. (1995). Race, Religious Involvement, and Depressive Symptomatology in Southeastern U.S. Community. *Social Science & Medicine*, *11*(11), 1561–1572. doi:10.1016/0277-9536(94)00273-V PMID:7667660

*Health, United States, 2015 with Special Feature on Racial and Ethnic Health Disparities.* (2015). Hyattsville: National Centre for Health Statistics.

Hogben, M., & Leichliter, J. S. (2008). Social determinants and sexually transmitted disease disparities. *Sexually Transmitted Diseases*, *35*(12Suppl), 13–18. doi:10.1097/OLQ.0b013e31818d3cad PMID:18936725

Howard, V. R. (2017). *Dharma: the Hindu, Jain, Buddhist and Sikh traditions of India*. London: I.B.Tauris.

Hummer, R. A., Roger, R. G., Nam, C. B., & Ellison, C. G. (1999). Religious Involvement and U.S. Adult Mortality. *Demography*, *36*(2), 273–285. doi:10.2307/2648114 PMID:10332617

Intergroup Resources. (2012). Retrieved August 1, 2019, from www.intergroupresources.com: http://www.intergroupresources.com/race-and-racism/

Jevning, R., Wilson, A. F., & Davison, J. M. (1978). Adrenocortical activity during meditation. *Hormones and Behavior*, *10*(1), 54–60. doi:10.1016/0018-506X(78)90024-7 PMID:350747

JohnsonB. R.TompkinsR. B.WebbD. (2017, September). https://www.manhattaninstitute.org/pdf/crrucs_objective_hope.pdf. Retrieved 2017, from www.manhattaninstitute.org

King, C. J., & Redwood, Y. (2016). The Health Care Institution, Population Health and Black Lives. *Journal of the National Medical Association*, *108*(2), 131–136. doi:10.1016/j.jnma.2016.04.002 PMID:27372475

King, G. (1996). Institutional Racism and the Medical/Health Complex: A Conceptual Analysis. *Ethnicity & Disease*, *6*, 30–46. PMID:8882834

Koenig, H. G. (2012). Religion, Spirituality, and Health: The Research and Clinical Implications. *ISRN Psychiatry*, *2012*, 1–33. doi:10.5402/2012/278730 PMID:23762764

Koenig, H. G. (2015). Spirituality, and health: A review and update. *Advances in Mind-Body Medicine*, *§§§*, 19–26. PMID:26026153

Kung, H. C., Hoyert, D. L., Xu, J., & Murphy, S. L. (2008). *Deaths: Final data for 2005*. National Vital Statistics Report America.

Learning Objectives. (n.d.). Retrieved from http://sphweb.bumc.bu.edu/otlt/MPH-Modules/PH/PublicHealthHistory/PublicHealthHistory_print.html

Marrone, S. (2007). Understanding barriers to health care: A review of disparities in health care services among indigenous populations. *International Journal of Circumpolar Health*, *66*(3), 188–198. doi:10.3402/ijch.v66i3.18254 PMID:17655060

Martsolf, D. S., & Mickley, J. R. (1998). The concept of spirituality in nursing theories: Differing world view and extend of focus. *Journal of Advanced Nursing*, *27*(2), 294–303. doi:10.1046/j.1365-2648.1998.00519.x PMID:9515639

Miller, W. R., & Thorensen, C. E. (2003). Spirituality, religion, and health: An emerging research field. *The American Psychologist*, *58*(1), 24–35. doi:10.1037/0003-066X.58.1.24 PMID:12674816

Newman, L. M., & Berman, S. M. (2008). Epidemology of STD Disparities in African Communities. *Sexually Transmitted Diseases*, *35*(12Supplement), 4–12. doi:10.1097/OLQ.0b013e31818eb90e

Pamuk, E., Makuk, D., Heck, K., & Reuben, C. (1998). *Socioeconomic status and health chart book*. Hyattsville, MD: National Center for Health Statistics.

Patel, C., Marmot, G. M., Terry, J. D., Carruthers, M., Hunt, B., & Patel, M. (1985). Trial of relaxation reducing coronary risk: Fouryear follow up. *British Medical Journal*, *290*(6475), 1103–1106. doi:10.1136/bmj.290.6475.1103 PMID:3921124

Pesut, B., Fowler, M., Taylor, J. E., Reimer-Kirkham, S., & Sawatzky, R. (2008). Conceptualising spirituality and religion for healthcare. *Journal of Clinical Nursing*, *17*(21), 2803–2810. doi:10.1111/j.1365-2702.2008.02344.x PMID:18665876

Pirl, W. F., Roth, A. J., Cotton, S. P., Levin, E. G., & Fitzpatrick, C. M. (2000). Exploring the relationship amongspiritual well-beingh, quality of life, and psychological adjustment in women with breast cancer. *Psycho-Oncology*, *8*(5), 429–438.

Rachel, R. H., Eduardo, M. M., & Katy, B. K. (2016). Structural Racism and Supporting Black Lives - The Role of Health Professionals. *The New England Journal of Medicine*, 2113-2115.

Rights, H. C. (2000). *Educating Children and Youth Against Racism*. Durban: United Nations High Commissioner for Human Rights.

Risch, N., Burchard, E., Ziv, E., & Tang, H. (2007). Categirization of humans in biomedical research: Genes, race and disease. *Genome Biology*, *3*(7), comment2007.1. doi:10.1186/gb-2002-3-7-comment2007

Romeo, N., Gallo, O., & Tagarelli, G. (2015). From Disease to Holiness: Religious-based health remedies of Italian folk medicine (XIX-XX century). *Journal of Ethnobiology and Ethnomedicine*, *11*(1), 50. doi:10.118613002-015-0037-z PMID:26048412

Saadi, A. (2016, February 7). *KevinMD*. Retrieved July 31, 2019, from www.kevinmd.com: https://www.kevinmd.com/blog/2016/02/muslim-american-doctor-racism-hospitals.html

Schneider, R. H., Nidich, S. I., Salerno, J. W., Sharma, H. R., Robinson, C. E., Nidich, R. J., & Alexander, C. N. (1998). Lower lipid peroxide levels in practitioners of the transcendental mediation program. *Psychosomatic Medicine*, *60*(1), 38–41. doi:10.1097/00006842-199801000-00008 PMID:9492237

Shafranske, E. P., & Malony, H. N. (1990). Clinical psychologists' religious and spiritual orientation and their practice of psychotherapy. *Psychotherapy (Chicago, Ill.)*, *27*(1), 72–78. doi:10.1037/0033-3204.27.1.72

Skevington, S. M., Carse, M. S., & Williams, A. C. (2001). Validation of the WHOQOL-100: Pain management improves quality of life for chronic pain patients. *The Clinical Journal of Pain*, *17*, 264–275. doi:10.1097/00002508-200109000-00013 PMID:11587119

Stark, R. (2012). *America's Blessings: How Religion Benefits Everyone, Including Atheists* (Vol. 1). Templeton Foundation Press.

Tovato, F. (1990). Domestic/Religious Individualism and Youth Sicide in Canada. *Family Perspective, 24*(1), 69–81.

VanderWeele, T. J., Balboni, T. A., & Koh, H. K. (2017). Health and spirituality. *Journal of the American Medical Association, 318*(6), 519–520. doi:10.1001/jama.2017.8136 PMID:28750127

Walton, K. G., Pugh, N. D., Gerderloos, P., & Macrae, P. (1995). Stress reduction and preventing hypertension: Preliminary support for a psychoneuroendocrine mechanism. *Journal of Alternative and Complementary Medicine (New York, N.Y.), 1*(3), 263–283. doi:10.1089/acm.1995.1.263 PMID:9395623

White, P. (2015). The concept of diseases and health care in African traditional religion in Ghana. *Hervormde Teologiese Studies, 71*(3). doi:10.4102/hts.v71i3.2762

Wuthnow, R. (1998). *After heaven: Spirituality in America scince the 1950s.* University of California Press. doi:10.1525/california/9780520213968.001.0001

# Chapter 14
# Assessment of Healthcare Service Quality:
## Tertiary Care Hospitals of Dhaka City

**Segufta Dilshad**
*North South University, Bangladesh*

**Afsana Akhtar**
*BRAC University, Bangladesh*

**S. S. M. Sadrul Huda**
*East West University, Bangladesh*

**Nandeeta Samad**
*North South University, Bangladesh*

## ABSTRACT

*The service quality measurement of healthcare services is always a big concern for the hospitals, patient rights activists, regulators, and general patients. This study deals with quality assessment of healthcare facilities concerning the private and public health facilities of Dhaka, Bangladesh. This study follows the survey research approach. Using the purposive sampling method, the individuals have been selected from households who have received healthcare services from public or private hospitals of Dhaka city in last year. The study collected data among 410 respondents. Standard statistical software (i.e., SPSS and STATA) have been used to analyze the data. This study confirms existing evidence that Bangladeshi patients have a growing concern with lower level of satisfaction in public healthcare services. The respondents faced multi-dimensional problems, characterized by a low level of overall service quality, interpersonal service quality, and technical or treatment-related quality at public hospitals. Further research is recommended to analyze the issues further.*

DOI: 10.4018/978-1-7998-3576-9.ch014

# INTRODUCTION

The health care sector of Bangladesh improved significantly after the independence and the recent improved health indicators have been received well by the international development organizations (IMF, 2015; BDHS 2014). The statistical evidence shows that the key indicators such as life expectancy at birth has been raised from 44 years to 72 years during the period 1970 to 2014; all basic vaccination coverage by 12 months of age raised up to 78 per cent, contraceptive usage has increased from 8% to 62% among married women and fertility rate has decreased from 6.3 per family to 2.3 (BDHS, 2014). Even though the country needs to continue the progress in order to achieve international and national development goals.

The quality of health care traditionally has been measured by criteria such as morbidity and mortality. Despite all the good progress, the health and population sector is still often blamed to be poorly coordinated and inefficient in the delivery of health care services. Although these different hard indicators are still useful and essential to measure clinical outcome, a more subjective and softer assessment is still often left overlooked. In reality, many other service sectors and industries likewise have been slow to move beyond a supply side approach to quality assessment (Cf. Dagger el al, 2007). Patient's perspective in defining the quality is becoming the competitive advantage with the changing trends of health care sector (Donabedian, 1980). As a consequence, health care service providers are struggling to provide patient centered quality assessment approach.

Andaleeb et al, identified few factors such as doctors' treatment, the behaviour of nurses/boys and their services to patient are significantly influence patients' satisfaction. (Andaleeb et al.: 2007). However, these were some good attempts to point out some factors linked with patient's satisfaction. Therefore, the proposed study was an attempt to explore the patients' expectations and perception of quality health care and experiences of receiving public health care services, their condition of satisfaction etc.

# HEALTHCARE SYSTEM PROFILE

The health system of Bangladesh exists with pluralistic governance i.e. different stakeholders with their respective roles are working in various competitive and collaborative combinations. There are four such stakeholders working in this field of Bangladesh. The government sector playing a dominant role with a mandate to not only set policy and regulations, but also providing comprehensive health services. Almost two-thirds of total health expenditure is household expenditure in the private including formal and informal sectors. In addition, the vibrant and large non-government organization (NGO) sector that focuses on the health needs of the poor, often as part of a broad array of development interventions. Finally, the donor community that exercises disproportionate influences in determining policy and programmatic priorities, with technical assistance, and directs delivery strategies.

Bangladesh has low ratios of credentialed professionals- only 0.5 doctors and 0.2 nurses per 1000 people, far less than the minimum standard of 2.28 per 1000 recommended by WHO (Bangladesh Health System Review, 2015). Bangladesh also has a shortage of skilled health workers, with twice as many doctors as nurses clustered disproportionately in urban areas. At present there are 64,434 registered doctors, 6,034 registered dentists, 30,516 registered nurses, (MoHFW, 2013) and 27,000 midwives (MoHFW, 2012). There are high levels of out-of-pocket and informal payments for health services and medicines that are exhausting millions of households. Despite these endemic shortfalls in key areas of the health

*Box 1.*

---
**HEALTH SERVICE AND MEDICAL EDUCATION**
Total number of government facilities under the DGHS: 2,258
Total primary level facilities (except community clinic): 2,004
Total secondary & tertiary level facilities: 254
Total number of registered private hospitals and clinics: 5,054
Total number of registered private diagnostic centers: 9,529
No. of hospital beds under the DGHS: 52,807
No. of hospital beds in private hospitals registered by the DGHS: 90,587
Total in the DGHS-run facilities and registered private hospitals: 143,394

---

Source: *Health Bulletin, 2018, Ministry of Health and Family Welfare, Bangladesh*

system, pronounced and rapid progress in the most important health measurements - e.g. infant and child mortality, maternal mortality, fertility, and contraceptive prevalence- are remarkable.

In Bangladesh, depending upon the type of services patients required, the level of health care divided into three broader categories: primary care - where basic and general healthcare traditionally provided by community workers, family health practitioners, occasionally gynecologist; secondary care - the medical care provided by a physician who acts as a consultant at the request of the primary physician; tertiary care - specialized curative care, usually on referral from primary and secondary medical care personnel by specialists working in a center that has personnel and facilities for special investigation and treatment.

Currently, Bangladesh has been making significant socio-economic developments in recent years. GDP has been growing at an average rate of 6-7% over the past decade. However, despite improving healthcare indicators such as decline in mortality rates and increase in average life expectancy, the health sector of the country is yet to reach its full potential. Bangladesh government's share of spending on healthcare is 37% of the total healthcare expenditure. Public spending on health is financed from the non-development or revenue budget and the development budget or Annual Development Program (ADP) in the form of national tax, foreign development funds, and corporations and autonomous bodies. Tax and non-tax revenue and foreign loans and grants are channeled by the Ministry of Finance to the Ministry of Health and Family Welfare and other ministries.

Health services in Bangladesh remained predominantly financed by households' Out-of-Pocket-Payments (OOPP). Direct payment for the purchase of pharmaceuticals and medical goods is the predominant contributor to OOPP, either through self-purchase or on the advice of a formal or informal health-care provider. OOPPs are mostly direct payments made at private and NGO facilities and also to informal providers. The growing reliance on OOPP leaves the population at risk.

## Public Sector Health Services

The Ministry of Health and Family Welfare (MoHFW) of Bangladesh has an extensive health infrastructure. The service delivery structure follows the country's administrative pattern, starting from the national to the district, *upazila*, union and finally to the ward levels. It provides promotive, preventive, and curative services such as outdoor (outpatient), indoor (inpatient), and emergency care at different levels – primary, secondary and tertiary.

## Private Sector Health Services

In the private sector, providers can be grouped into two main categories. First, the organized private sector (both for-profit and nonprofit), which includes qualified practitioners of medicine. Second, the private informal sector, which consists of providers practicing in rural areas not having any formal qualifications such as untrained allopaths, homeopaths and *kobiraj* (practitioner using locally produced herbal medicines). Along with private clinics and hospitals, the number of diagnostic centers in the private sector is growing. Approximately 9,529 laboratories and other diagnostic centers were registered in 2018 with the Ministry of Health and Family Welfare (MoHFW, 2018). In the private for-profit sector, there are some large diagnostic centers in the cities (Lab Aid, Ibn Sina, Popular and Medinova) providing laboratory and specialized radiological tests. Some of these facilities maintain a high standard. In the nonprofit private sector, there are centers like the International Centre for Diarrhoeal Diseases and Research, Bangladesh (ICDDR,B), which has modern laboratories providing research facilities and extends laboratory services to the general community.

## Donors, NGOs, and Professional Groups

Bangladesh is known worldwide for having one of the most dynamic NGO sectors, with 2,471 NGOs registered with NGO Affairs Bureau working in the population, health and nutrition sector (2014). NGOs have been active in health promotion and prevention activities, particularly at the community level, and in family planning, maternal and child health areas.

Multiple donors, both multilateral and bilateral, have been actively engaged in health-care financing and planning. The main bilateral donors to the health and population sector in Bangladesh are the governments of Australia, Belgium, Canada, Germany, Japan, Netherlands, Norway, Sweden, the United Kingdom and the United States. The multilateral donors include the World Bank, European Union, UNICEF, ADB, Global Fund to Fight AIDS, Tuberculosis and Malaria (GFATM), and the GAVI Alliance. Moreover, there are a number of professional organizations, which address the rights of medical professionals at different levels, such as the Bangladesh Medical Asssociation (BMA), Bangladesh Private Medical Practitioners Association (BPMPA), Public Health Association of Bangladesh, Bangladesh Paediatric Society and the Nephrology Society of Bangladesh.

## HEALTH SECTOR CHALLENGES

The health system in Bangladesh is characterized by a massive shortage of skilled health workers clustered disproportionately in urban areas while rural facilities are overburdened, understaffed and insufficiently equipped. According to the Asia Pacific Observatory on Public Health Systems and Policies, several factors played important roles in hindering expected improvement in the overall health status of the country:

## OBJECTIVES

The general aim of the study is to know the patient's satisfaction over different health care services provided by government and private hospitals. Therefore, the specific objectives are:

## Assessment of Healthcare Service Quality

*Box 2.*

| |
|---|
| • The complexities of the mixed health systems and poor governance |
| • Inadequacy of health system resources and impact on quality of care |
| • Inadequate and uneven health system coverage |
| • Healthcare financing through massive OOPP by households |
| • Inequitable access to health services hindering universal health coverage |

- To explore patients self-perception of healthcare services, understand expectation and demands of health care towards public and private service providers;
- To assess the patients experiences of getting health care services provided by the public and private hospitals; identify patients' level of satisfaction or dissatisfaction in getting those services;
- To assess the issues and difficulties they face in receiving services.

## PERCEIVED SERVICE QUALITY

Service quality perceptions are generally defined as a consumer's judgment of, or impression about, an entity's overall excellence or superiority (Bitner and Hubbert 1994; Boulding et al. 1993; Cronin and Taylor 1992; Parasuraman, Zeithaml, and Berry 1985, 1988). This judgment is often described in terms of the discrepancy between consumers' expectations of service and actual service performance. Grönroos (1984), for example, emphasized the use of expectations as a standard of reference against which performance can be judged, and Parasuraman, Zeithaml, and Berry (1985) put forward service quality as the gap between expected and perceived service.

It's found that developing countries are under-represented in health care research though significant research initiatives are noticed to assess the healthcare scenario of the developed world. Murti, Deshpande and Srivastava (2013) conducted an empirical research on the impact of service quality on patient satisfaction and consumer behavioral intentions of private hospitals in Bhopal city of central India. They have adopted the definition of customer satisfaction from Rust and Oliver (1994), which reflects the degree to which a customer believes that the use of a service evokes positive feelings. According to prominent marketing scholar Philip Kotler (1991), satisfaction is the post-purchase evaluation of products or services based on the expectations of the customers before purchase. In healthcare industry, patient satisfaction should be indispensable in assessing service quality (Donabedian, 1982).

High service quality is also vital for positive behavioral intentions of the consumers. Some of the major components of behavioral intentions are loyalty, recommending the product or service to others, spending more on the company's products, less vulnerability to price premiums, complaining behavior and intention to repurchase (Cronin et al., 2000; Zeithaml et al., 1996).

Although commonly applied, this approach has been the subject of substantial criticism and debate. Babakus and Boller (1992), for example, suggested that the measurement of expectations adds limited information beyond what is gained from measuring service perceptions alone. Similarly, Dabholkar, Shepherd, and Thorpe (2000) found that perceptions performed better than difference measures when comparing these approaches, and both Cronin and Taylor (1992) and Brady and Cronin (2001) focused on performance-only measures (i.e., perceptions rather than expectations) when modeling service quality perceptions.

The widely applied SERVQUAL scale (Parasuraman, Zeithaml, and Berry 1985, 1988), for example, has been criticized insofar as its five dimensions, namely, reliability, empathy, tangibles, responsiveness, and assurance, are difficult to replicate across diverse service contexts (Buttle 1996). Furthermore, Brown, Churchill, and Peter (1993) found service quality to be unidimensional when applying the five-dimension SERVQUAL scale.

Brady and Cronin (2001) suggested that service quality comprises the dimensions of interpersonal quality, outcome quality, and environment quality. Semantic differences aside, these models suggest that service quality perceptions comprise four overarching dimensions, namely, interpersonal quality, technical quality, environment quality, and administrative quality. As well as providing a foundation for the development of our health service quality scale, the merging of these dimensions with SERVQUAL has most recently seen the SERVQUAL dimensions positioned as descriptors of these overarching dimensions.

Service quality perceptions have received the attractions of the researchers and considerable amount of research has been published in the area of service quality perceptions. However, the focus was developing generic service quality models (e.g., Brady and Cronin 2001; Parasuraman, Zeithaml, and Berry 1985). There is no specific study directly explored patient's assessment of health service quality. A significant number of study indications that service quality evaluations are likely to be context dependent (Babakus and Boller 1992; Carman 1990; Dabholkar, Thorpe, and Rentz 1996) and relatively few studies, in comparison, has focused on the development of context-specific service quality models.

## HEALTH SERVICE QUALITY

Turning attention to the health care literature, several conceptual frameworks for evaluating the quality of care are offered. Donabedian (1966, 1980, 1992) differentiated between two primary domains of managing health care quality, namely, technical and interpersonal processes. According to this framework, technical care refers to the application of medical science and technology to health care, while interpersonal care represents the management of the interaction that occurs between the service provider and consumer.

Within this conceptualization, a third element, the amenities of care, also contributes to health care quality. The amenities of care describe the intimate features of the environment in which care is provided. Brook and Williams (1975) put forward a conceptualization similar to that proposed by Donabedian (1966, 1980, 1992), in which technical care reflects how well diagnostic and therapeutic processes are applied and interactive care concerns the interactive behavior between the service provider and patient. Ware, Davies-Avery, and Stewart (1978) and Ware et al. (1983) also identified the interaction between a service provider and a patient, the technical quality of care, and the environment as important dimensions of patient satisfaction. These authors also provided support for the inclusion of a fourth dimension reflecting the administrative aspects of service provision. This dimension is similar to the enabling dimension proposed by McDougall and Levesque (1994). Finally, Wiggers et al. (1990) noted the importance of technical competence and interpersonal skills when assessing health care services.

More recently, Zineldin (2006) expanded these conceptualizations and found support for five quality dimensions: object or technical quality, quality processes or functional quality, quality infrastructure, quality interaction, and quality atmosphere. Similarly, Choi et al. (2005) forward a four-factor structure, including physician concern, staff concern, convenience of care process, and tangibles, which reflect aspects of technical, functional, environment, and administrative quality. Finally, Doran and Smith (2004) examined a model in which outcome is seen as a pivotal service quality dimension; empathy, assurance, responsiveness, and reliability as core aspects of quality; and tangibles or the physical aspects of the service as peripheral aspects. A comparison of the health care dimensions identified with those evident in the marketing literature indicates considerable overlap. That is, both literatures identify the importance of the technical, functional, environment, and administrative dimensions of the service experience.

## HIERARCHICAL MODEL OF HEALTH SERVICE QUALITY

Dagger, Sweeney and Johnson (2007) carried out a research to develop a multidimensional hierarchical scale for the measurement of service quality and examined whether the scale can appropriately predict major service outcomes – patient satisfaction and behavioral intentions. A qualitative study was conducted to collect data from oncology clinics and outdoor patients of a hospital. The research identified nine sub dimensions, namely interaction, relationship, outcome, expertise, atmosphere, tangibles, timeliness, operation, and support. These dimensions drive four primary dimensions: interpersonal, technical, environment, and administrative quality. These four primary dimensions are the drivers of the service quality perceptions of patients.

Through extensive literature review, the authors found that past researchers put much emphasis on developing a fundamental model of service quality (e.g., Brady & Cronin, 2001; Parasuraman, Zeithaml, & Berry, 1985). However, less attention has been given to develop situation centered framework to examine specific context, though service quality issues are fundamentally culture specific (Babakus & Boller, 1992; Carman 1990; Dabholkar, Thorpe, & Rentz 1996). Perceived service quality has been defined as "a consumer's judgment of, or impression about, an entity's overall excellence or superiority (Bitner & Hubbert, 1994; Boulding et al., 1993; Cronin & Taylor, 1992; Parasuraman, Zeithaml, & Berry, 1985, 1988).

To measure health service quality, the study began by investigating commonly cited primary dimensions of service quality in the marketing literature. Through this process, the study identified four primary dimensions that reflect service quality perceptions. The first of these dimensions, interpersonal quality, reflects the relationship developed and the dyadic interplay that occurs between a service provider and a user (Brady and Cronin 2001; Donabedian 1992; Grönroos 1984; Rust and Oliver 1994; Ware, Davies-Avery, and Stewart 1978). As services are produced, distributed, and consumed in the interaction between a service provider and a customer, the interpersonal process is crucial to the customer's ultimate perception of the service provider's performance. The second, technical quality, describes the outcome of the service process, or what a customer receives as a result of interacting with a service firm (Brady and Cronin 2001; Donabedian 1992; Grönroos 1984; Rust and Oliver 1994; Ware, Davies-Avery, and Stewart 1978). Technical quality reflects the expertise, professionalism, and competency of a service provider in delivering a service (Aharony and Strasser 1993; Zifko, Baliga and Krampf 1997). The third dimension is environment quality, which comprises a complex mix of environmental features (Baker 1986; Bitner 1992; Brady and Cronin 2001; Donabedian 1992). The final primary dimension we identified is administrative quality. Administrative service elements facilitate the production of the core service while adding value to a customer's use of a service (Grönroos 1990; Lovelock, Patterson, and Walker 2001; Ware, Davies-Avery, and Stewart 1978).

## Interpersonal Quality

Interpersonal quality reflects the relationship developed and the dyadic interplay between a service provider and a user (Brady and Cronin 2001; Grönroos 1984). Three core themes were found to constitute customers' perceptions of interpersonal quality; these were termed manner, communication, and relationship. The first, manner, describes the attitude and behavior of a service provider in the service setting (Bitner, Booms, and Tetreault 1990; Brady and Cronin, 2001).

## Technical Quality

Technical quality involves the outcomes achieved (Grönroos 1984; McDougall and Levesque 1994) and the technical competence of a service provider (Ware, Davies-Avery, and Stewart 1978). Two core themes underpinned customers' perceptions of technical quality: expertise and outcome. We believe that these themes are salient indicators of technical quality in the context of our study, in which service provision was both complex and ongoing. That is, customers evaluated technical quality on the basis of service provider expertise and the outcomes achieved over multiple service encounters. The first theme, expertise, reflects a provider's competence, knowledge, qualifications, or skill (Aharony and Strasser 1993). Expertise reflects the ability of a service provider to adhere to high standards of service provision (Zifko-Baliga and Krampf 1997).

## Environment Quality

The environment defines the complex mix of environmental features that shape consumer service perceptions (Gotlieb, Grewal, and Brown 1994). Atmosphere and tangibles were the key themes underlying customers' perceptions of environment quality. The first theme of atmosphere refers to the intangible, background characteristics of the service environment (Baker 1986; Bitner 1992). These elements generally exist below consumers' level of awareness, thus affecting the pleasantness of the surroundings (Kotler 1974).

## Administrative Quality

Administrative service elements facilitate the production of a core service while adding value to a customer's use of the service (Grönroos 1990; McDougall and Levesque 1994). Facilitating services are essential to the delivery and consumption of a core service, while supporting elements augment the service but are not necessary to core service delivery (Grönroos 1990; Lovelock, Patterson, and Walker 2001). Three themes comprised customers' perceptions of administrative quality: timeliness, operation, and support. The first, timeliness, refers to the factors involved in arranging to receive medical services, such as appointment waiting lists, waiting time, the ease of changing appointments, and hours of operation (Thomas, Glynne-Jones, and Chaiti 1997).

The reliable and valid scales of hierarchical model to measure service quality perceptions from the patients perspective examines the salience of health service quality in predicting important health service outcomes, namely, customer/patents satisfaction and behavioral intentions. The outcomes were selcted on the basis of the weight of research suggesting their importance as outcomes of service quality (e.g.,

Bitner and Hubbert 1994; Brady and Robertson 2001; Cronin, Brady, and Hult 2000; Cronin and Taylor 1992; Gotlieb, Grewal, and Brown 1994; Mohr and Bitner 1995).

## METHODOLOGY

Present study was an exploratory, observational, individual based cross-sectional study. Based on purposive sampling, the individuals have been selected from household level who has taken healthcare services from public or private hospitals of Dhaka city in last one year. The study collected data among 410 respondents. One to one interviews of each respondent were conducted with a structured questionnaire. The patients were asked to participate in a survey anonymously, and the study designed to explore patients' opinions about the services provided by the health service providers.

## RESULTS

A descriptive analysis has been conducted to explain the data. The below table represents, majority of the respondents i.e. 63.3% of the respondents were male over 36.4% female. In case of professional distribution, it is observed 22.9% of the respondents were private service holder, 18.6% were businessman and 13.3% were government service holders. It is observed that 50.8% of respondents were received services from private hospitals and 49.2% were received services from public hospitals. Among the respondents, 73% of the respondents taken the services from indoor, 18.4% were from outdoor and 8.7% were from emergency.

Present study identified four primary dimensions that reflect service quality perceptions and these four dimensions have been described according to patient's satisfaction level.

Here, we used logit link function for the flexibility of data interpretation. All analysis was carried out using STATA 14.0. Pearson's Chi-square test was performed and as our response variable is binary in nature, logistic regression is used. Based on the data collected through the survey, association between exposure and outcome was drawn through developing logistic regression model, which was also effective to control potential confounders. However, we considered a cut of value of 50%, which indicates that less than 50% denotes non-satisfactory hospital service while greater than 50% implies satisfactory service.

In the above exploratory analysis, we considered 10% significance level. In Pearsons's chi-squared test, we found 4 variables, which were significantly ($p < 0.1$) associated with hospital service satisfactory level. Here, we found significant association between income and hospital service satisfaction. It was evident from the data that, when there was an increment in income, hospital service satisfaction was growing. However, hospital service satisfaction was higher among doctors (100%) followed by lawyer and accountant (75%), businessmen (71.88%) and private service holder (65.71%). On the other hand, 52.38% government service holders expressed dissatisfaction towards hospital services. Satisfactory service had been given by private hospitals (86.39%) while 65.85% dissatisfaction had been expressed from those who received services from public hospitals.

After fitting logistic regression model with the variables that were significant in chi-squared test at 10% significance level, we found 1 variable that was significantly associated with satisfactory level of hospital service. Here, we considered 5% significance level. Here, respondents are 90% less satisfied with service provided by public hospitals than that of private hospitals.

*Table 1. Respondent's profile*

| | |
|---|---|
| **Gender** | |
| Male | 63.3 |
| Female | 36.4 |
| **Profession** | |
| Government service holder | 13.3 |
| Private service holder | 22.9 |
| Doctor | 7.8 |
| Businessman | 18.6 |
| Lawer/Accountant | 3.5 |
| Housewives | 3.8 |
| Others | 20.1 |
| **Types of hospital he/she took service** | |
| Private | 50.8 |
| Public | 49.2 |
| **Where he/she took services** | |
| Indoor | 73.0 |
| Outdoor | 18.4 |
| Emergency | 8.7 |

*Table 2. Descriptive statistics of the variable perceived service quality*

| | | Not Satisfactory | | Satisfactory | |
|---|---|---|---|---|---|
| | | Private | Public | Private | Public |
| 1 | The overall quality of the service provided by the hospital is excellent | 50.6% | 49.4% | 0% | 0% |
| 2 | The quality of the service provided at the hospital is impressive | 25% | 75% | 71.5% | 28.5% |
| 3 | The service provided by the hospital is of a high standard | 23.6% | 76.4% | 76.2% | 23.8% |
| 4 | I believe the hospital offers service that is superior in every way | 29.2% | 70.8% | 74.6% | 24.9% |

*Table 3. Descriptive statistics of the variable service satisfaction*

|   |   | Not Satisfactory | | Satisfactory | |
|---|---|---|---|---|---|
|   |   | Private | Public | Private | Public |
| 1 | My feelings towards the hospital are very positive | 50.5% | 49.5% | 0% | 0% |
| 2 | I feel good about coming to this hospital for my treatment | 50.4% | 49.6% | 0% | 0% |
| 3 | Overall I am satisfied with the hospital and the service it provides | 28.2% | 71.8% | 65.8% | 33.8% |
| 4 | I feel satisfied that the results of my treatment are the best that can be achieved | 28.4% | 71.6% | 68.5% | 31.5% |
| 5 | The extent to which my treatment has produced the best possible outcome is satisfying | 30.1% | 69.9% | 68.5% | 31% |

*Table 4. Descriptive statistics of the variable behavioral intentions*

|   |   | Not Satisfactory | | Satisfactory | |
|---|---|---|---|---|---|
|   |   | Private | Public | Private | Public |
| 1 | If I had to start treatment again I would want to come to this hospital | 26.2% | 73.8% | 72.7% | 27.3% |
| 2 | I would highly recommend the hospital to other patients | 26.7% | 73.8% | 71.8% | 27.7% |
| 3 | I have said positive things about the hospital to my family and friends | 23.5% | 76.5% | 70.4% | 29.2% |
| 4 | I intend to continue having treatment, or any follow-up care I need, at this hospital | 26.2% | 73.8% | 67.4% | 31.9% |
| 5 | I have no desire to change hospitals | 37.0% | 63.0% | 69.7% | 29.6% |
| 6 | I intend to follow the medical advice given to me at the hospital | 25.9% | 74.1% | 60.4% | 39.2% |
| 7 | I am glad I have my treatment at this hospital rather than somewhere else | 27.5% | 72.5% | 71.2% | 28.3% |

## DISCUSSION

The results illustrate patient's level of satisfaction with the public and private health care services they received from different hospitals Dhaka city. Among the male respondents 60.27% and among the female respondents 60.27% of them were satisfied with the services of healthcare services of Bangladesh and dissatisfaction was high among public hospitals comparing with private hospitals. The data represents, government service holders are less satisfied with the health care services. However, doctors were satisfied with the health care services. It is also observed that low-income group people are less satisfied with the services comparing with the high income group people. In addition, the found 71% of the emergency students are satisfied with the health care services of Bangladesh.

*Table 5. Descriptive statistics of the variable interpersonal quality*

|   |   | Not Satisfactory | | Satisfactory | |
|---|---|---|---|---|---|
|   |   | Private | Public | Private | Public |
| 1 | The interaction I have with the staff at the hospital is of a high standard | 27.0 | 73.0 | 77.7 | 21.7 |
| 2 | The interaction I have with the staff at the hospital is excellent | 27.1 | 72.9 | 73.6 | 25.9 |
| 3 | I feel good about the interaction I have with the staff at the hospital | 26.5 | 73.5 | 73.3 | 26.2 |

*Table 6. Descriptive statistics of the variable technical quality*

|   |   | Not Satisfactory | | Satisfactory | |
|---|---|---|---|---|---|
|   |   | Private | Public | Private | Public |
| 1 | The quality of the care I receive at the hospital is excellent | 22.4 | 77.6 | 72.0 | 27.6 |
| 2 | The care provided by the hospital is of a high standard | 24.2 | 75.8 | 72.4 | 27.2 |
| 3 | I am impressed by the care provided at the hospital | 23.7 | 76.3 | 70.8 | 28.8 |

*Table 7. Descriptive statistics of the variable environment quality*

|   |   | Not Satisfactory | | Satisfactory | |
|---|---|---|---|---|---|
|   |   | Private | Public | Private | Public |
| 1 | I believe the physical environment at the hospital is excellent | 20.1 | 79.9 | 76.5 | 23.0 |
| 2 | I am impressed with the quality of the hospital's physical environment | 18.5 | 81.5 | 76.2 | 23.3 |
| 3 | The physical environment at the hospital is of a high standard | 22.5 | 77.5 | 79.7 | 19.8 |

*Table 8. Descriptive statistics of the variable administrative quality*

|   |   | Not Satisfactory | | Satisfactory | |
|---|---|---|---|---|---|
|   |   | Private | Public | Private | Public |
| 1 | The administration system at the hospital is excellent | 22.7 | 77.3 | 73.7 | 25.8 |
| 2 | The administration at the hospital is of a high standard | 22.8 | 77.2 | 76.6 | 22.9 |
| 3 | I have confidence in the hospital's administration system | 26.3 | 73.7 | 72.4 | 27.1 |

*Table 9. Associated factors with hospital service satisfaction*

| Variables | Categories | Hospital service - Not Satisfactory (< 50%) | Hospital service - Satisfactory (≥ 50%) | P-value (< 0.3) |
|---|---|---|---|---|
| Age | 11-20 | 46.88 | 53.13 | 0.273 |
|  | 21-30 | 38.66 | 61.34 |  |
|  | 31-40 | 32.26 | 67.74 |  |
|  | 41-50 | 42.86 | 57.14 |  |
|  | 51-60 | 39.29 | 60.71 |  |
|  | 61-70 | 33.33 | 66.67 |  |
|  | 71-80 | 100.00 | 0.00 |  |
|  | 81-90 | 100.00 | 0.00 |  |
| Gender | Male | 39.73 | 60.27 | 0.889 |
|  | Female | 38.94 | 61.06 |  |
| Income (BDT) | <20,000 | 50.00 | 50.00 | 0.001 |
|  | 20,000-40,000 | 28.26 | 49.12 |  |
|  | 40,000-60,000 | 28.26 | 71.74 |  |
|  | 60,000-80,000 | 30.77 | 69.23 |  |
|  | 80,000-120,000 | 23.53 | 76.47 |  |
|  | >120,000 | 17.39 | 82.61 |  |
| Profession | Government service holder | 52.38 | 47.62 | 0.027 |
|  | Private service holder | 34.29 | 65.71 |  |
|  | Doctor | 0.00 | 100.00 |  |
|  | Businessmen | 28.13 | 71.88 |  |
|  | Lawyer/ Accountant | 25.00 | 75.00 |  |
|  | Housewives | 42.55 | 57.45 |  |
|  | Others | 48.53 | 51.47 |  |
| Types of hospital he/she took service from | Private | 13.61 | 86.39 | 0.001 |
|  | Public | 65.85 | 34.15 |  |
| Service taken | Indoor | 41.18 | 58.82 | 0.416 |
|  | Outdoor | 38.10 | 61.90 |  |
|  | Emergency | 29.03 | 70.97 |  |
| Duration of hospital stay | 1 week | 39.09 | 60.91 | 0.110 |
|  | 2 weeks | 53.33 | 46.67 |  |
|  | 3 weeks | 25.00 | 75.00 |  |
|  | >3 weeks | 40.00 | 60.00 |  |

*Table 10. Logistic regression model*

| Variables | Categories | Adjusted Odds Ratio | P-value <0.05 | 95% Conf. Interval |
|---|---|---|---|---|
| Income (BDT) | <20,000 | 1 | | |
| | 20,000-40,000 | 1.091645 | 0.836 | 0.4761613-2.5027 |
| | 40,000-60,000 | 1.319662 | 0.585 | 0.487249-3.574165 |
| | 60,000-80,000 | 0.6194398 | 0.443 | 0.182144-2.106606 |
| | 80,000-120,000 | 1.638188 | 0.442 | 0.4650054-5.771241 |
| | >120,000 | 1.701833 | 0.506 | 0.3555806-8.145087 |
| Profession | None | 1 | | |
| | Government service holder | 0.9267658 | 0.910 | 0.2461408-3.489446 |
| | Private service holder | 1.769719 | 0.366 | 0.5129972-6.105109 |
| | Doctor | 1 | | |
| | Businessmen | 2.194328 | 0.234 | 0.6007306-8.015369 |
| | Lawyer/Accountant | 1.302979 | 0.811 | 0.1495835-11.34987 |
| | Housewives | 1.007401 | 0.991 | 0.2860697-3.547586 |
| | Others | 1.248764 | 0.701 | 0.4017663-3.881387 |
| Type of hospitals | Private | 1 | | |
| | Public | 0.0949227 | 0.001 | 0.049489-0.1820669 |

The evidence further reveals that most of the participants had experienced facing deteriorated health condition at the public hospitals. Among the respondents, 76.2% of them were satisfied with the service quality of private hospitals and on the other hand, 76.4% of the respondents are not satisfied with the service quality of the public hospitals. Regarding treatment service satisfaction, the public hospitals services were less satisfactory comparing private hospitals. Despite some positive experiences, majority participants opine that they are less satisfied with public services specially the ways the services are operated and delivered to them. The situation of dissatisfaction arises due to many factors, which include administrative, interpersonal, environmental and technical service quality. The common expectations of patients are to get better quality of care, supportive staff members, and standard physical environment

*Figure 1. Bar diagram represents the distribution of health service quality satisfaction according to age*

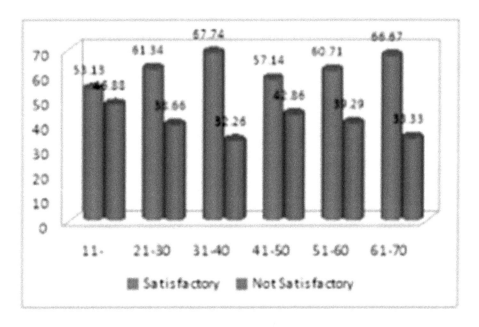

*Figure 2. Bar diagram represents the distribution of health service quality satisfaction according to gender*

*Figure 3. Bar diagram represents the distribution of health service quality satisfaction according to income*

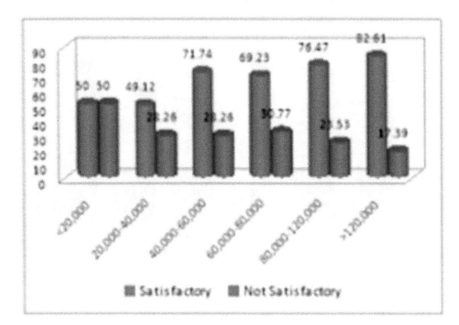

*Figure 4. Bar diagram represents the distribution of health service quality satisfaction according to types of hospitals*

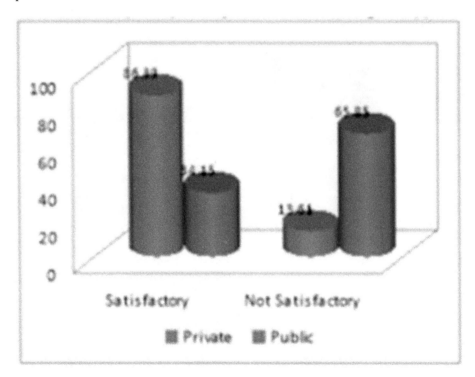

*Figure 5. Bar diagram represents the distribution of health service quality satisfaction according to profession*

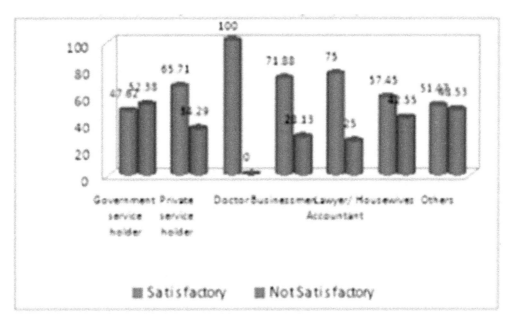

*Figure 6. Bar diagram represents the distribution of health service quality satisfaction according to types of service*

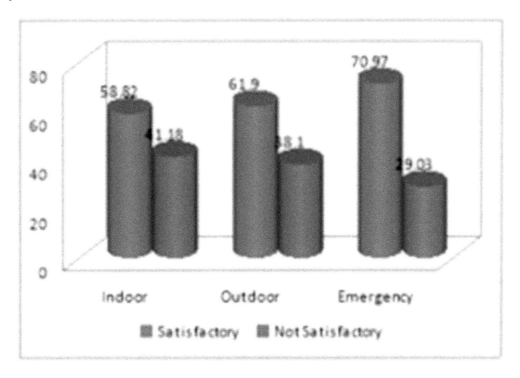

of hospitals. This study is consistent with Hussain, (2018) in demonstrating how the majority of Bangladeshi patients expect positive attitude and behavior of health service providers, particularly respect and politeness, privacy, short waiting time etc. in using public health services. The respondents provided much importance on those issues like technical and environmental issue including physical examination and explain the nature of the problem etc. Present study points out that care that meets all medical needs may fail to meet client's emotional or social needs due to ignoring the cultural background of patients and suggest for more culture specific in-depth studies.

## CONCLUSION

This study confirms existing evidence that Bangladeshi patients' have a growing concern with lower level of satisfaction in public health care services. The respondents faced multi-dimensional problems, characterized by low level of overall service quality, low level of interpersonal service quality, low level of technical or treatment related quality at public hospitals, low level of confidence among the administrative services. The problems are compounded by differential expectations and experiences of service users, which reveal that patients demand quality health care services. As the literature represents that there are a serious shortage of doctors, paramedics, nurses and midwives. The average doctor patients' ratio in Bangladesh is 4000: 1. There are 43 thousands registered doctor against 21 thousand nurses. Nurse patient ratio is 1: 15 in Bangladesh as against international standard for 1: 4. It is very difficult to provide quality health care services for the vast majority patients with limited number of health care professionals. Health is a basic human right. Therefore, the policy makers and service providers should be more responsive to initiate effective policy measures and program to improve the quality of health care services.

## REFERENCES

Aharony, L., & Strasser, S. (1993). Patient satisfaction: What we know about and what we still need to explore. *Medical Care Review*, *50*(1), 49–79. doi:10.1177/002570879305000104 PMID:10125117

Babakus, E., & Boller, G. W. (1992). An empirical assessment of the SERVQUAL scale. *Journal of Business Research*, *24*(3), 253–268. doi:10.1016/0148-2963(92)90022-4

Baker, J. (1986). The Role of the Environment in Marketing Services: The Consumer Perspective. In *The Service Challenge: Integrating for Competitive Advantage* (pp. 79–84). Chicago: American Marketing Association.

*Bangladesh Demographic and Health Survey Report*. (2014). Dhaka: Ministry of Planning, Planning Commission.

Bitner, M. J. (1992). Servicescapes: The impact of physical surroundings on customers and employees. *Journal of Marketing*, *56*(2), 57–71. doi:10.1177/002224299205600205

Bitner, M. J., Booms, B. H., & Tetreault, M. S. (1990). The service encounter: Diagnosing favorable and unfavorable incidents. *Journal of Marketing*, *54*(1), 71–84. doi:10.1177/002224299005400105

Bitner, M. J., & Hubbert, A. R. (1994). Encounter satisfaction versus overall satisfaction versus quality. *Service quality: New directions in theory and practice, 34*(2), 72-94.

Bitner, M. J., Hubbert, A. R., Rust, R., & Oliver, R. (1994). Service quality: New directions in theory and practice. *Encounter satisfaction versus overall satisfaction versus quality, 77*.

Boulding, W., Kalra, A., Staelin, R., & Zeithaml, V. A. (1993). A Dynamic Process Model of Service Quality: From Expectations to Behavioral Intentions. *JMR, Journal of Marketing Research, 30*(February), 7–27. doi:10.1177/002224379303000102

Brady, M. K., & Cronin, J. J. Jr. (2001). Some new thoughts on conceptualizing perceived service quality: A hierarchical approach. *Journal of Marketing, 65*(3), 34–49. doi:10.1509/jmkg.65.3.34.18334

Brady, M. K., & Robertson, C. J. (2001). Searching for a consensus on the antecedent role of service quality and satisfaction: An exploratory cross-national study. *Journal of Business Research, 51*(1), 53–60. doi:10.1016/S0148-2963(99)00041-7

Brook, R. H., & Williams, K. N. (1975). Quality of health care for the disadvantaged. *Journal of Community Health, 1*(2), 132–156. doi:10.1007/BF01319207 PMID:777052

Brown, T. J., Churchill, G. A. Jr, & Peter, J. P. (1993). Research note: Improving the measurement of service quality. *Journal of Retailing, 69*(1), 127–139. doi:10.1016/S0022-4359(05)80006-5

Buttle, F. (1996). SERVQUAL: Review, critique, research agenda. *European Journal of Marketing, 30*(1), 8–32. doi:10.1108/03090569610105762

Carman, J. M. (1990). Consumer perceptions of service quality: An assessment of T. *Journal of Retailing, 66*(1), 33.

Choi, K.-S., Lee, H., Kim, C., & Lee, S. (2005). The Service Quality Dimensions and Patient Satisfaction Relationships in South Korea: Comparisons across Gender, Age and Types of Service. *Journal of Services Marketing, 19*(3), 140–150. doi:10.1108/08876040510596812

Cronin, J. J. Jr, Brady, M. K., & Hult, G. T. M. (2000). Assessing the effects of quality, value, and customer satisfaction on consumer behavioral intentions in service environments. *Journal of Retailing, 76*(2), 193–218. doi:10.1016/S0022-4359(00)00028-2

Cronin, J. J. Jr, & Taylor, S. A. (1992). Measuring service quality: A reexamination and extension. *Journal of Marketing, 56*(3), 55–68. doi:10.1177/002224299205600304

Dabholkar, P. A., Shepherd, C. D., & Thorpe, D. I. (2000). A comprehensive framework for service quality: An investigation of critical conceptual and measurement issues through a longitudinal study. *Journal of Retailing, 76*(2), 139–173. doi:10.1016/S0022-4359(00)00029-4

Dabholkar, P. A., Thorpe, D. I., & Rentz, J. O. (1996). A measure of service quality for retail stores: Scale development and validation. *Journal of the Academy of Marketing Science, 24*(1), 3–16. doi:10.1007/BF02893933

Dagger, T. S., Sweeney, J. C., & Johnson, L. W. (2007). A hierarchical model of health service quality: Scale development and investigation of an integrated model. *Journal of Service Research*, *10*(2), 123–142. doi:10.1177/1094670507309594

Directorate General Health Service. (2018). *Health Bulletin*. Ministry of Health and Family Welfare.

Donabedian, A. (1966). Evaluating the quality of medical care. *The Milbank Memorial Fund Quarterly*, *44*(3), 166–206. doi:10.2307/3348969 PMID:5338568

Donabedian, A. (1980). A guide to medical care administration: medical care appraisal: quality and utilization. In *A guide to medical care administration: medical care appraisal: quality and utilization*. The American Public Health Association.

Donabedian, A. (1992). The role of outcomes in quality assessment and assurance. *QRB. Quality Review Bulletin*, *18*(11), 356–360. doi:10.1016/S0097-5990(16)30560-7 PMID:1465293

Doran, D., & Smith, P. (2004). Measuring service quality provision within an eating disorders context. *International Journal of Health Care Quality Assurance*, *17*(7), 377–388. doi:10.1108/09526860410563186 PMID:15552393

Gotlieb, J. B., Grewal, D., & Brown, S. W. (1994). Consumer satisfaction and perceived quality: Complementary or divergent constructs? *The Journal of Applied Psychology*, *79*(6), 875–885. doi:10.1037/0021-9010.79.6.875

Grönroos, C. (1984). *A service quality model and its marketing implications*. Academic Press.

Grönroos, C. (1990). *Service management and marketing: Managing the moments of truth in service competition*. Jossey-Bass.

International Monetary Fund. (2013). *Bangladesh: Poverty Reduction Strategy Paper*. IMF, Country Report N0. 13/ 63.

Koerner, M. M. (2000). The Conceptual Domain of Service Quality for Inpatient Nursing Services. *Journal of Business Research*, *48*(3), 267–283. doi:10.1016/S0148-2963(98)00092-7

Kotler, P. (1974). Atmospherics as a Marketing Tool. *Journal of Retailing*, *49*(4), 48–64.

Lovelock, C. H., Patterson, P. G., & Walker, R. (2001). *Services Marketing*. Frenchs Forest.

McDougall, G. H., & Levesque, T. J. (1994). A revised view of service quality dimensions: An empirical investigation. *Journal of Professional Services Marketing*, *11*(1), 189–209. doi:10.1080/15332969.1994.9985149

Meterko, M., Nelson, E. C., Rubin, H. R., Batalden, P., Berwick, D. M., Hays, R. D., & Ware, J. E. (1990). Patient judgments of hospital quality: Report of a pilot study. *Medical Care*, *28*(9), S1–S56. PMID:2214895

Mohr, L. A., & Bitner, M. J. (1995). The Role of Employee Effort in Satisfaction with Service Transactions. *Journal of Business Research*, *32*(2), 239–252. doi:10.1016/0148-2963(94)00049-K

Parasuraman, A., Zeithaml, V. A., & Berry, L. L. (1985). A conceptual model of service quality and its implications for future research. *Journal of Marketing, 49*(4), 41–50. doi:10.1177/002224298504900403

Parasuraman, A., Zeithaml, V. A., & Berry, L. L. (1988). Servqual: A multiple-item scale for measuring consumer perc. *Journal of Retailing, 64*(1), 12.

Stewart, A. L., Ware Jr, J. E., Brook, R. H., & Davies-Avery, A. (1978). *Conceptualization and Measurement of Health for Adults in the Health Insurance Study: Vol. II, Physical Health in Terms of Functioning.* The Rand Corporation. R-1987/2-HEW.

Taylor, S. A., & Baker, T. L. (1994). An assessment of the relationship between service quality and customer satisfaction in the formation of consumers' purchase intentions. *Journal of Retailing, 70*(2), 163–178. doi:10.1016/0022-4359(94)90013-2

Thomas, S., Glynne-Jones, R., & Chaiti, I. (1997). Is It Worth the Wait? A Survey of Patients' Satisfaction with an Oncology Outpatient Clinic. *European Journal of Cancer Care, 6*(1), 50–58. doi:10.1111/j.1365-2354.1997.tb00269.x PMID:9238930

Ware, J. E. Jr, Snyder, M. K., Wright, W. R., & Davies, A. R. (1983). Defining and measuring patient satisfaction with medical care. *Evaluation and Program Planning, 6*(3-4), 247–263. doi:10.1016/0149-7189(83)90005-8 PMID:10267253

Wensing, M., Grol, R., & Smits, A. (1994). Quality judgements by patients on general practice care: A literature analysis. *Social Science & Medicine, 38*(1), 45–53. doi:10.1016/0277-9536(94)90298-4 PMID:8146714

Wiggers, J. H., Donovan, K. O., Redman, S., & Sanson-Fisher, R. W. (1990). Cancer patient satisfaction with care. *Cancer, 66*(3), 610–616. doi:10.1002/1097-0142(19900801)66:3<610::AID-CNCR2820660335>3.0.CO;2-T PMID:2364373

World Health Organization. (2015). *Bangladesh health system review*. Manila: WHO Regional Office for the Western Pacific.

Zifko-Baliga, G. M., & Krampf, R. F. (1997). Managing perceptions of hospital quality. *Marketing Health Services, 17*(1), 28. PMID:10169030

Zineldin, M. (2006). The quality of health care and patient satisfaction. An Exploratory Investigation of the 5Q Model at Some Egyptian and Jordanian Medical Clinics. *International Journal of Health Care Quality Assurance, 19*(1), 60–92. doi:10.1108/09526860610642609 PMID:16548402

# Chapter 15
# Application of Nanotechnology in Global Issues

**Ranjit Barua**
*OmDayal Group of Institutions, India*

**Sudipto Datta**
*Indian Institute of Engineering Science and Technology, Shibpur, India*

**Jonali Das**
*Raja Peary Mohan College, Calcutta University, India*

## ABSTRACT

*Nanotechnology basically means any kind of technology in a nanoscale, which can be applied in the existent world. It is a comparatively new research field, but it is not a completely new area and the research draws insights from many other research areas. It is generally considered that nanotechnology makes possible the coming of the new Industrial Renaissance since it has the potential for a reflective impact on modern society and economy in the early 21$^{st}$ century, similar to that of information technology (IT), electronics technology, especially in semiconductor technology or molecular and cellular biology. The purpose of this chapter is to look into the present aspects of nanotechnology. In this chapter, the authors discuss a variety of applications of nanotechnology in recent decades like modern engineering, robotics, food technology, medicine, etc., and also they indicate the current and potential uses of nanoscience and nanotechnologies. Social and ethical impacts as well as health and environmental impacts will be highlighted.*

## INTRODUCTION

Nanotechnology is already serving significant role in fueling progress and transformation in numerous fields of technology, natural sciences and engineering divisions such as energy, information and communication technology, medicine, environmental technology, home security, transportation, and food safety, with lots of others. Existing nanotechnology has greatly transformed physics, chemistry,

DOI: 10.4018/978-1-7998-3576-9.ch015

## Application of Nanotechnology in Global Issues

biotechnology and material science and as a result we already know how to make innovative materials that have distinctive characteristics as their arrangements are resolute on the nanometer scale. Science and technology continue to progress forward in creating the fabrication of micro and nano-devices and systems that have major significance for a variety of consumer, biomedical and industrial applications (Morrison et al., 2008). Nanotechnology works at the first level of arrangement of atoms or molecules for both anthropogenic and other living systems. This is where the characteristics and purposes of all systems are described. Nanotechnology assures the capability to make specific machine and apparatus of molecular dimension. In its actual sense, "Nanotechnology" refers to the estimated capability to build things from the bottom up, applying tools and techniques, which are being developed to create high performance produces. This theoretical potential was first clearly imagined as near to our times as 1959 by physicist Richard Feynman. According to NNI and NSF (National Science Foundation), Nano-technology is the capability to recognize, manage and manipulate substance at the point of individual molecules and atoms. Nanotechnology is often suggested to as a common purpose technique because in its modern version it will have major impact on almost all industries and also all over areas of civilization (Mansoori GA et al., 2005). Science and technology are the main elements of worldwide technological contest. Combining science based on the combining characteristics of nature at the nanoscale offers a new base for integration of technology, innovation, and knowledge. There is a longitudinal progression of divergence and convergence in main regions of science and technology (Iverson NM et al., 2013). Such as at macro scale, the convergence of sciences was suggested during the recovery and it was pursued by constricted disciplinary specialization in science and technology in the 18th-20th centuries. The convergence at the nanoscale achieved its potency in about year 2000, and one may approximate a divergence of the nano-formation of architectures in the coming decades. Figure 1 shows application of nanotechnology in various fields.

*Figure 1. Nanotechnology in various fields*

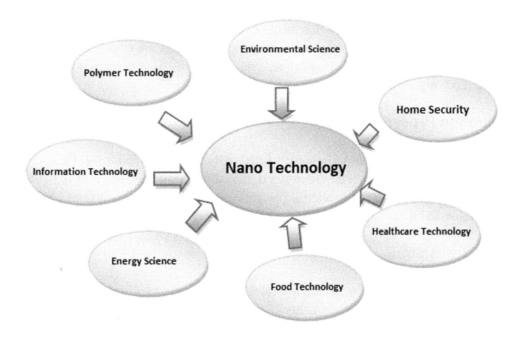

## FUNCTION OF NANOTECHNOLOGY IN VARIOUS FIELDS

Nanotechnology is doing to notably progress, numerous technology, constant transform and different engineering divisions like medicine, energy, environmental science, transportation, and food safety, information technology with lots of others field. Recently nanotechnology connect with existing advancement in biotechnology, chemistry, physics and materials science to create modern materials, which have unique properties as their arrangements are determined on the nanometer scale (Huang et al., 2009, Morrison et al., 2008). Engineers and researchers have been associating the particulars of functioning with nanoscale elements. At the present they have an outstanding idea of how to create a nanoscale materials by means of characteristics on no account imagined before. Manufactured goods applying nanoscale materials and methods are now accessible. Nanoscale silver also applies in anti bacterial wound healing treatment. A gas can be neutralized by nanoscale dry powder. Batteries for tools are being done with nanoscale materials in order to perform more quickly with less heat and more power (Mansoori GA et al., 2005). To prevent sunburns from ultraviolet sunlight, a sunscreens containing with nanoscale Zinc Oxide or Titanium Dioxide is used, which can reflect the UV light. Next session a variety of nanoscale based techniques and products based are explained in detail (Wesam Al-Mufti M et al., 2015).

### Nano Films

Various types nanoscale material can be applied in thin films to build the films as anti-reflective, ultraviolet resistant, water repellent, infrared-resistant, self-cleaning, anti-fog, anti-microbial, good conductivity or Scratch resistant (Mansoori GA et al., 2005). Nano films are applied now on computer display, eyeglasses and cameras to guard or care for the surfaces.

### Application of Drug Delivery Systems

Dendrimers are a one kind of nanostructure, which can be accurately considered and manufactured for a wide range of functions, including the treatment of complex diseases like cancer. This type of nanostructure can carry different materials and also their branches can perform numerous works at one time, for instance identifying affected cells, drug delivery, diagnosing diseased states, reporting outcomes of treatment and reporting location (Yi et al., 2005).

### Water Filtration Procedure

Scientists are researching with nanoscale sensors for identifying the contaminants in water system and also carbon nanotubes based membranes for water desalination. At present, this type of nano-materials which have huge prospective to purify and filter water contain with titanium dioxide, which is applied in sunscreen lotion and it has a great effect to neutralize bacteria.

### Nanoscale Transistors

Transistors are basically electronic switching apparatus where a small quantity of electricity is applied like a gateway to manage the flow of larger quantity of electricity. In any electronics device like computers, the more transistors are used for the greater power. Recently, the sizes of transistors have been decreasing,

*Application of Nanotechnology in Global Issues*

so computer or CPU has become more powerful. Most recently, the market's best commercial technology manufactured computer chips with transistors having 45 nanometer features. Latest announcements specify that 32 nanometer feature technology will be coming soon.

## Application of Carbon Nanotubes

Carbon nanotubes (CNTs) are basically allotropes of carbon with a long thin cylindrical nanostructure. This was discovered by Sumio Iijima in the year 1991. Carbon nanotubes are great macromolecules, which are exclusive for their geometrical dimensions like shape, size and amazing substantial characteristics. Basically it can be considered as a graphite sheet, which appears as a hexagonal lattice of carbon rolled into a cylinder. These interesting geometrical structures have sparked much stimulation in current years and a huge amount of research has been directed to their recognizing. This tube has been constructed with length-to-diameter ratio of up to 28,000,000:1, which is extensively larger than any other material. Cylindrical carbon molecules have a great characteristic, which make them prospectively useful in many applications in electronics, nanotechnology and other fields of materials science, in addition to probable applies in architectural areas (Dimitrios et al., 2014; Wesam Al-Mufti M et al., 2015). They show unexpected strength and exclusive electrical properties, and are proficient conductors of heat. Their final practice, though, may be restricted by their probable toxicity. Figure 2 shows the carbon nano tubes structure.

*Figure 2. Carbon nanotube structure*

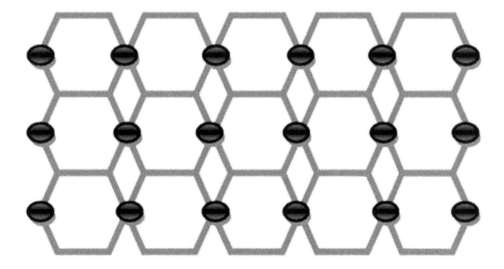

## APPLICATIONS OF NANOTECHNOLOGY

The application of nanotechnology are being expanded that could drive the worldwide market for mineral, agricultural and further non-fuel commodities. At present, depending on the strong inter disciplinary nature of nanotechnology there are numerous research areas and various prospective applications that engages nanotechnology (Yi et al., 2005). Here we present a concise summary about various nano science and nanotechnology in progress expansions.

### Nanorobots in Biomedical Area

The nanorobots containing nano biosensors and nano-actuators can provide a new prospect for supplying innovative medical equipment especially for the medical staffs and doctors. Microscopic mechanical environment is totally different from the conventional techniques methods which are using in the feed forward control of event-based are required to successfully spread new the medical technologies (Adam et al., 2013). In 1980s microelectronics were invented and utilized in various fields, in medical field microelectronics led to the manufacturing of biomedical instrumentation. In the similar way nano-electronics will also further minimize the medical systems giving more efficiency for pathological forecast. The micro devices used for medical device and surgery is authentic which came with many improvements in clinical practices in the current years. Like example catheterization is an important procedure among other biomedical instrumentation for intracranial and heart surgery. Currently, researchers are working hard to convert the manufacturing technology from microelectronica to nanoelectronics for miniaturization of medical instruments (Bae et al., 2011). Applicable biomedical sensors are progressing by pervasive medicine and tele-operated surgery by this same technology the development of biomolecular actuators are also made. For molecular machines a nanotechnology prototypes are being investigated as well as some exciting instruments for sensing and propulsion are also being presented. Also for medical practices more complicated nanorobots or molecular machines having inbuilt nano-scopic specification signify new gears for medical practices.

### Nanorobot Development for Defense

In defense area integration of nano-biotechnology systems will highly be benefit from the recent trends and achievements. Because of such technology trends the scientific international community which included pharmaceutical and medical sectors showed more interest in the development of molecular machines.

### Application in Spacecraft

Using carbon nanotubes to make the cable needed for the space elevator, a system which could significantly reduce the cost of sending material into orbit. To reduce the weight of spaceships, the parts are made by carbon and it is also even increase the structural strength. Producing thrusters for spacecraft that use MEMS devices to accelerate nanoparticles. This should reduce the weight and complexity of thruster systems used for interplanetary missions. Carbon nanotubes are also use to construct the lightweight solar sails where the pressure of light from the sun reflecting on the mirror-like solar cell to propel a spacecraft (Silva, J. et al., 2014). The main benefit of this solves the problem of having to lift enough fuel into orbit to power spacecraft for the period of interplanetary missions. Nanosensors network are also apply to explore a large areas of other planets such as Mars for traces of water. Improvement in nanomaterials makes less weight of solar sails and a cable wire for the space elevator feasible (Mukherjee, B. et al., 2014). Additionally, new materials combined with nanorobots and nanosensors could develop the performance of spacesuits, spaceships, and the apparatus applied to discover moons or other planets, making nanotechnology a significant element of the "final frontier".

## Nanotechnology in Electronics: Nanoelectronics

How can nanoelectronics improve the capabilities of electronic components? Nanoelectronics holds some answers for how we might increase the capabilities of electronics devices while we reduce their weight and power consumption. Some of the nanoelectronics areas under development include: a. Progressing display screens on electronics appliances. This engages reducing power consumption while decreasing the weight and thickness of the screens. b. Increasing the density of memory chips. Researchers are expanding a type of memory chip with an anticipated density of 1TB of memory per square inch or greater. c. Minimizing the dimension of transistors applied in integrated circuits. It is also believed that it may be able to "put the power of all of today's present computers in the palm of your hand".

## Nanotechnology in Medicine

Applications of nanotechnology in medicine currently being developed involve employing nano-elements to bring heat, drugs, light or other matters to exact cells in the human body. A nano particle to be applied in this way permits recognition and treatment of diseases or injuries inside the targeted cells, thus reducing the harm to healthy cells in the body (Kompella et al., 2013). Researchers are looking into the following nanoelectronics projects: a. Using electrodes made from nanowires that would enable flat panel displays to be flexible as well as thinner than current flat panel displays. b. Using carbon nanotubes to direct electrons to illuminate pixels, resulting in a lightweight, millimeter thick "nanoemmissive" display panel. c. Making integrated circuits with features that can be measured in nanometers (nm), such as the process that allows the production of integrated circuits with 45 nm wide transistor gates. d. Using nanosized magnetic rings to make Magneto Resistive Random Access Memory (MRAM) which research has indicated may allow memory density of 400 GB per square inch. e. Developing molecular-sized transistors which may allow us to shrink the width of transistor gates to approximately one nm which will significantly increase transistor density in integrated circuits. While most purposes of nanotechnology in medicine are still under development, nano crystalline silver is already being used as a antimicrobial agent in the treatment of wounds (Adam et al., 2013; Zamani et al., 2013). Following applications also will be released very soon: a. Qdots that recognize the cancer cells in the body, b. Nano particles deliver chemotherapy drugs applied directly to cancer cells to minimize injure to healthy cells, c. Nanotubes used in broken bones to provide a structure for new bone material to grow, d. Nanoshells that concentrate the heat from infrared light to destroy cancer cells with minimal damage to surrounding healthy cells, e. Nanoparticles that can attach to cells infected with various diseases and allow a doctor to identify, in a blood sample, the particular disease (Stevensona, C.L. et al., 2012).

## RISKS OF NANOTECHNOLOGY

These days it seems you need the prefix "nano" for products or applications if you want to be either very trendy or incredibly scary. If it were able of existing outdoors, and using biomass as raw material, it could severely damage the environment. If wrong people increased the capability to manufacture any preferred product, they could imperative the world, or reason massive demolition in the effort. Definite products like as prevailing aerospace weapons, and also microscopic antipersonnel devices, supply special cause of concern (Cagdas, M. et al., 2014), (Liu Y et al., 2015). Grey goo is applicable here as well. Obviously, the unobstructed accessibility of nanotechnology creates serious risks, which may well be more important than the profits of cheap, clean, convenient, self included manufacturing. Immature utilize of molecular manufacturing could guide to black markets, uneven arms races ending in enormous destruction, and perhaps a release of grey goo. Another important aspect of the nanoscale is that the smaller a nanoparticle gets, the larger its relative surface area becomes. Its electronic structure changes dramatically, too. Both effects lead to greatly improved catalytic activity but can also lead to aggressive chemical reactivity.

### Environmental Issues

In free form nanoelements can be discharged in the air or water during production (or else production mistakes) or as dissipate byproduct of production, and also finally gather in the water, soil or plant life. On the other hand, in fixed form, wherever they are component of a fabricated product or substance, they will finally have to be disposed of as waste or recycled. It is not known yet if certain nano particles will compose a absolutely new class of non-biodegradable pollutant. It could be removed from water or air because most conventional filters are not appropriate for such responsibilities

### Health Issues

Health and environmental issues combine in the workplace of companies engaged in producing or using nanomaterials and in the laboratories engaged in nanoscience (Oupicky D et al., 2014). To accurately evaluate the health hazards of engineered nano-elements, the entire life cycle of these particles require to be estimated (Halo TL et al., 2014), as well as their storage, fabrication and circulation, application and potential abuse, and also removal. The impact on humans or the environment may differ at dissimilar steps of the life cycle (Lauriano et al., 2016).

## CONCLUSION

It is widely felt that nanotechnology will be the next industrial revolution. Today, many of world's most creative scientists and engineers are finding new ways to use nanotechnology to improve the world in which we live. We see doctors detecting disease at its earliest stages and treating illness such as heart disease, cancer and diabetes with further effective and safer medicines. Although there are many research challenges ahead, nanotechnology already is producing a wide range of beneficial materials and pointing to breakthrough in many fields.

## REFERENCES

Adam, T., Hashim, U., Ali, M. E., & Leow, P. L. (2013). The electroosmosis mechanism for fluid delivery in PDMS multi-layer microchannel. *Advanced Science Letters, 19*(1), 12–15. doi:10.1166/asl.2013.4670

Bae, Y. H., & Park, K. (2011). Targeted drug delivery to tumors: Myths, reality and possibility. *Journal of Controlled Release, 153*(3), 198–205. doi:10.1016/j.jconrel.2011.06.001 PMID:21663778

Cagdas, M., Sezer, A.D., & Bucak, S. (2014). *Application of nanotechnology in drug delivary- Liposomes as Potential Drug Carrier Systems for Drug Delivery*. Academic Press.

Halo, T. L., McMahon, K. M., Angeloni, N. L., Xu, Y., Wang, W., Chinen, A. B., ... Thaxton, C. S. (2014). NanoFlares for the detection, isolation, and culture of live tumor cells from human blood. *Proceedings of the National Academy of Sciences of the United States of America, 111*(48), 17104–17109. doi:10.1073/pnas.1418637111 PMID:25404304

Huang, J.-W., Chang, C.-K., Lu, K., Huang, J.-T., & Lin, C.-R. (2009). Novel approach for nanoporous gas sensor fabrication using anodic aluminum oxidation and MEMS process, Nanotechnology, 2009. *IEEE-NANO 2009. 9th IEEE Conference.*

Iverson, N. M., Barone, P. W., Shandell, M., Trudel, L. J., Sen, S., Sen, F., ... Strano, M. S. (2013). In vivo biosensing via tissue localizable near-infrared-fluorescent single-walled carbon nanotubes. *Nature Nanotechnology, 8*(11), 873–880. doi:10.1038/nnano.2013.222 PMID:24185942

Kompella, U. B., Aniruddha, C., Amrite, A. C., Ravia, R. P., & Durazoa, S. A. (2013). Nanomedicines for Back of the Eye Drug Delivery, Gene Delivery, and Imaging. *Progress in Retinal and Eye Research, 36*, 172–198. doi:10.1016/j.preteyeres.2013.04.001 PMID:23603534

Lauriano Souza, V. G., & Fernando, A. L. (2016). Nanoparticles in food packaging: Biodegradability and Potential migration to food—A review. *Food Packaging and Shelf Life, 2016*(8), 63–70. doi:10.1016/j.fpsl.2016.04.001

Liu, Y., Zhang, L., Wei, W., Zhao, H., Zhou, Z., Zhang, Y., & Liu, S. (2015). Colorimetric detection of influenza A virus using antibody-functionalized gold nanoparticles. *Analyst (London), 140*(12), 3989–3995. doi:10.1039/C5AN00407A PMID:25899840

Mansoori, G. A., Soelaiman, T. A. F., & Soelaiman, T. A. F. (2005). Nanotechnology-An Introduction for the Standards Community. *Journal of ASTM International, 2*, 1–22.

Morrison, D., Dokmeci, M., Demirci, U., & Khademhosseini, A. (2008). *Biomedical Nanostructures* (G. Kenneth, H. Craig, T. L. Cato, & N. Lakshmi, Eds.). John Wiley & Sons, Inc.

Mukherjee, B., Dey, N. S., Maji, R., Bhowmik, P., Das, P. J., & Paul, P. (2014). Current Status and Future Scope for Nanomaterials in Drug Delivery (pp. 525–539). Academic Press. doi:10.5772/58450

Nikolelis, D. P., Arzum Erdem, T. V., & Nikoleli, G.-P. (Eds.). (2014). *Portable biosensing of food toxicants and environmental pollutants* (Vol. 1). New York: CRC Press.

Oupicky, D., & Li, J. (2014). Bioreducible polycations in nucleic acid delivery: Past, present, and future trends. *Macromolecular Bioscience, 14*(7), 908–922. doi:10.1002/mabi.201400061 PMID:24678057

Silva, J., Fernandes, A. R., & Baptista, P. V. (2014). Application of Nanotechnology in Drug Delivery. Academic Press.

Stevensona, C. L., Santini, A. J. T. Jr, & Langerb, R. (2012). Reservoir-Based Drug Delivery Systems Utilizing Microtechnology. *Advanced Drug Delivery Reviews*, *64*(14), 1590–1602. doi:10.1016/j.addr.2012.02.005 PMID:22465783

Wesam Al-Mufti, M., Hashim, U., Rahman, M., Adam, T., & Arshad, M. (2015). Studying Effect Dimensions of Design and Simulation Silicon Nanowire Filed Effect Biosensor. *Applied Mechanics and Materials*, *754*, 854–858. doi:10.4028/www.scientific.net/AMM.754-755.854

Yi, M., Jeong, K.-H., & Lee, L. P. (2005). Theoretical and experimental study towards a nanogap dielectric biosensor. *Biosensors & Bioelectronics*, *20*(7), 1320–1326. doi:10.1016/j.bios.2004.05.003 PMID:15590285

Zamani, M., Prabhakaran, M. P., & Ramakrishna, S. (2013). Advances in drug delivery via electrospun and electrosprayed nanomaterials. *International Journal of Nanomedicine*, *8*, 2997–3017. PMID:23976851

# Chapter 16
# The Social Significance of Visual Arts:
## State, Nation, and Visual Art in Japan and Finland

**Mika Markus Merviö**
*Kibi International University, Japan*

## ABSTRACT

*This chapter focuses on visual arts in Japan and Finland and the ways that social and political developments have been linked to reconstruction of cultural traditions. In particular, the most important turning point appears to be the establishment of modern state. This chapter follows the circumstances that prevented Japanese state for a long time during the era of feudal society from giving culture and visual arts more freedom and larger role in the process that leads to modernization and the nation state. After this, the author moves to discuss the parallel developments in Finland, where the process is very different. The idea of this chapter is to look these parallel developments both in terms of contrasts and similarities. In both societies, the visual arts have a long history, and often in national history writing, the cultural traditions are presented as a long unbroken narrative. However, the social and political developments have been greatly affected by cultural developments, but also culture has played a major role in social development.*

## INTRODUCTION

The Japanese state has been reluctant and slow to realize how important visual arts for the Japanese society and culture and that visual arts have a huge potential to make Japan and Japanese ideas known abroad, in addition, to contributing to global culture. The idea of state being responsible for promoting the cultured life of people or the idea of cultural rights of people both are rather foreign to Japanese society. However, the whole idea of 'being Japanese' is very much a social and cultural reconstruction and the Japanese state has been actively encaged in shaping that reconstruction together with other pillars

DOI: 10.4018/978-1-7998-3576-9.ch016

of establishment in Japan, such as the educational and economic institutions. In visual arts the Western influences seeped in gradually during the Edo period and well before Japanese politival elite was ready to contemplate real opening of the society to the Western culture. The hybridization in arts had been going on for a long time before the actual Western painting landed in Japan. This chapter follows the circumstances that prevented Japanese state for a long time during the era of feudal society from giving culture and visual arts more freedom and larger role in the process that lead to modernization and nation state. After this I move to discuss the parallel developments in Finland. The idea of this chapter is to look these parallel developments both in terms of contrasts and similarities. In both societies the visual arts have a long history and often in national history writing the cultural traditions are presented as a long unbroken narrative. However, the social and political developments have been creatly affected cultural developments, but also culture has played a major role in social development.

Japan has a long history as an island territory that is easily distinguishable from its neighbours and the long history of more or less centralized rule under the emperors and shôguns. However, Japanese modern nationalism is largely a product of Meiji period political program and it is wise to regard Japanese nationalism as very different ideological construction than most other nationalisms of the world. Even the Japanese language is very much part of the standarization process of Meiji Period when the new standard Japanese *(hyôjungo)* was created on the basis of Kantô dialect by the government in order facilitate smoother governance and education. In feudal Japan the rulers did not care that much about the commoners or their culture or communication. However, the Japanese population and cities were already large enough to support elite culture that later has surved as the basis of both literary and visual culture. How much it reflects the histories of the past generations in Japan is open to debate but it certainly has become a central element in the construction of the narrative of Japanese culture and its continued tradition.

Meanwhile in Finland the area has been inhabited for a very long time by a population primarily using Finnish, a Finno-Ugrian language, a non-Indo-European language, which continues to mark their difference from both the Swedes and Russians. The Swedes gradually were able to bring Finland under their domination after the first crusade in the 1150s. In medieval times Finland was regarded in Sweden as a part of the electoral kingdom and was represented along the other regions at the stones of Mora. However, much of the current area of Finland remained as a wilderness where the Swedish crown had little influence. The effective rule of Sweden covered mostly the areas in the southwest and south. What also mattered was that in all parts of Sweden the peasants remained independent and were never brought under feudalism, which suited well the Finnish population. However, the small political and cultural elite had a reason to use also Swedish and quite a few also studied at universities elsewhere in Europe. Only after the establishment of the Vasa dynasty in Sweden in 1523 the Swedish crown started more ambitious program of central government, which soon turned Sweden into a major military power and multinational empire. The empire-building wars usually meant hardship to the Finnish population. King Gustav Vasa also brought Lutheran Reformation to Sweden and Finland as the state church and this way strengthened the role of King. For Finnish language and future cultural development this was helpful as Finnish Lutheran clergyman and scholar Mikael Agricola translated the New Testament (1548) and wrote several religious books in Finnish in order to teach the Bible. However, he also created singlehandedly the basis for written Finnish language and quite masterfully coined new words on basis of existing Finnish words to express foreign ideas and phenomena. Agricola had studied in Wittenberg under Philipp Melanchton, who was a known specialist of Greek language. Agricola was himself an extraordinary talented linguist. There is a good reason to argue that he played the most important individual role in

## The Social Significance of Visual Arts

the Finnish cultural history (for more on Agricola, Heininen 2019). The Lutheran church required everyone to learn to read and write, which also created the basis for national educational system for years to come. However, in terms of secular literature it was only in the 19th century with a surge of books and newspapers published in Finnish and the modern compulsory educational system when the Finnish language took off as a medium of literary expression on par with Swedish.

The whole idea of 'Finland' as a nation state started in the 19th century very much as a high culture project invented by rather small cultural and political elite that understood very well that the Russian authorities might easily find the Finnish cultural Renaissance tolerable or even a suitable as a model for much of the rest of Russian Empire. In the Finnish Golden Age of National Romanticism visual arts, Finnish literarature (especially in Finnish language) and music played each a major role and all these further proliferated to play a major political role as well as bridged Finland to Western Europe by establishing and nurturing values that would help the process of nation building and integration into the European political system. This development was gradually distancing Finland ever further from the value systems and autocracy that served as the basis of Russian Empire. At the same time Finnish culture also distinquished itself from Swedish culture although the Swedish model of state-supported flourishing culture served as a major inspiration. The simple idea was to nurture culture that is worthy of a nation state.

The history of Finnish culture, state and nation is quite different from such cultures/nations where the present political state has a long history as the most important agent behind the cultural development. In Finland it was the cultural awakening itself that made the state possible in 1917, when Finland became independent. At that time there also was a rare political and practical opportunity to break free from the domination of larger neighbours. Of course, this was a part of a larger pattern and was helped by the Wilsonian principle of the Right of National Self-Determination, that influenced the thinking on international relations and international law during the World War I. However, those same neighbours that had so long dominated Finland politically have left their mark to Finnish culture and the existing cultural links nowadays serve as basis for peaceful cooperation. This history has made Finland particularly suitable for the European Union and for the type of peaceful integration and universal value systems that the European Union stands for. Visual art both in Finland and Japan has never become totally controlled by the state for its own interests. In fact, the aesthetic ideas sometimes have marked similarities, especially in the approach to nature.

## CHANGES IN VISUAL EXPRESSION IN JAPAN DURING THE TRANSFORMATION FROM EDO SOCIETY TO MODERN STATE

While the political elite invented traditions of modern Japan during the Meiji Period it also created at the same time narratives of Japanese history and culture to suit the uses of new state and nationalism that supported it – a process which is rather typical for nation-states everywhere. However, the invention of traditions or narratives does not mean that they are all false or invented from a scratch. Instead, in Japan the old cultural traditions of the samurai class provided a perfect arsenal from which the political elite could choose the most suitable ones for the whole nation. The refined tastes of upper classes served as a basis for artistic traditions that are officially associated with the essence of Japanese art and culture. The modern Japanese state presented the Japanese authentic cultural traditions as a proof of Japanese cultural superiority and, therefore, there has all along been a tendency also to protect the Japanese cultural traditions from foreign influences if those influences appeared to be harmful for Japanese traditions or question their right to live and flourish. *Nihonga* (Japanese style painting on silk or paper) was seen by the policy makers as more Japanese than *yôga* (Western painting), regardless of the theme of painting and without much of critical discussion whether *nihonga* really is that uniquely Japanese. After all, both Japanese *nihonga* and Western painting *(yôga)* during Meiji Period and afterwards were both deeply influenced by former foreign and Japanese art and by each other. There is also the fundamental issue of profound Chinese influence in Japanese cultural traditions, including all aspects of visual art.

The conservatism of Japanese old visual art traditions has also another dimension. The old cultural traditions of high culture enjoyed the patronage of the upper classes and, therefore, social status and arts were (/are) closely connected. The rigid social order of Edo Period was based on rather narrowminded control of the Tokugawa Bakufu. While the central government was rather undeveloped and did not even try to fullfil anything resembling the functions of modern state, the Bakufu nevertheless did not want to see any open challenge to social order. Social rank determined everything especially within the ruling samurai class: the shape and size of residences (and gardens alike) were closely regulated, so were the colour and designs of clothing and even many small but symbolically significant minute details were strictly regulated, such as the the borders of tatami straw mats reserved to various ranks of officials in Edo castle. In other words, the social order and control had also a very strong aspect of visual symbolism. This meant that there was very little room left for individual choice. As Nishiyama Matsunosuke points out, "knowing one's place" in Japanese feudal system did not only require everyone from avoiding rising above their place but also banned them from falling below one's status (Nishiyama 1997: 32).

In other words, political discourse could not be openly exercised through the means of art or any other means of free expression of ideas and the relative isolation of Japan as an island nation also prevented new and especially potentially subversive ideas from spreading. The realm of politics and Orthodox political discourse was reserved to a very small class of samurai and its top echelon. This class covered less than 7 percent of the population while the peasants officially were regarded as the second most importat class but were kept closely politically and economically controlled by the samurai. The artisans and merchants (and those falling outside classes) were in the lowest position and certainly knew better than openly raising their voices against the system. The samurai class itself was divided over such issues as hereditary or family support to Tokugawa rulers (*fudai* group) and plain subjugation to the strongest faction after the battle of Sekigahara (*tozama* group). In economic terms the fundamentally untrustworthy *tozama* group emerged as victors and destabilized the whole system towards the end of Edo Period.

Ideological foundation was provided by official recognition of a few simple tenets of Neo-Conficianism. Japan did have in the beginning of Edo Period learned scholars such as Hyashi Razan, Fujiwara Seika and Ogyû Sorai who understood well the Chinese thought and the possibilities for social thought. However, the Bakufu was able to hijack the Neo-Confucian thought and by appointing some of the leading scholars to official roles it effectively established an official simplified state ideology that also substituted Buddhist religious thought (and the role of Buddlish temples) by a thought system which appeared to be rational compared with previous and still competing religious doctrines (Cf. Ooms 1984, 27-61). However, the Bakufu support was a kiss of death in terms of free discussion and development of anything resembling 'rational thought'.

The Neo-Confucian discourses could not fully hide the fact that the small political clique that had managed to grab the power did everything it could to prevent open political and social discourse that could potentially exposed its own weaknesses. In other words, the system very much fostered an ethos that 'About what one can not speak, one must remain silent' (Cf. Ludwig Wittgenstein, Tractacus Logico-

*The Social Significance of Visual Arts*

Philosophicus where he is discussing the ideas outside the border of language being simply madness). It is not that all Japanese would have been blind to politics or that Japanese language would not have words for subversive or progressive political ideas. For instance, Ogyû Sorai advocated a serious study of Chinese classics with original sources and his ideas attracted a sizable following against the realities of Edo Period Japan (Cf. Nishiyama 1997: 67). However, the educational system and, for instance, publishing were not large enough to adequately support any broader attempts to spread such rather elitist ideas much beyond a relatively small groups of people, who themselves were very conscious of their social status.

There simply was no room left for alternative ways of expression and the virtually everyone were intentionally deprived of freedom to express themselves. In visual art the refined realm of aesthetic pleasure was the escape that was left open, especially among the new rich groups among merchant class where their tastes were less stringently controlled. For instance, the extravagance or Rimpa school tells much about this splendid escapism and how it relatively quickly became part of the mainstream, and even (part of) the upper classes could enjoy all the glitter, fancy details and exquisite craftsmanship that Rimpa offered, while few seemed to notice that exquisite beauty and impeccable taste easily risks rising above one's place. Both art and social thought easily lead to dangerous directions and the Edo high culture was quite a mine field, but, of course, the elite also acquired the survival skills to make the best of situation. However, there are also a good number of cases where learned men ended up becoming boiled alive, crucified or otherwise losing their lives for expressing their thoughts too freely. This kind of situation made it very difficult to spread culture or ideas to all segments of population and, therefore, Japanese nationalism and the whole idea of common Japanese culture starts from rather unique situation when Japanese modernisation started from the top with Meiji Restoration.

Japan with its rigid social order had a very different social development leading to modernization than most European countries: in Europe it was the middle class which created a mass market for arts (starting in the Netherlands) and served as a basis for new political and intellectual elite, The European middle class was also early to develop a vast array of political and cultural ideologies, many of which directly challenged old political leadership and cultural traditions. The meaning of visual arts and arts in general were reinvented and reinterpreted in Europe especially in the 19th century, a process that also created the foundation for modern art from the beginning of the 20th century. The artists, especially painters, musicians and writers, poets and novelists, were particularly active politically. The Finnish model fits this pattern.

In Europe, the most important development in art theory were the ideas of Immanuel Kant that would prove to be influential for centuries and which still are with us in new reinterpreted forms. Immanuel Kant's understanding about '*beauty*' was that it existed in the world without being contaminated by our ideas of purpose and usefulness (Kant 1799/1954). People with good aesthetic judgement according to Kant are able to appreciate art as it is, with innate or acquired disinterestedness in the mundane aspects of it. Meanwhile making art itself requires something like a '*genius*'. Kant saw *genius* very much in the context of the old Roman and Latin meaning for embodiment of a noble spirit rather than the modern stereotype of *genius* being something like a mad scientist or demigod churning out new things and products to the mesmerized public (For more on the origin and concept of the Genius, Murray, P. ed. 1989).

Both the art connoisseurs with good taste and the *genius* artists have often became something like caricatures when Kant's works were popularized and re-interpreted. However, Kantian art theory together with rather elitist approach to art education and to ideas about art appreciation were very influential for European art before the birth of modern art (around the end of 19the century) which would then diversify and democratise ideas about artistic expression and appreciation of art. For most modern people the elitism that is part of Kantian art theory (as it is often presented in its simple form) equals to dead art and snobbery and arrogant self-importance rather than something as elusive as 'good taste'. Moreover, another important development in European art was that public ownership and grand scale exhibition of art spread to wide public steadily after opening of the Royal collections of Louvre to the public during the French Revolution. The Dutch national collections that evolved into the Rijksmuseum were started in year 1800 by the Batavian Republik and its art loving merchant rulers to emulate the French example. Nowadays it is taken as granted that the best art should be in museums within an easy reach of everyone and this kind of thinking is becoming more and more global. This reflects both the idea of art belonging to everyone and the also a more patronasing approach according to which the plebs should be educated about their own culture and that national culture requires a constant effort to keep people informed about their past and use it as a springboard to future glory (Cf. Merviö 2017: 3-4).

In Japan, something like the Kantian art theory did not make any sense as especially its vocabulary was based on a totally different social order and cultural history. In visual arts the most important new development would be the introduction of Western painting *(yôga)* to Japan. However, there could be no jump to a fundamentally different mode of expression. During the Edo Period Western painting was effectively prevented by the Bakufu policies from spreading although the use of pespective in Japanese arts obviously benefited from the sporadic contacts with Western art. So called *Nishiki-e* style was spreading in Edo by print makers of relatively modest social standing and education. The Japanese traditions placed great importance to emptiness *(kû)* which translated into blank areas in paintings (on silk, paper or wood). These ideas stemmed from Buddhist philosophy of Chinese origin and in Japan they also were accompanied with ideas about space *(ma)*, and these ideas grew up to encompass significance in terms of natural philosophy (and having obvious connections with Shintô. However, in relatively lowly art of *ukiyo-e* in new capital Edo it was easier to break the traditions and start filling the whole surface with colour and innovative compositions. Moreover, these same artists also started employing Western techniques of perspective. The Western paintings and especially prints and etchings that were more readily available made a profound impression to Japanese artists and the Japanese artists soon were hungry for new ideas and during the 19th century these ideas transformed Japanese visual arts and soon grew to have a profound influence also among the more respected and traditional art genres, including the Japanese scroll painting on silk and paper *(kakejiku)* (Cf. Nishiyama: 69-71).

One interesting, and as such rare and unusual, document was produced by Shiba Kôkan (1747-1818), a painter within the influential Kanô school, who studied Western painting and write in 1799 a short *Discussion on Western Painting (Seiyô gadan)*, where he lists the weaknesses of Eastern painting starting from the inability of tradtional painters to recognise Western art as anything else than absurd and unworthy of study. He continues to tell that Western art can realistically depict such objects as human hair or an ox in their entirety rather than as collections of lines. Then he moves to discuss shading and perspective and the durality of oil painting in contrast to Eastern methods. He is most impressed by the ability of Western painting to recreate an accurate image in a way that a written word can never do and points out that the Eastern artists are not in such a position: the Japanese artists just draw a circle and call it a sphere whereas the Western artists actually paint a spherical object or have no problems depicting the height of a nose in their portraits. In the end of his writing Shiba Kôkan lists Japanese (serious) painting schools as those of Tosa, Kanô and the Chinese one and continues with the ultimate insult stating that none of them even know how to paint the Mount Fuji or any of the famous mountains of China: they just paint a nameless mountain and call it a (celebrated) landscape (Shiba Kôkan's full text translated in, Traditional Japanese Arts and Culture. An Illustrated Sourcebook, 2006: 173-176).

*The Social Significance of Visual Arts*

For the established masters of famous painting schools it was difficult or unheard to break traditions, even if the artists themselves were open to new ideas and modes of expression. Remember this was in a country where the whole social order rested on seniority and knowing one's place. Furthermore, the art education and finding work was based on a very close but hierarchical relationship between the master and student and most masters were not happy to see their students becoming rebels of any kind. However, for less esteemed art forms and artists the rules were more flexible and novelty could be turned into a business opportunity. The *ukiyo-e* artists discovered the same economic rule that the Western printmakers and artists especially in the Netherlands and Germany had discovered before: multiplying the number of prints and cultivating new art audiences brings more stabile income than sticking to unique art pieces or traditions. The rest is history: among the artists who attended the *Daishôkai* gatherings organised by Nishiki-e group were such names as Hokusai, Utamaro, Harunobu and Haruaki, who would hybridise Japanese art and Western art. Furthermore, such artists as Hokusai, who went the furthest in their experiments would also have a significant impact on Western art and in world art, and gained the well-earned respect and recognition in the end. Japanese art was imported *en masse* to European markets after such pioneering import-export businessmen as Siegfried Samuel Bing in France saw the opportunities from the 1870s on and went also to do business in Japan with art and artefacts that met the changing tastes of Japanese elite (For more on the role of Bing: Wichmann 1981: 8-9,12,21). Segments of the Japanese art establishment and manufacturing were quick to realise the new business opportunities and they started to churn out art that would sell well in Europe and America and, for instance, in the case of wood block prints it was relatively easy to make reprints. In Impressionism, Art Nouveau and, for instance, in the art of Vincent van Gogh the Japanese influences were crucial and very visible. Sometimes the compositions were directly copied, sometimes there were *ukiyo-e* prints hanging on the walls in European paintings and sometimes the Japanese animals and plants or colour combinations appeared to have made their journey to European art (See e.g. Wichmann 1981 and *Gohho to nihonten* 1992). These Japanese influences also directly and indirectly arrived just on time to influence Finnish Golden Age of painting when the national romanticism in art was exploring new forms of expression and when Japanese sensitivity to natural phenomena was received as godsent.

During the Bakumatsu years (before Japan fully embraced the West) the pioneers of Japanese *yôga* such as Takahashi Yûichi trained seriously and learned the Western methods – to be mixed with their profound knowledge of Japanese painting, *nihonga* methods. In early Meiji Period there was a brief period when Western painting had a chance to be fully introduced to Japan. *Kôbu bijutsu gakkô* (Technical Art School工部美術学校, 1876-1883) was established by the Meiji government and it employed Italian artists to teach western methods. Also Takahashi Yûichi was hired as a professor. However, the political currents soon turned against *yôga* and *nihonga* soon gained dominance and official support. Many *nihonga* artists adopted western techniques and practices, for instance, with perspective, shading, themes or composition. Western painting continued in Japan, but Western painters could not count on official support or even popular support and understanding (Cf. Hashimoto 2007: 14-66). Against all odds many *yôga* artists persevered in Japan and many of these artists went to study in Europe as Japan lacked opportunities. While some Japanese artists developed quite international outlook and taste, the Japanese *yôga* community as a whole shifted towards Japanization *(nihonka)* already during the Meiji Period and during the early Shôwa Period *yôga* was actively supporting militarism in Japan and was seen primarily as a tool of war propaganda.

In other words, the Japanese *yôga* artists were seldom testing the limits of freedom of expression and, instead, many of the *yôga* artists were quick to discover that *yôga* offered far better means for dramatic battlefield panorama paintings than *nihonga*. Lucky for *nihonga* most of the *nihonga* artists saw no reason to abandon their traditional themes. It is also likely that at least some officers knew better *nihonga* and had more respect for its artistic tradition. Moreover, no one suspected the Japanese pedigree of *nihonga* and, therefore, there was no need to prove anything. On the contrary, *yôga* was something that had to Japanized first in order to show its value to the new Japanese regime. In fact, the military leaders gave very precise instructions what kind of *yôga* they wanted, and that was what they got: they wanted realistic and accurate paintings, which recorded the bravery and suffering of the soldiers. Accuracy meant that such details as those of weaponry and uniforms as well as details relating to geography and foreign places had to be right. Interestingly, it was the military that was criticizing the weakness of a distinct Japanese style of *yôga*. They wanted *yôga* to surpass the western tradition (without even knowing much about the western tradition) and also record the establishment of a new Japanese imperial culture and saw no problem in the idea of instructing from above what the artists should do to achieve that artistic greatness. Of course, propaganda painting was something of a global trend at the same time with Nazi Germany, Italy and Soviet Union all pioneering the *genre,* while battlefield panoramas have a proud tradition within western tradition especially during the Renaissance and Baroque art.

In Japan the ideas of modernism in visual arts could not easily develop as the authorities were actively trying to destroy everything that was distantly connected with anarchism or socialism. Suspicious ideas often included democracy and Christianity, meaning that there was little room for anything directly connected with the western tradition, except for technical skills. This does not mean that the authorities could successfully weed out everything they wanted as Japan did already have its small communities of people who had gained a deep knowledge about all the forbidden fruits, from anarchism to communism and from Christianity to democracy. Taishô Period, after all, is strongly associated with the ultimately failed attempt to develop a functioning multiparty democracy and a pluralistic society – and a cultural renewal. Visual arts in Taishô democracy reflect some of those attempts to bring new ideas to Japan. Especially in the less ideological niches of popular culture Western ideas and techniques were wholeheartedly embraced, often in a delightful mixture with Japanese traditions. The Japanese illustrations and graphic design, as well as fashion of this era show that the Japanese had no problem turning away from Japanese traditions to adopt Western styles or mixed styles. The influences of *Art Nouveau* and *Art Deco* can easily be detected in Japanese styles of the late Meiji to early Shôwa, but in Japan it was difficult to raise this art to the level of high art as that position was already reserved to *nihonga*. For instance, *Art Nouveau* never attained the political significance of cultural awakening and national romanticism that it had in many parts of Europe, and, especially, in Finland. In Japan, *Art Nouveau* was mostly a harmless decorative element and had little to do with new representation of Japanese culture.

Meanwhile the modern styles of painting or symbolism in Japanese *yôga* could not hope to gain acceptance in the ideological climate of the time – and they also were far more difficult to utilize in commercial ways or turn to fancy decorative novelties. That left the artistically more ambitious and more internationally minded artists with the only choice of finding their supporters among the small community of people who fell under the radar of the military and who knew well enough the Western art world and its currents. Taishô democracy did nurture a small group of internationally minded people, some with considerable wealth to spend on art. The Japanese modernization also did, indeed, create a new middle class, but the rise of militarism soon made life very difficult for anyone who did not fully support the aims and dominance of the military. It seems that most *yôga* artists were quick to choose the side of the military and the works defying the new political order were quite rare in visual arts. However, in literature it is easier to find examples of artists who clearly shared different values and understood

*The Social Significance of Visual Arts*

the value of their own art (see e.g. Tanizaki Jun'ichirô and Yosano Akiko). Meanwhile, in Finland art moved on to serve the rather insular purposes of nation building while a small but significant minority followed closely the exciting new directions of modern art in Europe, especially France. Compared with Japan in Finland the modernization of visual arts could take place without too much suppression and the art establishment as well as buying middle class demonstrated more understanding to changing modes of expression. Furthermore, in Finland the art institutions and art establishment grew in size and influence and were able to establish certain independence from the state and politics in spite of becoming increasingly dependent on public funding. In Japan a similar development may be seen after the boom of art museums after the 1970s.

## THE GRADUAL RISE OF FINNISH VISUAL ARTS AS A SOCIALLY SIGNIFICANT ACTIVITY

Finland was for a long time a poor and backward part of the Swedish kingdom and visual art as well as art education were very much ignored during that long era. The market for visual art was small and was taken care by few individual who often were from Swedish mainland or from Central Europe. Even these painters usually did not have enough real artistic work available and they usually had to take any kind of painting jobs. Finnish peasants did decorate buildings (included wooden churches) and objects that they needed in their life. This art was primarily utilitarian and its aesthetic level often is even more primitive than that of pre-historical Finnish art. In fact, some of the art forms that can be found, for instance, in some Finnish medieval churces are almost unbelievably naivistic or clumsy – or original – showing that the artists (or the clergy behind them) had limited chances to copy the art of other parts of Europe. After the Reformation the Lutheran church did favour literacy but it did not do much to support the arts. Compared with the Swedish era, the early hunters in Finland actually demonstrated in their art very sensitive eye especially for the nature and animals, but this tradition did not fare too well in the agrarian society and Finnish folk art during the Swedish period follows often the Swedish models. There is a good reason to assume that in the pre-Christian Finnish society when the population depended more heavily on hunting and fishing the visual art and artists had a more significant social role to play than in agricultural society where the church as well as the Swedish state had a negative opinion about the traditional nature worship. In Finland the Christian art did not make particularly strong impact either during the Catholic or Lutheran period. In short, Finland in terms of visual art remained a backwater of Swedish empire, especially during the time that Sweden possessed parts ot Germany and the Baltic states that had much more developed art institutions and markets, that could not even be crushed by the Reformation and its ideas of piety and austerity. The first art guild in Finland was establish only in 1785 in Turku (Åbo) in the very end of Swedish era. The guilds were altogether abolished in 1868 when Finland rapidly modernized and industrialized as a result of domestic efforts as well as support from Alexander II, the most progressive and liberal of all Russian Czars. For the future development of visual arts it probably was a blessing that the guild system never became as strong as in many other parts of Europe as the guilds very not in place to control the new developments in the 19th century. This was particularly important as Finland suddenly became a major centre of female artists even by global standards. Of course, there were also other developments that helped to make it easier to enter visual arts. One major factor everwhere was that the synthetic organic pigments were invented largely in Germany in the early 19th century making possible the cheap production of oil paints. As a results it became far easier to practice oil painting and it also became possible to paint outside *en plain air*.

In the following I summarise the history of the establishment of art education in Finland in the 19th century to show how late Finland was to develop its art institutions and yet how important they became towards the end of century. The guild system was copied from Sweden and Germany, where larger cities and commercial base existed for such an institution to thrive and also to maintain certain standards of quality. Like in all other crafts the master craftsman in arts had to produce a showpiece, literally a masterpiece to demonstrate his skills. In the beginning in Turku it was an altarpiece but by the middle of the 19th century it was gradually changed to a secular work of painting a carriage or the interiour of a house with nice painted decorations. Since the art scene generally was so underdeveloped in Finland also the jealousies and frictions between different kinds of painters and artists were rare while those were quite common in European countries with bigger art markets and clearly distinquished art genres and hierarchies.

In Swedish mainland the painters' guilds were far better organised and ambitious and there were sometimes eminent teachers as well as students from Central Europe working in Stockholm and as a result there were even few such internationally known painters of Swedish origin as Michael Dahl (1659-1743) who ended up living in Britain and painting British, Danish and Swedish royals. In Finland, the visual arts in the early part of the 19th century had very little in common with the European art scene and Kant's aesthetic theories did not have much fertile soil in Finland. This situation actually shows the limits of a theory that treats good aesthetic judgement more or less as separate of the social atmosphere and the art scene around. The reality of many societies has been that good aesthetic judgement can be in most societies, but it needs a supporting community to leave its mark to the whole society. Both in Finland and Japan it is easy to identify processes that contributed to the development of aesthetic values in society. In Finland, the modest efforts in the 19th century helped to bring Fonland closer to the European cultural models. Particularly influential were the Swedish, German and Russian models that were within easy reach.

The art guild in Turku started in 1830 on its own initiative a drawing school for its own members, particularly to the apprentices. The most influential individual was a 30 year old painting master Carl Gustaf Södestrand, who during his apprentice years had received more ambitious artistic training at the Royal Swedish Academy of Fine Arts in Stockholm and who wanted to make use of his experience in Finland and took as granted that the European model of art education would be needed to raise the standards in Finland. The training in the beginning consisted entirely of drawing skills. In 1843 the school had advanced and was able to start its "Antiquities class" for which purpose they had acquired a good collection of gypsum sculpture copies from Stockholm and St Petersburg.

In this phase Robert Wilhelm Ekman (1808-1873) returned to Finland in 1846 and started teaching the Antiquities class while Söderstrand continued to take care of the rest. Ekman was born in Uusikaupunki in Finland and studied at the Royal Swedish Academy of Fine Arts and after winning a generous Royal scholarship (with a condition of taking the Swedish citizenship, as Finland was not any more part of Sweden) he studied and worked in Italy, France and the Netherlands (1837-1844). After that he returned to Finland with the Swedish title of court painter and, for instance, won a major task of decorating the Turku (medieval) cathedral with large frescos. In 1855 he regained his Finnish citizenship.

It was lucky that Södestrand and Ekman with their talent and knowledge were available and wanted to bring European painting tradition to Finland. Both of then had become orphans at young age and their artistic talents had helped them to secure a profession. Söderstrand came from a very modest social background and with no formal primary education and actually he first travelled to Stockholm as a translator to a Russian general (being a native Swedish speaker and having learned Russian as live-in male servant from fourteen years of age). While in Stockholm he befriended people in the painter shop and got interested in painting. In Finland he then received a master's qualification from the guild and after that he moved tonce again back to Stockholm to work and during his free time he received a permission to enroll to the Art Academy. Meanwhile, Ekman was from a wealthier family, but his father died when he was just ten years old. His mother could not help him with schooling but his oldest brother Fredrik who had already entered university noticed his brother's artistic talent and bought first private lessons from G.W.Finnberg who was the best available artist in the city. Next Fredrik took his brother to Stockholm and got him enrolled to the Royal Swedish Academy of Fine Arts and even started himself art studies at the same Academy for a while to stay with his brother. Ekman then became a great success story and won every possible honour that could have neen earned in that Academy. Södestrand and Ekman got to know each other at the academy and they had in common. Both of them had studied just art and had a very pragmatic attitude to art and life. The kind of teaching they brought to Finland in the mid 19th century was very close to the ideas of Italian Renaissance and Mannerism. In short, it was thought that anatomy, perspective and basic drawing skills would serve well any artist.

The drawing school activities were funded largely by Söderström himself, who would not take a salary and who used his own very limited financial resources to keep the school going, although there were also individuals who supported the school, the most notable having been Apothecary Carl Arvidson, who just liked art. Ekman received a salary and much of the honour for the school. Ekman also became the favourite painter of Turku's upper class (For the early years of art education in Finland a good comprehensive source is: Willner-Rönnholm 1996).

Ekman had ambitious plans to establish a Finnish National Art Academy. However, Finnish Art Society had become active in Helsinki (Helsingfors), the new capital, and started a drawing school in 1848 with one teacher, Bernt Abraham Godenhjelm, who had moved from St Petersburg to run the school. Godenhjelm also had quite a few private students and among them was Ida Silfverberg, who is known to have copied oil paintings. In short, Godenhjelm had far more modern approach than the pioneers in turku. By the 1860s the government started to support the Helsinki school and turned its back to Ekman and his plans. By this time also Ekman's artistic style started to be questioned. Finnish elite was well connected with the Central Europe and many understood that visual arts were changing rapidly elsewhere while Ekman represented a bygone world. When Ekman applied to a job at the Helsinki school in 1868 a much younger teacher, and his former student, Adolf von Becker (1831-1909) was chosen. Von Becker had continued his studies first in Copenhagen and then in Düssidolf and Paris. His connections with Paris would turn out to be a real blessing to the next generation of artists in Finland (For more, "Suomen Taideyhdistyksen piirustuskoulu Helsingissä" and "Suomen Taideyhdistyksen perustaminen").

Adolf von Becker also started his own private school in 1872 in addition to his formal teaching, most likely to able to use nude models. The list of his students reads as the honour roll of Finnish Golden Age of painting: Helene Schjerfbeck, Elin Danielson(-Gambogi), Helena Westermarck and Akseli Gallén-Kallela. Adolf von Becker kept travelling and also encouraged his students to travel and this turned out to be the secret of how to actually change and upgrade the Finnish art life. What is also remarkable is that there was a generation of female students who would start in Helsinki and continue in different parts of Europe and become among the best painters of their generation in the world in a time when this was rare. Adolf von Becker is known to have strongly disagreed with younger and more modern (and nowadays much better-known artists, such as Albert Edelfelt and Ville Vallgren) but, anyway, he played a major role in the internationalization and modernisation of Finnish visual arts. His realistic paintings, such as his cat paintings also tell about a great sense of humour, but they certainly have not become accepted as great paintings and did not keep up with rapidly changing European or Finnish artistic ideas.

## THE INTERNATIONALIZATION OF FINNISH ARTS AND CULTURE

For Finnish domestic political purposes the national romanticism in visual arts played an important role and supported other artistic pursuits. However, it is a small wonder how quickly Finnish artists started producing work that was well received both abroad and in Finland. As we have seen the people who organized the drawing schools started from the point zero and themselves had not much of an agenda nor funds or political support. Once arts started to get official and other recognition in Finland it was also quite a balancing act to produce art that would please two so different audiences: the international audience that was expecting new and original art and the domestic audience that was often more focused on the theme and (social & political) meaning of art rather than its aesthetic values. The Finnish artists who studied abroad loosely followed the artistic styles that were in fashion. For instance, Werner Holmberg (1830-1860) who studied after Helsinki in Düsseldorf applied the style of Düsseldorfer Malerschule (Düsseldorf School) to depict Finnish landscape in the light of German Romantik. Of course, in Finland it looks really exotic to see landscapes in bright sunlight. Holmberg was well received in Germany and offered a professorship in Weimar but died at the age of 30, which may partly explain why the German Romantic art with its fanciful yet detailed landscapes never really made it in Finland. One can also argue that many other aspects of of German Romantik remained foreign in Finland.

The most influential thinker of Romantic nationalism in Germany was Georg Wilhelm Friedrich Hegel (1770–1831), whose central idea was that "spirit of the age" (*Zeitgeist*) inhabited a particular people at a particular time, and that, when that people became the active determiner of history, it was when their cultural and political moment had come. In addition to Zeitgeist Hegel identified *Weltgeist* (world spirit) and *Volksgeist* (national spirit) and created a whole philosophical system on basis of these spirits (see e.g. Hegel 1807). Hegel argued that the Protestant Reformation had helped the *Zeitgeist* settle on the German-speaking peoples. Romantic nationalists along Hegel in Germany focused on language as a marker of ethnicity, race, culture, religion, and customs of the nation and equated German speakers as people who shared German culture. However, it did not take that long in Germany for the less cultured nationalists to establish links with the state and military. Towards the end of 19th century the nationalists had adopted a program of 'Drang nach Osten' (Drive to the East), which rewrote the German history and tried to legitimize expansion of Germany especially to the east. In the 20th century these ideas would soon lead to catastrophic choices and in the light of that history German nationalism or Hegel can not be disassociated from the militarism and brutal violence that it legitimized.

In Finland, the elite was largely Swedish-speaking or most often multilingual. However, Hegel's ideas had a major impact among the Finnish politicians and cultural elite. The most influential Finnish politician in the 19th century, Johan Vilhelm Snellman (1806-1881) used Hegel's ideas of national awakening to promote social, cultural and economic development in Finland. He actually had studied in Germany and wrote at length about Hegel's philosophy. Since the majority of Finns were Finnish speaking Snellman understood that speedy progress required that Finnish language would need to become the tool of national awakening and cultural renaissance. That would require major educational reforms and that people would turn their focus to their own traditions. The collection of Finnish oral poetry, most

*The Social Significance of Visual Arts*

famously Kalevala, helped to mobilize support to such a cultural program. Snellman also received the support of the Russian Czars for his efforts as the Russian Czars did not believe that national awakening or Finnish language would become a threat to Russia. The Russian leaders believed that a higher status to Finnish language would instead help to separate Finland from Sweden and foster loyalty towards the Russian Czar and stabilize the situation of Finland as a Grand Duchy of Russian Empire. It also helped that the Poles adopted an openly rebellious stance and served as an example of less reasonable people. In Finland, the national awakening was primarily focused on culture itself and the political, social and cultural realites of the time meant that the Finnish cultural program would become more tolerant toward other cultures than in most other parts of Europe. Of course, the reality was that a culturally and economically more advanced Finland would not only be better prepared to stand free from Sweden but also from Russia – if that was needed. Finnish cultural elite certainly was very selective when it came to German philosophical concepts regarding narion and nationalism. On the other hand, also the German cultural elite was very selective when it came to the actual cultural diversity in German lands and even among themselves.

The changes in politics and society had opened the doors to very different art forms in Finland. The artists were actively sought to create art for the nation. At least the elite realized well that the whole idea of Finnish nation required dynamic, vibrant and, first of all, original Finnish culture. Those artists who left an enduring mark to Finnish art and still rank high among the Finnish population as symbols of Finnish culture were crossing borders and styles in all different ways. Helene Schjerfbeck (1862-1946) enrolled at the age of eleven to Helsinki drawing school. Her hip injury prevented her to study at regular school but von Becker recognized her talent and paid her fees and introduced especially French styles to her.her works started to win prizes in competions when she was seventeen and next she received a travel grant from the Russian Senate to study in Paris. She and many of her friends from Finnish drawing schools attended more female- friendly and liberal art schools in Paris and quite a few of them, including Schjerfbeck, went on to paint in Brittany which was then a popular destination among painters. After winning a pronze medal in 1889 Paris World Fair Schjerfbeck returned to Finland and briefly taught at her old school. However, in 1902 she moved to live together with her mother in Hyvinkää and continued to live there for the rest of her life. She went through realism, naturalism and expressionism, but did all of them with her special sensitive touch and often returned to her old themes with a radically different approach. Schjerfbeck did follow her own way and did not attempt to play any public role. She lived for her art and nowadays on basis of exhibition visitor numbers as well as several surveys is the most respected and loved visual artist of all times in Finland.

The opposite of Schjerfbeck in terms of political role is Akseli Gallén-Kallela. He also moved to Paris after von Becker's education and also was closely involved with Edvard Munch with whom he had joint exhibitions in Berlin. The art of Gallén-Kallela was heavy in symbolism and it served well to illustrate the Kalevala mythology and the mystical Finnish past. However, Gallén-Kallela was a really innovative artist working with many different materials. He also took his whole family to Kenya to paint over 150 expressionistic paintings with bright African colours. However, he returned to Finland to remain the quintessential representative of Finnish national romanticism and new Finnish art. For the Finns his art represents the Golden Age of painting and is also closely associated with the development that lead to Finnish independence. When Schjerfbeck represents the human face of Finnish artistic expression, Gallén-Kallela is difficult to see reparate from politics and Finnish national history. He also served as the aide-de-camp of General Mannerheim and, for instance, designed flags, symbols and uniforms for the newly independent Finland and in this way became the closest thing to a state artist that Finland has ever seen.

Most other visual artists in Finland during this period fall somewhere between the above-mentioned two especially nationally celebrated greats. Among the female artists there are such great artists as Fanny Churberg (1845-1892, an artist whose style bears similarities with expressionist before anyone turned to these ideas in the Nordic countries – but she also played an important role in textile art), Ellen Thesleff (1869-1954, an expressionist), Elin Danielson-Gambogi (1861-1919, a realist with a very peculiar sense of humour) and Maria Wiik (1853-1928, a psychologically oriented painter) who all are part of world art and most likely will receive more attention in the years to come. Most of the Finnish female artists of that time were less political in the sense of national politics of the time and were more tuned to ageless art and basic human condition. This actually may help to make their art more interesting for a bigger audience.

Among the male artists such people as Pekka Halonen (1865-1933), Eero Järnefelt (1863-1937), Hjalmar Munsterhjelm (1940-1905) and Albert Edelfelt (1854-1905) defined very much the styles and new criteria set for visual arts. By doing so they also produced visual images that illustrate Finnish history, landscape and people of their age. They also managed to became part of the bedrock of Finnish nationalism while remaining cosmopolitan people with direct links to artists in different parts of Europe and sometimes even beyond Europe. Apparently the gender roles at that time made it easier for the larger audience in Finland to understand the social and political influence and significance of the work of male artists. They themselves were often very conscious of their role – although they also understood well that outside Finland it did not count much to be famous in Finland. However, nationalism was strong in the Nordic countries as well as in Germany where already then ideas of very narrowly defined ethnic nationalism had much support. In needs to be noted that in Finland the cultural elite was rather cosmopolitan due to its linguistic and cultural background. With the female artists there apparently was less of a pressure to become cultural icons and social influencers of the time. For Finnish art and artists in late 19th century: *Nordic Dawn. Modernism's Awakening in Finland 1890-1920* (2005) and *Pohjoismainen taide 1880- luvulla* (1986).

I conclude my analysis of Finnish visual art here with the political independence of Finland (1917). Once the goal of national independence was reached there was less of a reason to glorify national traditions in visual art in order to inspire people. This also meant that the Golden Age of Romantic Nationalism was over. Moreover, the modernism had already openened new perspectives and opportunities for the artists. The political independence meant freedom in more than one way to the artists, if that was what they wanted. Most artists followed closely the developments in European modern art. Art started to reflect more the diversity of social and political opinions of the time. Furthermore, the relationship between to state, nation and arts/ culture had become far more complicated and there would be no going back to the simple origins. As we have seen, Japanese state and art developed differently, but after the Pacific war the political situation in Japan finally allowed visual arts to break free from excessive political control. The present situation allows much freedom and global influences/ interaction for both artistic communities. Yet, the national traditions continue myriad ways.

## CONCLUSION

This chapter has focused on visual arts in Japan and Finland and the ways that social and political developments have been closely conncetcd to reconstruction of cultural traditions. The most important turning point appears to be the establishment of modern state. However, this chapter has illustrated how the particular strains of nationalisms in Japan and Finland were very different reflecting different historical and social backgrounds. In both cases the state and political elite played an important role when they tried to influence the development of visual art traditions. In the Finnish case the art was widely understood to be a very important tool of social development. In Japan, the political elite for a long time was suspicious of freedom of expression, which certainly had a major impact on the development of Japanese visual arts. The idea of this chapter was to look at parallel developments in these two countries both in terms of contrasts and similarities. In both societies the visual arts have a long history and often in national history writing the cultural traditions are presented as a long unbroken narrative. A more careful analysis or more nuanced reconstruction of past developments in these areas indicates that in both societies individual artists played a major role in changing the direction of art. In both societies art has survived very different challenges from strict censorship to ignorance of aesthetic values. However, the story, history, has continued. States and nations come and go, but Art remains.

## "Ars longa, vita brevis"

This well-known (broken) Latin aphorism quotes the first two lines of *Aphorismi* by Greek physician Hippocrates (Hippocrates, see below for the text in Greek and English). The Latin version actually goes: *Vīta brevis, ars longa, occāsiō praeceps, experīmentum perīculōsum, iūdicium difficile.*

It is written by a natural scientist and 'Art' term *tékhnē* (τέχνη) usually is understood/translated as 'craft' rather than Art in the sense of eternal accomplishment. The intention of Hippocrates apparently was to tell how futile all our earthly efforts are – and especially his own efforts. He was not really making a point about Art. However, the modern interpretation of this ancient textual fragment usually is seen more or less as a statement that Art lasts forever, whereas (human) life and less artistic endeavors will perish in no time. The fate of this aphorism already tells how fragile are the translations, interpretations and reconstructions of past. However, Art remains.

Life is short, and Art long; the crisis fleeting; experience perilous, and decision difficult. The physician must not only be prepared to do what is right himself, but also to make the patient, the attendants, and externals cooperate (Adams translation of the whole aphorism, 1868).

## REFERENCES

English translation by Charles Darwin Adams, 1868. (n.d.). http://www.perseus.tufts.edu/hopper/text?doc=Perseus%3atext%3a1999.01.0248%3atext%3dAph

Gohho to nihonten. (1992). *Vincent van Gogh and Japan*. The National Museum of Art, Kyôto, Setagaya Art Museum and TV Asahi. Nissha Printing.

Hashimoto, O. (2007). *Nihon bijutsushi 7*. Tôkyô: Shinchôsha.

Hegel, G. W. F. (1807). *Phaenomenologie des Geistes*. Gutenberg Projekt-DE. http://www.gutenberg.org/cache/epub/6698/pg6698-images.html

Heininen, S. (2019). *Kansallisbiografia. Agricola, Mikael (noin 1510-1557)*. http://urn.fi/urn:nbn:fi:sks-kbg-000014

Hippocrates. (c. 400 B.C.). *Aphorismi*. Greek original text edited by A. Littré. Perseus Digital Library. Tufts University. http://www.perseus.tufts.edu/hopper/text?doc=Perseus%3Atext%3A1999.01.0250%3Atext%3DAph.%3Achapter%3D1%3Asection%3D1

Kant, I. (1954). Kritik der Urteilskraft (K. Vorländer, Ed.). Hamburg: Verlag von Felix Meiner. (Originally published 1799)

Klinge, M. (2015). *Kansallisbiografia. Snellman, Johan Vilhelm (1806 - 1881)*. Online: http://urn.fi/urn:nbn:fi:sks-kbg-003639

Koja, S. (Ed.). (2005). *Nordic Dawn. Modernism's Awakening in Finland 1890-1920*. Prestel.

Kôkan, S. (2006). Discussion on Western Painting (Seiyô gadan). In S. Addiss, G. Groemer, & J. T. Rimer (Eds.), *Traditional Japanese Arts and Culture. An Illustrated Sourcebook* (pp. 173–176). Honolulu, HI: University of Hawai'i Press. (Original work published 1799)

Merviö, M. (2017). *Interpretation of Visual Arts Across Societies and Political Culture: Emerging Research and Opportunities*. Hershey, PA: IGI International. doi:10.4018/978-1-5225-2554-7

Murray, P. (Ed.). (1989). *Genius. The History of an Idea*. Oxford: Basil Blackwell.

Nishiyama, M. (1997). *Edo Culture. Daily Life and Diversions in Urban Japan, 1600-1868* (G. Groemer, Trans. & Ed.). Honolulu, HI: University of Hawai'i Press.

Ooms, H. (1984). Early Tokugawa Ideology. In P. Nosco (Ed.), *Confucianism and Tokugawa Culture* (pp. 27–61). Honolulu, HI: University of Hawai'i Press.

*Pohjoismainen taide 1880- luvulla*. (1986). Helsinki: Amos Andersonin Taidemuseo.

Suomen Taideyhdistyksen perustaminen. Lähteillä. (n.d.). *Suomen kansallisgalleria*. http://www.lahteilla.fi/fi/page/suomen-taideyhdistys-suomen-taideyhdistyksen-perustaminen

Suomen Taideyhdistyksen piirustuskoulu Helsingissä. Lähteillä. (n.d.). *Suomen kansallisgalleria*. http://www.lahteilla.fi/fi/page/suomen-taideyhdistys-suomen-taideyhdistyksen-piirustuskoulu-helsingissä

Wichmann, S. (1981). Japonisme. The Japanese Influence on Western Art since 1858. London: Thames & Hudson.

Willner-Rönnholm, M. (1996). *Taidekoulun arkea ja unelma. Turun piirustuskoulu 1830-1981*. Vammala: Turun taidemuseo.

Wittgenstein, L. (1922). *Tractacus Logico-Philosophicus*. London: Kegan Paul. https://people.umass.edu/klement/tlp/tlp.pdf

# Compilation of References

Abe, J. O., Popoola, A. P. I., Ajenifuja, E., & Popoola, O. M. (2019). Hydrogen energy, economy and storage: Review and recommendation. *International Journal of Hydrogen Energy*, *44*(29), 15072–15086. doi:10.1016/j.ijhydene.2019.04.068

Abernathy, W. J. (1978). *The Productivity Dilemma*. Baltimore, MD: Johns Hopkins Univ. Press.

Abrams, M., & Harpham, G. (2011). *A Glossary of Literary Terms*. Boston: Cengage Learning.

Abusada, H., & Elshater, A. (2018). *The Fifth Dimension: Urban design- The atmospheres of the City*. Cairo, Egypt: Academic Bookshop. (in Arabic)

Achard, F., & Banoin, M. (2003). Fallows, forage production and nutrient transfers by livestock in Niger. *Nutrient Cycling in Agroecosystems*, *65*(2), 183–189. doi:10.1023/A:1022111117516

Actualidad.com. (2009, November 25). *Clinic closed and surgeon arrested for malpractice*. Accessed March 3, 2019 retrieved from https://www.diariolibre.com/actualidad/clinic-closed-andsurgeon-arrested-for-malpractice-MIDL224822

Adams, W. (2006). *The future of sustainability, re-thinking environment and development in the twentyfirst century*. The World Conservation Union.

Adam, T., Hashim, U., Ali, M. E., & Leow, P. L. (2013). The electroosmosis mechanism for fluid delivery in PDMS multi-layer microchannel. *Advanced Science Letters*, *19*(1), 12–15. doi:10.1166/asl.2013.4670

Adhikari, K., & Hartemink, A. E. (2016). Linking soils to ecosystem services— A global review. *Geoderma*, *262*, 101–111. doi:10.1016/j.geoderma.2015.08.009

Agarwal, A. (1997). *Homicide by Pesticides*. New Delhi: Centre for Science and Environment.

Aggarwal, A. L. (1993). *Air Quality Management Goals in India*. Paper presented at International Workshop on Urban Air Quality Management, Mumbai, India.

Aguey-Zinsou, K.-F., & Ares-Fernandez, J.-R. (2010). Hydrogen in magnesium: New perspective toward functional stores. *Energy & Environmental Science*, *3*(5), 252. doi:10.1039/b921645f

Aguey-Zinsou, K.-F., Ares-Fernandez, J.-R., Klaasen, T., & Bormann, R. (2007). Effect of Nb2O5 on MgH2 properties during mechanical milling. *International Journal of Hydrogen Energy*, *32*(13), 2400–2407. doi:10.1016/j.ijhydene.2006.10.068

Aharony, L., & Strasser, S. (1993). Patient satisfaction: What we know about and what we still need to explore. *Medical Care Review*, *50*(1), 49–79. doi:10.1177/002570879305000104 PMID:10125117

Ahern, J. (2012). Urban landscape sustainability and resilience: The promise and challenges of integrating ecology with urban planning and design. *Landscape Ecology*, *28*(6), 1203-1212. doi:10.100710980-012-9799-z

Aherna, J., Cilliers, S., & Niemeläc, J. (2014). The concept of ecosystem services in adaptive urban planning and design: A framework for supporting innovation. *Landscape and Urban Planning*, *125*, 254–259. doi:10.1016/j.landurbplan.2014.01.020

Ahern, J. (2011). From fail-safe to safe-to-fail: Sustainability and resilience in the new urban world. *Landscape and Urban Planning*, *100*(4), 341–343. doi:10.1016/j.landurbplan.2011.02.021

Ahluwalia, R. K., Peng, J -K., & Hua, T. Q. (2015). Sorbent material property requirements for on board hydrogen storage for automotive fuel cell systems. *International Journal of Hydrogen Energy*, *40*, 673.

AHRQ. (2017). *Agency for Healthcare Research and Quality. National healthcare disparities report*. Washington, DC: Government Printing Office.

AICAF. (1994). *Report of the study for formulation of sustainable agriculture development plans for Africa – Valley bottom rice farming in West Africa (second phrase)*. Association for International Cooperation of Agriculture and Forestry.

Albright, M. (2016). Servant leadership: Not just buzzwords. *Strategic Finance*, *98*(4), 19-20.

Allred, A. (2015, July 30). *5 On Your Side Investigates: Going under the knife, "on the cheap"* Accessed on March 6, 2019 from https://www.usatoday.com/story/news/health/2015/07/30/5-onyour-side-investigates-going-under-on-cheap/30868471/

Almstead, A., Brouder, P., Karlsson, S., & Lundmark, L. (2014). Beyond post-productivism: From rural policy discourse to rural diversity. *European Countryside*, *4*(4), 297–306. doi:10.2478/euco-2014-0016

Alsan, M., Bloom, D. E., & Canning, D. (2006). The effect of population health on foreign direct investment inflows to low-and middle-income countries. *World Development*, *34*(4), 613–630. doi:10.1016/j.worlddev.2005.09.006

Alter, A. (2017). *Irresistible: The Rise of Addictive Technology and the Business of Keeping Us Hooked*. Harmondsworth: Penguin.

Altieri, M. A., Companioni, N., Cañizares, K., Murphy, C., Rosset, P., Bourque, M., & Nicholls, C. I. (1999). The greening of the "barrios": Urban agriculture for food security in Cuba. *Agriculture and Human Values*, *16*(2), 131–140. doi:10.1023/A:1007545304561

Altieri, M. A., & Toledo, V. M. (2011). The agroecological revolution in Latin America: Rescuing nature, ensuring food sovereignty and empowering peasants. *The Journal of Peasant Studies*, *38*(3), 587–612. doi:10.1080/03066150.2011.582947

Andersen, K. V., & Lorentzen, M. (2005). *The Geography of the Danish Creative Class – A Mapping and Analysis. Danish part of "Technology, Talent and Tolerance in European Cities: A comparative analysis"*. Cph Business Sch.

Andersen, L. B., Bjoernholt, B., Bro, L. B., & Holm-Petersen, C. (2016). Leadership and motivation: A qualitative study of transformational leadership and public service motivation. *Review of Adm. The Sciences*.

Anderson, J., & Gronkvist, S. (2019). Large scale storage of hydrogen. *International of Hydrogen Energy*, *44*(23), 11901–11919. doi:10.1016/j.ijhydene.2019.03.063

Anderson, P. A. (2008). Storage of hydrogen in zeolites. In G. Walker (Ed.), *Solid-state hydrogen storage: Materials and chemistry* (pp. 233–260). Woodhead Publishing.

Anon. (1996). *Report on Water Quality Monitoring of Yamuna River*. New Delhi: Central Pollution Control Board, mimeo.

Antolini, E. (2003). Formation of carbon supported PtM alloys for low temperature fuel cells: A review. *Materials Chemistry and Physics*, *78*(3), 563–573. doi:10.1016/S0254-0584(02)00389-9

*Compilation of References*

Antolini, E. (2009). Carbon supports for low temperature fuel cell catalysts. *Applied Catalysis B: Environmental*, *88*(1-2), 1–24. doi:10.1016/j.apcatb.2008.09.030

Aoyagi, H., Aoki, K., & Masumoto, T. (1995). Effect of ball milling on hydrogen absorption properties of FeTi, Mg2Ni and LaNi5. *Journal of Alloys and Compounds*, *231*(1-2), 804–809. doi:10.1016/0925-8388(95)01721-6

Aprea, C., & Greco, A. (1998). An experimental evaluation of the greenhouse effect in R22 substitution. *Energy Conversion and Management*, *39*(9), 877–887. doi:10.1016/S0196-8904(97)10058-9

Aprea, C., Greco, A., & Maiorino, A. (2013). The use of the first and of the second order phase magnetic transition alloys for an AMR refrigerator at room temperature: A numerical analysis of the energy performances. *Energy Conversion and Management*, *70*, 40–55. doi:10.1016/j.enconman.2013.02.006

Aprea, C., Greco, A., Maiorino, A., & Masselli, C. (2015). A comparison between rare earth and transition metals working as magnetic materials in an AMR refrigerator in the room temperature range. *Applied Thermal Engineering*, *91*, 767–777. doi:10.1016/j.applthermaleng.2015.08.083

Aprea, C., Greco, A., Maiorino, A., & Masselli, C. (2018a). The drop-in of HFC134a with HFO1234ze in a household refrigerator. *International Journal of Thermal Sciences*, *127*, 117–125. doi:10.1016/j.ijthermalsci.2018.01.026

Aprea, C., Greco, A., Maiorino, A., & Masselli, C. (2018b). Solid-state refrigeration: A comparison of the energy performances of caloric materials operating in an active caloric regenerator. *Energy*, *165*, 439–455. doi:10.1016/j.energy.2018.09.114

Aprea, C., Greco, A., Maiorino, A., & Masselli, C. (2018c). Energy performances and numerical investigation of solid-state magnetocaloric materials used as refrigerant in an active magnetic regenerator. *Thermal Science and Engineering Progress*, *6*, 370–379. doi:10.1016/j.tsep.2018.01.006

Aprea, C., Greco, A., Maiorino, A., & Masselli, C. (2019). The Employment of Caloric-Effect Materials for Solid-State Heat Pumping. *International Journal of Refrigeration*. doi:10.1016/j.ijrefrig.2019.09.011

Aprea, C., Greco, A., Maiorino, A., Masselli, C., & Metallo, A. (2016a). HFO1234yf as a drop-in replacement for R134a in domestic refrigerators: A life cycle climate performance analysis. *International Journal of Heat and Technology*, *34*(2), S212–S218. doi:10.18280/ijht.34S204

Aprea, C., Greco, A., Maiorino, A., Masselli, C., & Metallo, A. (2016b). HFO1234ze as drop-in replacement for R134a in domestic refrigerators: An environmental impact analysis. *Energy Procedia*, *101*, 964–971. doi:10.1016/j.egypro.2016.11.122

Armitage, D., Marschke, M., & Plummer, R. (2008). Adaptive comanagement and the paradox of learning. *Global Environmental Change*, *18*(1), 86–98. doi:10.1016/j.gloenvcha.2007.07.002

Asalatha (n.d.).

Asanuma. S. (2004). Desertification in Burkina Faso in West Africa and Farmer's Coping Strategy – Life-scale Analysis. *Expert bulletin for International Cooperation of Agriculture and Forestry, 25*(2), 1-18. (in Japanese)

Association for Emissions Control by Catalyst (AECC). (n.d.). *Adsorbers*. Retrieved from https://www.aecc.eu/technology/adsorbers/

Ataca, C., Akturk, E., & Ustunel, H. (2009). High-capacity hydrogen storage by metallized graphene. *Applied Physics Letters*, *93*(4), 043123. doi:10.1063/1.2963976

Athanasopoulou, A., Moss-Cowan, A., Smets, M., & Morris, T. (2018). Claiming the Corner Office: Female CEO Careers and Implications for Leadership Development. *Human Resource Management Journal, 57*(2), 617–639. doi:10.1002/hrm.21887

Atkin, C. K. (1973). Instrumental utilities and information seeking. In P. Clarke (Ed.), *New models for communication research* (pp. 205–242). Beverly Hills, CA: Sage.

Atkinson, R., & Castro, D. (2008). *Digital quality of life.* The Information Technology and Innovation Foundation.

Atun, Gurol-Urganci & Sheridan. (2007a). Uptake and diffusion of pharmaceutical innovations in health systems. *International Journal of Innovation Management, 11*(2), 299-321.

Atun, R. A., & Sheridan, D. (2007). Editorial Innovation In Healthcare: The Engine Of Technological Advances. *International Journal of Innovation Management, 11*(2), v–x. doi:10.1142/S1363919607001692

Avolio, B. J., & Bass, B. M. (Eds.). (2001). *Developing potential across a full range of Leadership Tm: Cases on transactional and transformational leadership.* Psychology Press. doi:10.4324/9781410603975

Awogbade, M. O. (1983). *Fulani Pastoralism – Jos Case Study.* Ahmade Bello University Press Limited.

Babakus, E., & Boller, G. W. (1992). An empirical assessment of the SERVQUAL scale. *Journal of Business Research, 24*(3), 253–268. doi:10.1016/0148-2963(92)90022-4

Bae, Y. H., & Park, K. (2011). Targeted drug delivery to tumors: Myths, reality and possibility. *Journal of Controlled Release, 153*(3), 198–205. doi:10.1016/j.jconrel.2011.06.001 PMID:21663778

Baines, C., & Smart, J. (1991). *A Guide to Habitat Creation.* London Ecology Unit.

Baker, J. (1986). The Role of the Environment in Marketing Services: The Consumer Perspective. In *The Service Challenge: Integrating for Competitive Advantage* (pp. 79–84). Chicago: American Marketing Association.

Balcerzak, M., Nowak, M., & Jurczyk, M. (2012). Nanocrystalline TiNi, Ti2Ni alloys for hydrogen storage. *Materials Science and Engineering, 33*, 370–373.

Baldi, A., Palsson, G. K., Gonzalez-Silveira, M., Schreuders, H., Slaman, M., Rector, J. H., ... Griessen, R. (2015). Mg/Ti multilayers: Structural, optical and hydrogen absorption properties. *Physical Review. B, 81*(22), 224203. doi:10.1103/PhysRevB.81.224203

Bandura, A. (1999). Moral disengagement in the perpetration of inhumanities. *Personality and Social Psychology Review, 3*(3), 193–209. doi:10.120715327957pspr0303_3 PMID:15661671

Banerjee, S., Secchi, S., Fargione, J., Polasky, S., & Kraft, S. (2013). How to sell ecosystem services: A guide for designing new markets. *Frontiers in Ecology and the Environment, 11*(6), 297–304. doi:10.1890/120044

*Bangladesh Demographic and Health Survey Report.* (2014). Dhaka: Ministry of Planning, Planning Commission.

Barbujani, G., Ghirotto, S., & Tassi, F. (2013). Nine things to remember about human genome diversity. *Tissue Antigens, 82*(3), 155–164. doi:10.1111/tan.12165 PMID:24032721

Barcelona. (2011). *Smart city 22.* http://www.22barcelona.com

Barclay, J. A. (1982). Theory of an active magnetic regenerative refrigerator (No. LA-UR-83-1251; CONF-821237-1). Los Alamos National Lab.

Barclay, J. A. (1994). Active and passive magnetic regenerators in gas/magnetic refrigerators. *Journal of Alloys and Compounds, 207*, 355–361. doi:10.1016/0925-8388(94)90239-9

*Compilation of References*

Barefoot, J. K. (2012). *US Multinational Companies – Operations of US Parents and their Affiliates in 2010*. Bur. US Survey of Current Business.

Barkhordarian, G., Klaasen, T., & Bormann, R. (2006). Catalytic mechanism of transition-metal compounds on Mg hydrogen sorption reaction. *The Journal of Physical Chemistry B*, *110*(22), 11020–11024. doi:10.1021/jp0541563 PMID:16771356

Bassan, S., & Michaelsen, M. A. (2013). Honeymoon, medical treatment or big business? an analysis of the meanings of the term "reproductive tourism" in German and Israeli public media discourses. *Philosophy, Ethics, and Humanities in Medicine; PEHM*, *8*(1), 1–8. doi:10.1186/1747-5341-8-9 PMID:23962355

Bass, B. (1988). The inspirational processes of leadership. *Journal of Management Development*, *7*(5), 21–31. doi:10.1108/eb051688

Batchelor, C., & Butterworth, J. (2008). *Learning Alliance Briefing Note 9: Visioning* (draft). http://www.switchurbanwater.eu/la_guidance.php

Batty, M. (2008). The size, scale, and shape of cities. *Science*, *319*(5864), 769–771. doi:10.1126cience.1151419 PMID:18258906

Bavec, M., Mlakar, S. G., Rozman, C., Pazek, K., & Bavec, F. (2009). Sustainable agriculture based on integrated and organic guidelines: Understanding terms—the case of Slovenian development and strategy. *Outlook on Agriculture*, *38*(1), 89–95. doi:10.5367/000000009787762824

Beattie, A. (2005). *Cairo: A Cultural and Literary History*. Oxford: Signal Books.

Beattie, S. D., Setthanan, U., & McGrady, G. S. (2011). Thermal desorption of hydrogen from magnesium hydride (MgH2): An in situ microscopy study by environmental SEM and TEM. *International Journal of Hydrogen Energy*, *36*(10), 6014–6021. doi:10.1016/j.ijhydene.2011.02.026

Becker, F. D., & Steele, F. (1995). *Workplace by design: mapping the high-performance workscape*. New York: Jossey-Bass.

Belisent, J. (2010). *Getting clever about smart cities: new opportunities require new business models*. Forester for Ventor Strategy Professionals.

Bell, B. (1914, Oct. 2). Health Seekers Bureau. *Summit City Journal*.

Bell, R., Jung, J., & Zacharilla, L. (2009). *Broadband economies: creating the community of the 21st century*. Intelligent Community Forum.

Benoit, M., Rizzo, D., Marraccini, E., Moonen, A. C., Galli, M., Lardon, S., ... Bonari, E. (2012). Landscape agronomy: A new field for addressing agricultural landscape dynamics. *Landscape Ecology*, *27*(10), 1385–1394. doi:10.100710980-012-9802-8

Benson, H., Beary, J. F., & Carol, M. P. (1974). The relaxation response. *Psychiatry*, *37*(1), 37–46. doi:10.1080/00332747.1974.11023785 PMID:4810622

Benson, H., & Klipper, M. (1975). *The Relaxation Response*. W Morrow & Co.

Berchicci, L., & Bodewes, W. (2005). Bridging Environmental Issues with New Product Development. *Business Strategy and the Environment*, *14*(5), 272–285. doi:10.1002/bse.488

Berger, I. E., Cunningham, P. H., & Drumwright, M. E. (2007). Mainstreaming Corporate Social Responsibility: Developing Markets for Virtue. *California Management Review*, *49*(4), 132–157. doi:10.2307/41166409

Berkes, F., Colding, J., & Folke, C. (2000). Rediscovery of traditional ecological knowledge as adaptive management. *Ecological Applications*, *10*(5), 1251–1262. doi:10.1890/1051-0761(2000)010[1251:ROTEKA]2.0.CO;2

Berkman, N.D., Sheridan, S.L., Donahue, K.E., Halpern, D.J., Viera, A., Crotty, K., ... Viswanathan, M., (2011, March). *Health literacy interventions and outcomes: An updated systematic review*. Evidence Report/Technology Assessment No. 199.

Berkman, N. D., Davis, T. C., & McCormack, L. (2010). Health literacy: What is it? *Journal of Health Communication*, *15*(S2), 9–19. doi:10.1080/10810730.2010.499985 PMID:20845189

Berman, D. K. (2008, October 28). The game: Post-Enron crackdown comes up woefully short. *The Wall Street Journal*, p. C2.

Bhat, V. V., Rougier, A., Aymard, L., Nazri, G. A., & Tarascon, J. M. (2008). High surface area niobium oxides as catalysts for improved hydrogen sorption properties of ball milled MgH2. *Journal of Alloys and Compounds*, *406*(1-2), 507–512. doi:10.1016/j.jallcom.2007.05.084

Biswas, A. K., & Hartly, K. (2017). Asist K Biswas & Kris Hartly. Delhi should follow Beijing example in tackling air pollution. *The Wire*.

Bitner, M. J., & Hubbert, A. R. (1994). Encounter satisfaction versus overall satisfaction versus quality. *Service quality: New directions in theory and practice, 34*(2), 72-94.

Bitner, M. J., Hubbert, A. R., Rust, R., & Oliver, R. (1994). Service quality: New directions in theory and practice. *Encounter satisfaction versus overall satisfaction versus quality, 77*.

Bitner, M. J. (1992). Servicescapes: The impact of physical surroundings on customers and employees. *Journal of Marketing*, *56*(2), 57–71. doi:10.1177/002224299205600205

Bitner, M. J., Booms, B. H., & Tetreault, M. S. (1990). The service encounter: Diagnosing favorable and unfavorable incidents. *Journal of Marketing*, *54*(1), 71–84. doi:10.1177/002224299005400105

Bizer, C., Heath, T., & Berners-Lee, T. (2009). Linked data – the story so far. *International Journal on Semantic Web and Information Systems*. http://linkeddata.org/docs/ijswis-special-issue

Bjørk, R., Bahl, C. R. H., & Katter, M. (2010a). Magnetocaloric properties of $LaFe_{13-x-y}Co_xSi_y$ and commercial grade Gd. *Journal of Magnetism and Magnetic Materials*, *322*(24), 3882–3888. doi:10.1016/j.jmmm.2010.08.013

Blench, R. (2010). *The transformation of conflict between pastoralists and cultivators in Nigeria, Review Paper prepared for DFID, Nigeria*. Review Paper prepared for DFID, Nigeria. Unpublished paper. Online. http://www.rogerblench.info/Development/Nigeria/Pastoralism/Fadama%20II%20paper.pdf

Blewitt, J. (2008). *Understanding sustainable development*. London: Earthscan.

Bogdanovic, B., Felderhoff, M., Pommerin, A., Schuth, F., & Spielkamp, N. (2006). Advanced hydrogen-storage materials based on Sc-, Ce-, and Pr-doped NaAlH4. *Advanced Materials*, *18*(9), 1198–1201. doi:10.1002/adma.200501367

Böhme, G. (1993). Atmosphere as the Fundamental Concept of a New Aesthetics. *Thesis Eleven*, *36*, 113-126.

Böhme, G. (2005). Atmosphere as the subject matter of architecture. In H. Meuron & P. Ursprung (Eds.), Natural History (pp. 398–407). London: Springer Science & Business Media.

Böhme, G. (2013). Atmosphere as Mindful Physical Presence in Space. *OASE Journal for Architecture*, 21-32.

Böhme, G. (1998). Atmosphere as an aesthetic concept. In G. Confurius (Ed.), *Constructing Atmospheres* (Vol. 68, pp. 112–115). Daidalos.

Böhme, G. (2000). Acoustic Atmospheres: A Contribution to the Study of Ecological Aesthetics. *The Journal of Acoustic Ecology, 1*(1), 14–18.

Böhme, G. (2010). On beauty. *The Nordic Journal of Aesthetics, 39*, 22–33.

Böhme, G. (2014). Urban Atmospheres: Charting New Directions for Architecture and Urban Planning. In C. Borch (Ed.), *Architectur, Architectural Atmospheres – On the Experience and Politics of Architecture* (pp. 42–59). Basel: Birkhäuser. doi:10.1515/9783038211785.42

Bohn, K., & Viljoen, A. (2005). More city with less space: Vision for lifestyle. In A. Viljoen (Ed.), *Continuous Productive Urban Landscapes: Designing Urban Agriculture for Sustainable Cities* (pp. 251–264). Burlington, MA: Architectural Press.

Bolund, P., & Hunhammar, S. (1999). Ecosystem services in urban areas. *Ecological Economics, 29*(2), 293–301. doi:10.1016/S0921-8009(99)00013-0

Boonyabancha, S. (2003). A decade of change: From the Urban Community Development Office (UCDO) to the Community Organizations Development Institute (CODI) in Thailand. Poverty Reduction in Urban Areas working paper 12. International Institute for Environment and Development.

Bostonian, N. J., Goulet, H., O'Hara, J., Masner, L., & Racette, G. (2004). Towards insecticide free orchards: Flowering plants to attract beneficial arthropods. *Biocontrol Science and Technology, 14*(1), 25–37. doi:10.1080/09583150310001606570

Boulding, W., Kalra, A., Staelin, R., & Zeithaml, V. A. (1993). A Dynamic Process Model of Service Quality: From Expectations to Behavioral Intentions. *JMR, Journal of Marketing Research, 30*(February), 7–27. doi:10.1177/002224379303000102

Bouoiyour, J. (2007). The determining factors of foreign direct investment in Morocco. *Savings and Development*, 91-106.

Boyd, C. E., & Gross, A. (2000). Water use and conservation for inland aquaculture ponds. *Fisheries Management and Ecology, 7*(1-2), 55–63. doi:10.1046/j.1365-2400.2000.00181.x

Brady, M. K., & Cronin, J. J. Jr. (2001). Some new thoughts on conceptualizing perceived service quality: A hierarchical approach. *Journal of Marketing, 65*(3), 34–49. doi:10.1509/jmkg.65.3.34.18334

Brady, M. K., & Robertson, C. J. (2001). Searching for a consensus on the antecedent role of service quality and satisfaction: An exploratory cross-national study. *Journal of Business Research, 51*(1), 53–60. doi:10.1016/S0148-2963(99)00041-7

Breese, G. (1974). *Urban and Regional for the Delhi-N. Delhi area, Capital for conquerors and Country*. Academic Press.

Brondolo, E., Rieppi, R., Kelly, K. P., & Gerin, W. (2003). Perceived racism and blood pressure: A review of the literature and conceptual and methodological critique. *PubMed, 25*(1), 55–56. doi:10.1207/S15324796ABM2501_08 PMID:12581937

Brook, R. H., & Williams, K. N. (1975). Quality of health care for the disadvantaged. *Journal of Community Health, 1*(2), 132–156. doi:10.1007/BF01319207 PMID:777052

Brourwer, J., & Powell, J. M. (1995). Soil aspects of nutrient cycling in a manure experiment in Niger. In *Livestock and sustainable nutrient cycling in mixed farming systems of sub-Saharan Africa. Vol. 11: Technical papers. Proceedings of an International Conference held in Addis Ababa, Ethiopia, 22-26 November 1993*. ILCA.

Brourwer, J., & Powell, J. M. (1998). Micro-topography, water balance, millet yield and nutrient leaching in a manuring experiment on sandy soil in south-west Niger. In *Soil fertility management in West African land use systems. Proceedings of a workshop held in Niamey, 4-8 March 1997*. Margarf.

Brown, J. (2018) Hypernatural monitoring: The real reason we're so addicted to our smartphone. *Independent*. https://www.independent.co.uk/news/long_reads/smartphone-addiction-hypernatural-monitoring-technology-nomophobia-social-media-a8253191.html

Brown, P.C., Roediger, H. L., & McDaniel M. A. (2014). *Make it stick*. Harvard University Press 2014.

Brown, G. V. (1976). Magnetic heat pumping near room temperature. *Journal of Applied Physics*, *47*(8), 3673–3680. doi:10.1063/1.323176

Brown, J. S., & Domanski, P. A. (2014). Review of alternative cooling technologies. *Applied Thermal Engineering*, *64*(1-2), 252–262. doi:10.1016/j.applthermaleng.2013.12.014

Brown, T. J., Churchill, G. A. Jr, & Peter, J. P. (1993). Research note: Improving the measurement of service quality. *Journal of Retailing*, *69*(1), 127–139. doi:10.1016/S0022-4359(05)80006-5

Bruins, H. J., Evenari, M., & Nessler, U. (1986). Rainwater-harvesting agriculture for food production in arid zones: The challenge of the African famine. *Applied Geography (Sevenoaks, England)*, *6*(1), 13–32. doi:10.1016/0143-6228(86)90026-3

Burbidge, J. L. (1971). Production flow analysis. *Production Engineering*, *50*(4-5), 139–152. doi:10.1049/tpe.1971.0022

Burra, S., & Patel, S. (2002). Community toilets in Pune and other Indian Cities. *Participatory Learning and Action*, *44*, 43–45.

Busch, L., & Bain, C. (2004). New! Improved? The Transformation of the Global Agrifood System. *Rural Sociology*, *69*(3), 321–346. doi:10.1526/0036011041730527

Business Dictionary. (2018). *Ambiance*. Retrieved from www.businessdictionary.com: http://www.businessdictionary.com/definition/ambiance.html

Buttle, F. (1996). SERVQUAL: Review, critique, research agenda. *European Journal of Marketing*, *30*(1), 8–32. doi:10.1108/03090569610105762

CAG Report. (2005). *CAG (The Comptroller & Auditor General of India) Report on Government of Delhi*. Author.

Cagdas, M., Sezer, A.D., & Bucak, S. (2014). *Application of nanotechnology in drug delivery- Liposomes as Potential Drug Carrier Systems for Drug Delivery*. Academic Press.

Caldwell, D. F., & Moberg, D. (2007). An exploratory investigation of the effect of ethical culture in activating moral imagination. *Journal of Business Ethics*, *73*(2), 193–204. doi:10.100710551-006-9190-6

Caligiuri, P. (2013, March). Develop your cultural agility. *TD Magazine*. Retrieved from https://www.td.org/magazines/td-magazine/develop-your-cultural-agility

Cameron, K., Crooks, V. A., Chouinard, V., Snyder, J., Johnston, R., & Casey, V. (2014). Motivation, justification, normalization: Talk strategies used by Canadian medical tourists regarding their choices to go abroad for hip and knee surgeries. *Social Science & Medicine*, *106*, 93–100. doi:10.1016/j.socscimed.2014.01.047 PMID:24556288

Cardella, U., Decker, L., & Klein, H. (2017). Roadmap to economically viable hydrogen liquefaction. *International Journal of Hydrogen Energy*, *42*(19), 13329–13338. doi:10.1016/j.ijhydene.2017.01.068

Carman, J. M. (1990). Consumer perceptions of service quality: An assessment of T. *Journal of Retailing*, *66*(1), 33.

*Compilation of References*

Carmona, M., Heath, T., Oc, T., & Tiesdell, S. (2003). *Public Places, Urban Spaces: The Dimensions of Urban Design.* Oxford: Architectural Press.

Carr, N. (2017, Oct. 6). How Smartphones Hijack Our Minds. *Wall Street Journal.*

Cavazza, L. (1996). Agronomia aziendale e agronomia del territorio. *Riv Agron, 30,* 310–319.

Cayne, B. S. (1998). *The New Lexicon Webster's Dictionary Of The English Language.* New York: Lexicon Publications Inc.

Census. (2011). *Govt. of India.*

Chambers, A., Park, C., Baker, R. T. K., & Rodriguez, N. M. (1998). Hydrogen storage in graphite nanofibers. *Journal of Physical Chemistry, 102*(22), 4253–4256. doi:10.1021/jp9801141

Chandrakumar, K. R. S., & Ghosh, S. K. (2008). Hydrogen adsorption on Na-SWCNT systems. *Nano Letters, 8,* 13–19. doi:10.1021/nl071456i PMID:18085807

Chattopadhyay, S. (2007). Religion, spirituality, health and medicine: Why should Indian physicians care? *Journal of Postgraduate Medicine, 53*(4), 262. doi:10.4103/0022-3859.33967 PMID:18097118

Chee-Yee, C., & Kumar, S. (2003). Sensor networks: Evolution, opportunities, and challenges. *Proceedings of the IEEE, 91*(8), 1247–1256. doi:10.1109/JPROC.2003.814918

Chen, C.-Y., & Lin, S.-K. (2007). Improvement of hydrogen absorption performance of mechanically alloyed Mg2Ni powders by silver doping. *Materials Transactions, 48*(5), 1113–1118. doi:10.2320/matertrans.48.1113

Cheng, L. K., & Kwan, Y. K. (2000). What are the determinants of the location of foreign direct investment? The Chinese experience. *Journal of International Economics, 51*(2), 379–400. doi:10.1016/S0022-1996(99)00032-X

Chen, J. M., & Hamilton, D. L. (2015). Understanding diversity: The importance of social acceptance. *Personality and Social Psychology Bulletin, 41*(4), 586–598. doi:10.1177/0146167215573495 PMID:25713169

Chen, J., & Yan, Z. (1998). The effect of thermal resistances and regenerative losses on the performance characteristics of a magnetic Ericsson refrigeration cycle. *Journal of Applied Physics, 84*(4), 1791–1795. doi:10.1063/1.368349

Chen, P., & Zhu, M. (2008). Recent progress in hydrogen storage. *Materials Today, 11*(12), 36–43. doi:10.1016/S1369-7021(08)70251-7

Chen, Y. S., Lai, S. B., & Wen, C. T. (2006). The influence of green innovation performance on corporate advantage in Taiwan. *Journal of Business Ethics, 67*(4), 331–339. doi:10.100710551-006-9025-5

Choi, K.-S., Lee, H., Kim, C., & Lee, S. (2005). The Service Quality Dimensions and Patient Satisfaction Relationships in South Korea: Comparisons across Gender, Age and Types of Service. *Journal of Services Marketing, 19*(3), 140–150. doi:10.1108/08876040510596812

Chokshi, Patil, Khanna, Neogi, Sharma, Paul, & Zodpey. (2016). Health Systems in India. *J Perinatol., 36*(3), S9–S12. doi:10.1038/jp.2016.184

Choudhary, A. A. (2017). Amit Anand Choudhary. Yamuna pollution: Supreme Court seeks report on sewage treatments plants. Economic Times.

Chowdhury, A. R. (1997). Linking Threat. Down to Earth, Society for Environmental Communications, 16(14), 20-24.

Chowdhury, A. R. (1998). Breathing Benzene. Down to Earth, Society for Environmental Communications, 7(1), 15.

Christian, M., & Aguey-Zinsou, K. (2010). Destabilization of complex hydrides through size effects. *Nanoscale*, *2*(12), 2587–2590. doi:10.1039/c0nr00418a PMID:20886168

Cleveland, M., & And Cleveland, S. (2018). Building engaged communities—A collaborative leadership approach. *Smart Cities*, *1*(1), 155–162. doi:10.3390martcities1010009

Coey, J. M. D. (2012). Permanent magnets: Plugging the gap. *Scripta Materialia*, *67*(6), 524–529. doi:10.1016/j.scriptamat.2012.04.036

Collette. (2011). Thunnus obesus The IUCN Red List of Threatened Species 2011: e.T21859A9329255. doi:10.2305/IUCN.UK.2011-2

Companioni, N., Rodríguez Nodals, A., Carrión, M., Alonso, R. M., Ojeda, Y., & Peña, E. (1997). La Agricultura Urbana en Cuba. Su participación en la seguridad alimentaria. Conferencias. III Encuentro Nacional de Agricultura Orgánica. Villa Clara, Cuba: Central University of Las Villas.

Comstock, G. W., & Patridge, K. B. (1972). Church Attendance and Health. *Journal of Chronic Diseases*, *25*(12), 665–672. doi:10.1016/0021-9681(72)90002-1 PMID:4648512

Confederation of Indian Industry. (2017). https://cii.in/WebCMS/Upload/Economic_Overview_Delhi.pdf

Costanza, R., d'Arge, R., De Groot, R., Farber, S., Grasso, M., Hannon, B., … Paruelo, J. (1998). The value of the world's ecosystem 2 services and natural capital. *Ecological Economics*, *25*(1), 3-15.

Cote, M., & Nightingale, A. J. (2012). Resilience thinking meets social theory: Situating social change in socio-ecological systems (SES) research. *Progress in Human Geography*, *36*(4), 475–489. doi:10.1177/0309132511425708

Cozens, M. (2008). New urbanism, crime and the suburbs: A review of the evidence. *Urban Policy and Research*, *26*(4), 429–444. doi:10.1080/08111140802084759

Crawford, M. (2010). *Creating a forest garden: working with nature to grow edible crops*. Green Books.

Crimmins, E. M., Kim, J. K., Alley, D. E., Karlamangla, A., & Seeman, T. (2007). Hispanic paradox in biological risk profiles. *American Journal of Public Health*, *97*(7), 1305–1310. doi:10.2105/AJPH.2006.091892 PMID:17538054

Croix, St. (1972). *The Fulani of Northern Nigeria: Some general notes*. Gregg International Publishers Limited.

Cronin, J. J. Jr, Brady, M. K., & Hult, G. T. M. (2000). Assessing the effects of quality, value, and customer satisfaction on consumer behavioral intentions in service environments. *Journal of Retailing*, *76*(2), 193–218. doi:10.1016/S0022-4359(00)00028-2

Cronin, J. J. Jr, & Taylor, S. A. (1992). Measuring service quality: A reexamination and extension. *Journal of Marketing*, *56*(3), 55–68. doi:10.1177/002224299205600304

Crooks, V. A., Turner, L., Snyder, J., Johnston, R., & Kingsbury, P. (2011). Promoting medical tourism to India: Messages, images, and the marketing of international patient travel. *Social Science & Medicine*, *72*(5), 726–732. doi:10.1016/j.socscimed.2010.12.022 PMID:21310519

Crouzat, E., Mouchet, M., Turkelboom, F., Byczek, C., Meersmans, J., Berger, F., & Lavorel, S. (2015). Assessing bundles of ecosystem services from regional to landscape scale: Insights from the French Alps. *Journal of Applied Ecology*, *52*(5), 1145–1155. doi:10.1111/1365-2664.12502

CSA America, Inc. (n.d.). *ANSI HGV 2-2014*. ANSI Webstore.

*Compilation of References*

Cuéllar-Padilla, M., & Calle-Collado, Á. (2011). Can we find solutions with people? Participatory action research with small organic producers in Andalusia. *Journal of Rural Studies*, *27*(4), 372–383. doi:10.1016/j.jrurstud.2011.08.004

Cullen, G. (1961). *The Concise Townscape*. Architectural Press.

Cullen-Lester, K. L., Maupin, C. K., & Carter, D. R. (2017). Incorporating social networks into leadership development: A conceptual model and evaluation of research and practice. *The Leadership Quarterly*, *28*(1), 130–152. doi:10.1016/j.leaqua.2016.10.005

Cunliffe, A. L., & Eriksen, M. (2011). Relational leadership. *Human Relations*, *64*(11), 1425–1449. doi:10.1177/0018726711418388

Dabholkar, P. A., Shepherd, C. D., & Thorpe, D. I. (2000). A comprehensive framework for service quality: An investigation of critical conceptual and measurement issues through a longitudinal study. *Journal of Retailing*, *76*(2), 139–173. doi:10.1016/S0022-4359(00)00029-4

Dabholkar, P. A., Thorpe, D. I., & Rentz, J. O. (1996). A measure of service quality for retail stores: Scale development and validation. *Journal of the Academy of Marketing Science*, *24*(1), 3–16. doi:10.1007/BF02893933

Daft, R. L., & Becker, S. W. (1978). *The Innovative Organization*. New York: Elsevier.

Dagger, T. S., Sweeney, J. C., & Johnson, L. W. (2007). A hierarchical model of health service quality: Scale development and investigation of an integrated model. *Journal of Service Research*, *10*(2), 123–142. doi:10.1177/1094670507309594

Daily Excelsior. (2000). *As many as 15,000 units were given licences even after the Supreme Court's 1996 Order*. Author.

Daloz Parks, S. (2005). *Leadership Can be Taught*. Boston, MA: Harvard Business School Press.

Damanpour, F. (1991, September). Organizational innovation: A meta-analysis of effects of determinants and moderators. *Academy of Management Journal*, *34*, 555–590.

Dan'Kov, S. Y., Tishin, A. M., Pecharsky, V. K., & Gschneidner, K. A. (1998). Magnetic phase transitions and the magnetothermal properties of gadolinium. *Physical Review. B*, *57*(6), 3478–3490. doi:10.1103/PhysRevB.57.3478

Daryani, M., Simchi, A., Sadati, M., Hosseini, H. M., Targholizadeh, H., & Khakbiz, M. (2014). Effects of Ti-based catalysts on hydrogen desorption kinetics of nanostructured magnesium hydride. *International Journal of Hydrogen Energy*, *39*(36), 21007–21014. doi:10.1016/j.ijhydene.2014.10.078

Davies, K. K., Fisher, K. T., Dickson, M. E., Thrush, S. F., & Le Heron, R. (2015). Improving ecosystem service frameworks to address wicked problems. *Ecology and Society*, *20*(2), art37. doi:10.5751/ES-07581-200237

Day, D. (2000). Leadership development: A review in context. *The Leadership Quarterly*, *11*(4), 581–611. doi:10.1016/S1048-9843(00)00061-8

Daylight, B.T. (2017, Dec. 23). 'The difficulty is the point': Teaching spoon-fed students how to really read. *Reading Australia Guardian*.

De Molina, M. G. (2012). *Agro ecology and politics. How to get sustainability?* About the necessity for a political agroecologist. *Agroecology and Sustainable Food Systems*, *37*, 45–59. doi:10.1080/10440046.2012.705810

de Oliveira, N. A., & von Ranke, P. J. (2005). Theoretical calculations of the magnetocaloric effect in MnFeP0.45As0.55: A model of itinerant electrons. *Journal of Physics Condensed Matter*, *17*(21), 3325–3332. doi:10.1088/0953-8984/17/21/025

De Pree, M. (2002). Servant-leadership: Three things necessary. *Focus on leadership: Servant leadership for the*, *21*, 89-100.

De Schutter, O. (2010). *Report submitted by the special rapporteur on the right to food*. United Nations Human Rights Council.

De Zeeuw, H., Van Veenhuizen, R., & Dubbeling, M. (2011). The role of urban agriculture in building resilient cities in developing countries. *J Agric Sci*, *149*(S1), 153–163. doi:10.1017/S0021859610001279

Deb, S., Barbhuiya, A. R., Arunachalam, A., & Arunachalam, K. (2008). Ecological analysis of traditional agroforest and tropical forest in the foothills of Indian eastern Himalaya: Vegetation, soil and microbial biomass. *Tropical Ecology*, *49*, 73–78.

Debye, P. (1926). Einige bemerkungen zur magnetisierung bei tiefer temperatur. *Annalen der Physik*, *386*(25), 1154–1160. doi:10.1002/andp.19263862517

Dein, S. (2010, January 10). *Religion, Spirituality, and Mental Health*. Retrieved from https://www.psychiatrictimes.com/schizophrenia/religion-spirituality-and-mental-health

Dewar, R. D., & Dutton, J. E. (1986). The adoption of radical and incremental innovations: An empirical analysis. *Management Science*, *32*(11), 1422–1433. doi:10.1287/mnsc.32.11.1422

Dewey, J. (2005). *Art as experience* (1934 ed.). Penguin.

Dewey, J. (2008). *The Later Works of John Dewey, 1925-1953: 1934, Art as Experience* (A. Boydston, Ed.; Vol. 10). Carbondale, IL: Southern Illinois University Press.

Dictionary.com. (2018). *Ambience*. Retrieved from www.dictionary.com: https://www.dictionary.com/browse/ambience

Diesel Technology Forum. (n.d.). *About clean diesel: What is SCR?* Retrieved from https://www.dieselforum.org/about-clean-diesel/what-is-scr

Dillon, A. C., Jones, K. M., Bekkedahl, T. A., Kiang, C. H., Bethune, D. S., & Heben, M. J. (1997). Storage of hydrogen in single walled carbon nanotubes. *Nature*, *386*(6623), 377–379. doi:10.1038/386377a0

Dinca, M., Dailly, A., Liu, Y., Brown, C. M., Neumann, D. A., & Long, J. R. (2006). Hydrogen storage in a microporous metal organic framework with exposed Mn2+ coordination sites supporting information. *Journal of the American Chemical Society*, *128*(51), 16876–16883. doi:10.1021/ja0656853 PMID:17177438

DIPP. (n.d.). *Government of India*. Retrieved from https://dipp.gov.in/

Directorate General Health Service. (2018). *Health Bulletin*. Ministry of Health and Family Welfare.

Dohmen, T., & Falk, A. (2012). *Individual Risk Attitudes: Measurement, Determinants and Behavior. Consequences*. JEEA.

Doley, K. (2016). *How a botched hair transplant put rs 1.25 lakh cr beauty biz in india under scanner*. Financial Express.

Donabedian, A. (1966). Evaluating the quality of medical care. *The Milbank Memorial Fund Quarterly*, *44*(3), 166–206. doi:10.2307/3348969 PMID:5338568

Donabedian, A. (1980). A guide to medical care administration: medical care appraisal: quality and utilization. In *A guide to medical care administration: medical care appraisal: quality and utilization*. The American Public Health Association.

Donabedian, A. (1992). The role of outcomes in quality assessment and assurance. *QRB. Quality Review Bulletin*, *18*(11), 356–360. doi:10.1016/S0097-5990(16)30560-7 PMID:1465293

Doran, D., & Smith, P. (2004). Measuring service quality provision within an eating disorders context. *International Journal of Health Care Quality Assurance*, *17*(7), 377–388. doi:10.1108/09526860410563186 PMID:15552393

*Compilation of References*

Dornheim, M., Doppiu, S., Barkhordarian, G., Boesenberg, U., Klaasen, T., & Gutfleisch, O. (2007). Hydrogen storage in magnesium-based hydrides and hydride composites. *Scripta Materialia*, *56*(10), 841–846. doi:10.1016/j.scriptamat.2007.01.003

Dow, M. G., Kearns, W., & Thornton, D. H. (1996). The internet II: Future effects on cognitive behavioral practice. *Cognitive and Behavioral Practice*, *3*(1), 137–157. doi:10.1016/S1077-7229(96)80035-6

Downs, G. W., & Mohr, L. B. (1976). Conceptual issues in the study of innovation. Admin. Sci. Quart., 21.

Dubbeling, M., Campbell, M. C., Hoekstra, F., & van Veenhuizen, R. (2009). Building resilient cities. *Urban Agriculture Magazine*, *22*, 3–11.

Dufour, J., & Huot, J. (2007). Rapid activation, enhanced hydrogen sorption kinetics and air resistance in laminated Mg-Pd 2.5 at%. *Journal of Alloys and Compounds*, *439*(1-2), L5–L7. doi:10.1016/j.jallcom.2006.08.264

Dufrenne, M. (1973). *The Phenomenology of Aesthetic Experience*. Evanston, IL: Northwestern University Press.

Durbin, D. J., & Malardier-Jugroot, C. (2013). Review of hydrogen storage techniques for on board vehicle applications. *International Journal of Hydrogen Energy*, *38*(34), 14595–14617. doi:10.1016/j.ijhydene.2013.07.058

Duru, M., Thérond, O., & Fares, M. (2015). Designing agroecological transitions: A review. *Agronomy for Sustainable Development*, *35*(4), 1237–1257. doi:10.100713593-015-0318-x

Duus, R., & Cooray, M. (2014). Together we Innovate: Cross-cultural teamwork through virtual platforms. *Journal of Marketing Education*, *26*(3), 244–257. doi:10.1177/0273475314535783

Economic Survey of India. (1999-2000). Planning Department, Govt. of India.

Egan, K. (2014). Film Production Design: Case Study of The Great Gatsby. *Elon Journal of Undergraduated Research in Communications*, *5*(1), 5–14.

Eich, D. (2008). A Grounded Theory of High-Quality Leadership Program: Perspectives from Student Leadership Development Programs in Higher Education. *Journal of Leadership & Organizational Studies*, *15*, 176–187. doi:10.1177/1548051808324099

Ellis, M. A., Ferree, D. C., Funt, R. C., & Madden, L. V. (1998). Effects of an apple scab-resistant cultivar on use patterns of inorganic and organic fungicides and economics of disease control. *Plant Disease*, *82*(4), 428–433. doi:10.1094/PDIS.1998.82.4.428 PMID:30856893

Ellison, C. G. (1995). Race, Religious Involvement, and Depressive Symptomatology in Southeastern U.S. Community. *Social Science & Medicine*, *11*(11), 1561–1572. doi:10.1016/0277-9536(94)00273-V PMID:7667660

Elmqvist, T., Colding, J., Barthel, S., Borgström, S., Duit, A., Lundberg, J., ... Bengtsson, J. (2004). The dynamics of social–ecological systems in urban landscapes: Stockholm and the National Urban Park, Sweden. *Annals of the New York Academy of Sciences*, *1023*(1), 308–322. doi:10.1196/annals.1319.017 PMID:15253913

Elshater, A., & Abusaada, H. (2017). *What Is Urban Design? Learning and Thought*. Partridge Publishing.

English translation by Charles Darwin Adams, 1868. (n.d.). http://www.perseus.tufts.edu/hopper/text?doc=Perseus%3atext%3a1999.01.0248%3atext%3dAph

Ergen, M. (2009). *Mobile broadband—including WiMAX and LTE*. Springer.

Erk, S., Spitzer, M., Wunderlich, A. P., Galley, L., & Walter, H. (2002). Cultural Objects modulate reward circuitry. *Neureport*, *13*(18), 2499–2503. doi:10.1097/00001756-200212200-00024 PMID:12499856

Ernstson, H., Sörlin, S., & Elmqvist, T. (2008). Social movements and ecosystem services–the role of social network structure in protecting and managing urban green areas in Stockholm. *Ecology and Society, 13*(2), 39. https://www.ecologyandsociety.org/vol13/iss2/art39/

Ettlie, J. E., Bridges, W. P., & O'Keefe, E. D. (1984, June). Organization strategy and structural differences for radical versus incremental innovation. *Management Science, 30*(6), 682495. doi:10.1287/mnsc.30.6.682

Ettlie, J. E., & Rubenstein, A. H. (1987, June). Firm size and product innovation. *Journal of Product Innovation Management, 4*(2), 89–108. doi:10.1111/1540-5885.420089

EU No 517/2014, European Regulation No 517/2014 on fluorinated greenhouse gases and repealing Regulation (EC) No 842/2006, 2014, *Off. J. Eur. Union*.

Evans, R. G., & Stoddart, G. L. (1990). Producing health, consuming health care. *Social Science & Medicine, 31*(12), 1347–1363. doi:10.1016/0277-9536(90)90074-3 PMID:2126895

Evenari, M., Shanan, L., & Tadmor, N. (1982). *The Negev: the challenge of a desert*. Cambridge: Harvard University Press. doi:10.4159/harvard.9780674419254

Fabusoro, E. (2006). *Property rights, access to natural resources and livelihood security among settled Fulani agropastoralists in Southwestern Nigeria* (Unpublished research report). University of Agriculture, Abeokuta, Nigeria.

Fairhurst, G. (2016). Reflections on Leadership and Ethics in Complex Times. *Atlantic Journal of Communication, 24*(1). . doi:10.1080/15456870.2016.1113612

Fang, Z.-Z., Kang, X.-D., & Wang, P. (2010). Improved hydrogen storage properties of LiBH4 by mechanical milling with various carbon additives. *International Journal of Hydrogen Energy, 35*(15), 8247–8252. doi:10.1016/j.ijhydene.2009.12.037

Felipe-Lucia, M. R., Martín-López, B., Lavorel, S., Berraquero-Díaz, L., Escalera-Reyes, J., & Comín, F. A. (2015). Ecosystem services flows: Why stakeholders' power relationships matter. *PLoS One, 10*(7), e0132232. doi:10.1371/journal.pone.0132232 PMID:26201000

Ferguson, B. G., & Morales, H. (2010). Latin American agroecologists build a powerful scientific and social movement. *Journal of Sustainable Agriculture, 34*(4), 339–341. doi:10.1080/10440041003680049

Fernandez, M., Goodall, K., Olson, M., & Mendez, E. (2012). Agroecology and alternative agrifood movements in the United States: Towards a sustainable agrifood system. *Agroecology and Sustainable Food Systems, 37*, 115–126. doi:10.1080/10440046.2012.735633

Ferrer, R. D., Sridharan, M. G., Garcia, G., Pi, F., & Viejo, J. R. (2007). Hydrogenation properties of pure magnesium and magnesium-aluminum thin films. *Journal of Power Sources, 169*(1), 117–122. doi:10.1016/j.jpowsour.2007.01.049

FICCI. (2015). *Healthcare: The neglected GDP driver Need for a paradigm shift*. FICCI.

Fisher, L., Knapman, C., Luhrmann, B., Martin, C., Wick, D. (Producers), & Luhrmann, B. (Director). (2013). *The Great Gatsby* [Motion Picture]. Retrieved 10 30, 2018, from https://www.youtube.com/watch?v=PNLoFqVXoqg

Florida, R. (2003). *The rise of the creative class and how it's transforming work, leisure, community and everyday life*. Perseus Books Group.

Florida, R. (2012). *The Rise of the Creative Class-Revisited:10th Anniversary Edition*. Basic Books.

Flyvbjerg, B., Garbuio, M., & Lovallo, D. (2009). Delusion and Deception in Large Infrastructure Projects: Two Models for Explaining and Preventing Executive Disaster (February 2009). *California Management Review*, *51*(2), 170–193. doi:10.2307/41166485

Forde, D. (1970). The Nupe. In D. Forde (Ed.), *Peoples of the Niger-Benue Confluence* (pp. 15–52). London: International African Institute.

Francese, M., & Mulas-Granados, C. (2015). *Functional Income Distribution and Its Role in Explaining Inequality.* IMF. doi:10.5089/9781513549828.001

Francis, C. A., & Porter, P. (2011). Ecology in sustainable agriculture practices and systems. *Critical Reviews in Plant Sciences*, *30*(1-2), 64–73. doi:10.1080/07352689.2011.554353

Franco, V., Blázquez, J. S., Ingale, B., & Conde, A. (2012). The magnetocaloric effect and magnetic refrigeration near room temperature: Materials and models. *Annual Review of Materials Research*, *42*(1), 305–342. doi:10.1146/annurev-matsci-062910-100356

French, R. P., & Raven, B. (1959). The bases of social power. In D. Cartwright (Ed.), *Studies in social power* (pp. 150–167). Ann Arbor: University of Michigan, Institute for Social Research.

Frey, D. (2011). *Bioshelter market garden: a permaculture farm*. New Society.

Frolich, T. C., Sauter, M. B., & Stebbins, S. (2015, June 29). The worst companies to work for. *24/7 Wall St/Yahoo Finance*.

Frontline. (2000a, Feb. 4). Article. *Frontline*, 82.

Frontline. (2000b). Article. *Frontline*, *17*(25), 1-7.

Frontline. (2000c). Article. *Frontline*, *17*(4), 92.

Frontline. (2000d). Article. *Frontline*, *17*(2).

Frost & Sullivan. (n.d.). *Healthcare sector growth trends (US $Billion)*. Retrieved from https://www.researchgate.net/figure/Healthcare-sector-growth-trend-Source-Frost-and-Sullivan-LSI-Financial-Services_fig1_328540551

Frost, J. K., Honeycutt, J. M., & Heath, S. K. (2017). Relational maintenance and social support in the aftermath of sudden and expected death. *Communication Research Reports*, *34*(4), 326–334. doi:10.1080/08824096.2017.1350573

Galafassi, D., Daw, T., Munyi, L., Brown, K., Barnaud, C., & Fazey, I. (2017). Learning about social-ecological trade-offs. *Ecology and Society*, *22*(1), 2. doi:10.5751/ES-08920-220102

Galloway, S.M. (2016). What makes a leader transformational? *Leadership Excellence*, *33*(5), 20-21.

Galloway, R. F. (1997, Summer). Community Leadership Programs: New Implications for Local Leadership Enhancement, Economic Development, and Benefits for Regional Industries. *Economic Devekopment Review*, *15*(2), 6–9.

Gandy, M. (2017). Urban atmospheres. *Cultural Geographies*, *24*(3), 353–374. doi:10.1177/1474474017712995 PMID:29278257

Gardiner, B. (2017). China's Surprising Solutions to Clear Killer Air. *National Geographic*.

Gauthier, C. (2005). Measuring Corporate Social and Environmental Performance: The Extended Life-Cycle Assessment. *Journal of Business Ethics*, *59*(1/2), 199–206. doi:10.100710551-005-3416-x

Geels, F. W., & Kemp, R. (2007). Dynamics in socio-technical systems: Typology of change processes and contrasting case studies. *Technology in Society*, *29*(4), 441–455. doi:10.1016/j.techsoc.2007.08.009

Geels, F. W., & Schot, J. (2007). Typology of sociotechnical transition pathways. *Research Policy*, *36*(3), 399–417. doi:10.1016/j.respol.2007.01.003

Gefu, J. O. (1989). The dynamics of Nigerian pastoralism: An overview. In *Pastoralism in Nigeria: Past, Present & Future. Proceedings of the national conference on pastoralism in Nigeria*. Ahmadu Bello University.

Gefu, J. O. (1992). *Pastoralist Perspectives in Nigeria: The Fulbe of Udubo Grazing Reserve*. Uppsala: Scandinavin Institute of African Studies.

Gehl, J. (1987). *Life Between Buildings*. New York: Van Nostrand Reinhold.

Gehl, J., & Svarre, B. (2013). *How to Study Public Life*. Island Press. doi:10.5822/978-1-61091-525-0

Giauque, W. F. (1927). A thermodynamic treatment of certain magnetic effects. A proposed method of producing temperatures considerably below 1 absolute. *Journal of the American Chemical Society*, *49*(8), 1864–1870. doi:10.1021/ja01407a003

Gillett, E., Lehr, H., & Osorio, C. (2004). Local government broadband initiatives. *Telecommunications Policy*, *28*(7), 537–558. doi:10.1016/j.telpol.2004.05.001

Glaeser, E., & Gottlieb, J. D. (2006). Urban Resurgence and the Consumer City. *Urban Studies (Edinburgh, Scotland)*, *43*(8), 1275–1299. doi:10.1080/00420980600775683

Glenn, N. M., McGannon, K. R., & Spence, J. C. (2013). Exploring media representations of weight-loss surgery. *Qualitative Health Research*, *23*(5), 631–644. doi:10.1177/1049732312471731 PMID:23282795

Gliessman, S. R. (2007). *Agroecology: The ecology of sustainable food systems* (2nd ed.). New York: CRC Press.

Gliessman, S. R., & Rosemeyer, M. (Eds.). (2009). *The conversion to sustainable agriculture: principles, processes, and practices*. Boca Raton, FL: CRC Press; doi:10.1201/9781420003598

Global Top 100. (2017). *Companies by market capitalisation. 31 March 2017 update*. PWC: www.pwc.com/top100

Goetzler, W., Zogg, R., Young, J., & Johnson, C. (2014). Alternatives to vapor-compression HVAC technology. *ASHRAE Journal*, *56*(10), 12.

Gohho to nihonten. (1992). *Vincent van Gogh and Japan*. The National Museum of Art, Kyôto, Setagaya Art Museum and TV Asahi. Nissha Printing.

Goldberg, L. R. (1993). The structure of phenotypic personality traits. *The American Psychologist*, *48*(1), 26–34. doi:10.1037/0003-066X.48.1.26 PMID:8427480

Gomiero, T., Pimentel, D., & Paoletti, M. G. (2011). Is there a need for a more sustainable agriculture? *Critical Reviews in Plant Sciences*, *30*(1-2), 6–23. doi:10.1080/07352689.2011.553515

González, J. A., & Rossi, A. (2011). *New trends for smart cities, open innovation Mechanism in Smart Cities. European commission within the ICT policy support programme*. http://opencities.net/sites/opencities.net/ðles/content-ðles/repository/D2.2.21%20New%20trends%20for%20Smart%20Cities.pdf

Gotlieb, J. B., Grewal, D., & Brown, S. W. (1994). Consumer satisfaction and perceived quality: Complementary or divergent constructs? *The Journal of Applied Psychology*, *79*(6), 875–885. doi:10.1037/0021-9010.79.6.875

Gountas. J, Ciorciari J (2010). Inside the Minds of Trendsetters. *Aus Magazine Sci/Tech*, 14-17.

Government of India. (2009). *National Health Accounts – India (2004–05) –with provisional estimates from 2005-06 to 2008-09). National Health Accounts Cell*. New Delhi: Ministry of Health and Family Welfare.

*Compilation of References*

Graetz, J. (2012). Metastable metal hydrides for hydrogen storage. *ISRN Materials Science*, 18.

Grayzel, J. (1990). Markets and migration: A Fulbè pastoral system in Mali. In J. Galaty & D. Johnson (Eds.), *The world of pastoralism* (pp. 35–67). New York: Guilford Press.

Greco, A., Aprea, C., Maiorino, A., & Masselli, C. (2019). A review of the state of the art of solid-state caloric cooling processes at room-temperature before 2019. *International Journal of Refrigeration*, *106*, 66–88. doi:10.1016/j.ijrefrig.2019.06.034

Greco, A., & Vanoli, G. P. (2005). Flow boiling heat transfer with HFC mixtures in a smooth horizontal tube. Part II: Assessment of predictive methods. *Experimental Thermal and Fluid Science*, *29*(2), 199–208. doi:10.1016/j.expthermflusci.2004.03.004

Green Climate Fund. (2019). *Process for readiness support*. Retrieved from https://www.greenclimate.fund/gcf101/empowering-countries/readiness-support

Griffel, M., Skochdopole, R. E., & Spedding, F. H. (1954). The heat capacity of gadolinium from 15 to 355 K. *Physical Review*, *93*(4), 657–661. doi:10.1103/PhysRev.93.657

Griffero, T. (2016). Atmospheres and Felt-bodily Resonances. *Studi di estetica*, *49*(4), 1-49.

Griffero, T. (2014). *Atmospheres: Aesthetics of Emotional Spaces* (S. D. Sanctis, Trans.). Ashgate Publishing Limited.

Griffin, R. J., Dunwoody, S., & Neuwirth, K. (1999). Proposed model of the relationship of risk information seeking and processing to the development of preventive behaviors. *Environmental Research*, *80*(2), S230–S245. doi:10.1006/enrs.1998.3940 PMID:10092438

Grönroos, C. (1984). *A service quality model and its marketing implications*. Academic Press.

Grönroos, C. (1990). *Service management and marketing: Managing the moments of truth in service competition*. Jossey-Bass.

Groves, K. (2007). Integrating leadership development and succession planning best practices. *Journal of Management Development*, *26*(3), 239–260. doi:10.1108/02621710710732146

Guillou, F., Legait, U., Kedous-Lebouc, A., & Hardy, V. (2012). Development of a new magnetocaloric material used in a magnetic refrigeration device. In *EPJ Web of Conferences* (Vol. 29, p. 21). EDP Sciences. 10.1051/epjconf/20122900021

Guoxian, L., Erde, W., & Shoushi, F. (2000). Hydrogen absorption and desorption characteristics of mechanically milled Mg-35 wt% FeTi1.2 powders. *Journal of Alloys and Compounds*, *297*, 222–231.

Gurin, P., Dey, E., Hurtado, S., & Gurin, G. (2002). Diversity and higher education: Theory and impact on educational outcomes. *Harvard Educational Review*, *72*(3), 330–367. doi:10.17763/haer.72.3.01151786u134n051

Hage, J. (1980). *Theories of Organization: Form, Process and Transformation*. New York: Wiley.

Hall, J., & Vredenburg, H. (2003, Fall). The Challenges of Innovating for Sustainable Development. *MIT Sloan Management Review*, 61–68.

Halo, T. L., McMahon, K. M., Angeloni, N. L., Xu, Y., Wang, W., Chinen, A. B., ... Thaxton, C. S. (2014). NanoFlares for the detection, isolation, and culture of live tumor cells from human blood. *Proceedings of the National Academy of Sciences of the United States of America*, *111*(48), 17104–17109. doi:10.1073/pnas.1418637111 PMID:25404304

Ham, B., Junkaew, A., Arroyave, R., Park, J., Zhou, H. C., Foley, D., ... Zhang, X. (2014). Size and stress dependent hydrogen desorption in metastable Mg hydride films. *International Journal of Hydrogen Energy, 39*(6), 2597–2607. doi:10.1016/j.ijhydene.2013.12.017

Hampwaye, G. (2013). Benefits of urban agriculture: Reality of illusion? *Geoforum, 49*, R7–R8. doi:10.1016/j.geoforum.2013.03.008

Hanada, N., Ichikwa, T., & Fujii, H. (2005). Catalytic effect of nanoparticles 3D-transition metals on hydrogen storage properties in magnesium hydride MgH2 prepared by mechanical milling. *Journal of Physical Chemistry, 109*(15), 7188–7194. doi:10.1021/jp044576c PMID:16851820

Harmer, R. (1999). Creating New Native Woodlands: Turning Ideas into Reality. Forestry Commission Information Note 15.

Harrison, D., Welchman, E., & Thonhauser, T. (2017). H-4 alkanes: A new class of hydrogen storage material? *International Journal of Hydrogen Energy, 42*(4), 1–6. doi:10.1016/j.ijhydene.2016.12.144

Hasan, A. (1997). *Working with government: The story of OPP's collaboration with state agencies for replicating its Low Cost Sanitation Programme*. Karachi, Pakistan: City Press.

Hashimoto, O. (2007). *Nihon bijutsushi 7*. Tôkyô: Shinchôsha.

Haslam, J. (2008). Pits, Pendulums, and Penitentiaries: Reframing the Detained Subject. *Texas Studies in Literature and Language, 50*(3), 268–284. doi:10.1353/tsl.0.0004

Hatfield, J. (2007) Beyond the edge of the field. *Arch. Soil Sci. Soc. Am.* https://www.soils.org/about-society/presidents-message/archive/13

*Health, United States, 2015 with Special Feature on Racial and Ethnic Health Disparities.* (2015). Hyattsville: National Centre for Health Statistics.

Heasley, L., & Delehanty, J. (1996). The Politics of Manure: Resource Tenure and the Agropastoral Economy in Southwestern Niger. *Society & Natural Resources, 9*(1), 31–46. doi:10.1080/08941929609380950

Hegel, G. W. F. (1807). *Phaenomenologie des Geistes*. Gutenberg Projekt-DE. http://www.gutenberg.org/cache/epub/6698/pg6698-images.html

Heifetz, R. A., Grashow, A., & Linsky, M. (2009). *The practice of adaptive leadership: Tools and tactics for changing your organization and the world*. Harvard Business Press.

Heininen, S. (2019). *Kansallisbiografia. Agricola, Mikael (noin 1510-1557)*. http://urn.fi/urn:nbn:fi:sks-kbg-000014

Hellström, T. (2007). Dimensions of Environmentally Sustainable Innovation: The Structure of Eco-Innovation Concepts. *Sustainable Development, 15*(3), 148–159. doi:10.1002d.309

Hickson, G. B., Federspiel, C. F., Pichert, J. W., Miller, C. S., Gauld-Jaeger, J., & Bost, P. (2002). Patient complaints and malpractice risk. *Journal of the American Medical Association, 287*(22), 2951–2957. doi:10.1001/jama.287.22.2951 PMID:12052124

Hiernaux, P., Fernàndez-Rivera, S., Schlecht, E., Turner, M. D., & Williams, T. O. (1997). Les transferts de fertilité per la betail dans les agro-écosystèmes du Sahel (Nutrient transfers of animals in Sahelian agroecosystems). In *Proceedings of Regional Workshop on Soil fertility management in West Africa land use systems*. Margraf Verlag.

Higgs, M. (2009). The good, the bad and the ugly: Leadership and narcissism. *Journal of Change Management, 9*(2), 165–178. doi:10.1080/14697010902879111

Hindustan Times. (2018). *Delhi world's most polluted city, Mumbai worse than Beijing: WHO.* Author.

Hippocrates. (c. 400 B.C.). *Aphorismi*. Greek original text edited by A. Littré. Perseus Digital Library. Tufts University. http://www.perseus.tufts.edu/hopper/text?doc=Perseus%3Atext%3A1999.01.0250%3Atext%3DAph.%3Achapter%3D1%3Asection%3D1

Hirose, S. (2002). Rice and upland farming in the Nupe community. In S. Hirose & T. Wakatsuki (Eds.), *Restoration of Inland Valley Ecosystems in West Africa* (pp. 183–212). Association of Agriculture and Forest Statistics.

Hirscher, M. (2009). Handbook of hydrogen storage: New materials for future energy storage. Weinheim, Germany: Wiley-VCH Verlag GmbH & Co. KGaA.

Hogben, M., & Leichliter, J. S. (2008). Social determinants and sexually transmitted disease disparities. *Sexually Transmitted Diseases, 35*(12Suppl), 13–18. doi:10.1097/OLQ.0b013e31818d3cad PMID:18936725

Hohm, C., & Snyder, J. (2015). "It was the best decision of my life": A thematic content analysis of former medical tourists' patient testimonials. *BMC Medical Ethics, 16*(1), 1–7. doi:10.1186/1472-6939-16-8 PMID:25614083

Holen, A. (1976). *The Psychology of Silence*. Oslo: Dyade Publishing.

Holling, C. S. (1978). *Adaptive environmental assessment and management*. New York: John Wiley.

Holt, A. R., Alix, A., Thompson, A., & Maltby, L. (2016). Food production, ecosystem services and biodiversity: We can't have it all everywhere. *The Science of the Total Environment, 573*, 1422–1429. doi:10.1016/j.scitotenv.2016.07.139 PMID:27539820

Holt-Gimenez, E. (2006). *Campesino a campesino: voices from Latin America's farmer to farmer movement for sustainable agriculture*. Oakland: Food First.

Holtzberg, F., Gambino, R. J., & McGuire, T. R. (1967). New ferromagnetic 5: 4 compounds in the rare earth silicon and germanium systems. *Journal of Physics and Chemistry of Solids, 28*(11), 2283–2289. doi:10.1016/0022-3697(67)90253-3

Hood, C. (1991). Public Management for All Seasons. *Public Administration, 6*(3).

Hood, C., & Dixon, R. (2015). What We Have to Show for 30 Years of New Public Management: Higher Costs, More Complaints. *Governance: An International Journal of Policy, Administration and Institutions, 28*(3), 265–267. doi:10.1111/gove.12150

Hopen, C. E. (1958). *The pastoral Fulbe Family in Gwandu*. London: OUP for IAI.

Hopkins, L., Labonté, R., Runnels, V., & Packer, C. (2010). Medical tourism today: What is the state of existing knowledge? *Journal of Public Health Policy, 31*(2), 185–198. doi:10.1057/jphp.2010.10 PMID:20535101

Howard, V. R. (2017). *Dharma: the Hindu, Jain, Buddhist and Sikh traditions of India*. London: I.B.Tauris.

Huang, H. J., & Wang, X. (2012). Pd nanoparticles supported on low-defect graphene sheets for use as high performance electrocatalysts for formic acid and methanol oxidation. *Journal of Materials Chemistry. A, Materials for Energy and Sustainability, 22*, 22533–22541.

Huang, J.-W., Chang, C.-K., Lu, K., Huang, J.-T., & Lin, C.-R. (2009). Novel approach for nanoporous gas sensor fabrication using anodic aluminum oxidation and MEMS process, Nanotechnology, 2009. *IEEE-NANO 2009. 9th IEEE Conference.*

Hua, T. Q., Ahluwalia, R. K., Peng, J.-K., Kromer, M., Lasher, S., McKenney, K., ... Sinha, J. (2011). Technical assessment of compressed hydrogen storage tank systems for automotive application. *International Journal of Hydrogen Energy, 36*(4), 3037–3049. doi:10.1016/j.ijhydene.2010.11.090

Hu, F. X., Shen, B. G., Sun, J. R., Cheng, Z. H., Rao, G. H., & Zhang, X. X. (2001). Influence of negative lattice expansion and metamagnetic transition on magnetic entropy change in the compound LaFe 11.4 Si 1.6. *Applied Physics Letters*, *78*(23), 3675–3677. doi:10.1063/1.1375836

Hult, F., & Sivanesan, G. (2014). What good cyber resilience looks like. *Journal of Business Continuity & Emergency Planning*, *7*(2), 112–125. PMID:24457323

Hummer, R. A., Roger, R. G., Nam, C. B., & Ellison, C. G. (1999). Religious Involvement and U.S. Adult Mortality. *Demography*, *36*(2), 273–285. doi:10.2307/2648114 PMID:10332617

Huot, J., Akiba, E., & Takasa, T. (1995). Mechanical alloying of Mg-Ni compounds under hydrogen and inert atmosphere. *Journal of Alloys and Compounds*, *231*(1-2), 815–819. doi:10.1016/0925-8388(95)01764-X

Huot, J., Liang, G., Boily, S., Neste, A. V., & Schulz, R. (1999). Structural study and hydrogen sorption kinetics of ball milled magnesium hydride. *Journal of Alloys and Compounds*, *293-295*, 495–500. doi:10.1016/S0925-8388(99)00474-0

Hurk, P.A.M., Wingens, T., & Giomni, F. (2011). On the Relationship Between the Practice of Meditation and Personality – an Exploratory Analysis of the Mediating Role of Mindfulness Skills. *MF*, *2*, 194-200.

Hwang, H. T., & Varma, A. (2014). Hydrogen storage for fuel cell vehicles. *Current Opinion in Chemical Engineering*, *5*, 42–48. doi:10.1016/j.coche.2014.04.004

Ibrahim, S. (1992). *The Nupe and their neighbours from the 14th century*. Ibadan, Nigeria: Heinemann Educational Books.

Ichikwa, T., Hanada, N., Isobe, S., Leng, H. Y., & Fujii, H. (2005). Hydrogen storage properties in Ti catalyzed Li-N-H system. *Journal of Alloys and Compounds*, *404-406*, 435–438. doi:10.1016/j.jallcom.2004.11.110

Iijima, S. (1991). Helical microtubes of graphitic carbon. *Nature*, *354*(6348), 56–58. doi:10.1038/354056a0

Imison, M., & Schweinsberg, S. (2013). Australian news media framing of medical tourism in low- and middle-income countries: A content review. *BMC Public Health*, *13*(1), 1–12. doi:10.1186/1471-2458-13-109 PMID:23384294

India Services Medical Value Travel Services. (2018). *Sector Overview*. Retrieved from https://www.indiaservices.in/medical

India Today. (2018). *Eastern Peripheral Expressway inaugurated by PM Modi is likely to decrease Delhi pollution by 27 per cent*. Author.

Intergroup Resources. (2012). Retrieved August 1, 2019, from www.intergroupresources.com: http://www.intergroupresources.com/race-and-racism/

International Monetary Fund. (2013). *Bangladesh: Poverty Reduction Strategy Paper*. IMF, Country Report N0. 13/ 63.

International Organisation of Standardization (ISO). (2018). *ISO 19881: Gaseuos hydrogen – Land vehicle fuel containers*. ISO.

Ismaila, D. (2002). *Nupe in history (1300 to date)*. Jos, Nigeria: Olawale Publishing Company Ltd.

Iverson, N. M., Barone, P. W., Shandell, M., Trudel, L. J., Sen, S., Sen, F., ... Strano, M. S. (2013). In vivo biosensing via tissue localizable near-infrared-fluorescent single-walled carbon nanotubes. *Nature Nanotechnology*, *8*(11), 873–880. doi:10.1038/nnano.2013.222 PMID:24185942

J.P. (2018, Jan. 25). How China cut its air pollution. *The Economist*.

Jaaniste, L. O. (2010). The Ambience of Ambience. *A Journal of Media and Culture*, *13*(2). Retrieved from http://www.journal.media-culture.org.au/index.php/mcjournal/article/view/238

*Compilation of References*

Jacke, D., & Toensmeier, E. (2005). *Edible forest gardens: ecological design and practice for temperate climate permaculture* (Vol. 2). Chelsea Green Publishing Company.

Jacke, D., & Toensmeier, E. (2005). *Edible forest gardens: ecological design and practice for temperate-climate permaculture*. White River Junction: Chelsea Green.

Jain, I. P., Jain, P., & Jain, A. (2010). Novel hydrogen storage materials: A review of lightweight complex hydrides. *Journal of Alloys and Compounds*, *503*(2), 303–339. doi:10.1016/j.jallcom.2010.04.250

Jain, I. P., Lal, C., & Jain, A. (2010). Hydrogen storage in Mg: A most promising material. *International Journal of Hydrogen Energy*, *35*(10), 5133–5144. doi:10.1016/j.ijhydene.2009.08.088

James, W. (1907). *Pragmatism's Conception of Truth. Lecture 6 in Pragmatism: A New Name for Some Old Ways of Thinking*. New York: Longman Green and Co. doi:10.1037/10851-000

Jenner, R. (2013). Ambient atmospheres: Exhibiting the immaterial in works by Italian Rationalists Edoardo Persico and Franco Albini. *Interstices: Journal of Architecture and Related Arts*, 13-24.

Jeon, K.-J., Moon, H. R., Ruminski, A. M., Jiang, B., Kisielowski, C., Bardhan, R., & Urban, J. J. (2011). Air-stable magnesium nanocomposites provide rapid and high capacity hydrogen storage without using heavy-metal catalysts. *Nature Materials*, *97*(4), 083106–083103. doi:10.1038/nmat2978 PMID:21399630

Jevning, R., Wilson, A. F., & Davison, J. M. (1978). Adrenocortical activity during meditation. *Hormones and Behavior*, *10*(1), 54–60. doi:10.1016/0018-506X(78)90024-7 PMID:350747

JohnsonB.R.TompkinsR.B.WebbD. (2017, September). https://www.manhattaninstitute.org/pdf/crrucs_objective_hope.pdf. Retrieved 2017, from www.manhattaninstitute.org

Johnson, C. E. (2017). *Meeting the ethical challenges of leadership: Casting light or shadow*. Sage Publications.

Johnson, P. (1990). *A History of Christianity*. Harmondsworth: Penguin Books.

Johnston, H. A. S. (1967). *The Fulani Empire of Sokoto*. London: Oxford University Press.

Jordan, N. R., Bawden, R. J., & Bergmann, L. (2008). Pedagogy for addressing the worldview challenge in sustainable development of agriculture. *Journal of Natural Resources and Life Sciences Education*, *37*, 92–99.

Jun, J. (2016). Framing service, benefit, and credibility through images and texts: A content analysis of online promotional messages of Korean medical tourism industry. *Health Communication*, *31*(7), 845–852. doi:10.1080/10410236.2015.1007553 PMID:26644259

Jun, J., & Oh, K. M. (2015). Framing risks and benefits of medical tourism: A content analysis of medical tourism coverage in Korean American community newspapers. *Journal of Health Communication*, *20*(6), 720–727. doi:10.1080/10810730.2015.1018574 PMID:25942506

Kagawa, A., Ono, E., Kusakabe, T., & Sakamoto, Y. (1991). Absorption of hydrogen by vanadium-rich V-Ti based alloys. *Journal of the Less Common Metals*, *172-174*, 64–70. doi:10.1016/0022-5088(91)90433-5

Kahneman, D., Tversky, A. (1979). Prospect Theory: an analysis of decision under risk. *Econom*, 263-91.

Kahneman, D. (2003). Maps of bounded rationality: Psychology for behavioral economics. *The American Economic Review*, *93*(5), 1449–1475. doi:10.1257/000282803322655392

Kant, I. (1954). Kritik der Urteilskraft (K. Vorländer, Ed.). Hamburg: Verlag von Felix Meiner. (Originally published 1799)

Kaplan, S. (1999). Discontinuous Innovation and the Growth Paradox. *Strategy and Leadership, 27*(2), 16–21. doi:10.1108/eb054631

Karaman, M., Çatalkaya, H., & Aybar, C. (2016). Institutional Cybersecurity from Military Perspective. *International Journal of Information Security Science, 5*(1), 1–7.

Kathuria, V. (2001). *Relocation, Ban or Boon*. Chennai: Hindu Survey of Environment.

Kato, S., & Ahern, J. (2008). Learning by doing: Adaptive planning as a strategy to address uncertainty in planning. *Environment & Planning, 51*(4), 543–559. doi:10.1080/09640560802117028

Kegan, R., Lahey, L. L., Miller, M. L., & Fleming, A. (2016). *An everyone culture*. Becoming a Deliberately Developmental Organization.

Kemp, E., Williams, K. H., & Porter, M. III. (2015). Hope across the seas: The role of emotions and risk propensity in medical tourism advertising. *International Journal of Advertising, 34*(4), 621–640. doi:10.1080/02650487.2015.1024385

Keuken, M.P. (1998). *Result of Preliminary Measurements and Segregations for a cost Effective Air Pollution Assessment*. Paper presented at workshop on Integrated Approach to Vehicular Pollution Control in Delhi.

Keynes, J. M. (1936). *The General Theory of Employment, Interest and Money*. Palgrave MacMillan.

Kezar, A., & Lester, J. (2010). Breaking the barriers of essentialism in leadership research: Positionality as a promising approach. *Feminist Formations, 22*(1), 163–185. doi:10.1353/nwsa.0.0121

Khafidz, N. Z. A., Yaakob, Z., & Lim, K. L. (2016). The kinetics of lightweight solid-state hydrogen storage materials: A review. *International Journal of Hydrogen Energy, 41*, 13131–13151. doi:10.1016/j.ijhydene.2016.05.169

Khan, M. J., Chelliah, S., Haron, M. S., & Ahmed, S. (2017). Role of travel motivations, perceived risks and travel constraints on destination image and visit intention in medical tourism: Theoretical model. *Sultan Qaboos University Medical Journal, 17*(1), e11–e17. doi:10.18295qumj.2016.17.01.003 PMID:28417022

Kim, S. K., Hong, S. A., Son, H. J., Han, W. S., Michalak, A., Hwang, S. J., & Kang, S. O. (2015). Dehydrogenation of ammonia borane by cation Pd(II) and Ni(II) complexes in a nitromethane medium: Hydrogen release and spent fuel characterization. *Dalton Transactions (Cambridge, England), 44*(16), 7373–7381. doi:10.1039/C5DT00599J PMID:25799252

King, C. J., & Redwood, Y. (2016). The Health Care Institution, Population Health and Black Lives. *Journal of the National Medical Association, 108*(2), 131–136. doi:10.1016/j.jnma.2016.04.002 PMID:27372475

King, G. (1996). Institutional Racism and the Medical/Health Complex: A Conceptual Analysis. *Ethnicity & Disease, 6*, 30–46. PMID:8882834

Kitanovski, A., & Egolf, P. W. (2006). Thermodynamics of magnetic refrigeration. *International Journal of Refrigeration, 29*(1), 3–21. doi:10.1016/j.ijrefrig.2005.04.007

Kitanovski, A., Plaznik, U., Tomc, U., & Poredoš, A. (2015). Present and future caloric refrigeration and heat-pump technologies. *International Journal of Refrigeration, 57*, 288–298. doi:10.1016/j.ijrefrig.2015.06.008

Klein, H. J., Simic, D., Fuchs, N., Schweizer, R., Mehra, T., Giovanoli, P., & Plock, J. A. (2017). Complications after cosmetic surgery tourism. *Aesthetic Surgery Journal, 37*(4), .474-482.

Kleperis, J., Lesnicenoks, P., Grinberga, L., Chikvaidze, G., & Klavins, J. (2013). Zeolite as material for hydrogen storage in transport applications. *Latvian Journal of Physics and Technical Sciences, 50*(3), 59–64. doi:10.2478/lpts-2013-0020

*Compilation of References*

Klinge, M. (2015). *Kansallisbiografia. Snellman, Johan Vilhelm (1806 - 1881)*. Online: http://urn.fi/urn:nbn:fi:sks-kbg-003639

Knapp, S. (2015). Lean Six Sigma implementation and organizational culture. *International Journal of Health Care Quality Assurance*, 28(8), 855–863. doi:10.1108/IJHCQA-06-2015-0079 PMID:26440487

Knight, K. E. (1967, October). A descnptive model of the intra-firm innovation process. *The Journal of Business*, 40(4), 478496. doi:10.1086/295013

Knox, P. L. (1994). *Urbanization: introduction to urban geography*. New Jersey: Prentice Hall.

Koenig, H. G. (2012). Religion, Spirituality, and Health: The Research and Clinical Implications. *ISRN Psychiatry*, 2012, 1–33. doi:10.5402/2012/278730 PMID:23762764

Koenig, H. G. (2015). Spirituality, and health: A review and update. *Advances in Mind-Body Medicine*, §§§, 19–26. PMID:26026153

Koerner, M. M. (2000). The Conceptual Domain of Service Quality for Inpatient Nursing Services. *Journal of Business Research*, 48(3), 267–283. doi:10.1016/S0148-2963(98)00092-7

Koh, G., Zhang, Y.-W., & Pan, H. (2012). First-principles study on hydrogen storage by graphitic carbon nitride nanotubes. *International Journal of Hydrogen Energy*, 37(5), 4170–4178. doi:10.1016/j.ijhydene.2011.11.109

Koja, S. (Ed.). (2005). *Nordic Dawn. Modernism's Awakening in Finland 1890-1920*. Prestel.

Kôkan, S. (2006). Discussion on Western Painting (Seiyô gadan). In S. Addiss, G. Groemer, & J. T. Rimer (Eds.), *Traditional Japanese Arts and Culture. An Illustrated Sourcebook* (pp. 173–176). Honolulu, HI: University of Hawai'i Press. (Original work published 1799)

Komninos, N. (2008a). *Intelligent Cities and Globalisation of Innovation Networks*. London: Routledge. doi:10.4324/9780203894491

Komninos, N., & Sefertzi, E. (2009). *Intelligent cities: R&D offshoring, web 2.0 product development and globalization of innovation systems*. Second Knowledge Cities.

Kompella, U. B., Aniruddha, C., Amrite, A. C., Ravia, R. P., & Durazoa, S. A. (2013). Nanomedicines for Back of the Eye Drug Delivery, Gene Delivery, and Imaging. *Progress in Retinal and Eye Research*, 36, 172–198. doi:10.1016/j.preteyeres.2013.04.001 PMID:23603534

Konijnendijk, C., & Gauthier, M. (2006) Urban forestry for multifunctional land use. In *Cities farming for the future: urban agriculture for green and productive cities*. International Development Research Centre, Ottawa. https://www.idrc.ca/en/ev-103884-201-1-DO_TOPIC.html.

Kosiński, J., & Jones, R. C. (1979). *Being There* [Screenplay]. United Artists.

Kosiński, J. (1970). *Being There*. Harcourt Brace.

Kotler, P. (1974). Atmospherics as a Marketing Tool. *Journal of Retailing*, 49(4), 48–64.

Kremen, C., & Miles, A. (2012). Ecosystem services in biologically diversified versus conventional farming systems: Benefits, externalities, and trade-offs. *Ecology and Society*, 17(4), 40. doi:10.5751/ES-05035-170440

Kuhn, E. P., Colberg, P. J., Schnoor, J. L., Wanner, O., Zehnder, A. J. B., & Schwarzenbach, R. P. (1985). Microbial transformation of substitute benzenes during infiltration of river water to groundwater: Laboratory column studies. *Environmental Science & Technology*, 19(10), 961–967. doi:10.1021/es00140a013

Kumar, A.K., Chen, L.C., Choudhury, M., Ganju, S., Mahajan, V., Sinha, A., & Sen, A. (2011). *Financing healthcare for all: challenges and opportunities.* Doi:10.1016/S0140-6736(10)61884-3

Kumaran, P., Giridharan, V., Gokulakrishnan, R., & Hariharan, M. (2018). A Review of room temperature magnetocaloric materials for home appliance. *International Journal of Pure and Applied Mathematics, 119*(16), 2053–2059.

Kumar, B. M. (2006). Agroforestry: The new old paradigm for Asian food security. *Journal of Tropical Agriculture, 44*, 1–14.

Kumar, S., Jain, A., Ichikawa, Y., Kojima, Y., & Dey, G. K. (2017). Development of vanadium based hydrogen storage material: A review. *Renewable & Sustainable Energy Reviews, 72*, 791–800. doi:10.1016/j.rser.2017.01.063

Kumar, S., & Krishnamurthy, N. (2011). Synthesis of V-Ti-Cr alloys by aluminothermy co-reduction of its oxide. *International Journal of Applied Ceramic Technology, 5*, 181–186.

Kumar, S., Taxak, M., & Krishnamurthy, N. (2013a). Hydrogen absorption kinetics of V-Al alloy. *Journal of Thermal Analysis and Calorimetry, 112*(1), 5–10. doi:10.100710973-012-2558-1

Kumar, S., Taxak, M., & Krishnamurthy, N. (2013b). Synthesis and hydrogen absorption in V-Ti-Cr alloy. *Journal of Thermal Analysis and Calorimetry, 112*(1), 51–57. doi:10.100710973-012-2643-5

Kumar, S., Taxak, M., Krishnamurthy, N., Suri, A. K., & Tiwari, G. P. (2012). Terminal solid solubility of hydrogen in V-Al solid solutions. *International Journal of Refractory Metals & Hard Materials, 31*, 76–81. doi:10.1016/j.ijrmhm.2011.09.009

Kung, H. C., Hoyert, D. L., Xu, J., & Murphy, S. L. (2008). *Deaths: Final data for 2005.* National Vital Statistics Report America.

Kurtishi-Kastrati, S., Ramadani, V., Dana, L. P., & Ratten, V. (2016). Do foreign direct investments accelerate economic growth? The case of the Republic of Macedonia. *International Journal of Competitiveness, 1*(1), 71–98. doi:10.1504/IJC.2016.075903

Kyoto Protocol to the United nation Framework Convention on climate change. (1997). United Nation Environment Program (UN).

Labonté, R., Crooks, V. A., Valdés, A. C., Runnels, V., & Snyder, J. (2018). Government roles in regulating medical tourism: Evidence from Guatemala. *International Journal for Equity in Health, 17*(1), 1–10. doi:10.118612939-018-0866-1 PMID:30236120

LaMotte, S. (2019, January 10). *Surgeries in Mexico linked to antibiotic-resistant infections in US, CDC says.* Retrieved March 3, 2019 from https://www.cnn.com/2019/01/10/health/mexicosurgery-antibiotic-resistant-infection-cdc/index.html

Lancaster, B., & Marshall, J. (2008). *Water-harvesting earthworks.* Tucson: Rainsource.

Langmi, H., Book, D., Walton, A., Johnson, S. R., Al-Mamouri, M. M., Speight, J. D., ... Anderson, P. A. (2005). Hydrogen storage in ion-exchanged zeolites. *Journal of Alloys and Compounds, 404-406*, 637–642. doi:10.1016/j.jallcom.2004.12.193

Langmi, H., Walton, A., Al-Mamouri, M. M., Johnson, S. R., Book, D., Speight, J. D., ... Harris, I. R. (2003). Hydrogen adsorption in zeolites A,X,Y and RHO. *Journal of Alloys and Compounds, 356-357*, 256–357, 710–715. doi:10.1016/S0925-8388(03)00368-2

Larsen, T. (2006). *Neuropsychology for Economists.* Lulu.com.

Larsen, T. (2017). Homo Neuroeconomicus – a Neuroeconomic Review of Functional Magnetic Resonance Imaging of Economic Choice. *IJUHD, 7*(1), 44–57.

*Compilation of References*

Larsson, G., Sandahl, C., Söderhjelm, T., Sjövold, E., & Zander, A. (2017). Leadership behavior changes following a theory-based development intervention: A longitudinal study of subordinates' and leaders' evaluations. *Scandinavian Journal of Psychology*, *58*(1), 62–68. doi:10.1111jop.12337 PMID:27859309

Lauriano Souza, V. G., & Fernando, A. L. (2016). Nanoparticles in food packaging: Biodegradability and Potential migration to food—A review. *Food Packaging and Shelf Life*, *2016*(8), 63–70. doi:10.1016/j.fpsl.2016.04.001

Lawton, L. M., Jr., Zimm, C. B., & Jastrab, A. G. (1999). *U.S. Patent No. 5,934,078*. Washington, DC: U.S. Patent and Trademark Office.

Leakey, R. R. B. (2012). *Multifunctional agriculture and opportunities for agroforestry: implications of IAASTD. In Agroforestry: the future of global land use* (pp. 203–214). Dordrecht: Springer. doi:10.1007/978-94-007-4676-3_13

Learning Objectives. (n.d.). Retrieved from http://sphweb.bumc.bu.edu/otlt/MPH-Modules/PH/PublicHealthHistory/PublicHealthHistory_print.html

Lee, H., Wright, K. B., O'Connor, M., & Wombacher, K. (2014). Framing medical tourism: An analysis of persuasive appeals, risks and benefits, and new media features of medical tourism broker websites. *Health Communication*, *29*(7), 637–645. doi:10.1080/10410236.2013.794412 PMID:24138286

Lee, J. H., Ryu, J., Kim, J. Y., Nam, S.-W., Han, J. H., Lim, T.-H., ... Yoon, C. W. (2014). Carbon dioxide mediated, reversible chemical hydrogen storage using Pd nanocatalyst supported on mesoporous graphitic carbon nitride. *Journal of Materials Chemistry. A, Materials for Energy and Sustainability*, *2*(25), 9490. doi:10.1039/c4ta01133c

Lefebvre, H. (2004). *Rhythmanalysis, space, time and everyday life*. London: Continuum.

Leon, N. (2006). *The well connected city A report on municipal networks Supported by The Cloud*. Imperial College London.

Levidow, L., Pimbert, M., & Vanloqueren, G. (2014). Agroecological research: conforming—or transforming the dominant agro-food regime? *Agroecology and Sustainable Food Systems*, *38*(10), 1127–1155. doi:10.1080/21683565.2014.951459

Lev, L., & Stevenson, G. W. (2011). Acting collectively to develop midscale food value chains. *Journal of Agriculture, Food Systems, and Community Development*, *1*(4), 119–128. doi:10.5304/jafscd.2011.014.014

Levy, D. J., & Glimcher, P. W. (2012). The root of all value: A neural common currency. *Current Opinion in Neurology*, *22*(6), 1027–1038. doi:10.1016/j.conb.2012.06.001 PMID:22766486

Levy, J. (2014). *Accounting for ROI and the History of Capital*. U. of Chicago Press.

Liang, J. G., & Sandmann, L. R. (2015). Leadership for community engagement: A distributive leadership perspective. *Journal of Higher Education Outreach & Engagement*, *19*(1), 35–64.

Li, Q., & Thonhauser, T. (2012). A theoretical study of the hydrogen storage potential of $(H_2)_4 CH_4$ in metal organic framework materials and carbon nanotubes. *Journal of Physics Condensed Matter*, *24*(42), 424204. doi:10.1088/0953-8984/24/42/424204 PMID:23032298

Lister, N. M. (2007). Sustainable large parks: Ecological design or designer ecology? In G. Hargreaves & J. Czerniak (Eds.), *Large parks* (pp. 35–54). New York: Architectural Press.

Liu, W., Setijadi, E., Crema, L., Bartali, R., Laidani, N., Aguey-Zinsou, K. F., & Speranza, G. (2018). Carbon nanostructures/Mg hybrid materials for hydrogen storage. *Diamond and Related Materials*, *82*, 19–24. doi:10.1016/j.diamond.2017.12.003

Liu, Y., Infante, I. C., Lou, X., Bellaiche, L., Scott, J. F., & Dkhil, B. (2014). Giant room-temperature elastocaloric effect in ferroelectric ultrathin films. *Advanced Materials*, *26*(35), 6132–6137. doi:10.1002/adma.201401935 PMID:25042767

Liu, Y., Ren, L., He, Y., & Cheng, H. P. (2010). Titanium-decorated graphene for high capacity hydrogen storage studied by density functional simulations. *Journal of Physics Condensed Matter*, *22*(44), 445301. doi:10.1088/0953-8984/22/44/445301 PMID:21403342

Liu, Y., Zhang, L., Wei, W., Zhao, H., Zhou, Z., Zhang, Y., & Liu, S. (2015). Colorimetric detection of influenza A virus using antibody-functionalized gold nanoparticles. *Analyst (London)*, *140*(12), 3989–3995. doi:10.1039/C5AN00407A PMID:25899840

Li, W., Li, C., Ma, H., & Chen, J. (2007). Supporting information-magnesium nanowires: Enhanced kinetics for hydrogen absorption and desorption. *Journal of the American Chemical Society*, *129*(21), 6710–6711. doi:10.1021/ja071323z PMID:17488082

Li, Y., & Yang, R. T. (2006). Hydrogen storage in low silica type X zeolites. *The Journal of Physical Chemistry B*, *110*(34), 17175–17181. doi:10.1021/jp0634508 PMID:16928014

Loofboro, L. (1993). *Tenure Relations in Three Agropastoral Villages: A Framework for Analyzing Natural Resource Use and Environmental Change in the Arrondissement of Boboye, Niger*. Discussion Paper No. 4. Land Tenure Center.

Lototsky, M. V., Yartys, V. A., & Zavaliy, I. Y. (2005). Vanadium based BCC alloys phase structural characteristics and hydrogen sorption properties. *Journal of Alloys and Compounds*, *404-406*, 421–426. doi:10.1016/j.jallcom.2005.01.139

Lovell, S. (2010). Multifunctional urban agro ecology for sustainable land use planning in the United States. *Sustainability*, *2*(8), 2499–2522. doi:10.3390u2082499

Lovell, S. T., DeSantis, S., Nathan, C. A., Olson, M. B., Méndez, V. E., Hisashi, C., & ... . (2010). Integrating agroecology and landscape multi-functionality in Vermont: An evolving framework to evaluate the design of agroeco-systems. *Agricultural Systems*, *103*(5), 327–341. doi:10.1016/j.agsy.2010.03.003

Lovelock, C. H., Patterson, P. G., & Walker, R. (2001). *Services Marketing*. Frenchs Forest.

Lu, K. (1996). Nanocrystalline metals crystallized from amorphous solids: Nanocrystallization, structure and properties. *Materials Science and Engineering*, *16*(4), 161–221. doi:10.1016/0927-796X(95)00187-5

Luo, Q., Zhao, D. Q., Pan, M. X., & Wang, W. H. (2007). Magnetocaloric effect of Ho-, Dy-, and Er-based bulk metallic glasses in helium and hydrogen liquefaction temperature range. *Applied Physics Letters*, *90*(21), 211903. doi:10.1063/1.2741120

Lu, S. G., Rožič, B., Zhang, Q. M., Kutnjak, Z., Li, X., Furman, E., ... Blinc, R. (2010). Organic and inorganic relaxor ferroelectrics with giant electrocaloric effect. *Applied Physics Letters*, *97*(16), 162904. doi:10.1063/1.3501975

Lynch, K. (1972). *What Time is this Place?* MIT Press.

M.C. Mehta V. Union of India and Others, (1985). *Case No. LA. No. 1254, Writ Petition (C), No. 4677*.

M.C. Mehta V. Union of India and Others, (1985). *Writ Petition 4677/1985 (1996. 07.08)*

M.C. Mehta V. Union of India, (1985). *Writ Petition (C) No. 13029*.

Maddison, A. (2003). *The World Economy: Historical Statistics*. OECD Data 1928-98.

Ma, H., & Cheng, F. (2012). Nickel-metal hydride (Ni-MH) rechargeable batteries. In J. Zhang, L. Zhang, H. Liu, A. Sun, & R.-S. Liu (Eds.), *Electrochemical technologies for energy storage and conversion* (pp. 175–237). Wiley-VCH Verlag. doi:10.1002/9783527639496.ch5

Mahapatra, D. (2012). Half of Delhi's Population lives in slums. *The Times of India*.

*Compilation of References*

Mair, G. W., Scherer, F., & Hoffman, M. (2015). Type approval of composite gas cylindres-probabilistic analysis and standards requirements concerning minimum burst pressure. *International Journal of Hydrogen Energy*, *40*(15), 5359–5366. doi:10.1016/j.ijhydene.2015.01.161

Mangen, A. (2014). *Evolution of Reading in the Age of Digitisation: European Cooperation in the field of Scientific and Technical Research*. COST IS1404 E-READ.

Mangen, A., & van der Weel, A. (2016). The evolution of reading in the age of digitization: An integrative framework for reading research. *Literacy*.

Mansoori, G. A., Soelaiman, T. A. F., & Soelaiman, T. A. F. (2005). Nanotechnology-An Introduction for the Standards Community. *Journal of ASTM International*, *2*, 1–22.

Mao, W. L., Koh, C. A., & Sloan, E. D. (2007). Clathrate hydrates under pressure. *Physics Today*, *60*(10), 42–47. doi:10.1063/1.2800096

Markus, M., Yasami, M. T., & Ommeren, M. (2012). *DEPRESSION – A Global Public Health Concern*. WHO. doi:10.1037/e517532013-004

Marques, F. (2010) Constructing sociotechnical transitions toward sustainable agriculture. In *Proc. Symp. Innov. Sustain. Dev. Agric*. Food ISDA.

Marrone, S. (2007). Understanding barriers to health care: A review of disparities in health care services among indigenous populations. *International Journal of Circumpolar Health*, *66*(3), 188–198. doi:10.3402/ijch.v66i3.18254 PMID:17655060

Martín-López, B., Gómez-Baggethun, E., García-Llorente, M., & Montes, C. (2014). Trade-offs across value-domains in ecosystem services assessment. *Ecological Indicators*, *37*, 220–228. doi:10.1016/j.ecolind.2013.03.003

Martsolf, D. S., & Mickley, J. R. (1998). The concept of spirituality in nursing theories: Differing world view and extend of focus. *Journal of Advanced Nursing*, *27*(2), 294–303. doi:10.1046/j.1365-2648.1998.00519.x PMID:9515639

Mason, A. (2014). Overcoming the dual-delivery stigma: A review of patient-centeredness within the Costa Rican medical tourism industry. *International Journal of Communication and Health*, *4*, 1–9.

Mason, A., & Wright, K. B. (2011). Framing medical tourism: An examination of appeal, risk, convalescence, accreditation, and interactivity in medical tourism web sites. *Journal of Health Communication*, *16*(2), 163–177. doi:10.1080/10810730.2010.535105 PMID:21161812

Mason, M. (1981). *Foundations of the Bida Kingdom*. Zaria, Nigeria: Ahmadu Bello University Press.

Masuda, M. (2002). People and Forests in Guinea Savanna. In S. Hirose & T. Wakatsuki (Eds.), *Restoration of Inland Valley Ecosystems in West Africa* (pp. 233–302). Association of Agriculture and Forest Statistics.

Maznevski, M. L., & DiStefano, J. J. (2000). Global Leaders are Team Players: Developing Global Leaders through Membership on Global Teams. *Human Resource Management*, *39*(2 & 3), 195–208. doi:10.1002/1099-050X(200022/23)39:2/3<195::AID-HRM9>3.0.CO;2-I

McDougall, G. H., & Levesque, T. J. (1994). A revised view of service quality dimensions: An empirical investigation. *Journal of Professional Services Marketing*, *11*(1), 189–209. doi:10.1080/15332969.1994.9985149

McGeough, U., & Newman, D. (2004). *Model for sustainable urban design with expanded sections on distributed energy resources*. Sustainable Energy Planning Office Gas Technology Institute and Oak Ridge National Laboratory GTI Project # 30803-23/88018/65952.

McGrath, I. (1997). Information superhighway or information traffic jam for healthcare consumers? *Clinical Performance and Quality Health Care, 5*(2), 90–93. PMID:10167219

McKay, L. (2016). Generating ambivalence: Media representations of Canadian transplant tourism. *Studies in Social Justice, 10*(2), 322–341. doi:10.26522sj.v10i2.1421

Meadows, D.H., Meadows, D.L., Randers, J., Behrens, W.W. (1972). *The Limits to Growth*. Univ. Books.

Mehta, M.C. (2000). Mashesh Chander Mehta in an interview with V. Venkatesan. *Frontline*.

Méndez, V. E., Bacon, C. M., & Cohen, R. (2013). Agroecology as a transdisciplinary, participa-tory, and action-oriented approach. *Agroecology and Sustainable Food Systems, 37*(1), 3–18.

Merckx, T., & Pereira, H. M. (2015). Reshaping agrienvironmental subsidies: From marginal farming to large-scale rewilding. *Basic and Applied Ecology, 16*(2), 95–103. doi:10.1016/j.baae.2014.12.003

Merlock Jackson, K., Lyon Payne, L., & Shepherd Stolley, K. (2013). Celebrity treatment: The intersection of star culture and medical tourism in American society. *Journal of American Culture, 36*(2), 124–134. doi:10.1111/jacc.12019

Merriam Webster. (2018). *Ambience*. Retrieved from www.merriam-webster.com: https://www.merriam-webster.com/dictionary/ambience

Merviö, M. (2017). *Interpretation of Visual Arts Across Societies and Political Culture: Emerging Research and Opportunities*. Hershey, PA: IGI International. doi:10.4018/978-1-5225-2554-7

Metcalf, G. E. (2019). On the Economics of a Carbon Tax for the United States. *BPEA Conference Draft*.

Meterko, M., Nelson, E. C., Rubin, H. R., Batalden, P., Berwick, D. M., Hays, R. D., & Ware, J. E. (1990). Patient judgments of hospital quality: Report of a pilot study. *Medical Care, 28*(9), S1–S56. PMID:2214895

Meyer, M. H., & Roberts, E. B. (1986). New product strategy in small cal issues and applications of stabstical techniques, technology-based firms A pilot study. *Management Science, 32*. doi:10.1287/mnsc.32.7.806

Midilli, A., Ay, M., Dincer, I., & Rosen, M. (2005). On hydrogen and hydrogen energy strategies I: Current status and needs. *Renewable & Sustainable Energy Reviews, 9*(3), 255–271. doi:10.1016/j.rser.2004.05.003

Millenium Ecosystem Assessment. (2005). Ecosystems and human well-being: synthesis; Mukul A. Lack of funds hampers social science research. *Times of India*. Retrieved from https://timesofindia.indiatimes.com/india/Lack-of-funds-hamperssocial-science-research/articleshow/10237494.cms

Miller, W. R., & Thorensen, C. E. (2003). Spirituality, religion, and health: An emerging research field. *The American Psychologist, 58*(1), 24–35. doi:10.1037/0003-066X.58.1.24 PMID:12674816

Milman, O. (2018, January 1). Vehicles are now America's biggest CO2 source but EPA is tearing up regulations. *The Guardian*. Retrieved from https://www.theguardian.com/environment/2018/jan/01/vehicles-climate-change-emissions-trump-administration

Milner, H. R. IV. (2007). Race, culture, and researcher positionality: Working through dangers seen, unseen, and unforeseen. *Educational Researcher, 36*(7), 388–400. doi:10.3102/0013189X07309471

Ministry of Health and Family Welfare Government of India. (2005). *National Family Health Survey (NFHS-3)*. Retrieved from https://dhsprogram.com/pubs/pdf/frind3/frind3-vol1andvol2.pdf

MOEF. (1997). *White Paper on Pollution in Delhi*. Author.

Mohr, L. A., & Bitner, M. J. (1995). The Role of Employee Effort in Satisfaction with Service Transactions. *Journal of Business Research*, *32*(2), 239–252. doi:10.1016/0148-2963(94)00049-K

Mollison, B., & Holmgren, D. (1978). *Permaculture one: A perennial agricultural system for human settlements*. Tagari.

Mollison, B. (1979). *Permaculture: a designer's manual*. Tagari Publications.

Montreal Protocol on substances that deplete the ozone layer. (1987). United Nation Environment Program (UN).

Mooij, L. (2013). Destabilization of magnesium hydride through interface engineering. *LPA Mooij*. Retrieved at https://d1rkab7tlqy5f1.cloudfront.net/TNW/Afdelingen/ChemE/PIs/Dam%2C%20Bernard/Supervised%20theses/PhD_Thesis_L.P.A._Mooij_1_October_2013_small.pdf

Morrison, D., Dokmeci, M., Demirci, U., & Khademhosseini, A. (2008). *Biomedical Nanostructures* (G. Kenneth, H. Craig, T. L. Cato, & N. Lakshmi, Eds.). John Wiley & Sons, Inc.

Morrison, K., Sandeman, K. G., Cohen, L. F., Sasso, C. P., Basso, V., Barcza, A., ... Gutfleisch, O. (2012). Evaluation of the reliability of the measurement of key magnetocaloric properties: A round robin study of La (Fe, Si, Mn) H$\delta$ conducted by the SSEEC consortium of European laboratories. *International Journal of Refrigeration*, *35*(6), 1528–1536. doi:10.1016/j.ijrefrig.2012.04.001

Mota-Babiloni, A., Navarro-Esbrí, J., Molés, F., Cervera, Á. B., Peris, B., & Verdú, G. (2016). A review of refrigerant R1234ze (E) recent investigations. *Applied Thermal Engineering*, *95*, 211–222. doi:10.1016/j.applthermaleng.2015.09.055

Mouchet, M. A., Lamarque, P., Martín-López, B., Crouzat, E., Gos, P., Byczek, C., & Lavorel, S. (2014). An interdisciplinary methodological guide for quantifying associations between ecosystem services. *Global Environmental Change*, *28*, 298–308. doi:10.1016/j.gloenvcha.2014.07.012

MTA. (2017). *Medical Tourism Association 2016-2017 Global Buyers Guide*. Retrieved online March 12, 2019 from https://www.medicaltourismassociation.com/en/prod40_global-buyersreport-2016-2017.html

Mukherjee, B., Dey, N. S., Maji, R., Bhowmik, P., Das, P. J., & Paul, P. (2014). Current Status and Future Scope for Nanomaterials in Drug Delivery (pp. 525–539). Academic Press. doi:10.5772/58450

Murray, P. (Ed.). (1989). *Genius. The History of an Idea*. Oxford: Basil Blackwell.

Murty, M. N., & Kumar, S. (2002). Measuring Cost of Environmentally Sustainable Industrial Development in India: A Distance Function Approach. *Environment and Development Economics*, *7*(3), 467–486. doi:10.1017/S1355770X02000281

Mussery, A., Leu, S., Lensky, I., & Budovsky, A. (2013). The effect of planting techniques on arid ecosystems in the northern Negev. *Arid Land Research and Management*, *27*(1), 90–100. doi:10.1080/15324982.2012.719574

Muyia, H., & Kacirek, K. (2009, December). An Empirical Study of a Leadership Development Training Program and Its Impact on Emotional Intelligence Quotient (EQ) Scores. *Advances in Developing Human Resources*, *11*(6), 703–718. doi:10.1177/1523422309360844

Nadel, S. F. (1942). *A black Byzantium: The kingdom of Nupe in Nigeria*. London: Oxford University Press.

Nagy, G. (2020). A Minoan-Mycenaean scribal legacy for converting rough copies into fair copies. *3rd January 2020; Classical Inquiries*. Harvard's Center for Hellenic Studies. Harvard University. https://classical-inquiries.chs.harvard.edu/a-minoan-mycenaean-scribal-legacy-for-converting-rough-copies-into-fair-copies/

Nagy, J. G., & Edun, O. (2002). *Assessment of Nigerian government fertilizer policy and suggested alternative market-friendly policies*. Unpublished IFDC consultancy paper.

Nair, P. K. R. (1993). *An introduction to agroforestry*. Kluwer Academic Publishers. doi:10.1007/978-94-011-1608-4

Nair, P. K. R. (2007). The coming age of agroforestry. *Journal of the Science of Food and Agriculture, 87*(9), 1613–1619. doi:10.1002/jsfa.2897

Narayan, V., & Debroy, S. (2019). Hair transplant death: Procedure went on for 12 hours till man complained of pain. *The Times of India*.

National Capital Region Planning Board. (1995). *Status of the Environment: NCT*. Author.

National Livestock Development Project. (2003). Accelerated development of grazing reserves and stock routes for settlement of pastoralists. Report submitted by the Sub-committee on grazing reserves and stock routes to the National Livestock Development Project, Kaduna, Nigeria.

National Livestock Development Project. (2007). *Report of the pastoral resolve (pare) committee on pastoral development and empowerment*. Kaduna, Nigeria: National Livestock Development Project.

NCR Planning Board. (1999). *Delhi 1999, A Fact Sheet*. New Delhi: NCR Planning Board, India Habitat Centre.

Neef, A. (1997). Le contrat de parcage, fumure pour les riches? Une étude de cas au sud-ouest du Niger. In *Soil fertility management in West African land use systems. Proceedings of a workshop held in Niamey, 4-8 March 1997*. Margarf.

Neef, A. (2001). Land Tenure and Soil Conservation Practices – Evidence From West Africa and Southeast Asia. In *Sustaining the Global Farm. Selected papers from the 10th International Soil Conservation Organization Meeting held on May 24-29, 1999*. Purdue University and the USDA-ARS National Soil Erosion Research Laboratory.

Nehls, T., Jiang, Y., Dennehy, C., Zhan, X., & Beesley, L. (2015). From waste to value: urban agriculture enables cycling of resources in cities. In Urban Agriculture Europe (pp. 170–173). Berlin: Jovis.

Nelson, E., Scott, S., Cukier, J., & Galán, Á. L. (2009). Institutionalizing agroecology: Successes and challenges in Cuba. *Agriculture and Human Values, 26*(3), 233–243. doi:10.100710460-008-9156-7

Newman, L. M., & Berman, S. M. (2008). Epidemology of STD Disparities in African Communities. *Sexually Transmitted Diseases, 35*(12Supplement), 4–12. doi:10.1097/OLQ.0b013e31818eb90e

NFHS. (1993). *Population Research Centre*. Delhi: Institute of Economic Growth.

Ngamvichaikit, A., & Beise-Zee, R. (2014). Communication needs of medical tourists: An exploratory study in Thailand. *International Journal of Pharmaceutical and Healthcare Marketing, 8*(1), 98–117. doi:10.1108/IJPHM-10-2012-0010

Niederdeppe, J., Hornik, R., Kelly, B., Frosch, D., Romantan, A., Stevens, R., ... Schwarz, S. (2007). Examining the dimensions of cancer-related information scanning and seeking behavior. *Health Communication, 22*(2), 153–167. doi:10.1080/10410230701454189 PMID:17668995

Niemelä, J., Saarela, S. -R., Söderman, T., Kopperoinen, L., Yli-Pelkonen, V., Väre, S., & Kotze, D. (2010). Using the ecosystem services approach for better planning and 13 conservation of urban green spaces: A Finland case study. *Biodiversity and Conservation, 19*(11), 3225-3243.

Nikolelis, D. P., Arzum Erdem, T. V., & Nikoleli, G.-P. (Eds.). (2014). *Portable biosensing of food toxicants and environmental pollutants* (Vol. 1). New York: CRC Press.

Nishiyama, M. (1997). *Edo Culture. Daily Life and Diversions in Urban Japan, 1600-1868* (G. Groemer, Trans. & Ed.). Honolulu, HI: University of Hawai'i Press.

Nissen, L., Merrigan, D., & Kraft, M. K. (2005, March). Moving mountains together: Strategic community leadership and systems change. *Child Welfare, 84*(2), 123–140. PMID:15828404

Nonaka, I., & Takeuchi, H. (1995). *The Knowledge-Creating Company*. Oxford, UK: Oxford University Press.

Norberg-Schulz, C. (1988). *Architecture: Meaning and Place, Selected Essays (Architectural Documents)*. New York: Rizzoli International Publications.

Nordhaus, W. (2007). *The Challenge of Global Warming: Economic Models and Environmental Policy*. Yale Univ.

Nye, J. S. (2008). *The powers to lead*. Oxford, UK: Oxford University Press.

NYSC. (2010). *Nigeria - Bida Agricultural Development in Niger State Project*. http://documents.worldbank.org/curated/en/309101468099257815/pdf/multi-page.pdf

O'Connor, A. (2015, August 9). Coca-Cola funds scientists who shift blame for obesity away from bad diets. *The New York Times*, p. A1.

O'Malley, K., Ordaz, J., Adams, J., Randolph, K., Ahn, C., & Stetson, N. T. (2015). Applied hydrogen storage research and development a perspective from the US department of energy. *Journal of Alloys and Compounds, 645*(1), 419–422. doi:10.1016/j.jallcom.2014.12.090

Oaten, M., & Cheng, K. (2006). Longitudinal gains in self-regulation from regular physical exercise. *British Journal of Health Psychology, 11*(4), 717–733. doi:10.1348/135910706X96481 PMID:17032494

Ogawa, R. (1998). Agriculture in Pastoral Fulbe Society – Re-thinking of the "traditional" production system. In Y. Takamura & Y. Shigeta (Eds.), *The problems of Agriculture in Africa* (pp. 88–113). Kyoto University Press. (in Japanese)

Oh, H., Chung, M. H., & Labianca, G. (2004). Group social capital and group effectiveness: The role of informal socializing ties. *Academy of Management Journal, 47*(6), 860–875. doi:10.5465/20159627

Olomola, A. S. (1998). *Pastoral Development and Grazing Resource Management in Nigerian Savannah Area*. Paper presented at "Crossing Boundaries, the 7th annual conference of the International Association for the Study of Common Property, Vancouver, Canada.

Omotayo, A. M. (2002). A land-use system and the challenge of sustainable agro-pastoral production in southwestern Nigeria. *International Journal of Sustainable Development and World Ecology, 9*(3), 369–382. doi:10.1080/13504500209470131

Ooms, H. (1984). Early Tokugawa Ideology. In P. Nosco (Ed.), *Confucianism and Tokugawa Culture* (pp. 27–61). Honolulu, HI: University of Hawai'i Press.

Oshun, M., Ardoin, N., & Ryan, S. (2011). *Use of the planning outreach liaison model in the neighborhood planning process: a case study in Seattle's Rainier Valley neighborhood*. Urban Stud Res.

Osty, P.-L., Le Ber, F., & Lieber, J. (2008). Raisonnement à partir de cas et agronomie des territoires. *Rev Anthr Connaissances, 2*(2), 169–193. doi:10.3917/rac.004.0169

Ottman, J. A., Stafford, E. R., & Hartman, C. L. (2006). Green Marketing Myopia. *Environment, 48*(5), 22–36. doi:10.3200/ENVT.48.5.22-36

Otto, U. (1998). Innovative Qualität statt neues Etikett: Erste Erfahrungen mit einer konsequenten Programmatik Bürgerschaftlichen Engagements. In J. Braun & O. Klemmert (Eds.), *ISAB-Schriftenreihe: Vol. 54. Selbsthilfeförderung und bürgerschaftliches Engagement in Städten und Kreisen. Fachtagung des Bundesministeriums für Familie, Senioren, Frauen und Jugend am 16./17. Februar 1998 in Bonn* (pp. 215–235). Köln: ISAB.

Oupicky, D., & Li, J. (2014). Bioreducible polycations in nucleic acid delivery: Past, present, and future trends. *Macromolecular Bioscience*, *14*(7), 908–922. doi:10.1002/mabi.201400061 PMID:24678057

Pahl-Wostl, C., Craps, M., Dewulf, A., Mostert, E., Tabara, D., & Taillieu, T. (2007). Social learning and water resources management. *Ecology and Society, 12*(2), 5. https://www.ecologyandsociety.org/vol12/iss2/art5/

Pallasmaa, J. (2014). Space, Place, and Atmosphere: Peripheral Perception in Existential Experience. In C. Borch (Ed.), *Architectural Atmospheres: On the Experience and Politics of Architecture* (pp. 18–41). Basel: Birkhauser. doi:10.1515/9783038211785.18

Pamuk, E., Makuk, D., Heck, K., & Reuben, C. (1998). *Socioeconomic status and health chart book*. Hyattsville, MD: National Center for Health Statistics.

Pang, Y., & Li, Q. (2016). A review on kinetic models and corresponding analysis methods for hydrogen storage materials. *International Journal of Hydrogen Energy*, *41*(40), 18072–18087. doi:10.1016/j.ijhydene.2016.08.018

Parasuraman, A., Zeithaml, V. A., & Berry, L. L. (1985). A conceptual model of service quality and its implications for future research. *Journal of Marketing*, *49*(4), 41–50. doi:10.1177/002224298504900403

Parasuraman, A., Zeithaml, V. A., & Berry, L. L. (1988). Servqual: A multiple-item scale for measuring consumer perc. *Journal of Retailing*, *64*(1), 12.

Parikh, J. (2004). *Environmentally Sustainable Development in India.* Available at http://scid.stanford.edu/events/India2004/JParikh.pdf

Parkes, C. M. (1975). Determinants of outcome following bereavement. *Journal of Death and Dying*, *6*(4), 303–323. doi:10.2190/PR0R-GLPD-5FPB-422L

Parkes, C. M., & Weiss, R. S. (1983). *Recovery from bereavement*. New York, NY: Basic Book.

Patah, A., Takasaki, A., & Szmyd, J. S. (2009). Influence of multiple oxide (Cr2O3/Nb2O5) addition on the sorption kinetics of MgH2. *International Journal of Hydrogen Energy*, *34*(7), 3032–3037. doi:10.1016/j.ijhydene.2009.01.086

Patel, C., Marmot, G. M., Terry, J. D., Carruthers, M., Hunt, B., & Patel, M. (1985). Trial of relaxation reducing coronary risk: Four year follow up. *British Medical Journal*, *290*(6475), 1103–1106. doi:10.1136/bmj.290.6475.1103 PMID:3921124

Paunovic, P., Popovski, O., & Dimitrov, A. T. (2011). Hydrogen economy: The role of nano-scaled support materials for electrocatalysts aimed for water electrolysis. In J. P. Reithmaier, P. Paunovic, W. Kulisch, C. Popov, & P. Petkov (Eds.), NATO science for peace and security series B, Physics and biophysics (pp. 545-563). Academic Press.

Paxton, J. (2019). Economics training and hyperbolic discounting: Training versus selection eff. *Applied Economics*, *51*(55), 5891–5899. doi:10.1080/00036846.2019.1631439

PBB. (2017). *Patients Beyond Borders. Medical tourism statistics and facts*. Accessed March 1, 2019 from www.patientsbeyondborders.com/medical-tourism-statistics-facts

Pecharsky, V. K., & Gschneidner, K. A. Jr. (1997). Giant magnetocaloric effect in Gd 5 (Si 2 Ge 2). *Physical Review Letters*, *78*(23), 4494–4497. doi:10.1103/PhysRevLett.78.4494

Peck, J. (2005). Struggling with the creative class. *International Journal of Urban and Regional Research*, *29*(4), 740–770. doi:10.1111/j.1468-2427.2005.00620.x

Penney, K., Snyder, J., Crooks, V. A., & Johnston, R. (2011). Risk communication and informed consent in the medical tourism industry: A thematic content analysis of Canadian broker websites. *BMC Medical Ethics*, *12*(1), 1–9. doi:10.1186/1472-6939-12-17 PMID:21943392

Peoples Democracy. (2001). Weekly Organ of the Communist Party of India (Marxist), 25(18).

Pereira, J., & Bruera, E. (1998). The Internet as a resource for palliative care and hospice: A review and proposals. *Journal of Pain and Symptom Management*, *16*(1), 59–68. doi:10.1016/S0885-3924(98)00022-0 PMID:9707658

Perkins, A. (2018, Mar. 28). Jeremy Corbyn decries abuse of antisemitism protest MPs. *Guardian*.

Pesonen, O., & Alakunnas, T. (2017). *Energy storage: A missing piece of the puzzle for the self-sufficient living*. Lapland University of Applied Sciences.

Pesut, B., Fowler, M., Taylor, J. E., Reimer-Kirkham, S., & Sawatzky, R. (2008). Conceptualising spirituality and religion for healthcare. *Journal of Clinical Nursing*, *17*(21), 2803–2810. doi:10.1111/j.1365-2702.2008.02344.x PMID:18665876

Petersen, O. H., Hjelmar, U., & Vrangbæk, K. (2015). Is contracting out of public services still the great panacea? A systematic review of international studies from 2000-2014. *Proceedings NormaCare Meeting*.

Petersen, P., Mussoi, E. M., & Dalsoglio, F. (2012). Institutionalization of the agroecological approach in Brazil: Advances and challenges. *Journal of Sustainable Agriculture*, *37*, 103–114. doi:10.1080/10440046.2012.735632

Peterson, D. T., & Nelson, S. O. (1985). The isopiestic solubility of hydrogen in vanadium alloys at low temperatures. *Metallurgical and Materials Transactions. A, Physical Metallurgy and Materials Science*, *16A*, 36–374.

Petty, T. (2014). *Capital in the Twenty-First Century*. Belknap/Harvard University Press.

Phan, M. H., & Yu, S. C. (2007). Review of the magnetocaloric effect in manganite materials. *Journal of Magnetism and Magnetic Materials*, *308*(2), 325–340. doi:10.1016/j.jmmm.2006.07.025

Pigg, K. E. (1999). Community leadership and community theory: A practical synthesis. *Journal of the Community Development Society*, *30*(2), 196–212. doi:10.1080/15575339909489721

Pigou, A. C. (1920). *The Economics of Welfare*. Cambridge.

Piraux, M., Silveira, L., Diniz, P., Duque, G., Coudel, E., Devautour, H., ... Hubert, B. (2010) Agroecological transition as a socio-territorial innovation: the case of the territory of Borborema in Brazilian semi-arid. *Proc. Symp. Innov. Sustain. Dev. Agric.*

Pirl, W. F., Roth, A. J., Cotton, S. P., Levin, E. G., & Fitzpatrick, C. M. (2000). Exploring the relationship amongspiritual well-beingh, quality of life, and psychological adjustment in women with breast cancer. *Psycho-Oncology*, *8*(5), 429–438.

Poe, E. A. (2017). *Great Horror Stories*. Meniola, New York: Dover Publication.

*Pohjoismainen taide 1880- luvulla*. (1986). Helsinki: Amos Andersonin Taidemuseo.

Porter, M. (1990). *The Competitive Advantage of Nations*. New York: Free Press. doi:10.1007/978-1-349-11336-1

Porter, M., & Reinhardt, F. L. (2007). A Strategic Approach to Climate. *Harvard Business Review*, *85*(10), 22–26.

Powell, J. M., & Williams, T. O. (1993). Livestock, nutrient cycling and sustainable agriculture in the West African Sahel. IIED Gatekeeper Series No. 37. International Institute for Environment and Development.

Prabhukhot, P. R., Wagh, M. M., & Gangal, A. C. (2016). A review on solid storage of hydrogen material. *Advances in Energy and Power*, *4*(2), 11–22.

Prabhu, S. S., Nayak, S. N., Kapilan, N., & Hindasagen, V. (2017). An experiment and numeric study on the effects of exhaust gas temperature and flowrate on deposit formation in Urea Selective catalytic reduction (SCR) system of modern automobiles. *Applied Thermal Engineering*, *111*, 1211–1231. doi:10.1016/j.applthermaleng.2016.09.134

Prager, Reed, & Scott. (2012). Encouraging collaboration for the provision of ecosystem services at a landscape scale—rethinking agri-environmental payments. *Land Use Policy, 29*(1), 244–249. .landusepol.2011.06.012 doi:10.1016/j

Pranevicius, L., Milcius, D., Pranevicius, L. L., & Thomas, G. (2004). Plasma hydrogenation of Al, Mg and MgAl films under high-flux ion irradiation at elevated temperatures. *Journal of Alloys and Compounds, 373*(1-2), 9–15. doi:10.1016/j.jallcom.2003.10.029

Prein, P. (2002). Integration of aquaculture into crop–animal systems in Asia. *Agricultural Systems, 71*(1-2), 127–146. doi:10.1016/S0308-521X(01)00040-3

Pretty, J. (2005). Sustainability in agriculture: Recent progress and emergent challenges. Issues. *Environmental Science & Technology, 21*, 1–15.

Pretty, J. (2006). *Agroecological approaches to agricultural development*. Washington, DC: World Bank.

Pretty, J. (2008). Agricultural sustainability: Concepts, principles and evidence. *Philosoph Trans Royal Soc. Botanical Sciences, 363*(1491), 447–465. PMID:17652074

Przybylski, A. K., & Weinstein, N. (2012). Can you connect with me now? How the presence of mobile communication technology influences face-to-face conversation quality. *Journal of Social and Personal Relationships, 30*(3), 237–246. doi:10.1177/0265407512453827

Rachel, R. H., Eduardo, M. M., & Katy, B. K. (2016). Structural Racism and Supporting Black Lives - The Role of Health Professionals. *The New England Journal of Medicine*, 2113-2115.

Rafi-ud-din, Zhang, L., Ping, L., & Xuanhui, Q. (2010). Catalytic effects of nano-sized TiC additions on the hydrogen storage properties of LiAlH4. *Journal of Alloys and Compounds, 508*(1), 119–128. doi:10.1016/j.jallcom.2010.08.008

Raja Sabha. (2000). *Short duration discussion, shifting of industries from residential areas of Delhi*. Retrieved at www.rsdebate.nic.in

Rajalakshmi, T. K., & Venkatesan, V. (2001). Commuters Crisis. *Frontline, 18*(8), 14-27.

Rajput, A. (2017). MCD polls: Delhi slum clusters wait for promised development. *Hindustan Times*.

Ramachandran, R. (2000). The Lethal Zones. *Frontline, 17*(25).

Ratzan, S. C., & Parker, R. M. (2000). *Health literacy*. National Library of Medicine website. www.nlm.nih.gov/archive/20061214/pubs/cbm/hliteracy.html#15

Reda, M. R. (2009). The effect of organic additive in Mg/Graphite composite as hydrogen storage materials. *Journal of Alloys and Compounds, 480*(2), 238–240. doi:10.1016/j.jallcom.2009.02.021

Rights, H. C. (2000). *Educating Children and Youth Against Racism*. Durban: United Nations High Commissioner for Human Rights.

Risch, N., Burchard, E., Ziv, E., & Tang, H. (2007). Categirization of humans in biomedical research: Genes, race and disease. *Genome Biology, 3*(7), comment2007.1. doi:10.1186/gb-2002-3-7-comment2007

Rivera, M. A., Quigley, M. F., & Scheerens, J. C. (2004). Performance of component species in three apple-berry polyculture systems. *HortScience, 39*(7), 1601–1606. doi:10.21273/HORTSCI.39.7.1601

Roberts, E. B., & Berry, C. A. (1985). Entering new businesses: Selecting strategies for success. *Sloan Manage Rev*, 3–17.

Roberts, E. B., & Hauptman, O. (1987, March). The financing threshold effect on success and failure of biomedical and pharmaceutical star-ups. *Management Science, 33*(3), 381–394. doi:10.1287/mnsc.33.3.381

Roche, M., & Argent, N. (2012). The fall and rise of agricultural productivism? An Antipodean viewpoint. *Progress in Human Geography, 39*(5), 621–635. doi:10.1177/0309132515582058

Romeo, N., Gallo, O., & Tagarelli, G. (2015). From Disease to Holiness: Religious-based health remedies of Italian folk medicine (XIX-XX century). *Journal of Ethnobiology and Ethnomedicine, 11*(1), 50. doi:10.118613002-015-0037-z PMID:26048412

Rosset, P. M., & Martínez-Torres, M. E. (2012). Rural social movements and agroecology: Context, theory, and process. *Ecology and Society, 17*(3), 17. doi:10.5751/ES-05000-170317

Rosset, P. M., Sosa, B. M., Jaime, A. M. R., & Lozano, D. R. Á. (2011). The campesino-to-campesino agroecology movement of ANAP in Cuba: Social process methodology in the construction of sustainable peasant agriculture and food sovereignty. *The Journal of Peasant Studies, 38*(1), 161–191. doi:10.1080/03066150.2010.538584 PMID:21284238

Rottle, N., & Yocom, K. (2010). *Ecological design*. Lausanne: AVA Publishing.

Roy, R., Wield, D., Gardiner, J. P., & Potter, S. (1996). *Innovative Product Development*. Milton Keynes: The Open University.

Saadi, A. (2016, February 7). *KevinMD*. Retrieved July 31, 2019, from www.kevinmd.com: https://www.kevinmd.com/blog/2016/02/muslim-american-doctor-racism-hospitals.html

SAE International. (2018). *Standard for fuel systems in fuel cell and other hydrogen vehicles*. SAE International.

Sakintuna, B., Lamari-Darkrim, F., & Hirscher, M. (2007). Metal hydride materials for solid hydrogen storage: A review. *International Journal of Hydrogen Energy, 32*(9), 1121–1140. doi:10.1016/j.ijhydene.2006.11.022

Salameh, C. M. (2014). *Synthesis of boron or aluminum based functional nitrides for energy applications (hydrogen production and storage)*. Material chemistry. Université Montpellier II - Sciences et Techniques du Languedoc.

Sallah, M., & Perez, M. (2019, January 30). This business helped transform Miami into a national plastic surgery destination. Eight women died. *USA Today*. Retrieved February 20, 2019 from https://www.usatoday.com/in-depth/news/investigations/2019/01/31/miami-doctors-plasticsurgery-empire-becomes-floridas-deadliest-clinics/2729802002/?fbclid=IwAR2imyhjTKSLbXWGfTgPGg_KX7uBRAb9eEbqtdxZoDnl7RLhrw-qFt--iW0

Sandler, M. (2015, August 8). CEO pay soars at top not-for-profits. *Modern Healthcare*. PMID:26642549

Sangarè, M., Fernández-Rivera, S., Hiernaux, P., Bationo, A., & Pandey, V. (2002). Influence of dry season supplementation for cattle on soil fertility and millet (Pennisetum glaucum L.) yield in a mixed crop/livestock production system of the Sahel. *Nutrient Cycling in Agroecosystems, 62*(3), 209–217. doi:10.1023/A:1021237626450

Santos, B. (2007). *Crítica da razão indolente: contra o desperdício da experiência*. São Paulo: Cortez.

Sarasvathy, S., & Read, S. (2001). *Effectual Entrepreneurship*. Routledge.

Satell, G. (2017, Mar. 30). How to Win with Automation (Hint: It's Not Chasing Efficiency). *Harvard Business Review*.

Satyapal, S., Petrovic, J., Read, C., Thomas, G., & Ordaz, G. (2007). The US Department of Energy's national hydrogen storage project: Progress towards meeting hydrogen-powered vehicle requirements. *Catalysis Today, 120*(3-4), 246–256. doi:10.1016/j.cattod.2006.09.022

Säumel, I., Kotsyuk, I., Hölscher, M., Lenkereit, C., Weber, F., & Kowarik, I. (2012). How healthy is urban horticulture in high traffic areas? Trace metal concentrations in vegetable crops from plantings within inner city neighbourhoods in Berlin, Germany. *Environmental Pollution, 165*, 124–132. doi:10.1016/j.envpol.2012.02.019 PMID:22445920

Savelsbergh, C., Poell, R. F., & van der Heijden, B. (2015). Does team stability mediate the relationship between leadership and team learning? An empirical study among Dutch project teams. *International Journal of Project Management, 33*(2), 406–418. doi:10.1016/j.ijproman.2014.08.008

Schaffers, H., Garcia Guzmán, J., Navarro, M., & Merz, C. (Eds.). (2010). *Living Labs for Rural Development*. Madrid: TRAGSA. http://www.c-rural.eu

Scheffer, M., & Westley, F. (2007). The evolutionary basis of rigidity: locks in cells, minds, and society. *Ecology and Society, 12*(2), 36. https://www.ecologyandsociety.org/vol12/iss2/art36/

Schlecht, E., Hiernaux, P., Achard, F., & Turner, D. (2004). Livestock related nutrient budgets within village territories in western Niger. *Nutrient Cycling in Agroecosystems, 68*(3), 199–211. doi:10.1023/B:FRES.0000019453.19364.70

Schmitz, H. (2017). The Felt Body and Embodied Communication. *Yearbook for Eastern and Western Philosophy*, 9-19. doi:10.1515/yewph-2017-0004

Schneider, R. H., Nidich, S. I., Salerno, J. W., Sharma, H. R., Robinson, C. E., Nidich, R. J., & Alexander, C. N. (1998). Lower lipid peroxide levels in practitioners of the transcendental mediation program. *Psychosomatic Medicine, 60*(1), 38–41. doi:10.1097/00006842-199801000-00008 PMID:9492237

Schober, T. (1996). Vanadium, Niobium and tantalum-hydrogen. *Diffusion and Defect Data, Solid State Data. Part B, Solid State Phenomena, 49-50*, 357–422. doi:10.4028/www.scientific.net/SSP.49-50.357

Schoonhaven, C. B., Eisenhardt, K. M., & Lyman, K. (1990, March). Speeding degrees in organizational studies from the Univerproducts to market Waiting time to first product introductions in new sity of San Francisco firms. *Administrative Science Quarterly, 35*, 177–207. doi:10.2307/2393555

Schutt, F. (2003). *The Importance of Human Capital for Economic Growth*. IWIM. University of Bremen.

Schwarz, K., Cutts, B. B., London, J. K., & Cadenasso, M. L. (2016). Growing gardens in shrinking cities: A Solution to the Soil Lead Problem? *Sustainability, 8*(2), 1–11. doi:10.3390u8020141

Seemita, B., Dasgupta, K., Kumar, A., Ruz, P., Vishwanadh, B., Joshi, J. B., & Sudarsan, V. (2015). Comparative evaluation of hydrogen storage behavior of Pd doped carbon nanotubes prepared by wet impregnation and polyol methods. *International Journal of Hydrogen Energy, 40*(8), 3268–3276. doi:10.1016/j.ijhydene.2015.01.048

Segouin, C., Hodges, B., & Brechat, P.-H. (2005). Globalization in health care: Is international standardization of quality a step toward outsourcing? *International Journal for Quality in Health Care, 17*(4), 277–279. doi:10.1093/intqhc/mzi059 PMID:16033805

Selvam, P., Viswanathan, B., Swamy, C. S., & Srinivasan, V. (1986). Magnesium and magnesium alloy hydrides. *International Journal of Hydrogen Energy, 11*(3), 169–192. doi:10.1016/0360-3199(86)90082-0

Semper, G. (2004). *Style in the Technical and Tectonic Arts, Or, Practical Aesthetics*. Los Angeles: Getty Research Institute.

Servigne, P., & Stevens, R. (2015). *Comment tout peut s'effondrer. Petit manuel de collapsologie à l'usage des générations présentes*. Paris, France: Seuil.

Shafranske, E. P., & Malony, H. N. (1990). Clinical psychologists' religious and spiritual orientation and their practice of psychotherapy. *Psychotherapy (Chicago, Ill.), 27*(1), 72–78. doi:10.1037/0033-3204.27.1.72

Shahi, R. R., Tiwari, A. P., Shaz, M. A., & Srivastava, O. N. (2013). Studies on de/rehydrogenation characteristics of nanocrystalline MgH2 co-catalyzed with Ti, Fe, and Ni. *International Journal of Hydrogen Energy, 38*(6), 2778–2784. doi:10.1016/j.ijhydene.2012.11.073

Shapiro, J. (2003). *Smart cities: explaining the relationship between city growth and human capital*. Harvard University.

Shepard, M. (2013). *Restoration agriculture: real world permaculture for farmers*. Austin: Acres U.S.A.

Sherard, J. L., Decker, R. S., & Dunnigan, L. P. (1976). Identification and nature of dispersive soils. *Journal of the Geotechnical Engineering Division*, *102*, 287–301.

Sheriff, S. A., Yogi, G. D., Stefanakos, E., & Steinfield, A. (2014). *A handbook of hydrogen energy*. CRC Press. doi:10.1201/b17226

Shikano, K. (2002). Ecological anthropological study on daily herding activities of pastoral Fulani in central Nigeria. In S. Hirose & T. Wakatsuki (Eds.), *Restoration of Inland Valley Ecosystems in West Africa* (pp. 303–369). Association of Agriculture and Forest Statistics.

Shimada, M., Tamaki, H., Higuchi, E., & Inoue, H. (2014). Kinetic analysis for hydrogen absorption and desorption of MgH2-based composites. *Journal of Materials and Chemical Engineering*, *2*, 64–71.

Shimada, S. (1999). A study of increased food production in Nigeria: The effect of the Structural Adjustment Program of the local level. *African Study Monographs*, *20*(4), 175–227.

Silva, J., Fernandes, A. R., & Baptista, P. V. (2014). Application of Nanotechnology in Drug Delivery. Academic Press.

Silva, D. J., Ventura, J., Araújo, J. P., & Pereira, A. M. (2014). Maximizing the temperature span of a solid state active magnetic regenerative refrigerator. *Applied Energy*, *113*, 1149–1154. doi:10.1016/j.apenergy.2013.08.070

Simchi, H., Kaflou, A., & Simchi, A. (2009). Synergetic effect Ni and Nb2O5 on dehydrogenation properties of nanostructured MgH2 synthesized by high-energy mechanical alloying. *International Journal of Hydrogen Energy*, *34*(18), 7724–7730. doi:10.1016/j.ijhydene.2009.07.038

Simon Rojo, M., Moratalla, A. Z., Alonso, N. M., & Jimenez, V. H. (2014). Pathways towards the integration of peri-urban agrarian ecosystems into the spatial planning system. *Ecological Processes*, *3*(13), 16.

Simosi, M., & Xenikou, A. (2010). The role of organizational culture in the relationship between leadership and organizational commitment: An empirical study in a Greek organization. *International Journal of Human Resource Management*, *21*(10), 1598–1616. doi:10.1080/09585192.2010.500485

Simpson, J., & Proffitt, M. (1997). *Oxford English Dictionary Additions Series* (Vol. 3). Oxford: Oxford University Press.

Sinclair, A. (2013). Not just "adding women in": Women re-making leadership. *Seizing the Initiative: Australian Women Leaders in Politics, Workplaces and Communities*, 15-34.

Sinclair, A. (2009). Seducing leadership: Stories of leadership development. *Gender, Work and Organization*, *16*(2), 266–284. doi:10.1111/j.1468-0432.2009.00441.x

Singh, A. (1997). Ozone Distribution in Urban Environment of Delhi during Winter Months. In *Atmospheric Environment*. Elsevier Science Ltd. doi:10.1016/S1352-2310(97)00138-6

Skevington, S. M., Carse, M. S., & Williams, A. C. (2001). Validation of the WHOQOL-100: Pain management improves quality of life for chronic pain patients. *The Clinical Journal of Pain*, *17*, 264–275. doi:10.1097/00002508-200109000-00013 PMID:11587119

Smart city Edinburgh. (2011). http://www.edinburgh.gov.uk

Smith, A. (1776). *Inquiry into the Nature and Causes of the Wealth of Nations*. Oxford. doi:10.1093/oseo/instance.00043218

Smith, A., Voss, J. P., & Grin, J. (2010). Innovation studies and sustainability transitions: The allure of the multi-level perspective and its challenges. *Research Policy*, *39*(4), 435–448. doi:10.1016/j.respol.2010.01.023

Smith, Z. (2004). Preface. In *The Quiet American*. Penguin.

Smukler, S., Sánchez-Moreno, S., Fonte, S., Ferris, H., Klonsky, K., O'Geen, A., ... Jackson, L. (2010). Biodiversity and multiple ecosystem functions in an organic farmscape. *Agriculture, Ecosystems & Environment*, *139*(1-2), 80–97. doi:10.1016/j.agee.2010.07.004

Snyder, J., Johnston, R., Crooks, V. A., Morgan, J., & Adams, K. (2017). How medical tourism enables preferential access to care: Four patterns from the Canadian context. *Healthcare Analysis: HCA: Journal of Health Philosophy and Policy*, *25*(2), 138–150. doi:10.100710728-015-0312-0 PMID:26724280

Somayazulu, M. S., Finger, L. W., Hemley, R. J., & Mao, H. K. (1996). High pressure compounds in methane hydrogen mixtures. *Science*, *271*(5254), 1400–1402. doi:10.1126cience.271.5254.1400

Song, Y., Guo, Z. X., & Yang, R. (2004). Influence of selected alloying elements on the stability of magnesium dihydride for hydrogen storage applications: A first principles investigation. *Physical Review. B*, *69*(9), 094205. doi:10.1103/PhysRevB.69.094205

Souder, W. (1987). *Managing New Product Innovation*. Lexington: MA Lex.

South African Government. (2019, May 26). *President Cyril Ramaphosa signs 2019 Carbon Tax Act into law* [media statement]. Retrieved from https://www.gov.za/speeches/publication-2019-carbon-tax-act-26-may-2019-0000

Srinivasan, R. (2012). *Healthcare in India – 2025 Issues and prospects*. Retrieved from http://planningcommission.gov.in/reports/genrep/bkpap2020/26_bg2020.pdf

St. Johns herald and Apache news. *[volume]* (St. Johns, Apache Co., Ariz.), 11 April 1907. (n.d.). *Chronicling America: Historic American Newspapers*. Lib. of Congress. Retrieved from: https://chroniclingamerica.loc.gov/lccn/sn95060582/1907-04-11/ed-1/seq-8/

Stark, R. (2012). *America's Blessings: How Religion Benefits Everyone, Including Atheists* (Vol. 1). Templeton Foundation Press.

Staszewski, M., Osadnik, M., Czepelak, M., & Swoboda, P. (2011). Hydrogen storage alloys prepared by high energy milling. *Journal of Achievements in Materials and Manufacturing*, *44*, 154–160.

Statista. (n.d.). *Value of gross direct premiums from general insurance industry across India from FY 2002 to FY 2019*. Retrieved from https://www.statista.com/statistics/1075200/india-insurance-gross-direct-premium-value/

Stefansdottir, H. (2017). The role of urban atmosphere for non-work activity locations. *Journal of Urban Design*, 319–335. doi:10.1080/13574809.2017.1383150

Stein, A. (2001). *Participation and sustainability in social projects: The experience of the Local Development Programme (PRODEL) in Nicaragua*. IIED working paper 3 on poverty reduction in urban areas. International Institute for Environment and Development, London.

Stenning, D. (1959). *Savannah nomads: A Study of the Woodabe Pastoral Fulani of Western Bornu Province Northern Region*. International African Institute.

Stevenson, A. (2010). *Oxford Dictionary of English*. London: OUP Oxford.

Stevensona, C. L., Santini, A. J. T. Jr, & Langerb, R. (2012). Reservoir-Based Drug Delivery Systems Utilizing Microtechnology. *Advanced Drug Delivery Reviews*, *64*(14), 1590–1602. doi:10.1016/j.addr.2012.02.005 PMID:22465783

Stevenson, G. W., Clancy, K., King, R., Lev, L., Ostrom, M., & Smith, S. (2011). Midscale food value chains: An introduction. *Journal of Agriculture, Food Systems, and Community Development*, *1*(4), 1–8. doi:10.5304/jafscd.2011.014.007

Stewart, A. L., Ware Jr, J. E., Brook, R. H., & Davies-Avery, A. (1978). *Conceptualization and Measurement of Health for Adults in the Health Insurance Study: Vol. II, Physical Health in Terms of Functioning*. The Rand Corporation. R-1987/2-HEW.

Stoltz, S. E., & Popovic, D. (2007). A high-resolution core-level study of Ni-catalyzed absorption and desorption of hydrogen in Mg-films. *Surface Science*, *601*(6), 1507–1512. doi:10.1016/j.susc.2007.01.016

Struzhkin, V. V., Militzer, B., Mao, W. L., Mao, H.-K., & Hemley, R. J. (2007). Hydrogen storage in molecular clathrates. *Chemical Reviews*, *107*(10), 4133–4151. doi:10.1021/cr050183d PMID:17850164

Sturm, S., & Turner, S. (2012). "Built Pedagogy": The University of Auckland Business School as Crystal Palace. *Interstices*, *12*, 23–34.

Subramony, M., Segers, J., Chadwick, C., & Shayamsunder, A. (2018). Leadership development practice bundles and organizational performance: The mediating role of human capital and social capital. *Journal of Business Research*, *83*(C), 120–129. doi:10.1016/j.jbusres.2017.09.044

Suleiman, H. (1989). Policy issue in pastoral development in Nigeria. In *Pastoralism in Nigeria: Past, Present & Future. Proceedings of the national conference on pastoralism in Nigeria*. Ahmadu Bello University.

Summary for Policymakers. (2018). *Special Report on Global Warming of 1.5°C approved by governments*. IPCC.

Suomen Taideyhdistyksen perustaminen. Lähteillä. (n.d.). *Suomen kansallisgalleria*. http://www.lahteilla.fi/fi/page/suomen-taideyhdistys-suomen-taideyhdistyksen-perustaminen

Suomen Taideyhdistyksen piirustuskoulu Helsingissä. Lähteillä. (n.d.). *Suomen kansallisgalleria*. http://www.lahteilla.fi/fi/page/suomen-taideyhdistys-suomen-taideyhdistyksen-piirustuskoulu-helsingissä

Sustainable Development Goals. (2019). *Progress of goal 13 in 2019*. Retrieved from https://sustainabledevelopment.un.org/sdg13

Sustainable Development Goals. (n.d.). *Sustainable development goals*. United Nations Department of Information. Retrieved from https://sustainabledevelopment.un.org/?menu=1300

Swensen, G., & Jerpåsen, G. P. (2008). Cultural heritage in suburban landscape planning. A case study in Southern Norway. *Landscape and Urban Planning*, *87*(4), 289–300. doi:10.1016/j.landurbplan.2008.07.001

Taghavi, A., & Murat, A. (2011). A heuristic procedure for the integrated facility layout design and flow assignment problem. *Computers & Industrial Engineering*, *61*(1), 55–63. doi:10.1016/j.cie.2011.02.011

Tahir, M., Cao, C., Butt, F. K., Idrees, F., Mahmood, N., Ali, Z., ... Mahmood, T. (2013). Tubular graphitic-C3N4: A prospective material for energy storage and green photocatalysis. *Journal of Materials Chemistry. A, Materials for Energy and Sustainability*, *1*(44), 1. doi:10.1039/c3ta13291a

Taras, V., Caprar, D. V., Rottig, D., Sarala, R. M., Zakaria, N., Zhao, F., & Huang, V. Z. (2013). A global classroom? Evaluating the effectiveness of global virtual collaboration as a teaching tool in management education. *Academy of Management Learning & Education*, *12*(3), 414–435. doi:10.5465/amle.2012.0195

Taylor, S. A., & Baker, T. L. (1994). An assessment of the relationship between service quality and customer satisfaction in the formation of consumers' purchase intentions. *Journal of Retailing*, *70*(2), 163–178. doi:10.1016/0022-4359(94)90013-2

Terry, H., & Liller, K. D. (2014). The Doctoral student leadership institute: Learning to lead for the future. *Journal of Leadership Education, 13*(1), 126–135. doi:10.12806/V13/I1/IB2

Teyber, R., Trevizoli, P. V., Christiaanse, T. V., Govindappa, P., Niknia, I., & Rowe, A. (2018). Semi-analytic AMR element model. *Applied Thermal Engineering, 128*, 1022–1029. doi:10.1016/j.applthermaleng.2017.09.082

The Citizens Fifth Report. (1997). *State of India's Environment.* New Delhi: CSE.

The Economics of Ecosystems and Biodiversity (TEEB). (2010). *Mainstreaming the economics of nature: a synthesis of the approach, conclusions and recommendations of TEEB.* Geneva, Switzerland: TEEB.

The Guardian. (2017). Jamie Fullerton. Beijing hit by dirty smog but observers say air is getting better. *The Guardian.*

The Hindu Survey of Environment. (2001). Chennai: Academic Press.

The Times of India. (2001). *Naik takes a U-turn, says not enough CNG.* Author.

Thibaud, J.-P. (2002). From Situated Perception to Urban Ambiences. In *First international Workshop on Architectural and Urban Ambient Environment* (pp. 1-11). Nantes, France: Centre de recherche m'ethodologique d'architecture; Ecole d'architecture de Nantes.

Thomas, S., Glynne-Jones, R., & Chaiti, I. (1997). Is It Worth the Wait? A Survey of Patients' Satisfaction with an Oncology Outpatient Clinic. *European Journal of Cancer Care, 6*(1), 50–58. doi:10.1111/j.1365-2354.1997.tb00269.x PMID:9238930

Tidball, K. G., & Krasny, M. E. (2009). *From risk to resilience: what role for community greening and civic ecology in cities?* Wageningen: Wageningen Academic Publishers.

Till, J. (2009). *Architecture Depends.* MIT Press.

Torre Ugarte, D. G., & Hellwinckel, C. C. (2010). The problem is the solution: the role of biofuels in the transition to a regenerative agriculture. In P. Mascia, J. Scheffran, & J. Widholm (Eds.), *Plant biotechnology for sustainable production of energy and co-products* (pp. 365–384). New York: Springer. doi:10.1007/978-3-642-13440-1_14

Tovato, F. (1990). Domestic/Religious Individualism and Youth Sicide in Canada. *Family Perspective, 24*(1), 69–81.

Toxic Link. (2000). Cloning Bhopal: Exposing the Dangers in Delhi's Environment. Author.

Trevizoli, P. V., Nakashima, A. T., Peixer, G. F., & Barbosa, J. R. Jr. (2017). Performance assessment of different porous matrix geometries for active magnetic regenerators. *Applied Energy, 187*, 847–861. doi:10.1016/j.apenergy.2016.11.031

Trevizoli, P. V., Peixer, G. F., Nakashima, A. T., Capovilla, M. S., Lozano, J. A., & Barbosa, J. R. Jr. (2018). Influence of inlet flow maldistribution and carryover losses on the performance of thermal regenerators. *Applied Thermal Engineering, 133*, 472–482. doi:10.1016/j.applthermaleng.2018.01.055

TropSoils. (1991). *Integrated management of an agricultural watershed: Characterization of a research site near Hamdallaye, Niger.* TropSoils Bulletin No. 91-03.

Trygve, R., Sandrock, G., Ulleberg, O., & Vie, P. J. S. (2005). Hydrogen storage - Gaps and priorities. *HIA HCG*, 1-13.

Tsai, W., Prigerson, H. G., Li, C., Chou, W., Kuo, S., & Tang, S. T. (2016). Longitudinal changes and predictors of prolonged grief for bereaved family caregivers over the first 2 years after the terminally ill cancer patient's death. *Palliative Medicine, 30*(5), 495–503. doi:10.1177/0269216315603261 PMID:26311571

Tscharntke, T., Klein, A. M., Kruess, A., Steffan-Dewenter, I., & Thies, C. (2005). Landscape perspectives on agricultural intensification and biodiversity-ecosystem service management. *Ecology Letters*, *8*(8), 857–874. doi:10.1111/j.1461-0248.2005.00782.x

Tschumi, B. (1994). *Bernard Tschumi, architecture and event: April 21-July 5, 1994*. The Museum of Modern Art.

Tschumi, B. (1981). *The Manhattan Transcripts*. New York: St. Martin's Press.

Tubbs, S. L., & Schulz, E. (2006). Exploring a taxonomy of global leadership competencies and meta competencies. *The Journal of American Academy of Business, Cambridge*, *8*(2), 29–34.

Turner, L. (2013). *Patient mortality in medical tourism*. Oxford University Press; doi:10.1093/acprof:oso/9780199917907.003.0001

Turner, L. G. (2011). Quality in healthcare and globalization of health services: Accreditation and regulatory oversight of medical tourism companies. *International Journal for Quality in Healthcare: Journal of the International Society for Quality in Healthcare*, *23*(1), 1–7. doi:10.1093/intqhc/mzq078 PMID:21148210

Tyre, M. J., & Hauptman, O. (1992). Effectiveness of organizational responses to technological change in the production process. *Organization Science*, *3*(3), 301–320. doi:10.1287/orsc.3.3.301

Uhl-Bien, M. (2006). Relational Leadership Theory: Exploring the social processes of leadership and organizing. *The Leadership Quarterly*, *17*(6), 654–676. doi:10.1016/j.leaqua.2006.10.007

UN-Habitat. (2003). *Water and sanitation in the world's cities: Local action for global goals*. London: Earthscan.

United States Environment Protection Agency (EPA). (n.d.). *EPA history*. Retrieved from https://www.epa.gov/history

Upson, J. W., Damaraju, N. L., Anderson, J. R., & Barney, J. B. (2017). Strategic networks of discovery and creation entrepreneurs. *European Management Journal*, *35*(2), 198–210. doi:10.1016/j.emj.2017.01.001

USAID. (2007). *Nigeria food security update*. Famine early warning systems network paper May 2007. www.fews.net/south

Valenti, G. (2015). Hydrogen liquefaction and liquid hydrogen storage. *Compendium of Hydrogen Energy*, 27-51.

Van de Ven, A. H., Polley, D. E., Garud, R., & Venkataraman, S. (1999). *The Innovation Journey*. New York: Oxford University Press.

Van Herk, S., Zevenbergen, C., Rijke, J., Ashley, R., (2011). Collaborative research to support transition towards integrating flood risk management in urban development. *Journal of Flood Risk Management*, *4*(4), 306 - 317.

VanderWeele, T. J., Balboni, T. A., & Koh, H. K. (2017). Health and spirituality. *Journal of the American Medical Association*, *318*(6), 519–520. doi:10.1001/jama.2017.8136 PMID:28750127

VanVactor, J. D. (2012). Collaborative leadership model in the management of health care. *Journal of Business Research*, *65*(4), 555–561. doi:10.1016/j.jbusres.2011.02.021

Varshney, C. K. (1989). *VOC and Ozone Pollution: Health Implications*. New Delhi: Jawaharlal Nehru University, School of Environmental Sciences.

Vedel, A., & Thomsen, D. K. (2017). The Personality of Academic Majors. *Personality and Individual Differences*, *116*, 86–91. doi:10.1016/j.paid.2017.04.030

Veldkamp, A., Kok, K., De Koning, G. H. J., Schoorl, J. M., Sonneveld, M. P. W., & Verburg, P. H. (2001). Multi-scale system approaches in agronomic research at the landscape level. *Soil & Tillage Research*, *58*(3-4), 129–140. doi:10.1016/S0167-1987(00)00163-X

Veleckis, E., & Edwards, R. K. (1969). Thermodynamic properties on the system vanadium hydrogen, Niobium-hydrogen and tantalum-hydrogen. *Journal of Physical Chemistry*, *73*(3), 683–692. doi:10.1021/j100723a033

Venkatadri, U., Rardin, R. L., & Montreuil, B. (1997). A design methodology for fractal layout organization. *IIE Transactions*, *29*(10), 911–924. doi:10.1080/07408179708966411

von Kaufmann, R., Chater, S., & Blench, R. (1986). Livestock systems research in Nigeria's subhumid zone. In *Proceedings of the second ILCA/NAPRI symposium*. Addis Ababba: International Livestock Centre for Africa.

von Moos, L., Bahl, C. R. H., Nielsen, K. K., & Engelbrecht, K. (2014). The influence of hysteresis on the determination of the magnetocaloric effect in Gd5Si2Ge2. *Journal of Physics. D, Applied Physics*, *48*(2), 025005. doi:10.1088/0022-3727/48/2/025005

Walton, K. G., Pugh, N. D., Gerderloos, P., & Macrae, P. (1995). Stress reduction and preventing hypertension: Preliminary support for a psychoneuroendocrine mechanism. *Journal of Alternative and Complementary Medicine (New York, N.Y.)*, *1*(3), 263–283. doi:10.1089/acm.1995.1.263 PMID:9395623

Wang, D., Liao, B., Zheng, J., Huang, G., Hua, Z., Gu, C., & Xu, P. (2019). Development of regulations, codes and standards on composite tanks for on-board gaseous hydrogen storage. *International Journal of Hydrogen Energy*, *44*(40), 22643–22653. doi:10.1016/j.ijhydene.2019.04.133

Wang, H., Ouyang, L. Z., Peng, C. H., Zeng, M. Q., Chung, C. Y., & Zhu, M. (2004). MmM5/Mg multi-layer hydrogen storage thin films prepared by dc magnetron sputtering. *Journal of Alloys and Compounds*, *370*(1-2), L4–L6. doi:10.1016/j.jallcom.2003.09.019

Wang, L., Quadir, M. Z., & Aguey-Zinsou, K.-F. (2016). Ni coated LiH nanoparticles for reversible hydrogen storage. *International Journal of Hydrogen Energy*, *41*(15), 6376–6386. doi:10.1016/j.ijhydene.2016.01.173

Warburg, E. (1881). Über einige Wirkungen der Coërcitivkraft. *Annalen der Physik*, *13*, 141. doi:10.1002/andp.18812490510

Ware, J. E. Jr, Snyder, M. K., Wright, W. R., & Davies, A. R. (1983). Defining and measuring patient satisfaction with medical care. *Evaluation and Program Planning*, *6*(3-4), 247–263. doi:10.1016/0149-7189(83)90005-8 PMID:10267253

Webb, C. J. (2015). A review of catalyst-enhanced magnesium hydride as a hydrogen storage material. *Journal of Physics and Chemistry of Solids*, *84*, 96–106. doi:10.1016/j.jpcs.2014.06.014

Weiss, P., & Piccard, A. (1918). Sur un nouveau phénomène magnétocalorique. *CR (East Lansing, Mich.)*, *166*, 352–354.

Wensing, M., Grol, R., & Smits, A. (1994). Quality judgements by patients on general practice care: A literature analysis. *Social Science & Medicine*, *38*(1), 45–53. doi:10.1016/0277-9536(94)90298-4 PMID:8146714

Wesam Al-Mufti, M., Hashim, U., Rahman, M., Adam, T., & Arshad, M. (2015). Studying Effect Dimensions of Design and Simulation Silicon Nanowire Filed Effect Biosensor. *Applied Mechanics and Materials*, *754*, 854–858. doi:10.4028/www.scientific.net/AMM.754-755.854

Wezel, A., & Soldat, V. (2009). A quantitative and qualitative historical analysis of the scientific discipline agroecology. *International Journal of Agricultural Sustainability*, *7*(1), 3–18. doi:10.3763/ijas.2009.0400

White, P. (2015). The concept of diseases and health care in African traditional religion in Ghana. *Hervormde Teologiese Studies*, *71*(3). doi:10.4102/hts.v71i3.2762

WHO. (2008). *Commission on social determinants of health. Closing the gap in a generation: health equity through action on the social determinants of health: final report of the commission on social determinants of health*. Geneva: World Health Organization, 2008. Retrieved from https://www.who.int/social_determinants/thecommission/finalreport/en/

*Compilation of References*

WHO. (2010). *World health statistics 2010*. Retrieved from https://www.who.int/whosis/whostat/2010/en/

Wichmann, S. (1981). Japonisme. The Japanese Influence on Western Art since 1858. London: Thames & Hudson.

Wiggers, J. H., Donovan, K. O., Redman, S., & Sanson-Fisher, R. W. (1990). Cancer patient satisfaction with care. *Cancer*, *66*(3), 610–616. doi:10.1002/1097-0142(19900801)66:3<610::AID-CNCR2820660335>3.0.CO;2-T PMID:2364373

Willner-Rönnholm, M. (1996). *Taidekoulun arkea ja unelma. Turun piirustuskoulu 1830-1981*. Vammala: Turun taidemuseo.

Wilson, G. A. (2009). Post-Productivist and multifunctional agriculture. International Encyclopedia of Human Geography, 379–386. doi:10.1016/B978-008044910-4.00895-6

Wilson, W. (1984). *Resource management in a stratified Fulani community* (PhD dissertation). Howard University.

Wilson, G. A. (2008). From "weak" to "strong" multifunctionality: Conceptualising farm-level multifunctional transitional pathways. *Journal of Rural Studies*, *24*(3), 367–383. doi:10.1016/j.jrurstud.2007.12.010

Wiskerke, J. S. C. (2016). Urban food systems. In Cities and Agriculture. Developing resilient urban food systems (pp. 1–26). New York: Routledge.

Wittgenstein, L. (1922). *Tractacus Logico-Philosophicus*. London: Kegan Paul. https://people.umass.edu/klement/tlp/tlp.pdf

Woodward, D. P. (1992). Locational determinants of Japanese manufacturing start-ups in the United States. *Southern Economic Journal*, *58*(3), 690–708. doi:10.2307/1059836

World Health Organization. (1999). *The World Health Report 1999*. Geneva: WHO. Retrieved from https://apps.who.int/iris/handle/10665/42167

World Health Organization. (2015). *Bangladesh health system review*. Manila: WHO Regional Office for the Western Pacific.

World Population Review. (2018). https://worldpopulationreview.com/world-cities/delhi-population

Wroe, N. (2005). Crime Pays. Cruz-Smith M. Interview (Wroe N.). *The Guardian*. https://www.theguardian.com/books/2005/mar/26/featuresreviews.guardianreview15

Wuthnow, R. (1998). *After heaven: Spirituality in America scince the 1950s*. University of California Press. doi:10.1525/california/9780520213968.001.0001

Xia, G. (2015). *Light metal hydrides for reversible hydrogen storage applications*. University of Wollongong Research Online.

XiangDong, Y., & GaoQing, L. (2008). Magnesium-based materials for hydrogen storage: Recent advances and future perspectives. *Chinese Science Bulletin*, *53*, 2421–2440.

Yadav, S., Zhu, Z., & Singh, C. V. (2014). Defect engineering of graphene for effective hydrogen storage. *International Journal of Hydrogen Energy*, *39*(10), 4981–4995. doi:10.1016/j.ijhydene.2014.01.051

Yermakov, A. Y., Mushnikov, N. V., Uimin, M. A., Gaviko, V. S., Tankeev, A. P., Skripov, A. V., ... Buzlukov, A. L. (2006). Hydrogen reaction kinetics of Mg-based alloys synthesized by mechanical milling. *Journal of Alloys and Compounds*, *425*(1-2), 367–372. doi:10.1016/j.jallcom.2006.01.039

Yi, M., Jeong, K.-H., & Lee, L. P. (2005). Theoretical and experimental study towards a nanogap dielectric biosensor. *Biosensors & Bioelectronics*, *20*(7), 1320–1326. doi:10.1016/j.bios.2004.05.003 PMID:15590285

Yoon, M., Yang, S., Wang, E. Z. Z., & Zhang, Z. (2007). Charged fullerenes as high capacity storage media. *Nano Letters*, *7*(9), 2578–2258. doi:10.1021/nl070809a PMID:17718530

Yu, B. F., Gao, Q., Zhang, B., Meng, X. Z., & Chen, Z. (2003). Review on research of room temperature magnetic refrigeration. *International Journal of Refrigeration*, *26*(6), 622–636. doi:10.1016/S0140-7007(03)00048-3

Yu, B., Liu, M., Egolf, P. W., & Kitanovski, A. (2010). A review of magnetic refrigerator and heat pump prototypes built before the year 2010. *International Journal of Refrigeration*, *33*(6), 1029–1060. doi:10.1016/j.ijrefrig.2010.04.002

Zahiri, B., Amirkhiz, B. S., & Mitlin, D. (2010). Hydrogen storage cycling of MgH2 thin film nanocomposites catalyzed by bimetallic CrTi. *Applied Physics Letters*, *97*(8), 083106. doi:10.1063/1.3479914

Zaluska, A., Zaluski, L., & Strom-Olsen, J. O. (2000). Lithium-berylium hydrides: The lightest reversible metal hydride. *Journal of Alloys and Compounds*, *307*(1-2), 157–166. doi:10.1016/S0925-8388(00)00883-5

Zamani, M., Prabhakaran, M. P., & Ramakrishna, S. (2013). Advances in drug delivery via electrospun and electrosprayednanomaterials. *International Journal of Nanomedicine*, *8*, 2997–3017. PMID:23976851

Zhang, W., Ricketts, T. H., Kremen, C., Carney, K., & Swinton, S. M. (2007). Ecosystem services and dis-services to agriculture. *Ecological Economics*, *64*(2), 253–260. doi:10.1016/j.ecolecon.2007.02.024

Zheng, L., Liu, X., Xu, P., Liu, P., Zhao, Y., & Yang, J. (2012). Development of high pressure gaseous hydrogen storage technologies. *International Journal of Hydrogen Energy*, *37*(1), 1048–1057. doi:10.1016/j.ijhydene.2011.02.125

Zifko-Baliga, G. M., & Krampf, R. F. (1997). Managing perceptions of hospital quality. *Marketing Health Services*, *17*(1), 28. PMID:10169030

Zimm, C. B., Sternberg, A., Jastrab, A. G., Boeder, A. M., Lawton, L. M., & Chell, J. J. (2003). *U.S. Patent No. 6,526,759*. Washington, DC: U.S. Patent and Trademark Office.

Zineldin, M. (2006). The quality of health care and patient satisfaction. An Exploratory Investigation of the 5Q Model at Some Egyptian and Jordanian Medical Clinics. *International Journal of Health Care Quality Assurance*, *19*(1), 60–92. doi:10.1108/09526860610642609 PMID:16548402

Zumthor, P. (2006). *Atmospheres: Architectural Environments- Surrounding Objects*. Basel: Birkhauser.

# About the Contributors

**Mika Markus Mervïo** is Professor of International Relations at the Kibi International University, Okayama, Japan. Previously he has been working at the University of Tampere, Finland, from 1983-1995, and from which he also received his Ph.D. (International Relations). Then he moved to the International Christian University, Tokyo, Japan as a JSPS researcher and from there to the Miyazaki International University and the University of Shimane, where he worked as a Professor. From 2000 he has been working at the Kibi International University. His research interests include political and social issues, history, art and environment.

\* \* \*

**Hisham Abusaada** is a professor of architecture and urban design in the Housing and Building National Research Centre (HBRC), Egypt. During 1995-2015, he taught urban design course at universities in Saudi Arabia and Egypt. He is the author and co-authors of several scientific manuscripts and books in both Arabic and English published in scientific journals and conference proceedings. His area of interest is in urban design between theory and practice.

**Afsana Akhtar** is an Assistant Professor at BRAC University, Dhaka, Bangladesh.

**Siwela Jeffrey Baloyi** obtained his PhD in Catalysis at the University of the Witwatersrand. Dr Siwela has a strong background in material science, wastewater treatment technologies, reactor design, mathematical optimization, multivariate statistical methods, data analysis and new product development. Areas of interest are in material science, design and assembly of nanomaterials for energy conversion and storage, catalysis, wastewater treatment, optimization, statistical analysis of process data and computational fluid dynamics (CFD).

**Ranjit Barua** holds his master degree in manufacturing technology. His area of research interest is nanotechnology, additive manufacturing, nano-robotics, etc.

**Mohd. Yousuf Bhat** is Assistant Professor, Political Science, in Higher Education Department, J&K, India. He has Ph.D. in Political Science from Aligarh Muslim University (AMU) Aligarh, India (2003). He has published many research papers in reputed journals in areas of environment, human rights and human security. He has a special interest in International relations, Environment and Human Rights.

## About the Contributors

**Sakshi Bhati** is currently a graduate student in the Masters of Arts in Communication degree program in the Department of Communication at Pittsburg State University.

**Marisa Cleveland** is currently completing her doctorate in organizational leadership from Northeastern University, Marisa has a master's in educational administration, a bachelor's in speech communications from George Mason University, and is a 2015 Leadership Marco graduate.

**Simon Cleveland** is an associate professor and faculty director at Georgetown University. Dr. Cleveland is a graduate of George Mason University and The George Washington University, with Bachelor of Science degree in Business and a Master of Science degree in Project Management respectively, and has a Ph.D. in Information Systems from Nova Southeastern University's College of Engineering and Computing. He is also certified as a project management professional (PMP), Six Sigma Black Belt (SSBB) expert, and Information Technology Infrastructure Library (ITIL) foundations expert. Dr. Cleveland is the author of over 30 peer-reviewed publications in journals and conferences, and is a three-time recipient of the prestigious Dr. Harold Kerzner Scholarship which is awarded by the Project Management Institute Educational Foundation. Professionally, Dr. Cleveland has worked as both technology manager, program manager and PMO Director for organizations such as MD Anderson Cancer Center, Department of Homeland Security, NASA, Accenture, and AOL.

**Jonali Das** holds her bachelors degree in chemistry. Her area of interest is nano-material, material chemical composition study, etc.

**Sudipto Datta** holds a M.Tech in material science. His area of interest is bio-material, nano-material, etc.

**Segufta Dilshad** is a Lecturer, Department of Public Health, School of Health and Life Sciences, North South University.

**Regina Hoi Yee Fu** is an Associate Professor in the School of Economics, Senshu University, Japan. After obtaining her Ph.D. from The University of Tokyo, she has taught at The University of Tokyo and Hosei University. She joined Senshu University since 2018 and has been teaching courses on African Economy and Development Economics. Regina has published articles and book chapters on natural resource management, rural sociology, conflict resolution and economic development of Africa and Asia. Her academic interest is mainly on development of Africa and Asia, and on Economic Anthropology.

**Sadrul Huda** is an Assistant Professor at East West University, Dhaka, Bangladesh.

**Ran Jiang** is an undergraduate, School of Communication, Soochow University. Focus on persuasive communication in the area of marketing and health.

**Torben Larsen** is MSc Econ from Aarhus University with a degree in Strategic Management from University of Maryland/Tietgenskolen. He has many years of experience in the planning, research and development of health care systems including both Academic and Complementary Medicine. He has been project coordinator for the EC-research project 'Homecare' on behalf of University of Southern Denmark. He is guest professor at the International Institute of Advanced Studies and Cybernetics (IIAS).

*About the Contributors*

He received Academic Awards for his research: Lundbeck Prize 1991, Award from European Medical Informatics 1996, IIAS Book Award 2007. 2017 he published a neuroeconomic model (Larsen T: Homo Neuroeconomicus ..., IJUDH 2017).

**Alicia Mason** is a Professor in the Department of Communication at Pittsburg State University in Pittsburg, Kan. and teaches within the Strategic Communication program. Mason's academic interest in risk/crisis communication spans corporate, public health, and environmental contexts. Mason also teaches in the College of Arts & Sciences Interdisciplinary Studies Sustainability, Society and Resource Management undergraduate degree program.

**Claudia Masselli** was born in Naples (Italy) on May 13rd 1987. She is a Postdoctoral Researcher at Department of Industrial Engineering of University of Salerno (Italy). She earned the Ph.D title in Industrial Engineering, discussing the thesis entitled "Magnetic refrigeration an attraction toward our future". Among her main research topics there is caloric solid-state refrigeration both experimentally and numerically. Through the development of two-dimensional finite element models, she investigates the energy performances of different solid-state materials in solid-state refrigerators. Her researches embrace, also, the energy and environmental optimization of vapor compression cycles. She is winner of two Young Researcher Awards; she is member of the editorial board of 6 the international journals: among them, Annales de Chimie: Science des Matériaux, (ed. IIETA - ISSN: 0151-9107 print; 1958-5934 online; Scopus and Web of science indexed); she reviewer of 40 international journals related to heat transfer, refrigeration, energy and engineering science fields. She is Junior Member of the Commission B2 - Refrigerating equipment afferent to the Section B: Thermodynamics, equipment and systems of International Institute of Refrigeration (IIR). Furthermore, she belongs to many international scientific committees of international conferences focused on energy and engineering science fields.

**Athule Ngqalakwezi** is a Masters student at one of South Africa's leading research university (University of the Witwatersrand). She has experience working in Catalysis at one of South Africa's leading Mineral research company. She worked on various projects such CO oxidation, NOx reduction and Desulphurizing diesel. She attained her Post graduate Diploma with Cum Laude. She is a public speaker, debater,science communicator and a chess enthusiast. Her next step is to pursue her PhD outside South Africa.

**Diakanua Nkazi** is a Senior Lecturer at the Witwatersrand University. He published two patents for his PhD work at the Witswatersrand. Dr Nkazi is involved in a number of projects in the school.

**Thabang Abraham Ntho** is a principal research scientist in the Advanced Materials Division (AMD) at Mintek, which is a is an autonomous research and development (R&D) organization specializing in all aspects of mineral processing, extractive metallurgy and related technology. He gained a PhD degree in Chemistry in 2007 from the University of the Witwatersrand, Johannesburg, South Africa under the supervision of Prof Mike Scurrell and is a member of the South African Chemical Institute. His research interests range from computational design of new materials from first principles for application in energy, transport and health industries to materials discovery using DFT and machine learning tools.

**Nandeeta Samad** is a Lecturer, Department of Public Health, School of Health and Life Sciences, North South University.

**Elizabeth Spencer** is currently a doctoral candidate at the University of Kentucky's College of Communication and Information.

**José G. Vargas-Hernández**, M.B.A., Ph.D., Member of the National System of Researchers of Mexico and a research professor at University Center for Economic and Managerial Sciences, University of Guadalajara. Professor Vargas-Hernández has a Ph. D. in Public Administration and a Ph.D. in Organizational Economics. He has undertaken studies in Organisational Behaviour and has a Master of Business Administration, published four books and more than 200 papers in international journals and reviews (some translated to English, French, German, Portuguese, Farsi, Chinese, etc.) and more than 300 essays in national journals and reviews. He has obtained several international Awards and recognition.

# Index

## A

aesthetics 120, 122, 124-125
Agro-Ecology 14-25, 33
Air pollution 150-153, 158, 160, 163
AMR cycle 239-241

## B

behavioral economics 64, 75, 80-81

## C

Caloric effect 228
Corralling contract 34-37, 45, 47-55, 57-59
Cultural Agility 1-2, 9-10

## E

Ecosystem Services 14-17, 20-22, 25, 33, 104
Emissions 19, 21, 78, 150-152, 156-158, 160, 169-170, 172
Environmental Development 18, 33, 107
Environmental Protection Agency (EPA) 168
Environmental Remediation 168, 172

## F

Food System 19-21, 25, 33

## G

Gadolinium 241, 243-246
Global Warming 72, 158, 168, 227
Green economy 99, 105
green growth 99
Green Innovation 19, 100, 102-103, 106-108, 110-111, 113, 119
Greenhouse Gas 19-20, 78, 157, 172

## H

Health communication 206, 208, 210
Healthcare 74, 76, 133, 195-204, 207-210, 212, 215-218, 253-255, 257, 260-262, 265, 267, 271-273, 275, 279, 281
Hydrogen storage 168, 172-176, 178-184

## I

Industrialization 16, 65-66, 140, 168
interpretation 96, 209, 211, 279, 315

## L

Leadership Development 1-3, 6, 10

## M

Magnetocaloric materials 228, 236, 241, 243, 246-247
Medical tourism 197-198, 203, 206-219
meditation 74, 265-266
Migration 39, 44, 47, 49-50, 53, 137-139, 141, 161
M-Learning 84, 89, 91

## N

Naguib Mahfouz 121
Nano films 294
Nanotechnology 292-298
Nanotubes 178-179, 294-297
nationalism 302-303, 305, 312-314
neuroeconomics 64, 72, 76, 79-80
News media 206, 209
Nupe 34-35, 39-45, 47, 49-50, 52-55, 57-59

## O

on-board applications 169, 173, 175-176, 185

## P

painting 123, 302, 304, 306-311, 313
Patient mortality 206, 208-210, 216, 218-219
Pigovian Tax 70, 77
Political Will 133, 147, 203

## R

Relational Leadership 1-2, 6-8, 10
Religion 41, 253-255, 262-265, 267, 312
Resources entitlement 34-35
Risk communication 211, 218

## S

social significance 301
Sustainable City 100, 119
sustainable development 17, 33, 99-100, 102-103, 105, 171-172
SWOT-analysis 64-65, 67

## T

Traditional institution 35, 56, 59

## U

universal basic income 64, 76
urban agro-ecology 14-25, 33
urban design 100, 104, 120-121, 124-126, 130
Urban Space 99-101, 112-113, 119

## V

Vapor Compression 226-228, 247-248
Visual arts 124, 301-303, 305-306, 308-312, 314-315

## Purchase Print, E-Book, or Print + E-Book

IGI Global's reference books are available in three unique pricing formats:
Print Only, E-Book Only, or Print + E-Book.
Shipping fees may apply.

**www.igi-global.com**

# Recommended Reference Books

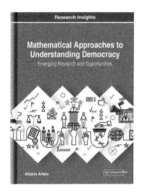

ISBN: 978-1-5225-7558-0
© 2019; 148 pp.
List Price: $165

ISBN: 978-1-5225-6918-3
© 2019; 740 pp.
List Price: $495

ISBN: 978-1-5225-7591-7
© 2019; 441 pp.
List Price: $215

ISBN: 978-1-5225-8160-4
© 2019; 450 pp.
List Price: $295

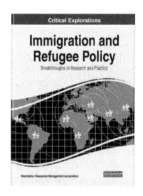

ISBN: 978-1-5225-8909-9
© 2019; 469 pp.
List Price: $330

ISBN: 978-1-5225-7429-3
© 2019; 453 pp.
List Price: $265

**Do you want to stay current on the latest research trends, product announcements, news and special offers?**
Join IGI Global's mailing list today and start enjoying exclusive perks sent only to IGI Global members.
Add your name to the list at **www.igi-global.com/newsletters**.

Publisher of Peer-Reviewed, Timely, and Innovative Academic Research

**IGI Global**
DISSEMINATOR OF KNOWLEDGE

www.igi-global.com   Sign up at www.igi-global.com/newsletters   facebook.com/igiglobal   twitter.com/igiglobal   linkedin.com/igiglobal

Ensure Quality Research is Introduced to the Academic Community

# Become an IGI Global Reviewer for Authored Book Projects

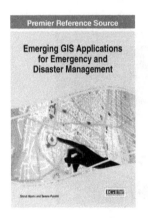 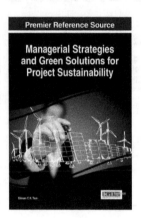  

## The overall success of an authored book project is dependent on quality and timely reviews.

In this competitive age of scholarly publishing, constructive and timely feedback significantly expedites the turnaround time of manuscripts from submission to acceptance, allowing the publication and discovery of forward-thinking research at a much more expeditious rate. Several IGI Global authored book projects are currently seeking highly-qualified experts in the field to fill vacancies on their respective editorial review boards:

### Applications and Inquiries may be sent to:
development@igi-global.com

Applicants must have a doctorate (or an equivalent degree) as well as publishing and reviewing experience. Reviewers are asked to complete the open-ended evaluation questions with as much detail as possible in a timely, collegial, and constructive manner. All reviewers' tenures run for one-year terms on the editorial review boards and are expected to complete at least three reviews per term. Upon successful completion of this term, reviewers can be considered for an additional term.

If you have a colleague that may be interested in this opportunity,
we encourage you to share this information with them.

# IGI Global Proudly Partners With eContent Pro International

## Receive a 25% Discount on all Editorial Services

## Editorial Services

IGI Global expects all final manuscripts submitted for publication to be in their final form. This means they must be reviewed, revised, and professionally copy edited prior to their final submission. Not only does this support with accelerating the publication process, but it also ensures that the highest quality scholarly work can be disseminated.

### English Language Copy Editing

Let eContent Pro International's expert copy editors perform edits on your manuscript to resolve spelling, punctuaion, grammar, syntax, flow, formatting issues and more.

### Scientific and Scholarly Editing

Allow colleagues in your research area to examine the content of your manuscript and provide you with valuable feedback and suggestions before submission.

### Figure, Table, Chart & Equation Conversions

Do you have poor quality figures? Do you need visual elements in your manuscript created or converted? A design expert can help!

### Translation

Need your documjent translated into English? eContent Pro International's expert translators are fluent in English and more than 40 different languages.

## Hear What Your Colleagues are Saying About Editorial Services Supported by IGI Global

"The service was very fast, very thorough, and very helpful in ensuring our chapter meets the criteria and requirements of the book's editors. I was quite impressed and happy with your service."

– Prof. Tom Brinthaupt,
Middle Tennessee State University, USA

"I found the work actually spectacular. The editing, formatting, and other checks were very thorough. The turnaround time was great as well. I will definitely use eContent Pro in the future."

– Nickanor Amwata, Lecturer,
University of Kurdistan Hawler, Iraq

"I was impressed that it was done timely, and wherever the content was not clear for the reader, the paper was improved with better readability for the audience."

– Prof. James Chilembwe,
Mzuzu University, Malawi

**Email: customerservice@econtentpro.com**          **www.igi-global.com/editorial-service-partners**

*Celebrating Over 30 Years* of Scholarly Knowledge Creation & Dissemination

www.igi-global.com

# InfoSci®-Books

A Database of Over 5,300+ Reference Books Containing Over 100,000+ Chapters Focusing on Emerging Research

GAIN ACCESS TO **THOUSANDS** OF REFERENCE BOOKS AT **A FRACTION** OF THEIR INDIVIDUAL LIST **PRICE**.

## InfoSci®-Books Database

The **InfoSci®-Books** database is a collection of over 5,300+ IGI Global single and multi-volume reference books, handbooks of research, and encyclopedias, encompassing groundbreaking research from prominent experts worldwide that span over 350+ topics in 11 core subject areas including business, computer science, education, science and engineering, social sciences and more.

### Open Access Fee Waiver (Offset Model) Initiative

For any library that invests in IGI Global's InfoSci-Journals and/or InfoSci-Books databases, IGI Global will match the library's investment with a fund of equal value to go toward **subsidizing the OA article processing charges (APCs) for their students, faculty, and staff** at that institution when their work is submitted and accepted under OA into an IGI Global journal.*

### INFOSCI® PLATFORM FEATURES

- No DRM
- No Set-Up or Maintenance Fees
- A Guarantee of No More Than a 5% Annual Increase
- Full-Text HTML and PDF Viewing Options
- Downloadable MARC Records
- Unlimited Simultaneous Access
- COUNTER 5 Compliant Reports
- Formatted Citations With Ability to Export to RefWorks and EasyBib
- No Embargo of Content (Research is Available Months in Advance of the Print Release)

*The fund will be offered on an annual basis and expire at the end of the subscription period. The fund would renew as the subscription is renewed for each year thereafter. The open access fees will be waived after the student, faculty, or staff's paper has been vetted and accepted into an IGI Global journal and the fund can only be used toward publishing OA in an IGI Global journal. Libraries in developing countries will have the match on their investment doubled.

**To Learn More or To Purchase This Database:**
www.igi-global.com/infosci-books

eresources@igi-global.com • Toll Free: 1-866-342-6657 ext. 100 • Phone: 717-533-8845 x100

Publisher of Peer-Reviewed, Timely, and
Innovative Academic Research Since 1988

# IGI Global's Transformative Open Access (OA) Model:
## How to Turn Your University Library's Database Acquisitions Into a Source of OA Funding

In response to the OA movement and well in advance of Plan S, IGI Global, early last year, unveiled their OA Fee Waiver (Offset Model) Initiative.

Under this initiative, librarians who invest in IGI Global's InfoSci-Books (5,300+ reference books) and/or InfoSci-Journals (185+ scholarly journals) databases will be able to subsidize their patron's OA article processing charges (APC) when their work is submitted and accepted (after the peer review process) into an IGI Global journal.*

## How Does it Work?

1. When a library subscribes or perpetually purchases IGI Global's InfoSci-Databases including InfoSci-Books (5,300+ e-books), InfoSci-Journals (185+ e-journals), and/or their discipline/subject-focused subsets, IGI Global will match the library's investment with a fund of equal value to go toward subsidizing the OA article processing charges (APCs) for their patrons.

   *Researchers:* Be sure to recommend the InfoSci-Books and InfoSci-Journals to take advantage of this initiative.

2. When a student, faculty, or staff member submits a paper and it is accepted (following the peer review) into one of IGI Global's 185+ scholarly journals, the author will have the option to have their paper published under a traditional publishing model or as OA.

3. When the author chooses to have their paper published under OA, IGI Global will notify them of the OA Fee Waiver (Offset Model) Initiative. If the author decides they would like to take advantage of this initiative, IGI Global will deduct the US$ 1,500 APC from the created fund.

4. This fund will be offered on an annual basis and will renew as the subscription is renewed for each year thereafter. IGI Global will manage the fund and award the APC waivers unless the librarian has a preference as to how the funds should be managed.

## Hear From the Experts on This Initiative:

"I'm very happy to have been able to make one of my recent research contributions, 'Visualizing the Social Media Conversations of a National Information Technology Professional Association' featured in the *International Journal of Human Capital and Information Technology Professionals*, freely available along with having access to the valuable resources found within IGI Global's InfoSci-Journals database."

– **Prof. Stuart Palmer**,
Deakin University, Australia

**For More Information, Visit:** www.igi-global.com/publish/contributor-resources/open-access or contact IGI Global's Database Team at eresources@igi-global.com.

Printed in the USA
CPSIA information can be obtained
at www.ICGtesting.com
LVHW081141141024
793749LV00006B/283